AF767697

Water Pollution and Conservation

Water Pollution and Conservation

Edited by **Sheryl McMillan**

Callisto Reference
New York

Published by Callisto Reference,
106 Park Avenue, Suite 200,
New York, NY 10016, USA
www.callistoreference.com

Water Pollution and Conservation
Edited by Sheryl McMillan

© 2016 Callisto Reference

International Standard Book Number: 978-1-63239-697-6 (Hardback)

This book contains information obtained from authentic and highly regarded sources. Copyright for all individual chapters remain with the respective authors as indicated. All chapters are published with permission under the Creative Commons Attribution License or equivalent. A wide variety of references are listed. Permission and sources are indicated; for detailed attributions, please refer to the permissions page and list of contributors. Reasonable efforts have been made to publish reliable data and information, but the authors, editors and publisher cannot assume any responsibility for the validity of all materials or the consequences of their use.

The publisher's policy is to use permanent paper from mills that operate a sustainable forestry policy. Furthermore, the publisher ensures that the text paper and cover boards used have met acceptable environmental accreditation standards.

Trademark Notice: Registered trademark of products or corporate names are used only for explanation and identification without intent to infringe.

Printed in China.

Contents

Preface

It is often said that books are a boon to mankind. They document every progress and pass on the knowledge from one generation to the other. They play a crucial role in our lives. Thus I was both excited and nervous while editing this book. I was pleased by the thought of being able to make a mark but I was also nervous to do it right because the future of students depends upon it. Hence, I took a few months to research further into the discipline, revise my knowledge and also explore some more aspects. Post this process, I began with the editing of this book.

Water pollution is a growing concern for today's generation. It refers to the contamination of water bodies like rivers, oceans, lakes, etc. Water gets polluted when chemicals, plastic, paint and other harmful products are released in it. Due to the growing industrial sector and human population, conservation of water has become an aim of regulatory bodies. This book will discuss the various ways in which water is being polluted and the processes to clean and conserve it. The detailed content provided in this book will prove to be useful for environmentalists, engineers, students, researchers and scientists. Readers will benefit from the number of concepts and theories that have been included in this book. It will serve as a resourceful guide and contribute to their knowledge of this subject.

I thank my publisher with all my heart for considering me worthy of this unparalleled opportunity and for showing unwavering faith in my skills. I would also like to thank the editorial team who worked closely with me at every step and contributed immensely towards the successful completion of this book. Last but not the least, I wish to thank my friends and colleagues for their support.

Editor

1

Groundwater vulnerability assessment in the vicinity of Ramtha wastewater treatment plant, North Jordan

Muheeb Awawdeh · Mutewekil Obeidat · Ghusun Zaiter

Abstract The main aim of this study was to evaluate the vulnerability of groundwater to contamination in the vicinity of Ramtha wastewater treatment plant using a modified DRASTIC method in a GIS environment. A groundwater pollution potential map was prepared using modified DRASTIC method by adding lineaments and land use/land cover parameters. The values of the modified DRASTIC index were classified into three categories: low, moderate and high. About 36.5 % of the study area is occupied by the high vulnerability class, 56.5 % is occupied by the moderate vulnerability class and 9 % is occupied by the low vulnerability class. Chemical analysis of the water samples collected from wells distributed in the study area and tapping Umm Rijam aquifer indicated that the nitrate concentration ranges from 20 to 193 mg/L with an average 65.5 mg/L. Nitrate exceeded the permissible limits of WHO and Jordanian standards in 69 and 54 % of the NO_3^- samples, respectively. The modified DRASTIC model was validated using nitrate concentration. Results showed a good match between nitrate concentrations level and the groundwater vulnerability classes.

Keywords Groundwater vulnerability · Ramtha wastewater treatment plant · DRASTIC · Geographic Information System · Nitrate

M. Awawdeh (✉) · G. Zaiter
Department of Earth and Environmental Sciences, Yarmouk University, Irbid, Jordan
e-mail: awawdeh@yu.edu.jo

M. Obeidat
Faculty of Science and Arts, Jordan University of Science and Technology, Irbid 22110, Jordan

Introduction

Jordan has very limited natural resources; it is ranked among the five most water-poor countries in the world (United Nations 2006). Current water availability in Jordan amounts to about 150 cubic meters per capita per year, but it is predicted to be about 91 cubic meters per capita per year by the year 2025, if no action has been made to meet the increasing need for water (Hadadin et al. 2010).

Water quality in Jordan has deteriorated due to various sources of pollution, and over-abstraction resulting in salinization. Over-abstraction of groundwater for irrigation has reduced the water table by 5 m in some aquifers and tripled salinity, and unregulated N-fertilizer application has increased nitrates in water supplies (Wardam 2004). Industrial discharges are improperly treated or untreated, which introduce toxicity into water resources (Mohsen 2007). It is found that about 70 % of spring water has biological contamination and surface water has high fecal coliform counts from non-point pollution sources, including point sources (wastewater treatment plants) operating over capacity (Wardam 2004). In recent years, nitrate has become a major source of groundwater pollution in Jordan (Margane et al. 1999). Therefore, groundwater pollution prevention and control are particularly critical. Hence, the research on groundwater vulnerability is essential to ensure the quality of groundwater and realize the sustainability of groundwater resources.

Nowadays, groundwater vulnerability assessment has become a useful tool for groundwater pollution prevention. Groundwater vulnerability mapping can indicate regions which are more vulnerable to contamination in the planning stage of socio-economic activities (Fritch et al. 2000; Huan et al. 2012). Vulnerability maps of aquifers are an important tool for the groundwater management and protection (Zwahlen 2004). Groundwater vulnerability maps provide useful information to protect groundwater resources and to evaluate the potential for water quality improvement with changes in agricultural practices and land use applications (Babiker et al. 2005; Burkart and Feher 1996; Connell and Daele 2003; Rupert 2001).

Vrba and Zaporozec (1994) distinguished intrinsic (or natural) vulnerability from specific (or integrated) vulnerability; the first term defined solely as a function of hydrogeological factors and the latter term defined by the potential impacts of specific land uses and contaminants. In other words, specific vulnerability integrates the contamination risk placed upon aquifers by human activities (Stigter et al. 2006). The intrinsic vulnerability of groundwater to contaminants takes into account the geological, pedological, hydrological and hydrogeological characteristics of an area (Vrba and Zaporozec 1994). Several methods have been developed to evaluate groundwater vulnerability, and choosing an appropriate one depends on many factors such as scale of the study area, data availability, and desired results (Al-Hanbali and Kondoh 2008). The most widely used method of vulnerability evaluation is DRASTIC model (Aller et al. 1987) due to its ease to use, minimum data requirement, and clear explanation of groundwater vulnerability.

The objectives of this study are (1) to assess the groundwater vulnerability to contamination in the study area using a modified DRASTIC method, and (2) to validate the groundwater vulnerability mapping using nitrate measurements.

Study area

The study area is a sub-basin of the Yarmouk River watershed, north Jordan (Fig. 1). Ramtha Wastewater Treatment Plant (RWWTP), located in the study area, was established in 1986 to serve the city of Ramtha. The plant is receiving wastewater from a population of about 70,000 people. Stabilization ponds were used in the treatment process. The activated sludge treatment process was used after upgrading the plant in 2003. Evaluation of several options for reuse of the reclaimed wastewater showed that agricultural use was the most suitable alternative. The agricultural land irrigated with the reclaimed wastewater is about 1 km^2, located around the plant. The climate of the study area has transitive climatic characteristics between the most humid zone of the country and the desert. The summer is hot and dry with mean daily temperature ranging from 19 to 31 °C. The mean annual minimum and maximum temperatures are 10.7 and 23.7 °C, respectively. Rainfall occurs mainly from November to March, with mean annual rainfall (1976–2002) at Ramtha rainfall station is 213.8 mm. Geologically, the study area consists of Oligocene Basalt (Ba), Wadi Shallala Chalk Formation (B5), which consists of chert and limestone of early Middle-early Late Eocene age, Umm Rijam Chert Limestone Formation (B4), which consists of chert and limestone of Paleocene age, and Amman Silicified Limestone (B2) of campanian age (Natural Resources Authority 1997) (Fig. 2). The subsurface geology comprises Mesozoic

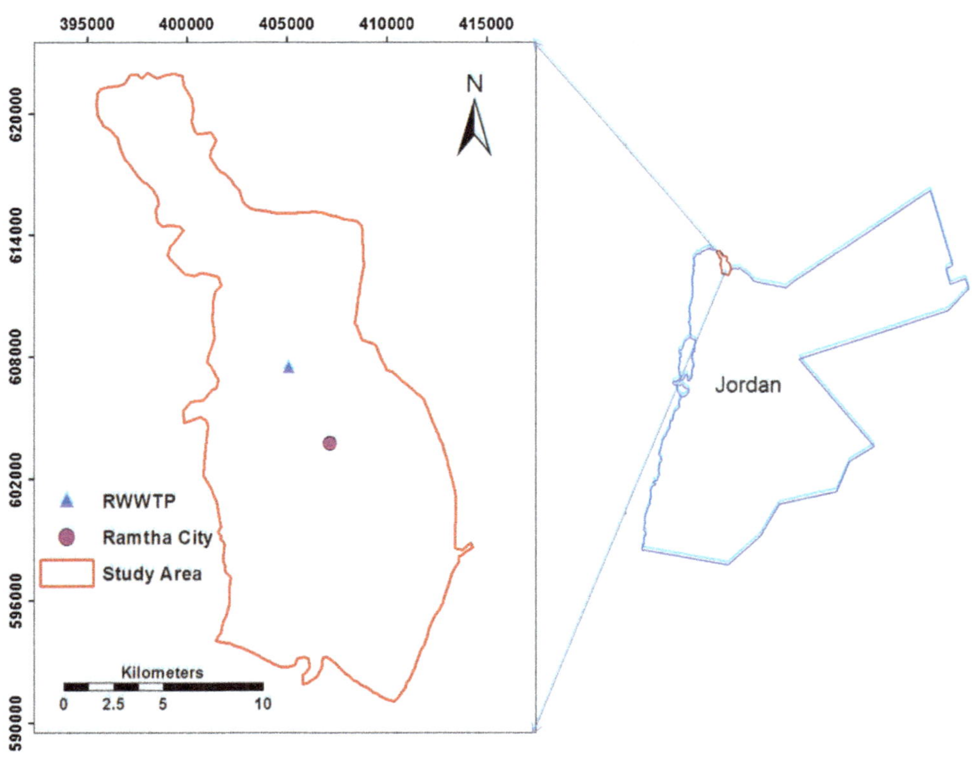

Fig. 1 Location map of the study area

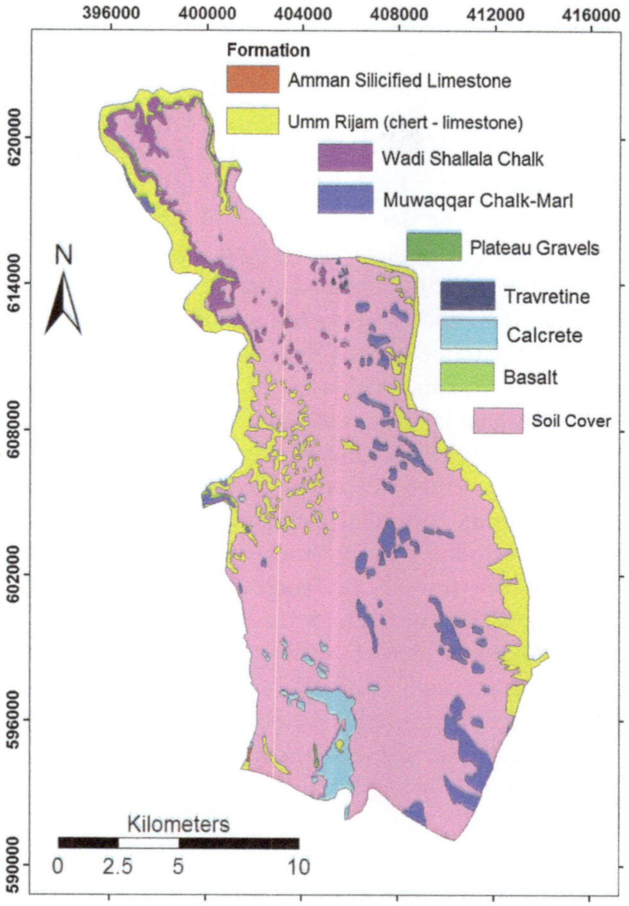

Fig. 2 Geology map of the study area

rocks that consist of Wadi Es Sir Limestone Formation (A7) of Turonian age and Muwaqqar Chalk Marl Formation (B3) of Maastrichtian age. From a hydrogeological point of view, the study area comprises two main aquifer systems: the Wadi Shallala/Umm Rijam (B5/B4). The second aquifer system consists of the Amman/Wadi Es Sir formations (B2/A7).

Materials and methods

Groundwater vulnerability assessment

DRASTIC is one of the most widely used models to assess groundwater vulnerability (Awawdeh and Jaradat 2010; Evans and Mayers 1990; Fritch et al. 2000; Knox et al. 1993; Nawafleh et al. 2011; Piscopo 2001; Rundquist et al. 1991; Secunda et al. 1998). In this model, spatial datasets on **D**epth to groundwater, Net **R**echarge, **A**quifer media, **S**oil media, **T**opography, **I**mpact of the vadose zone and the hydraulic **C**onductivity of the aquifer are combined (Navulur and Engel 1998). It is a numerical rating scheme,

Table 1 The DRASTIC model parameters

Factor	Description	Weight
Depth to water	Defined as the depth from the ground surface to the water level; deeper water level is less vulnerable to contamination	5
Impact of vadose zone	Represents the unsaturated zone, which attenuates the potential contamination of the aquifer	5
Net recharge	It is the amount of the water that penetrates the ground surface and reaches the water table. Recharge water transports the contaminants to the aquifer	4
Hydraulic conductivity	Defines the ability of the aquifer to transport water, which determines the flow rate of contaminants	3
Aquifer media	Refers to the saturated zone material properties, which controls the pollutant attenuation processes	3
Soil media	The unconsolidated or loose covering that covers the surface of the earth. It controls the amount of water infiltration	2
Topography	It is the slope of the land surface, which determines the amount of surface runoff; hence the amount of water percolation to the saturated zone	1

which was developed by the US Environmental Protection Agency (Aller et al. 1987; US EPA 1994). Determination of the DRASTIC index involves multiplying each factor weight by its point rating (Table 1) and summing the total (Knox et al. 1993). Rating (r) reflects the significance of classes which vary from 1 to 10 within each parameter and Weight (w) designates the importance for the parameter, and ranges between 1 and 5 (Fortin et al. 1997; Fritch et al. 2000; Knox et al. 1993).

In this study, two additional parameters were added to DRASTIC model to map the groundwater vulnerability in the study area more accurately: lineaments density and land use/land cover. Both parameters were assigned a weight value equal to 5, and the modified DRASTIC index is calculated using the following equation:

$$\text{Modified DRASTIC Index (MDI)} = D_r \times D_w + R_r \times R_w + A_r \times A_w + S_r \times S_w + T_r \times T_w + I_r \times I_w + C_r \times C_w + L_r \times L_w + \text{LULC}_r \times \text{LULC}_w$$

where D depth to groundwater, R recharge rate (net), A aquifer media, S soil media, T topography (slope), I impact of the vadose zone, C conductivity (hydraulic) of the aquifer, L lineaments density, LULC: land use/land cover.

ArcGIS 10 was used to process the datasets and to create the groundwater vulnerability map by overlaying the available data.

Sensitivity analysis

Sensitivity analysis (SA) is a significant component of any modeling project as it allows evaluation of the accuracy of the result (Baker 1992).Two sensitivity tests were performed; the map removal sensitivity analyses introduced by Lodwick et al. (1990) and the single-parameter sensitivity analysis introduced by Napolitano and Fabbri (1996). The map removal sensitivity analysis was performed to evaluate whether it was necessary to use all the parameters incorporated in the DRASTIC model (Hasiniaina et al. 2010). Lodwick et al. (1990) define a map removal sensitivity measure that represents the sensitivity associated with removing one parameter each time; it was computed using the equation:

$$S_x = \text{Absolute}\left((V_f/N) - (V_x/n)\right)$$

where S_x sensitivity associated with the removal of one parameter, V_f final model, V_x model excluding parameter, N number of parameter used in final model, n: number of parameter used in model without one parameter.

To assess the magnitude of variation created by removal of one parameter, the variation index V was calculated using the equation:

$$V = {}^{(V_f - V_x)}/_{V_f} \times 100$$

where V variation index of the removal parameter, V_f final model, V_x model excluding parameter.

The single-parameter sensitivity measure was developed to evaluate the impact of each of the DRASTIC parameters on the vulnerability index. It has been made to compare the effective or real weight of each input parameter in each polygon with the theoretical weight assigned by the analytical model (Babiker et al. 2005). It was computed using the equation:

$$W = {}^{(W_r \times W_w)}/_{V_f} \times 100$$

where W is the effective weight, W_r and W_w are the rating value and weight of each parameter, V_f is the final model.

Model validation

There are several validation methods of groundwater vulnerability mapping including correlation coefficient of the groundwater vulnerability and the actual pollution occurrence, the corresponding relationship between the extreme nitrate concentration and the vulnerability classes, analysis of variance of F statistic (Lake 2003) and level difference (Stigter et al. 2006). In most cases, correlation of vulnerability results with actual pollution occurrence is used (McLay et al. 2001; Rupert 2001), which was adopted for this study. The validation process is demonstrated by taking nitrate as the typical pollutant.

Nitrate (NO_3^-) is among the most important indicators of water resources contamination. Nitrate comes into water supplies through several sources such as septic systems (MacQuarrie et al. 2001), and fertilizers applied in agricultural fields, which contribute most to groundwater pollution (Baker 1992; Hubbard and Sheridan 1994; Postma et al. 1991). Urban development can also increase nitrate concentrations in groundwater (Ford and Tellam 1994; Lerner et al. 1999) through landfills, septic tanks and cesspools, domestic and industrial effluents, leaky sewage system and gasoline stations (Navarro and Carbonell 2007; Seiler 2005; Wakida and Lerner 2005).

Groundwater samples were collected during September and October 2011 from 12 wells (Fig. 3) tapping the B5/B4 aquifer. The methods described by APHA (1998) were followed during field and laboratory works. In addition, historical records of the hydrochemistry of groundwater were retrieved from Water Authority of Jordan (WAJ) archive from 1971 to 2008. At the time of sampling, basic information such as well location, depth and land use surrounding the wells was collected. Water samples were collected in clean polyethylene bottles washed previously with sampled water and tested in the field for the following

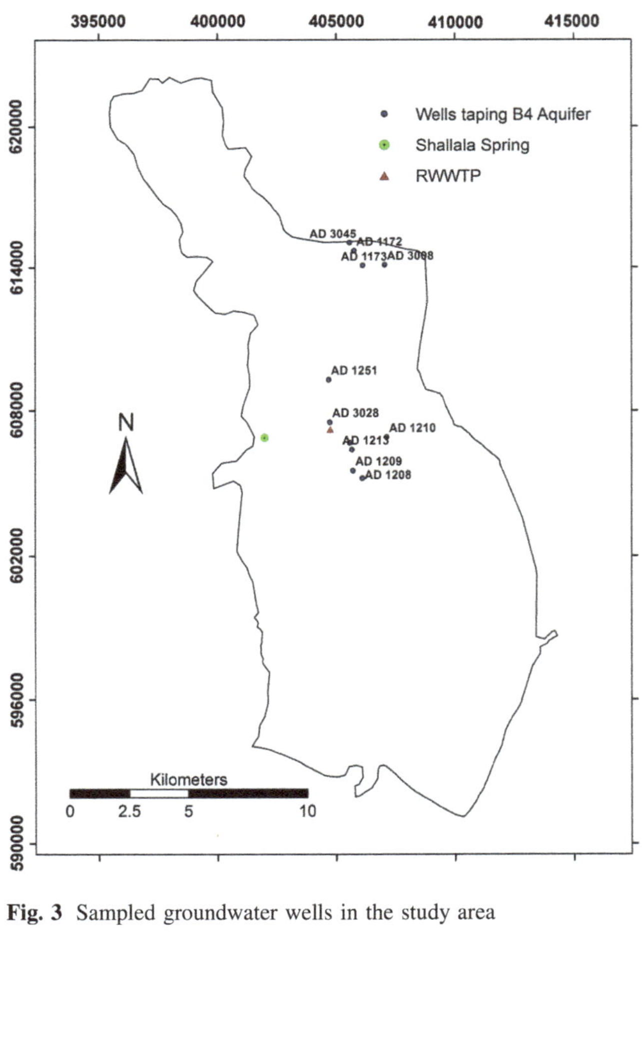

Fig. 3 Sampled groundwater wells in the study area

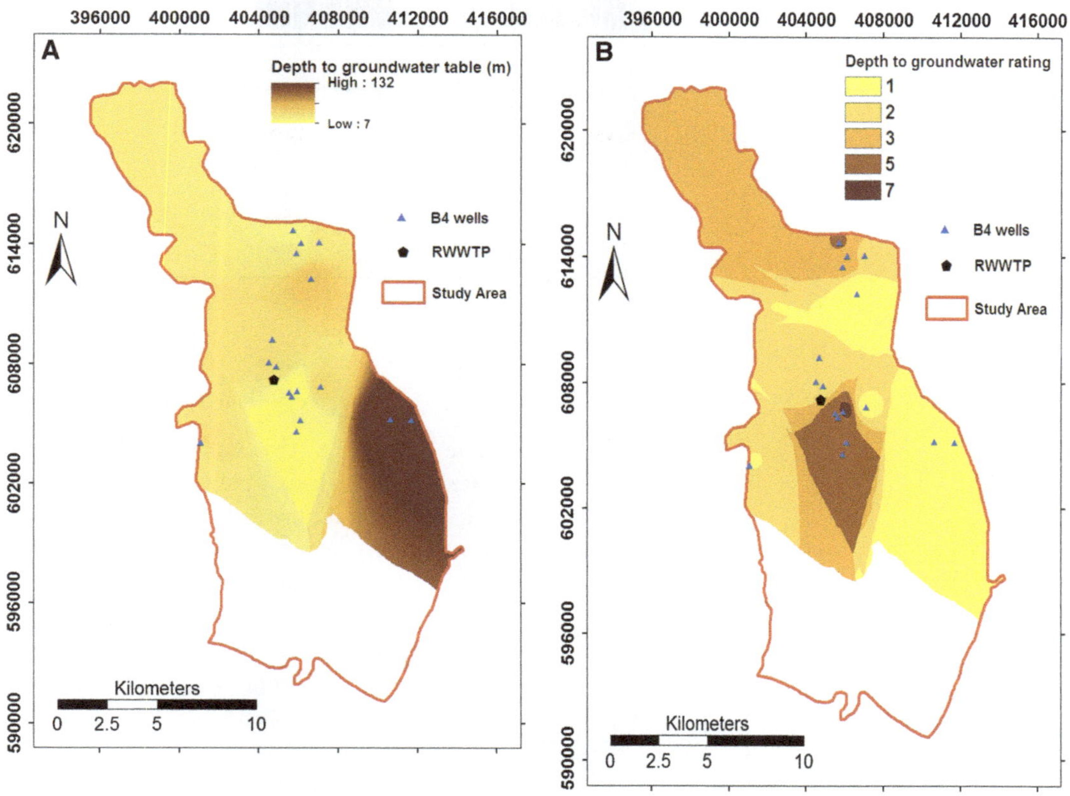

Fig. 4 Map of depth to groundwater contour map (**a**) and depth to groundwater rating (**b**) in the study area

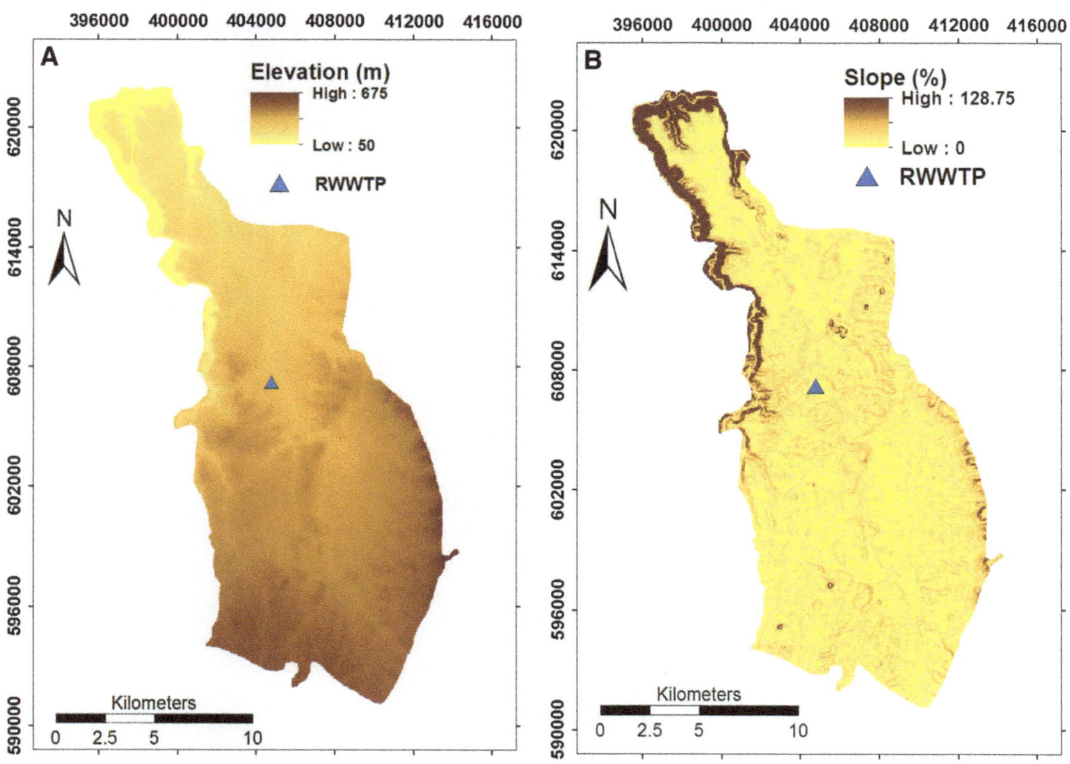

Fig. 5 Topography of the study area, DEM (**a**), and slope (**b**) in the study area

Table 2 Recharge ratings and weightings for the study area. (Piscopo 2001, Al-Adamat et al. 2003)

Slope %	Rating	Rainfall (mm)	Rating	Soil Permeability	Rating	Net recharge	Rating
<2	4	<500	1	Very slow	1	11–13	10
2–10	3	500–700	2	Slow	2	9–11	8
10–33	2	700–850	3	Moderate	3	7–9	5
>33	1	>850	4	Mod-high	4	5–7	3
				High	5	3–5	1

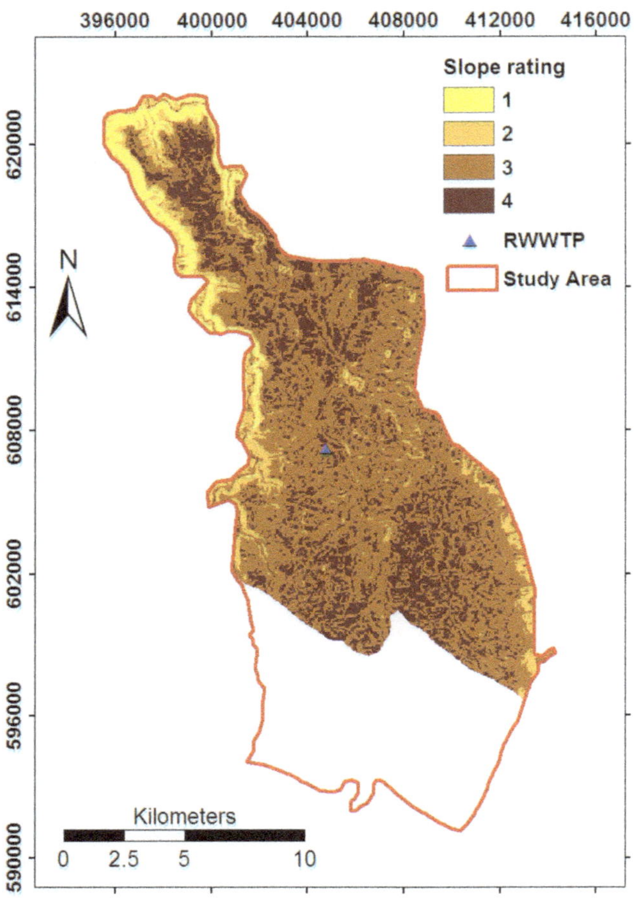

Fig. 6 Map of slope rating as factor in net recharge

properties: pH, electrical conductivity (EC), and temperature. The collected samples were transported to the Water Analysis Lab of Yarmouk University and stored in the refrigerator until analysis to prevent deterioration and changes in water quality.

Spectrophotometer (Thermo, Evolution100) was used to determine the concentration of NO_3^-. Nitrate concentration was determined by adding 25 ml from each sample in Erlenmeyer flask. Concentration 0.5 ml 1 N HCl was added to the sample and to the blank (distilled water). The absorbance was measured by the use of a spectrophotometer at 220 nm (APHA 1998).The concentration of NO_3^- was calculated according to the following equation:

$$NO_3^-(\mathrm{mg/L}) = F_{\mathrm{average}} \times (\text{Absorbance of Sample} - \text{Absorbance of Blank})$$

where $F = 42.52$, Absorbance of Blank = 0

Results of analysis were mapped spatially using ArcGIS 10 to demonstrate the distribution of pollution indicators.

For quality assurance and quality control, a sample blank, which consists of reagent water and all reagents that are normally in contact with a sample during the entire analytical procedure, was used to minimize the errors that come from any of the prepared chemical reagents. Standard blank was also used to remove the errors during standards preparation; it consists of reagent water and all reagents that are normally in contact with standard preparation. Calibration curve was used to determine the expected ion concentration of a measured sample. This curve contains at least three concentrations of each standard ion solution. By plotting the absorbance of the desired certified standard with its concentration, the curve must have a linear relationship with regression value (R^2) greater than 99 %. In addition, duplicate analyses were used, i.e., measure each sample twice under the same conditions to remove any errors during chemical analysis and to remove any personal errors. The average of the results is taken and compared with standard calibration curve.

Results and discussion

The modified DRASTIC vulnerability index

ArcGIS 10 was used to query, manipulate, and analyze the data for the assessment of groundwater vulnerability map. All maps are projected in JTM (Jordan Transverse Mercator).

Depth to groundwater

The shallower the water table, the more vulnerable is the groundwater to contamination. Depth to groundwater in the study area varies from 7 to 132 m (Fig. 4a), and its rating varies between 1 and 7 (Fig. 4b). The most vulnerable

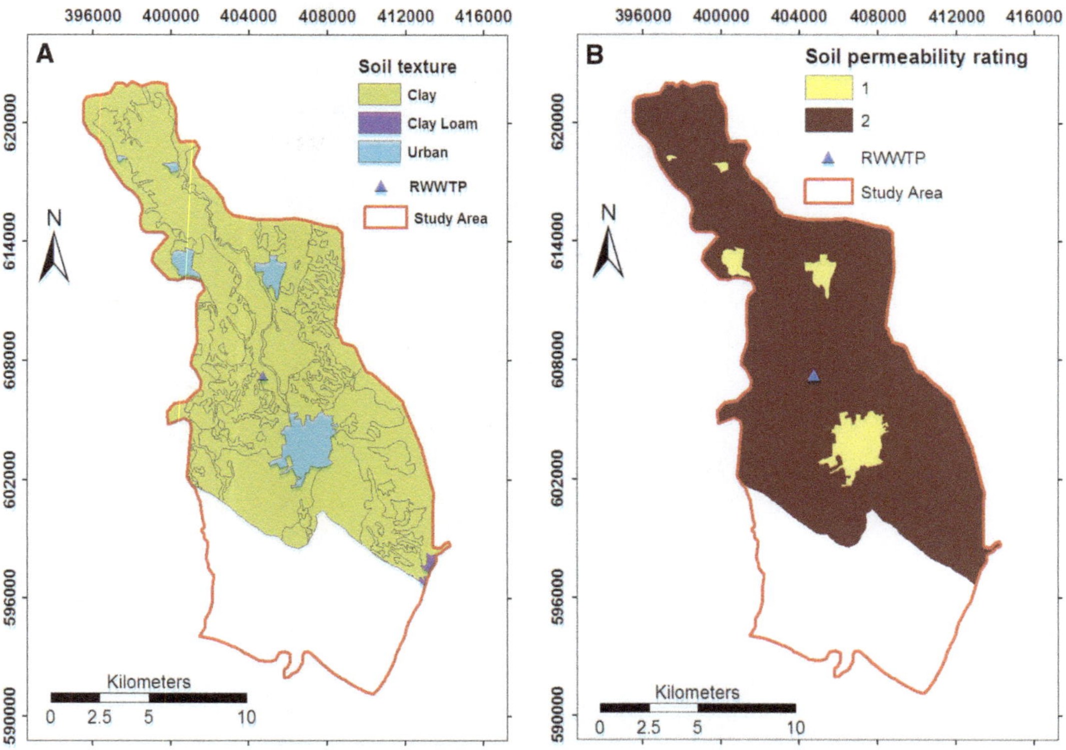

Fig. 7 Map of soil texture (**a**) and soil permeability rating (**b**) as a factor in net recharge in the study area

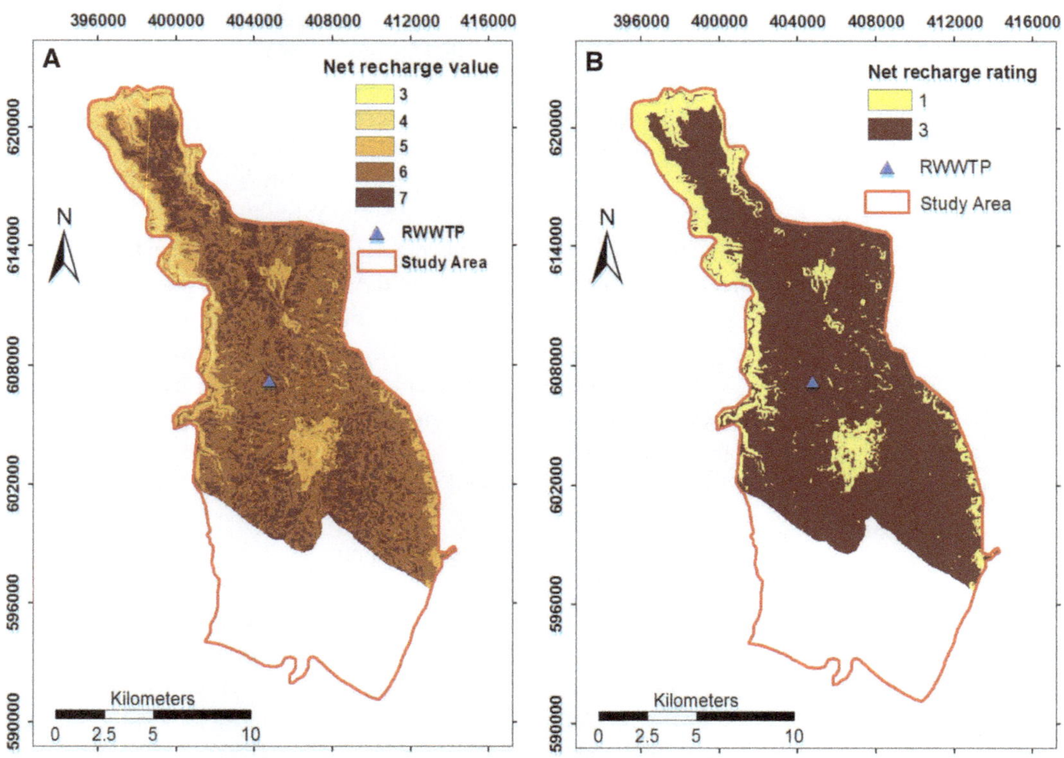

Fig. 8 Map of the net recharge value (**a**) and net recharge rating (**b**) in the study area

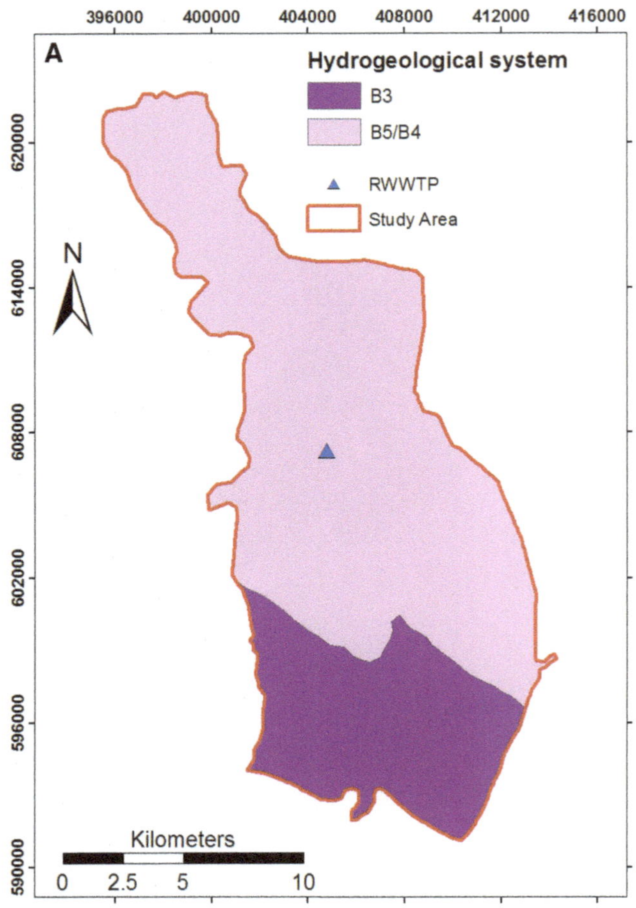

Fig. 9 Map of the aquifer media in the study area

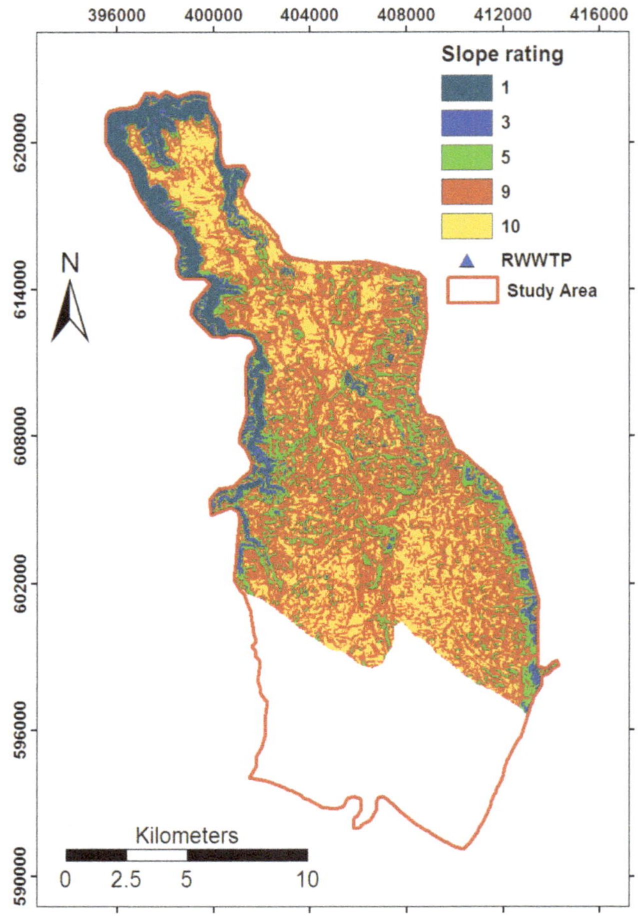

Fig. 10 Map of slope rating in the study area

groundwater was found in the central part of the study area surrounding the RWWTP.

Net recharge

Although net recharge dilutes the contaminant which enters the aquifer, it is the largest pathway for contaminant transport. Therefore, the amount of recharge is positively correlated with the vulnerability rating (Abdullahi 2009; Davis et al. 2002). Because there were no available data for net recharge in the study area, the following equation was used to calculate the net recharge value (Piscopo 2001):

$$\text{Recharge value} = \text{slope factor} + \text{rainfall factor (mm)} + \text{soil permeability factor}$$

The slope (%) in the study area (Fig. 5b) was derived from the ASTER digital elevation model (DEM) (Fig. 5a). The soil permeability map was generated from soil texture map (Fig. 7a) of the Ministry of Agriculture. The soil permeability classification and rating were based on the USDA (1994). Since there are only two rainfall stations in the study area, we used the mean value (256 mm) only because surface creation from two points is not statistically correct. Therefore, a fixed rating value of 1 was used according to Table 2 (Piscopo 2001). In addition, the slope and soil maps were rated (Figs. 6, 7b) according to the criteria given in Table 2.

The net recharge map was generated by stacking the factors of the net recharge parameters, which was then reclassified (Fig. 8) according to the values given in Table 2. The DRASTIC ratings of the net recharge parameter in the study area have two values (1 and 3), which indicate low net recharge rate.

Aquifer media

The greater permeability of the aquifer has a greater pollution potential rating thus the more vulnerable is the groundwater to contamination. Limestone, permeable sandstone, and unconsolidated sand and gravel have a relatively high rating aquifer media, whereas shale and unfractured igneous and metamorphic rocks have a lower rating (Hearne et al. 1992). The aquifer (B4/B5), which dominates most of the study area, is assigned a rating value 4, whereas the aquitard (B3) assigned a rating value 1 (Fig. 9).

Soil media

The soil map (Fig. 7) of scale 1:50,000 was obtained from the Ministry of Agriculture (2005). The study area is dominated by clayey soil (rating value 2), with very small area occupied by clay loam (rating value 3). Urban areas assigned a rating value of 1.

Topography

Slope map (Fig. 4b) was reclassified (Fig. 10) and rated in the range 1–10, with rating values 9 and 10 dominating the study area.

Impact of vadose zone

The geologic map (Fig. 2) is rated to reflect the impact of vadose zone: 1, 3, 4, 7 and 9 (Fig. 11). Most of the study area is covered by rating 7.

Hydraulic conductivity

Hydraulic conductivity map was generated based on the values (10 samples) taken from the open files of WAJ (2006) (Fig. 12). The values of hydraulic conductivity ranged from 2.99E-6 to 10E-5 m/s. DRASTIC ratings of the hydraulic conductivity in the study area were 1 and 2 only. The highly vulnerable groundwater according to hydraulic conductivity in the study area is found around and south east of the waste treatment plant.

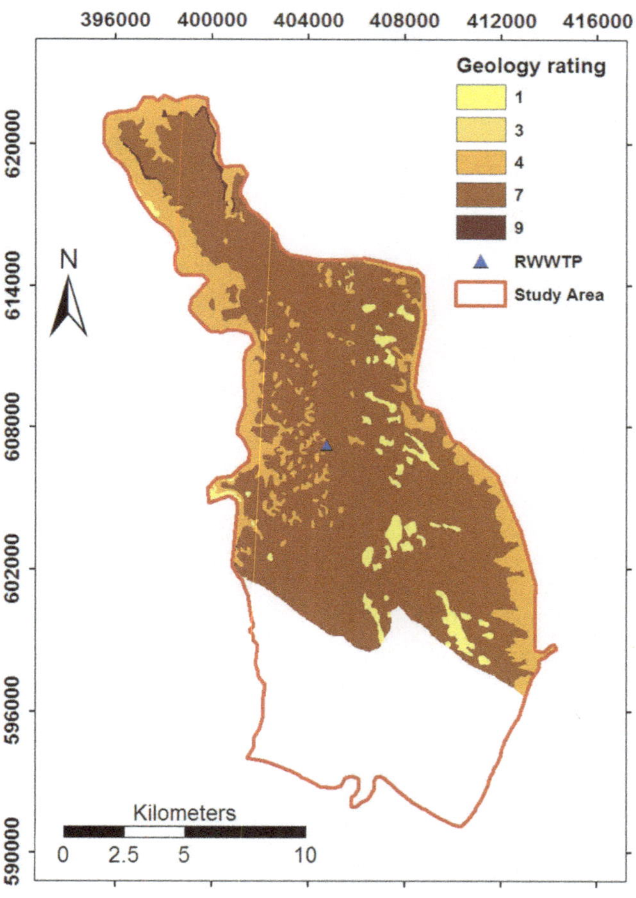

Fig. 11 Map of the impact of vadoze zone rating in the study area

Lineaments

The higher the lineament density means more vulnerable is the groundwater to contamination because lineaments are closely related to groundwater flow and contaminants migration (Mabee et al. 1994). Lineaments density was assigned a weight value of 5. Lineaments considered in the study area represent geological structures such as fractures and faults (Fig. 13a). A lineament density map was generated and rated into values 3 and 4 (Fig. 13b). The most vulnerable groundwater according to lineament density is found in the central parts of the study area.

Land use/land cover

The land use/land cover (LULC) in the study area comprises agriculture, rangeland, water, orchard, urban, and bareland (Fig. 14a). DRASTIC ratings of the LULC have the following values: 3,4,5,7 and 8 (Fig. 14b). Most of the

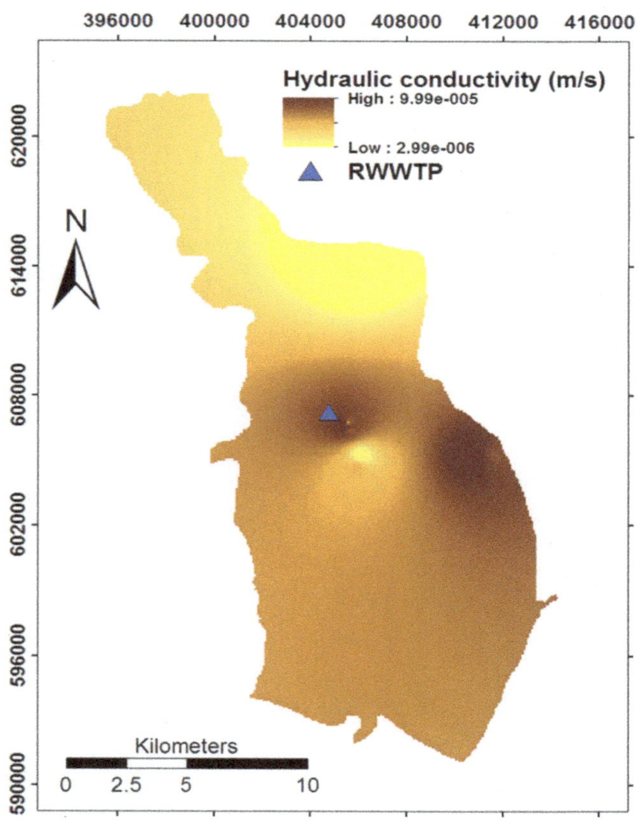

Fig. 12 Map of hydraulic conductivity (m/s) in the study area

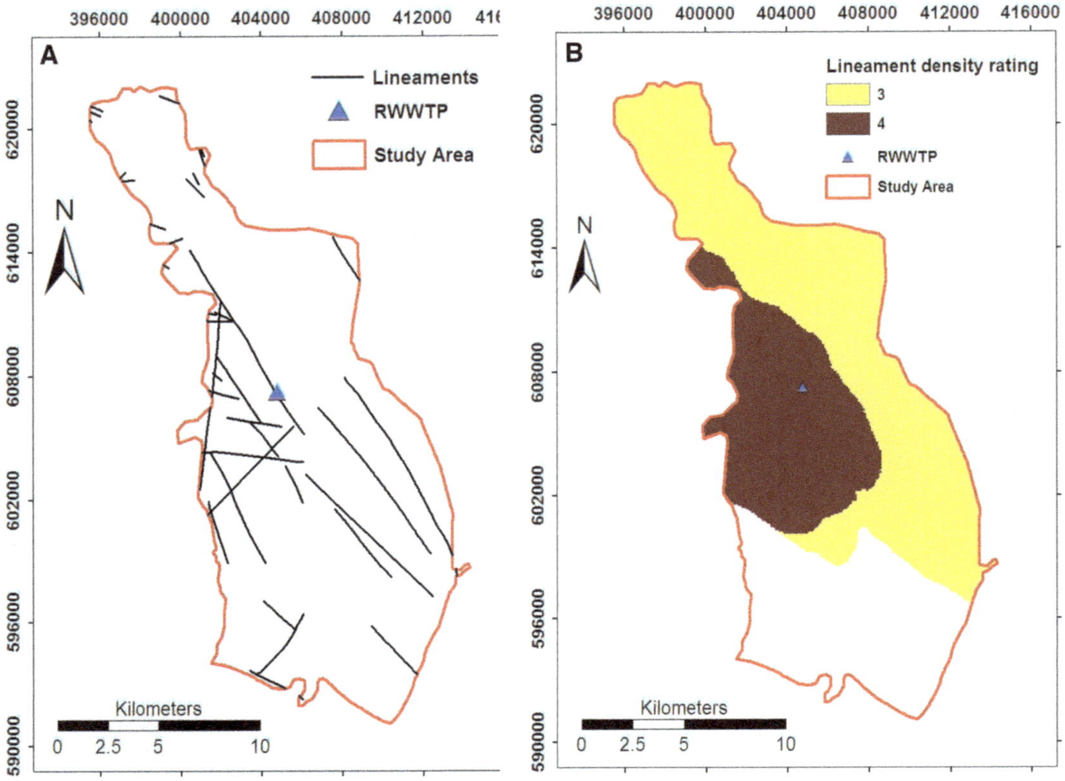

Fig. 13 Lineaments map (**a**) and lineament density rating (**b**) in the study area

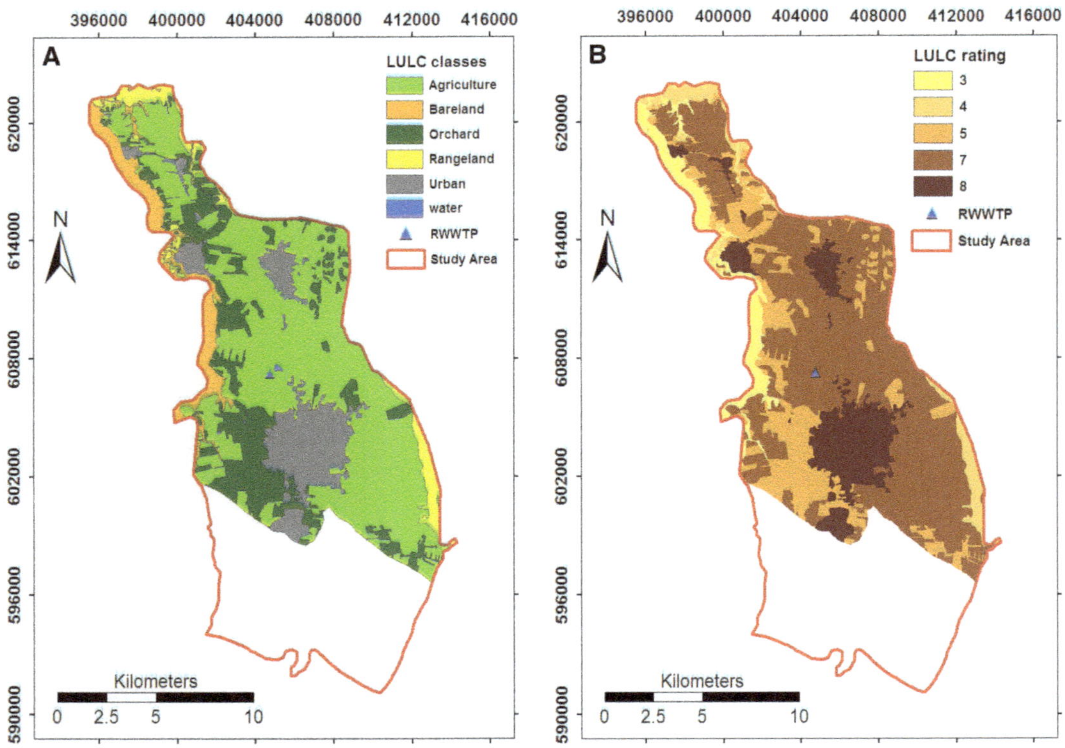

Fig. 14 Map of the major land use/land cover (**a**) in the study area and their ratings (**b**)

Table 3 DRASTIC Index and vulnerability class (Aller et al. 1987)

DRASTIC index	Vulnerability class
1–100	Low
101–140	Moderate
140–200	High
>200	Very high

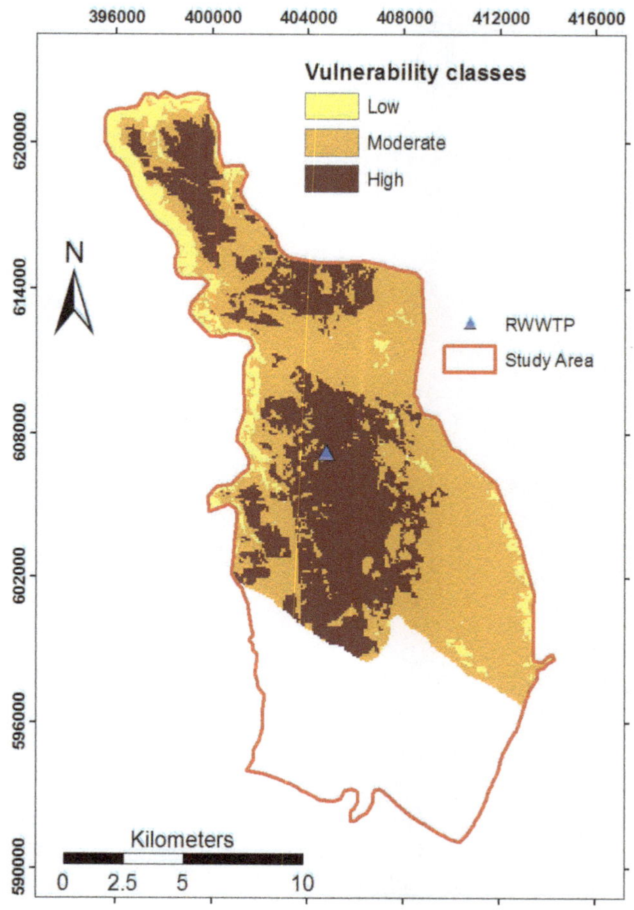

Fig. 15 The groundwater vulnerability map of the study area using the modified DRASTIC method

study area is covered by LULC rating 7, which represents agriculture.

Groundwater vulnerability modeling

The final modified DRASTIC vulnerability map was generated by overlying all the nine layers discussed earlier. The values of the DRASTIC index were classified according to the criteria given in Table 3 (Fig. 15). The vulnerability classes in the modified DRASTIC are low, moderate and high, which covered 9, 54.5, and 36.5 % of the study area, respectively.

As it is shown in Fig. 15a, the RWWTP is found within the high vulnerability class. This is due to the fact that

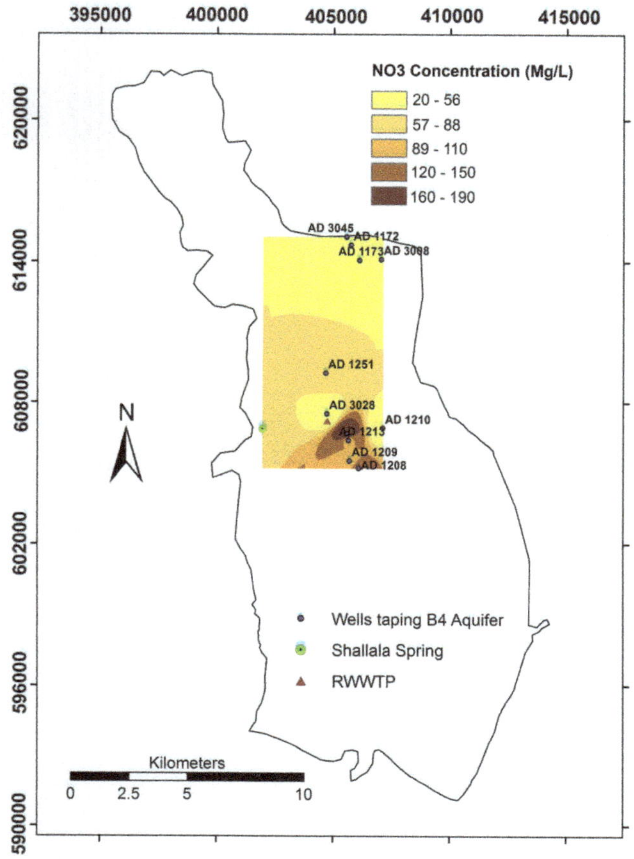

Fig. 16 Spatial distribution of the averages of the nitrate concentration for B4

depth to groundwater is shallow, with high net recharge value, high aquifer permeability, low slope, soil texture, high hydraulic conductivity, and high lineament density and land use (agriculture).

Sensitivity analysis

The highest variation index (27 %) is due to the removal of the LULC parameter from the model. This could mainly be attributed to the high theoretical weight assigned to this parameter (5). Hydraulic conductivity is the parameter that has the lowest effect on the variation in the final vulnerability (4 %). The vulnerability index also seems to be sensitive to the removal of the impact of vadose zone and lineament density (variation index = 20 and 13.5 %, respectively).

The single-parameter sensitivity analysis showed that the LULC parameter tends to be the most effective parameter in the vulnerability assessment (mean effective wt% is 23.2) in agreement with the result from map removal sensitivity analysis. Its effective weight exceeds the theoretical weight (15.2 %). The impact of vadose zone also shows high effective weight (18.7 %) that exceeds the theoretical weights (15.2 %). The rest of the parameters,

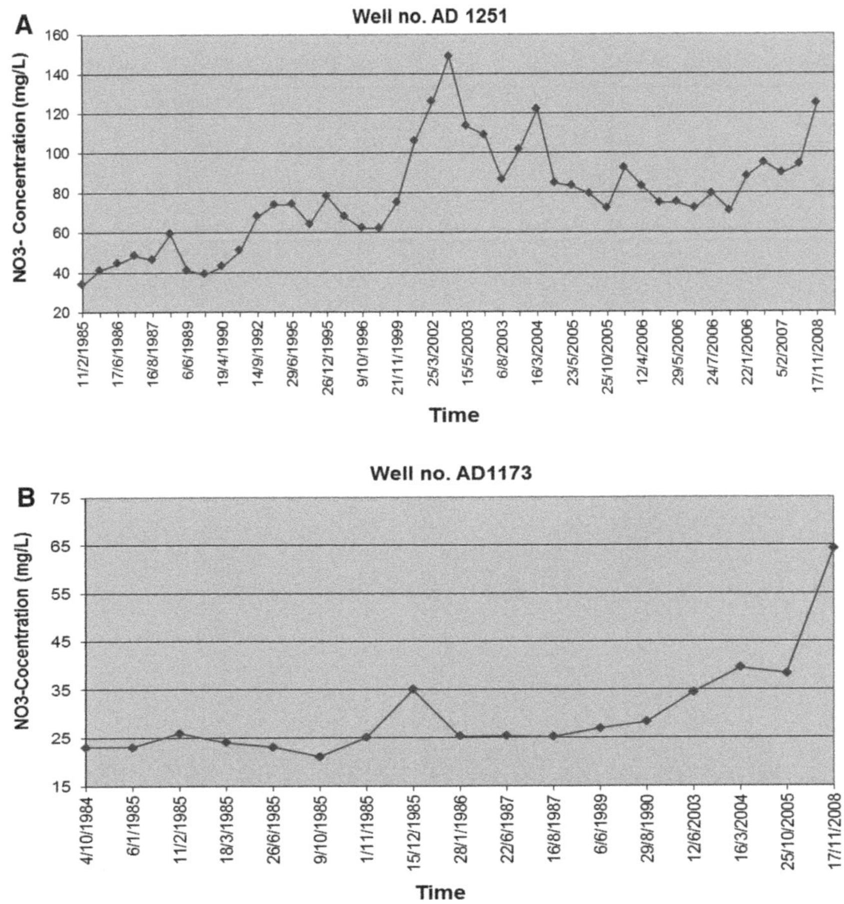

Fig. 17 Temporal variation in NO3-concentration for well no.AD1251 (**a**) and well no.AD1173 (**b**)

excluding topography have lower effective weights as compared to the theoretical weights. The significance of impact of vadose zone and LULC parameters highlights the importance of obtaining accurate, detailed, and representative information about these factors.

Model validation

Spatial concentrations of nitrate

Twelve samples of groundwater wells were sampled and analyzed for nitrate. Nitrate concentration in B4 aquifer was in the range of 20–193 mg/L with an average of 65.5 mg/L. The lowest value was recorded for well no. AD3028 and the highest value was recorded for well no. AD1296. Sixty-nine and 54 % of the B4 groundwater samples exceed the permissible limit of WHO (1993) and Jordanian (2001) standards of nitrate, respectively. Figure 16 shows that the highest concentration of NO_3^- for B4 aquifer was found toward the southeast of RWWTP.

Nitrate concentrations in the range 20–50 mg/L indicate the influence of human activities (Spalding and Exner 1993), whereas nitrate concentrations >50 mg/L are higher than the maximum acceptable concentration for drinking water (WHO 1993). It is found that 5 wells (83.5 %) of those having high nitrate concentration (>50 mg/L) are located in the highly vulnerability zone; and only the Shallala Spring is located in the moderate zone. Also, five of the wells (83.5 %) of those having a nitrate concentration between 20 and 50 mg/L are located in the moderate vulnerability zone. Only well no. AD3028 is found in the high vulnerability zone. These results show a very good match between nitrate concentrations level and the groundwater vulnerability model.

Nitrate concentration in the study area was correlated with depth to water table ($r = -0.51$). High concentration of nitrate is found in the area surrounding RWWTP, where depth to water table is <20 m below ground surface; this increases aquifer's vulnerability to contamination. Several studies showed a negative correlation between depth to groundwater and nitrate concentration (Changyuan et al. 2004; Enwright and Hudak 2009). A strong correlation ($r = 0.70$) exists between lineaments density and nitrate concentration. The pollution plume decreases away from the plant, towards the north, south, and east. A significant correlation exists between nitrate concentration and distance from the treatment plant ($r = -0.75$).

Temporal variation of nitrate

Nitrate concentration for well no. AD1251 (1985–2008) has increased from 34 mg/L in 1985 to more than 145 mg/L in 2002 (Fig. 17a). Starting 2003, the nitrate concentrations showed a clear decreasing trend, which is attributed to the upgrading of RWWTP from stabilization ponds to activated sludge treatment process.

Well no. AD1173 (1984-2005) showed a clear increase in nitrate concentration from an average of 25 mg/L before the existence of the treatment plant to 35 mg/L afterwards (Fig. 17b). However, the upgrading of the treatment plant did not reflect on the nitrate concentrations most probably because of the far distance between the plant and the groundwater well and topography. Statistical analysis by Obeidat et al. (2013) indicated significant differences between nitrate concentration before and after the construction of RWWTP.

Conclusions

The modified DRASTIC model developed a more accurate vulnerability map by adding the two parameters: land use and lineaments. Model validation proved good agreement between nitrate and the final vulnerability map. However, calibration of the DRASTIC model will depict groundwater vulnerability with higher accuracy in the view of model validation. Sensitivity analysis showed that the parameters LULC and the impact of vadose zone are most effective in vulnerability assessment. In B4 aquifer, there are 10 wells occurred in the zone of high vulnerability, and 1 well occurred in the zone of moderate vulnerability.

About 69 and 54 % of the B4 groundwater samples NO_3^- concentrations exceeded the permissible limit, the WHO (1993) and the Jordanian (2001) standards, respectively. Temporal data indicated the increase of nitrated concentrations due to the establishment of the RWWTP.

Acknowledgments Thanks are due to the Deanship of Scientific Research and Graduate Studies at Yarmouk University for providing the financial support to implement this work.

References

Abdullahi U (2009) Evaluation of models for assessing groundwater vulnerability to pollution in Nigeria. Bayero J Pure Appl Sci 2:138–142

Al-Adamat R, Foster I, Baban S (2003) Groundwater vulnerability and risk mapping for the Basaltic aquifer of the Azraq basin of Jordan using GIS, remote sensing and DRASTIC. Appl Geogr 23:303–324

Al-Hanbali A, Kondoh A (2008) Groundwater vulnerability assessment and evaluation of human activity impact (HAI) within the Dead Sea groundwater basin Jordan. Hydrogeol J 16(3):499–510

Aller L, Bennett T, Lehr J, Petty R, Hackett G (1987) DRASTIC: a standardized system for evaluating groundwater pollution potential using hydrogeologic settings: U.S. environmental protection agency report 600/2- 87/035, p 622

Awawdeh MM, Jaradat RA (2010) Evaluation of aquifers vulnerability to contamination in the Yarmouk River basin, Jordan, based on DRASTIC method. Arab J Geosci 3(3):273–282

Babiker I, Mohammed M, Hiyama T, Kato K (2005) A GIS-based DRASTIC model for assessing aquifer vulnerability in kakmigahara heights, Gifu Prefecture, central Japan. Sci Total Environ 345:127–140

Baker L (1992) Introduction to non-point source pollution in the United States and prospects for wetland use. Ecol Eng 1:1–26

Burkart MR, Feher J (1996) Regional estimation of ground water vulnerability to nonpoint sources of agricultural chemicals. Water Sci Technol 33(4–5):241–247

Changyuan T, Jianyao C, Njun S (2004) Long-term effect of wastewater irrigation onnitrate in groundwater in the North China Plain. In: Steenvoorden J, Endreny T (eds) Wastewater reuse and groundwater quality. Proceedings symposium HS04: IAHS Publications, Red Book Series, 285, pp 34–40

Connell L, Daele G (2003) A quantitative approach to aquifer vulnerability mapping. J Hydrol 276:71–88

Davis A, Long A, Wireman M (2002) A sensitivity method for carbonate aquifers in karst terrain. Environ Geol 42:65–72

Enwright N, Hudak P (2009) Spatial distribution of nitrate and related factors in the High Plains Aquifer, Texas. Environ Geol 58:1541–1548

Evans B, Mayers W (1990) A GIS-based approach to evaluating regional groundwater pollution potential with DRASTIC. J Soil Water Conserv 45:242–245

Ford M, Tellam J (1994) Source, type and extent of inorganic contamination within the Birmingham urban aquifer system, UK. J Hydrol 156:101–135

Fortin M, Thomson K, Edwards G (1997) The role of error propagation for integrating multisource data within spatial models: the case of the DRASTIC groundwater vulnerability model. Earth Surface Remote Sensing. London: Procedure SPIE Conference, pp 358–361

Fritch T, McKnight C, Yelderman J, Arnold J (2000) An aquifer vulnerability assessment of the paluxy aquifer, central Texas, USA, using GIS and a modified DRASTIC approach. Environ Manag 25:337–345

Hadadin N, Qaqish M, Akawwi E, Bdour A (2010) Water shortage in Jordan-sustainable solutions. Desalination 250:197–202

Hasiniaina F, Guoyi L, Zhou J (2010) Regional assessment of groundwater vulnerability in Tamtsag basin, Mongolia using DRASTIC model. J Am Sci 6(11):69

Hearne G, Wireman M, Campbell A, Turner S, Ingersoll G (1992) Vulnerability of the uppermost ground water to contamination in the greater Denver area. US Geological Survey, Water-Resources Investigations Report 92–4143

Huan H, Wang J, Teng Y (2012) Assessment and validation of groundwater vulnerability to nitrate based on a modified DRASTIC model: a case study in Jilin City of northeast China. Sci Total Environ 440:14–23

Hubbard R, Sheridan J (1994) Nitrates in groundwater in the southeastern USA. In: Adriano D, Iskadar A, Murarka I (eds)

Contamination of groundwater. Northwood, United Kingdom, Science Reviews, p 45–303
Jordanian Standards JS 286 (2001) Water—drinking water, 4th edn, Department of Standards and Specifications, Amman, The Hashemite Kingdom of Jordan
Knox R, Sabatini D, Canter L (1993) Subsurface transport and fate processes. Lewis Publishers, USA
Lake IR (2003) Evaluating factors influencing groundwater vulnerability to nitrate pollution: developing the potential of GIS. J Environ Manag 328(68):315–328
Lerner D, Yang Y, Barrett M, Tellam J (1999) Loading of non-agricultural nitrogen in urban groundwater. In: Ellis, JB (ed), Impacts of urban growth on surface and groundwater quality, IAHS publ. No. 259. IAHS Press, p 117–123
Lodwick W, Monson W, Svoboda L (1990) Attribute error and sensitivity analysis of map operations in geographical information system. Int J Geogr Inf Syst 4:13–28
Mabee S, Hardcastle K, Wise D (1994) A method of collecting and analyzing lineaments for regional-scale fractured-bedrock aquifer studies. Groundwater 32(6):884–894
MacQuarrie K, Sudicky E, Roberston W (2001) Numerical simulation of a fine-grained denitrification layer for removing septic system nitrate from shallow groundwater. J Hydrol 52:29–55
Margane A, Hobler M, Subah A (1999) Mapping of groundwater vulnerability and hazards to groundwater in the Irbid Area, N Jordan. Z Angew Geol 45(4):175–187
Mclay CD, Dragten R, Sparling G, Selvarajah N (2001) Predicting groundwater nitrate concentrations in a region of mixed agricultural land use: a comparison of three approaches. Environ Pollut 115(2):191–204
Mohsen M (2007) Water strategies and potential of desalination in Jordan. Desalination 203:27–46
Napolitano P, Fabbri A (1996) Single-parameter sensitivity analysis for aquifer vulnerability assessment using DRASTIC and SINTACS. Hydrol Water Resour Manag 235:559–566
Natural Resources Authority (1997) Geological Map of Irbid scale 1:50,000, sheet no. 15. Royal Jordanian Geographic Center, Amman
Navarro A, Carbonell M (2007) Evaluation of groundwater contamination beneath an urban environment: the Besòs river basin (Barcelona, Spain). J Environ Manag 85:259–269
Navulur K, Engel B (1998) Groundwater vulnerability assessment to non-point source nitrate pollution on a regional scale using GIS. Trans Am Soc Agric Eng 41:1671–1678
Nawafleh A, Awawdeh M, Salameh E (2011) Assessment of groundwater vulnerability to contamination in Irbid Governorate, North Jordan. DIRASAT 38(2)
Obeidat M, Awawdeh M, Al-Mughaid H (2013) Impact of a domestic wastewater plant on groundwater pollution, north Jordan. The Revista Mexicana de Ciencias Geológicas 30:371–384
Piscopo G (2001) Groundwater vulnerability map, explanatory notes, Castlereagh Catchment, NSW. Department of Land and Water Conservation, Australia, http://www.dlwc.nsw.gov.au/care/water/groundwater/reports/pdfs/castlereagh_map_notes.pdf. Accessed 20 Nov 2012
Postma D, Boesen C, Kristiansen H, Larsen F (1991) Nitrate reduction in an unconfined sandy aquifer: water chemistry, reduction processes and geochemical modeling. Water Resour Res 27:45–202
Rundquist DC, Peters AJ, Di L, Rodekohr DA, Ehrman RL, Murray G (1991) Statewide groundwater vulnerability assessment in Nebraska using the DRASTIC/GIS model. Geocarto Int 6:51–58
Rupert MG (2001) Calibration of the DRASTIC ground water vulnerability mapping method. Ground Water 39(4):625–630
Secunda S, Collin M, Melloul A (1998) Groundwater vulnerability assessment using a composite model combining DRASTIC with extensive agricultural land use. Environ Manag 54:39–57
Seiler R (2005) Combined use of 15 N and 18 O of nitrate and 11B to evaluate nitrate contamination in groundwater. Appl Geochem 20:1626–1636
Spalding R, Exner M (1993) Occurrence of nitrate in groundwater—a review. J Environ Qual 22:392–402
Stigter T, Almeida P, Carvalho D, Ribeiro L (2006) Evaluation of an intrinsic and a specific vulnerability assessment method in comparison on a regional scale. Hydrogeol J 14(3):79–99
United Nations (2006) Common Country Assessment: Jordan
United States Environmental Protection Agency (US EPA) (1994) Handbook: Groundwater and Wellhead Protection. US EPA Report No. EPA/625/R-94/001, Washington, DC, p 239
Vrba J, Zaporozec A (1994) Guidebook on mapping groundwater vulnerability—AH International Contributions to Hydrogeology 16, Hannover/FRG (Heise Publ.) p 131
Wakida T, Lerner D (2005) Non-agricultural sources of groundwater nitrate: a review and case study. Water Res 39:3–16
Wardam B (2004) More politics then water: water rights in Jordan. IUCN regional water program in west Asia and North Africa, Jordan. Global Issue paper No. 11
WHO (World Health Organization) (1993) Guidelines for drinking water quality, vol 1, Recommendations, 2nd edn, WHO, Geneva
Zwahlen F (ed) (2004) Vulnerability and risk mapping for the protection of carbonate (karst) aquifers. Final report (COST Action 620). European Commission, p 297

2

Hydro-geochemistry and application of water quality index (WQI) for groundwater quality assessment, Anna Nagar, part of Chennai City, Tamil Nadu, India

S. Krishna kumar · A. Logeshkumaran · N. S. Magesh · Prince S. Godson · N. Chandrasekar

Abstract In the present study, the geochemical characteristics of groundwater and drinking water quality has been studied. 24 groundwater samples were collected and analyzed for pH, electrical conductivity, total dissolved solids, carbonate, bicarbonate, chloride, sulphate, nitrate, calcium, magnesium, sodium, potassium and total hardness. The results were evaluated and compared with WHO and BIS water quality standards. The studied results reveal that the groundwater is fresh to brackish and moderately high to hard in nature. Na and Cl are dominant ions among cations and anions. Chloride, calcium and magnesium ions are within the allowable limit except few samples. According to Gibbs diagram, the predominant samples fall in the rock–water interaction dominance and evaporation dominance field. The piper trilinear diagram shows that groundwater samples are Na–Cl and mixed CaMgCl type. Based on the WQI results majority of the samples are falling under excellent to good category and suitable for drinking water purposes.

Keywords Gibbs and Piper diagram · Groundwater quality · Anna Nagar · Part of Chennai city

S. Krishna kumar (✉)
Department of Geology, University of Madras, Guindy campus, Chennai 600025, Tamil Nadu, India
e-mail: coralkrishna@yahoo.co.in

A. Logeshkumaran
Department of Civil Engineering, VKS College of Engineering and Technology, Desiyamagalam, Karur 639120, Tamil Nadu, India
e-mail: logucivil106@gmail.com

N. S. Magesh · P. S. Godson · N. Chandrasekar
Centre for GeoTechnology, Manonmaniam Sundaranar University, Tirunelveli 627 012, Tamil Nadu, India
e-mail: mageshissivan@gmail.com

P. S. Godson
e-mail: princegodsons@gmail.com

N. Chandrasekar
e-mail: profncsekar@gmail.com

Introduction

Ground water contamination in urban environment is a major issue and is complicated by large number of potential source of contamination (Jayaprakash et al. 2008). India has wide spectral variations of meteorological, topographical, geomorphological, hydrological, geological, and hydrogeological conditions. The chemistry of groundwater is an important factor determining its use for domestic, irrigation and industrial purposes. Utilization of land varies from place to place due to rapid urbanization and industrialization, without following the strict environmental norms, causing a lot of variation of quality of groundwater within a short distance, which constrains the developmental activities drastically everywhere (Subba Rao 1997, 2006). The urban aquifers are the only natural resource for drinking water supply, they are often perceived as of lesser relevance for the drinking water supply, leading to crisis in terms of drinking water scarcity, becoming increasingly polluted thereby decreasing their potability (Dixit et al. 2005; Tiwari et al. 2012). Once contamination of groundwater in aquifers occurs by means of industrial activities and urban development, it persists for hundreds of years because of very slow movement of water in them (Jerry 1986) and prompts investigations on their quality. The knowledge of hydrochemistry is important to assess the ground water quality in any area in which the ground water is used for both irrigation and drinking needs (Srinivas et al. 2013). The water quality assessment may give clear information about the subsurface geologic environments in which the water presents (Raju et al. 2011). The conventional techniques such as trilinear plots, statistical techniques are widely accepted methods to determine the quality of water. The problems of ground water quality are more acute in areas that are coastal, densely populated and thickly industrialized and have shallow groundwater tube wells (Ballukraya and Ravi 1999; Reddy and Subba Rao 2001; Malini et al. 2003; Krishna kumar et al. 2011; Padmalal et al. 2012; Bagyaraj et al. 2013; Selvam et al. 2013). The present study was carried out for qualitative analysis based on physico-chemical parameters of groundwater.

Study area

Chennai, formerly known as Madras, is the capital of the state of Tamil Nadu and India's fourth largest metropolitan city. The latitude of Chennai city is 13.04°N and longitude 80.17°E. The city covers an area of 174 km^2. The metropolitan area covers 1,177 km^2. The city is divided into four major parts like North, Central, South and West regions. The northern localities are devoted for the automobile industries, other medium- and small-scale industries. The study area has unmatched importance as a commercial and residential area in West Chennai. The study area extends from 13°5′31″ North to 80°11′57″ East (Fig. 1). The population of the Chennai city is 8,917,749 with a density of 26,702/km^2 (CDSH 2012).

The Northeast monsoon during the month of October, November and December chiefly contributes rainfall to the study area. Most of the precipitation occurs in the form of one or two cyclones caused due to depressions in Bay of Bengal. The average annual rainfall of the study area is 1,200 mm. The Chennai district enjoys a tropical climate with a mean annual temperature of 24.3–32.9 °C. The humidity is usually in the range of 6–84 %. Chennai is underlain by various geological formations from ancient Archaean to the Recent Alluvium. The geological formations can be grouped into three units, namely (1) the Archaean crystalline rocks, (2) consolidated Gondwana and Tertiary sediments, and (3) the Recent Alluvium. The study area chiefly consists of unconfined Tertiary sedimentary aquifers. The Archaean crystalline rocks comprise chiefly of Charnockites, gneisses and the associated basic and ultra basic intrusive. The alluvium consists of sand, silt

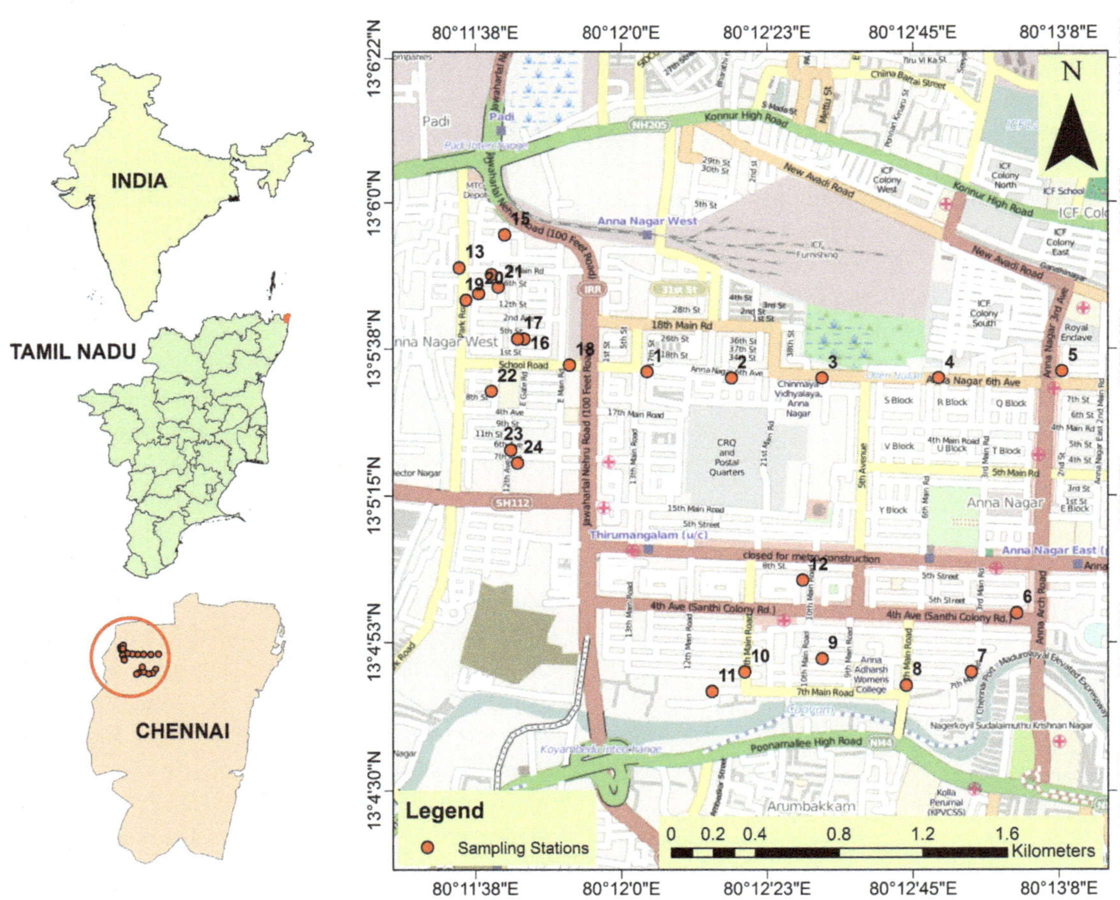

Fig. 1 Sampling location map of the study area

and clays. The thickness of alluvium varies from place to place and a maximum of 28 m is encountered in north Chennai. The computed transmissivity varied between 6.00 and 872 m^2/day and the storativity varied between 2.9×10^{-4} and 4.5×10^{-3} (CGWB Technical Report 2008).

Materials and methods

Representative ground water samples were obtained from 24 bore wells selected on the basis of geographical variation, the depth ranges from 80–150 m. The electrical conductivity, temperature and pH were measured in situ. Field measurements of pH and electrical conductivity (EC) were made using a handheld pH and EC meter (HANNA HI—9828, USA). Bicarbonate analysis was carried out using acid titration method; chloride concentration was measured by $AgNO_3$ titration method; sulphate by $BaCl_3$ turbidity method using a spectrophotometer. Sodium and potassium were analyzed using flame photometer; calcium and magnesium by the titration method. The analytical procedures are as suggested by the American Public Health Association (APHA 1995). The analytical precision and measurement reproducibility was less than 2 %. The ionic balance error for studied ions was within ±5 %. The base map of the study area was prepared using the Survey of India topographic sheets and digitized using Arc GIS 9.3 software. Trimble® Recon GPS was used to find the location of each sampling site and the coordinates were imported to GIS platform for preparation of the base map. The geochemical results are plotted on piper trilinear plot using AquaChem 4.0 software and Gibbs diagram is plotted to assess the quality controlling mechanism and dominated hydro-geochemical facies of the study area. Statistical analysis was performed using SPSS software package (SPSS, 2001). The physico-chemical parameters of the analytical results of groundwater were compared with standard guideline values recommended by the WHO.

Water quality index calculation

The water quality index (WQI) was calculated for evaluating influence of natural and anthropogenic activities based on several key parameters of groundwater chemistry. To calculate the WQI, the weight has been assigned for the physico-chemical parameters according to the parameters relative importance in the overall quality of water for drinking water purposes. The assigned weight ranges from 1 to 5. The maximum weight of 5 has been assigned for nitrate and TDS, 4 for pH, EC, SO_4, 3 for HCO_3, Cl, 2 for Ca, Na, K and weight 1 assigned for magnesium (Vasanthavigar et al. 2010). The relative weight is computed from the following equation.

$$Wi = wi / \sum_{i=1}^{n} wi, \qquad (1)$$

where

Wi is the relative weight

wi is the weight of each parameter

n is the number of parameters.

The quality rating scale for each parameter is calculated by dividing its concentration in each water sample by its respective standards (World Health Organization 2011) and multiplied the results by 100.

$$qi = (Ci/Si) \times 100, \qquad (2)$$

where

qi is the quality rating

Ci is the concentration of each chemical parameter in each sample in milligrams per liter

Si is the World Health Organization standard for each chemical parameter in milligrams per liter according to the guidelines of the (WHO 2011)

For computing the final stage of WQI, the SI is first determined for each parameter. The sum of SI values gives the water quality index for each sample.

$$SIi = Wi \times qi \qquad (3)$$

$$WQI = \sum SIi, \qquad (4)$$

where

SIi is the sub-index of *i*th parameter

qi is the rating based on concentration of *i*th parameter

n is the number of parameters

Results and discussion

The statistical parameters like minimum, maximum and mean concentration of physico-chemical parameters, major ion concentrations are tabulated in Table 1.

Temperature and pH

The temperature variation ranges from 26 to 27 °C with a mean value of 26.67 °C. The pH indicates the strength of the water to react with the acidic or alkaline material present in the water. It controls by carbon dioxide, carbonate and bicarbonate equilibrium (Hem 1985). The combination of CO_2 with water forms carbonic acid, which affects the pH of the water. The permissible limit of pH is 6.5–8.5. The pH in the groundwater is varied from 6.0 to 8.2. This may be attributed to the anthropogenic activities like sewage disposal and improper irrigation process and weathering process in the study area.

Table 1 Statistics of physico-chemical parameters, major ions ($n = 24$), World Health Organization (World Health Organization 2011) and Bureau of Indian Standards (BIS, 1991) for drinking waters of study area

Parameters	Minimum	Maximum	Mean	WHO standard (2011)	BIS standard (1991)
Temp (°C)	26 °C	27 °C	26.67 °C	–	–
pH (on scale)	6	8.2	7.03	6.5–8.5	6.5–8.5
EC (μS/cm)	580	7,250	1,828	1,500	–
TDS (mg/l)	373.5	4,669	1,177	500	500
HCO_3 (mg/l)	18.3	359.9	165.7	500	–
Cl (mg/l)	115.21	2,011.8	383.3	250	250
SO_4 (mg/l)	22.2	98.6	53.9	250	200
NO_3 (mg/l)	0.1	3.9	1.013	45	45
Ca^{2+} (mg/l)	26	130	53.5	75	75
Mg^{2+} (mg/l)	1.2	141.6	18.8	50	30
Na^+ (mg/l)	71	1,200	258.9	200	–
K^+ (mg/l)	3	36	9.33	12	–
TH (mg/l)	89.5	814.55	211.58	–	–

Electrical conductivity (EC)

Electrical conductivity is a measure of water capacity to convey the electrical current. The most desirable limit of EC in drinking water is prescribed as 1,500 μS/cm. The value of EC is between 580 and 7,250 μS/cm. EC measures the ability of a material to conduct an electric current such that the higher EC indicates enrichment of salts in the groundwater. Thus, the EC can be classified as type I, if the enrichments of salts are low (EC < 1,500 μS/cm); type II, if the enrichment of salts are medium (EC: 1,500 and 3,000 μS/cm); and type III, if the enrichments of salts are high (EC > 3,000 μS/cm; Sarath Prasanth et al. 2012). According to the classification of EC, 63 % of the total groundwater samples (1, 3–9, 12, 15, 18, 20–23) falling under the type I (low enrichment of salts), 25 % of the samples (13, 14, 16, 17, 19, 24) under the type II (medium enrichment of salts), and 12 % of the samples (2, 10, 11) under the type III (high enrichment of salts).

Total dissolved solids (TDS)

Total dissolved salts in the groundwater is between 373.52 and 4,669 mg/l. Low TDS (373.52–895.16 mg/l) is observed in the samples 1, 3–9, 15,18 and 23 which indicates the influence of rock–water interaction in relation to recharge water. The high TDS observed in the study area ranges from 959.56 to 4,669 mg/l; the occurrence of high TDS is due to the influence of anthropogenic sources, such as domestic sewage, septic tanks and agricultural activities. According to WHO specification TDS up to 500 mg/l is the highest desirable and up to 1,500 mg/l is maximum permissible.

Degree of groundwater quality can be classified as fresh, if the TDS is less than 1,000 mg/l; brackish, if the TDS is between 1,000 and 10,000 mg/l; saline, if the TDS is varied from 10,000 to 1,000,000 mg/l; and brine, if the TDS is more than 1,000,000 mg/l (Todd 1980). Accordingly, the quality of groundwater in the present study area is classified as fresh and brackish in 70 and 30 % of the total water samples.

Bicarbonate (HCO_3)

The concentration of carbonates in natural waters is a function of dissolved carbon dioxide, temperature, pH, cations and other dissolved salts. Bicarbonate concentration of natural waters generally held within a moderate range by the effects of the carbonate equilibrium. Most surface streams contained less than 200 mg/l of carbonate and bicarbonate, but in ground water somewhat higher. The concentration of bicarbonate is observed from 18.3 to 359.9 mg/l, few samples exceeding the permissible limits of bicarbonate.

Chloride (Cl)

The chloride ion is the most predominant natural form of the element chlorine and is extremely stable in water. The chloride in groundwater may be from diverse sources such as weathering, leaching of sedimentary rocks and soil, domestic and municipal effluents (Sarath Prasanth et al. 2012). The range of chloride is found to vary between 115.21 and 2,011.8 mg/l for water samples. As per (World Health Organization 2011) and Indian standards (BIS, 1991) the desirable limit for chloride is 250 mg/l. For the study area it has been found that in certain locations the chloride concentration exceeds the maximum permissible limit. The chloride ion concentration in groundwater of the study area exceeds the maximum allowable limit of 600 mg/l in two samples (2 and 11). They may be due to the lack of underground drainage system and bad maintenance of environment around the sources.

Sulphate (SO_4)

The concentration of sulphate is likely to react with human organs if the concentration exceeds the maximum allowable limit of 400 mg/l and causes a laxative effect on human system with the excess magnesium in groundwater. The content of SO_4^- is observed from 22.2 to 98.6 mg/l. However, the sulphate concentration in groundwater of the study area is within the maximum allowable limit in the entire sample.

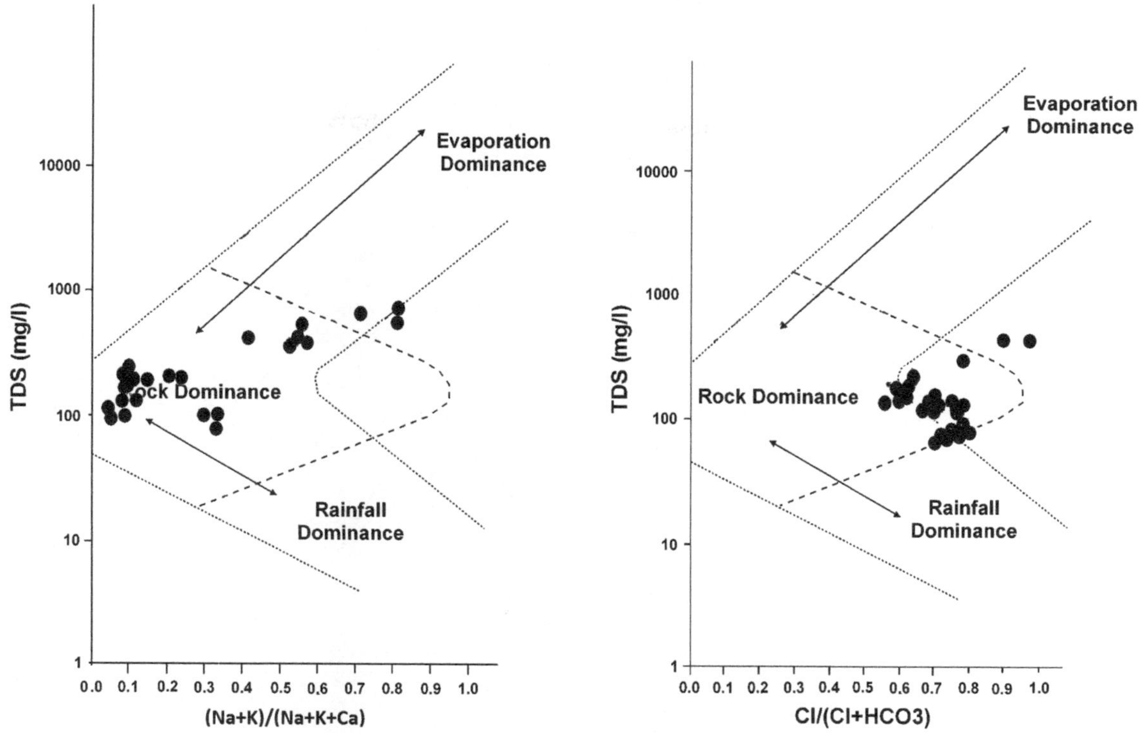

Fig. 2 Gibbs diagram

Nitrate (NO_3)

Nitrogen compounds are the most widespread contaminants in subsurface environments, mainly originating from non-point and multi-point agricultural sources (Pang et al. 2013). Epidemiological evidence suggests that nitrate-nitrogen exposure is strongly associated with several diseases, such as methemoglobinemia (blue baby syndrome), gastric cancer, thyroid disease and diabetes (Krishna Kumar et al., 2011). Hence, increasing nitrogen contamination seriously threatens public drinking water supply and human health. The value of NO_3^- in the groundwater is observed between 0.1 and 3.9 mg/l with an average value of 1.01 mg/l. All samples are not exceeding the permissible limit of 45 mg/l as per WHO standard.

Calcium and magnesium (Ca and Mg)

The calcium and magnesium in waters are generally used to classify the suitability of water. Calcium and magnesium are directly related to hardness of the water and these ions are the most abundant elements in the surface and groundwater and exist mainly as bicarbonates and to a lesser degree in the form of sulphate and chloride. The concentration of Ca^{2+} is between 26 and 130 mg/l, and concentration of Mg^{2+} is varied from 1.2 to 141.6 mg/l. The maximum concentration of calcium ions can cause abdominal ailments and is undesirable for domestic purposes as it causes encrustation and scaling. The higher concentration of Mg^{2+} (66 and 141.6 mg/l) is observed in the groundwater sample nos. 10 and 2. The rest of the groundwater samples show Mg^{2+} from 1.2 to 48 mg/l. The contribution of Ca^{2+} and Mg^{2+} to the total cations is approximately 7 to 19 % and 13 to 29 %, respectively.

Sodium and potassium (Na and K)

The concentration of Na^+ is varied from 71 to 1,200 mg/L. In general case, the Na^+ is the dominant ion among the cations and is present in most of the natural waters, which contributing approximately 53 to 69 % of the total cations. This is because of the silicate weathering and/or dissolution of soil salts stored by the influences of evaporation, anthropogenic activities, agricultural activities and poor drainage conditions. The higher contribution of Na^+ than that of the contribution of Ca^{2+} to the total cations is expected due to influence of ion exchange. However, the higher concentration of Na^+ observed from the groundwater samples contributes approximately 73 % to the total cations. K^+ is a naturally occurring element; but its concentration remains quite lower compared with Ca, Mg and Na. The average concentration of potassium is 9.33 mg/l and maximum value of 36 mg/l, which indicate the potassium form complexes in studied conditions.

Fig. 3 Piper trilinear diagram

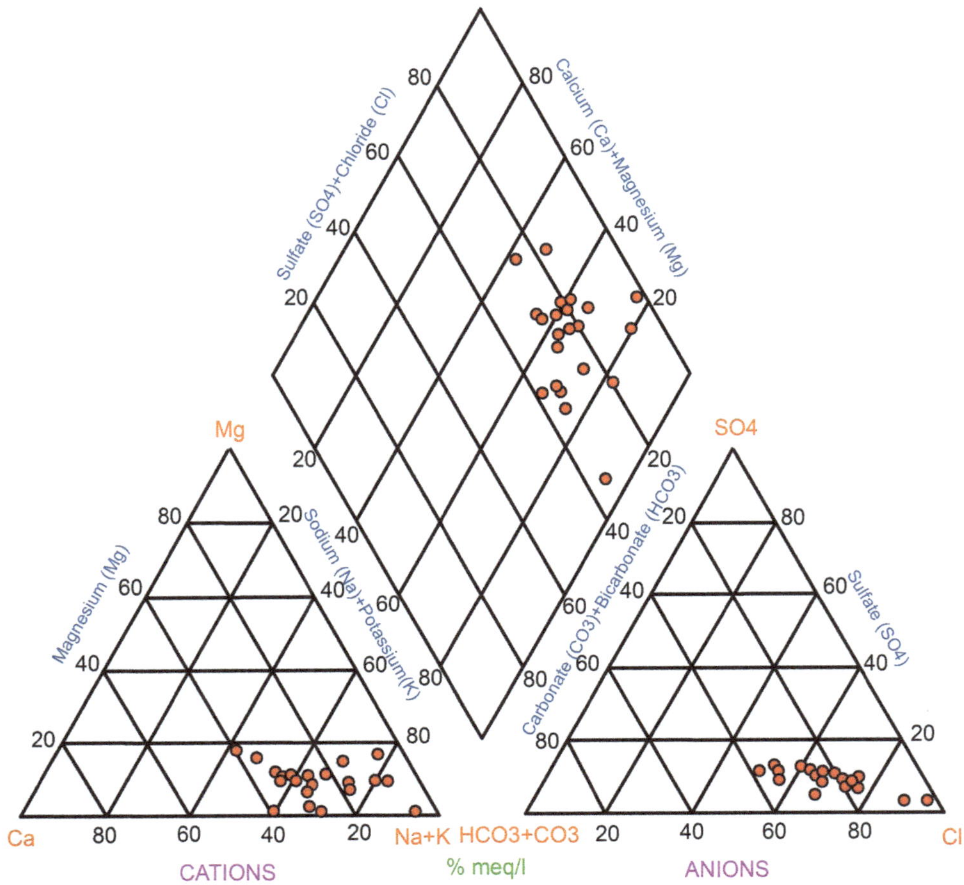

Table 2 Relative weight of chemical of physico-chemical parameters

Chemical parameters	WHO standards (2011)	Weight (wi)	Relative weight $Wi = wi/\sum_{i=1}^{n} wi$
pH (on scale)	6.5–8.5	4	0.114
EC (μS/cm)	500	4	0.114
TDS (mg/l)	500	5	0.142
HCO_3 (mg/l)	500	3	0.086
Cl (mg/l)	250	3	0.086
SO_4 (mg/l)	250	4	0.114
NO_3 (mg/l)	45	5	0.142
Ca (mg/l)	75	2	0.057
Mg (mg/l)	50	1	0.029
Na (mg/l)	200	2	0.057
K (mg/l)	200	2	0.057
		$\sum wi = 35$	$\sum wi = 0.998$

Total hardness (TH)

The classification of groundwater based on total hardness (TH) shows that a majority of the ground water samples fall in the hard water category. TH of the groundwater was calculated as suggested by Sawyer et al. 2003.

Table 3 Water quality classification ranges and types of water based on WQI values

Range	Type of water
<50	Excellent water
50–100	Good water
100–200	Poor water
200–300	Very poor water
>300	Water unsuitable for drinking purposes

$$TH(\text{as } CaCO_3)mg = Ca^{2+} + Mg^{2+}\left(\frac{\text{meq}}{1}\right) \times 50 \quad (5)$$

The hardness values range from 89.5 to 814.55 mg/l with an average value of 211.59 mg/l. The maximum allowable limit of TH for drinking purpose is 500 mg/l and the most desirable limit is 100 mg/l as per the WHO standard. Groundwater exceeding the limit of 300 mg/l is considered to be very hard. Sample nos. 2 and 11 exceed the maximum allowable limit of 500 mg/l. The hardness of the water is due to the presence of alkaline earths such as calcium and magnesium.

Table 4 Water quality index (WQI) classification for individual samples

S No.	WQI values	Water quality classification type
1	53.69	Good water
2	428.53	Water unsuitable for drinking purposes
3	81.71	Good water
4	48.63	Excellent water
5	49.41	Excellent water
6	44.28	Excellent water
7	80.87	Good water
8	80.47	Good water
9	80.14	Good water
10	226.75	Very poor water
11	399.18	Water unsuitable for drinking purposes
12	47.89	Excellent water
13	92.18	Good water
14	110.77	Poor water
15	92.93	Good water
16	101.30	Poor water
17	135.88	Poor water
18	48.54	Excellent water
19	162.39	Poor water
20	55.95	Good water
21	50.54	Good water
22	78.61	Good water
23	84.25	Good water
24	115.66	Poor water

Gibbs diagram

Gibbs diagram is widely used to establish the relationship of water composition and aquifer lithological characteristics. Three distinct fields such as precipitation dominance, evaporation dominance and rock–water interaction dominance areas are shown in the Gibbs diagram (Gibbs, 1970). The predominant samples fall in the rock–water interaction dominance and evaporation dominance field of the Gibbs diagram (Fig. 2). The rock–water interaction dominance field indicates the interaction between rock chemistry and the chemistry of the percolated waters under the subsurface.

$$\text{Gibbs ratio I (for anion)} = \frac{Cl^-}{(Cl^- + HCO_3^-)} \tag{6}$$

$$\text{Gibbs ratio II (for cation)} = \frac{Na^+ + K^+}{(Na^+ + K^+ + Ca^{2+})}, \tag{7}$$

whereas all the ionic concentration is expressed in meq/l.

Piper trilinear diagram

Piper diagram is extensively used to understand problems concerning the geochemical evolution of groundwater. This diagram consists of three distinct fields including two triangular fields and a diamond-shaped field. The cations expressed as percentage of total cations in meq/l as a single point on the left triangle while anions plot in the right triangle (Piper 1944). Each point is then projected into the upper field along a line parallel to the upper margin of the field and the point where the extension intersects indicates the character of the water as represented by the relationship among $Na^+ + K^+$, $Ca^{2+} + Mg^{2+}$, $CO_3^- + HCO_3^-$ and Cl^-, SO_4^{2-} ions. Similarities and differences among groundwater samples can be revealed from the trilinear diagram because water of similar qualities will tend to plot together as groups. Aquachem 4.0 scientific software is used for the plotting of piper trilinear diagram. The piper diagram is dominated by NaCl followed by mixed CaMgCl type facies (Fig. 3). This process indicates that alkalis ($Na^+ + K^+$) and strong acids ($Cl^-+SO_4^{2-}$) dominated over the alkaline earth ($Ca^{2+}+Mg^{2+}$) and weak acids. The elevated Na^+ concentrations coupled to low Ca^{2+} suggesting that Ca^{2+} and Na^+ ion exchange process is an important geochemical process for the Na–Cl type of groundwater.

Water quality index calculation (WQI)

The chemistry of groundwater is often used as a tool for discriminating the drinking and irrigation water quality (Subba Rao 2006; Vasanthavigar et al. 2010). Water quality index (WQI) is an important parameter for identifying the water quality and its sustainability for drinking purposes (Subba Rao, 1997; Magesh et al. 2013). WQI is defined as a technique of rating that provides the composite influence of individual water quality parameters on the overall water quality (Mitra and ASABE Member 1998). (World Health Organization 2011) standards for drinking water quality have been used to calculate the WQI. The relative weight (*wi*) was assigned for water quality parameters based on their relative importance on water quality for drinking purposes (Table 2). The water quality classification based on WQI values is shown in Table 3. The calculation of WQI for groundwater samples is shown in Table 4. A total of 24 samples were analyzed for WQI. Among these, 20.83 % of the samples showed excellent water, 45.83 % of the samples fell under good water category, 20.83 % of the samples showed poor water category, 4.16 and 8.33 % of the samples fell under very poor water quality and water unsuitable for drinking purposes, respectively. This may be due to effective leaching and dissolution process of rock salt and gypsum-bearing rock formations. High concentration of EC, chloride, sodium followed by calcium clearly suggests that rock–water interaction process is the main source for degrading the water quality in the study area.

Table 5 Correlation matrix (r_2) of studied physico-chemical parameters and major ions (N = 24) of groundwater

Parameters	Temp	pH	EC	TDS	CO_3	HCO_3	Cl	SO_4	NO_3	Ca	Mg	Na	K
Temp	1.000												
pH	0.244	1.000											
EC	−0.118	0.285	1.000										
TDS	−0.118	0.285	1.000	1.000									
CO_3	0.213	0.376	0.549	0.549	1.000								
HCO_3	0.013	0.605	0.255	0.255	0.076	1.000							
Cl	−0.125	0.158	0.952	0.952	0.471	0.114	1.000						
SO_4	0.131	0.536	0.777	0.777	0.454	0.601	0.638	1.000					
NO_3	0.227	0.362	0.297	0.297	−0.019	0.216	0.331	0.439	1.000				
Ca	−0.240	0.179	0.727	0.727	0.134	0.215	0.767	0.454	0.296	1.000			
Mg	−0.053	0.087	0.894	0.894	0.545	−0.107	0.872	0.565	0.231	0.621	1.000		
Na	−0.117	0.311	0.964	0.964	0.663	0.256	0.892	0.759	0.170	0.636	0.838	1.000	
K	−0.117	0.190	0.966	0.966	0.604	0.090	0.924	0.687	0.255	0.639	0.903	0.960	1.000

Correlation matrix

Statistical analysis was performed on the physico-chemical parameters and major ion concentration to detect the relationship and differences between the groundwater samples. In order to discuss the data, the values grouped with respect to the geochemical parameters. The average value of all the variables (temperature, pH, EC, TDS, CO_3^-, HCO_3^-, Cl^-, SO_4^{2-}, NO_3, Ca^{2+}, Mg^{2+}, Na^+, K^+) determined and tabulated as matrix (13 × 24) in Table 5. This matrix was analyzed with cluster analysis (hierarchical tree clustering, rescaled distance cluster combine analysis) using SPSS software (SPSS, 2001). The hierarchical cluster analysis was used to group water samples into significant clusters. The dendrogram analysis was performed using Ward method and the results of parameters are shown three groups in Fig. 4. Most of the samples were classified in group I and II with good correlation between SO_4, Ca, HCO_3, Cl and Na with EC and TDS. The group III with one sub-group constructed with EC and TDS. The possible salt combinations ($CaSO_4$, NaCl, and mixed $CaNaHCO_3$) are probably derived from weathering of rock salts, gypsum-bearing aquifers and irrigation return flow. The concentration of nitrate is probably derived from anthropogenic activities. The water quality management can be implemented by proper removal/recycling of waste water and installation of sewage treatment plants.

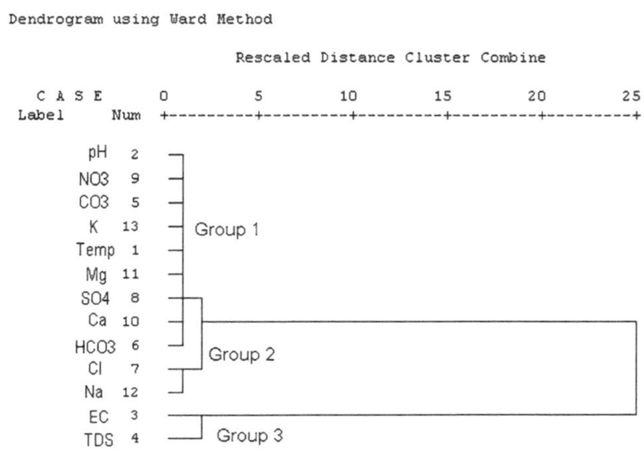

Fig. 4 Dendrogram for the groundwater grouping with respect to their physico-geochemical parameters

Conclusions

The study area is always under stress due to increasing population and more demand for water resources. The hydrogeochemical analysis of the study reveals that the groundwater is fresh to brackish and moderately high to hard. The order of the abundance of the major cation and anion is as follows: $Na^+ > Ca^{2+} > Mg^{2+} > K^+ = Cl^- > HCO_3^- > SO_4^{2-} > CO_3^{2-} > NO_3^{2-}$. Na and Cl are dominant ions among the studied cations and anions. The concentration of potassium ion is within the permissible limit for drinking purpose except few locations. 20 % of the groundwater samples have exceeded the permissible limit of sodium. Nitrate and sulphate ion concentrations are within the allowable limit for drinking purpose. Chloride, calcium and magnesium are within the allowable limit except few samples. According to Gibbs diagram, the predominant samples fall in the rock–water interaction dominance and evaporation dominance field. The piper trilinear diagram shows that groundwater in the study area is Na–Cl and mixed CaMgCl type. Based on the WQI classification majority of the samples are falling under excellent to good water category and suitable for drinking water purposes. To improve the quality of water, the Government and non-government organization should provide the moral support to design the rain water harvesting structures and artificial recharge methods for young generation especially budding civil engineers.

Acknowledgments KK thanks Prof. Ramachandramoorthy, Vice Chancellor, St. Peters University and Prof. P. Partheeban, Dean Academic, St. Peters College of Engineering and Technology, Avadi, Chennai, for their kind support and provide the permission to publish this work.

References

APHA (1995) Standard methods for the examination of water and wastewater, 19th edn. American Public Health Association, Washington, pp 1–467

Bagyaraj M, Ramkumar T, Venkatramanan S, Gurugnanam B (2013) Application of remote sensing and GIS analysis for identifying groundwater potential zone in parts of Kodaikanal Taluk, South India. Front Earth Sci 7(1):65–75

Ballukraya PN, Ravi R (1999) Characterisation of groundwater in the unconfined aquifer of Chennai city, India. J Geol Soc India 54:1–11

BIS (1991) (Bureau of Indian Standards) 10500 Indian Standard drinking water specification, 1st rev, 1–8

CDSH (2012) Chennai district statistical handbook. District Administration, Chennai

Dixit S, Gupta SK, Tiwari S (2005) Nutrients overloading of a freshwater lake in Bhopal, India. Electron Green J 21:2–6

Gibbs RJ (1970) Mechanism controlling world water chemistry Sciences 170: 795–840

Hem JD (1985) Study and interpretation of the chemical characteristics of natural water: US Geological Survey Water-Supply Paper 2254, 3rd edn, 263

Jayaprakash M, Giridharan L, Venugopal S, Krishna Kumar SP, Periyakali P (2008) Characterization and evaluation of the factors affecting the geochemistry of groundwater in Neyveli, Tamil Nadu, India. Environ Geol 54(4):855–867

Jerry AN (1986) Basic environmental technology (water supply, waste disposal and pollution control). Wiley, New York

Krishna Kumar S, Chandrasekar N, Seralathan P, Godson PS, Magesh NS (2011) Hydrogeochemical study of shallow carbonate aquifers, Rameswaram Island, India. Environ Monit Assess 184(7):4127–4139

Magesh NS, Krishnakumar S, Chandrasekar N, Soundranayagam JP (2013) Groundwater quality assessment using WQI and GIS techniques, Dindigul district, Tamil Nadu, India. Arab J Geosci 6(11):4179–4189

Malini S, Nagaiah N, Paramesh L, Venkataramaiah P, Balasubramanian A (2003) Groundwater quality around Mysore, Karnataka, India. Int J Environ Stud 60:87–98

Mitra BK, ASABE Member (1998) Spatial and temporal variation of ground water quality in sand dune area of Aomori Prefecture in Japan

Padmalal D, Maya K, Narendra Babu K, Baiju RS, Babura B (2012) Hydro chemical characterization and water quality assessment of the coastal springs of southern Kerala, India. J Appl Geochem 14(4):466–481

Pang Z, Yuan L, Huang T, Kong Y, Liu J, Li Y (2013) Impacts of human activities on the occurrence of groundwater nitrate in an alluvial plain: a multiple isotopic tracers approach. J Earth Sci 24(1):111–124

Piper AM (1944) A graphical interpretation of water—analysis. Trans Am Geophys Union 25:914–928

Raju NJ, Shukla UK, Ram P (2011) Hydrogeochemistry for the assessment of groundwater quality in Varanasi: a fast-urbanizing center in Uttar Pradesh, India. Environ Monit Assess 173:279–300

Reddy PM, Subba Rao N (2001) Effect of industrial effluents on the groundwater regime in Vishakhapatnam. Pollut Res 20(3):383–386

CGWB Technical Report (2008) District groundwater brochure Chennai district Tamil Nadu. pp 1–19

Sarath Prasanth SV, Magesh NS, Jitheshlal KV, Chandrasekar N (2012) Evaluation of groundwater quality and its suitability for drinking and agricultural use in the coastal stretch of Alappuzha District, Kerala, India. Appl Water Sci 2(3):165–175

Sawyer GN, McCartly DL, Parkin GF (2003) Chemistry for environmental engineering and science, 5th edn. McGraw Hill, New York, p 752

Selvam S, Manimaran G, Sivasubramanian P (2013) Hydrochemical characteristics and GIS-based assessment of groundwater quality in the coastal aquifers of Tuticorin corporation, Tamilnadu, India. Appl Water Sci 3:145–159

SPSS (2001) SPSS for Windows, Rel. 11.0.1. 2001. Chicago: SPSS Inc

Srinivas Y, Hudson Oliver D, Stanley Raj A, Chandrasekar N (2013) Evaluation of groundwater quality in and around Nagercoil town, Tamil Nadu, India: an integrated geochemical and GIS approach. Appl Water Sci 3:631–651

Subba Rao N (1997) Studies of water quality index in hard rock terrain of Guntur district, Andra Pradesh, India. In: National Seminar on Hydrology of Precambrian Terrains and hard rock areas, 129–134

Subba Rao N (2006) Seasonal variation of groundwater quality in a part of Guntur district, Andhra Pradesh, India. Environ Geol 49:413–429

Tiwari KK, Prasad RN, Chandra Ram, Mondal NC (2012) Geochemical parameters for assessment of groundwater quality around urban and suburban areas of Dausa city in Rajasthan, India. J Appl Geochem 14(2):184–193

Todd DK (1980) Groundwater hydrology. Wiley, New York

Vasanthavigar M, Srinivasamoorthy K, Vijayaragavan K, Rajiv Ganthi R, Chidambaram S, Anandhan P, Manivannan R, Vasudevan S (2010) Application of water quality for groundwater quality assessment: Thirumanimuttar Sub basin, Tamil Nadu, India. Environ Monit Assess 171(1–4):595–609

WHO (2011) WHO Guidelines for Drinking-water Quality, fourth ed. World Health Organization

3

Groundwater vulnerability to pollution mapping of Ranchi district using GIS

R. Krishna · J. Iqbal · A. K. Gorai · G. Pathak · F. Tuluri · P. B. Tchounwou

Abstract Groundwater pollution due to anthropogenic activities is one of the major environmental problems in urban and industrial areas. The present study demonstrates the integrated approach with GIS and DRASTIC model to derive a groundwater vulnerability to pollution map. The model considers the seven hydrogeological factors [Depth to water table (*D*), net recharge (*R*), aquifer media (*A*), soil media (*S*), topography or slope (*T*), impact of vadose zone (*I*) and hydraulic Conductivity(*C*)] for generating the groundwater vulnerability to pollution map. The model was applied for assessing the groundwater vulnerability to pollution in Ranchi district, Jharkhand, India. The model was validated by comparing the model output (vulnerability indices) with the observed nitrate concentrations in groundwater in the study area. The reason behind the selection of nitrate is that the major sources of nitrate in groundwater are anthropogenic in nature. Groundwater samples were collected from 30 wells/tube wells distributed in the study area. The samples were analyzed in the laboratory for measuring the nitrate concentrations in groundwater. A sensitivity analysis of the integrated model was performed to evaluate the influence of single parameters on groundwater vulnerability index. New weights were computed for each input parameters to understand the influence of individual hydrogeological factors in vulnerability indices in the study area. Aquifer vulnerability maps generated in this study can be used for environmental planning and groundwater management.

Keywords Groundwater vulnerability · Ranchi · DRASTIC · Sensitivity analysis · GIS

R. Krishna · J. Iqbal · A. K. Gorai (✉) · G. Pathak
Environmental Science and Engineering Group, Birla Institute of Technology Mesra, Ranchi 835215, India
e-mail: amit_gorai@yahoo.co.uk

F. Tuluri
Department of Technology, Jackson State University, Jackson, MS 39217, USA

P. B. Tchounwou
Department of Biology, Jackson State University, Jackson, MS 39217, USA

Introduction

Groundwater is the most important water resource on earth (Villeneuve et al. 1990). Groundwater quality is under considerable threat of contamination especially in agriculture-dominated areas due to intense use of fertilizers and pesticides (Giambelluca et al. 1996; Soutter and Musy 1998; Lake et al. 2003; Thapinta and Hudak 2003; Chae et al. 2004). Thus, the protection of groundwater against anthropogenic pollution is of crucial importance (Zektser et al. 2004). Assessment of groundwater vulnerability to pollution helps to determine the proneness of groundwater contamination and hence essential for managing and preserving the groundwater quality (Fobe and Goossens 1990; Worrall et al. 2002; Worrall and Besien 2004).

Groundwater vulnerability to pollution studies helps to categorize the land on the basis of its proneness to vulnerability (Gogu and Dassargues 2000a). That is, groundwater vulnerability assessment delineates areas that are more susceptible to contamination on the basis of the different hydrogeological factors and anthropogenic sources. In general, the study explains the estimation of the contaminants migration potential from land surface to groundwater through the unsaturated zones (Connell and

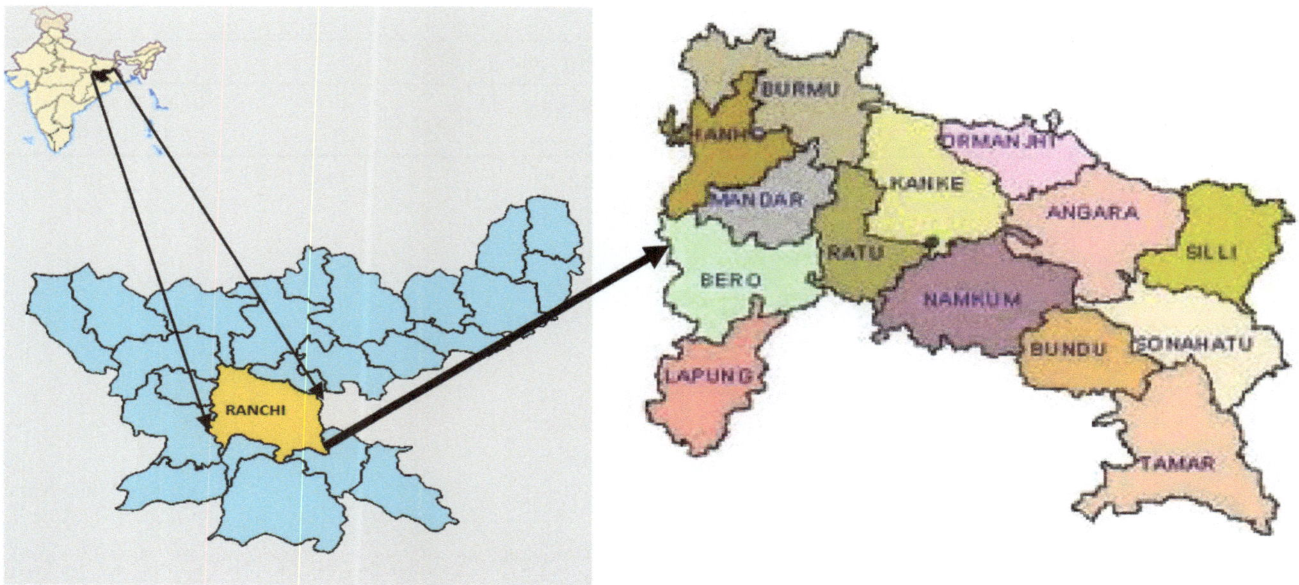

Fig. 1 Study area map

Fig. 2 Flow chart of the working methodology

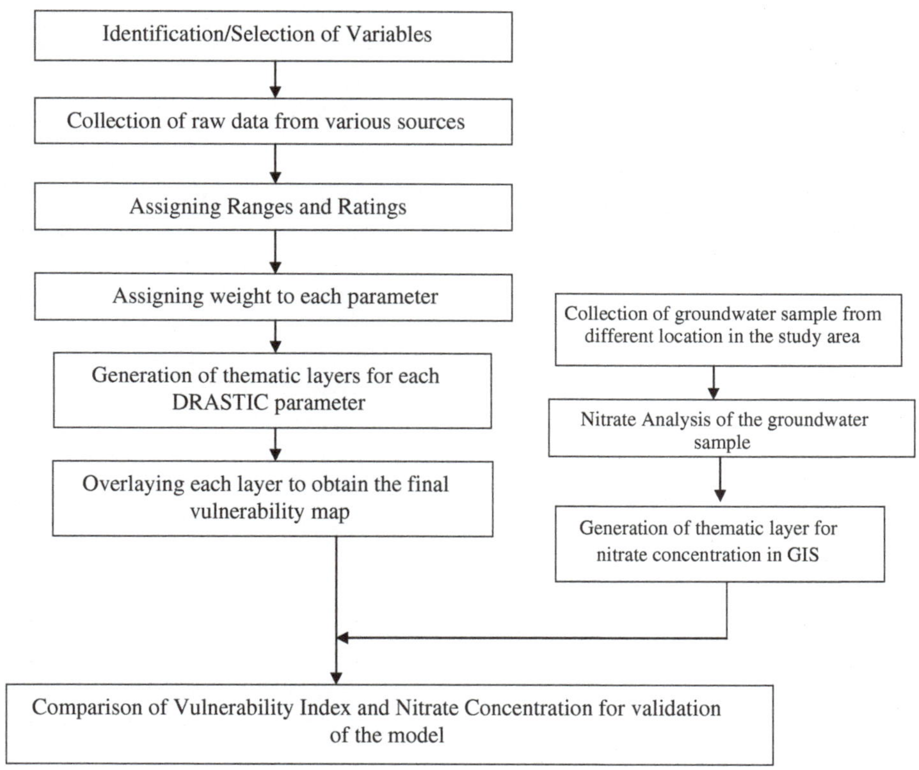

Daele 2003). Groundwater vulnerability assessment is essential for management of groundwater resources and subsequent land use planning (Rupert 2001; Babiker et al. 2005). Groundwater vulnerability maps provide visual information for more vulnerable zones which help to protect groundwater resources and also to evaluate the potential for water quality improvement by changing the agricultural practices and land use applications (Connell and Daele 2003; Rupert 2001; Babiker et al. 2005; Burkart and Feher 1996).

Groundwater vulnerability assessment can be used in planning, policy analysis, and decision making, viz., advising decision makers for adopting specific management options to mitigate the quality of groundwater

Table 1 Ranges and ratings for various hydrogeological settings (Aller et al. 1987)

Depth to groundwater			Net recharge		
Ranges (m)	Ratings (D_r)	Sub-index ($D_r \times D_w$)	Ranges (cm)	Ratings (R_r)	Sub-index ($R_r \times R_w$)
0–1.52	10	50	0–5.08	1	4
1.52–4.57	9	45	5.08–10.16	3	12
4.57–9.14	7	35	10.16–17.78	6	24
9.14–15.24	5	25	17.78–25.4	8	32
15.24–22.86	3	15	25.4+	9	36
22.86–30.48	2	10			
30.48+	1	5			
Weight (D_w)	5		Weight (R_w)	4	
Aquifer type			Soil type		
Type	Ratings (A_r)	Sub-index ($A_r \times A_w$)	Type	Ratings (S_r)	Sub-index ($S_r \times S_w$)
Massive shale	2	6	Thin or absent	10	20
Metamorphic/igneous	3	9	Gravel	10	20
Weathered metamorphic/igneous	4	12	Sand	9	18
Glacial till	5	15	Peat	8	16
Bedded sandstone, limestone and shale sequences	6	18	Shrinking and/or aggregated clay	7	14
Massive sandstone	6	18	Sandy loam	6	12
Massive limestone	6	18	Loam	5	10
Sand and gravel	8	24	Silty loam	4	08
Basalt	9	27	Clay loam	3	06
Karst limestone	10	30	Muck	2	04
Weight (A_w)	3		Non-shrinking and non-aggregated clay	1	02
Topography or slope			Weight (S_w)	2	
Ranges (in %)	Ratings (T_r)	Sub-index ($T_r \times T_w$)	Impact of vadose zone		
0–2	10	10	Type	Ratings (I_r)	Sub-index ($I_r \times I_w$)
2–6	9	9	Confining layer	1	5
6–12	5	5	Silt/clay	3	15
12–18	3	3	Shale	3	15
18+	1	1	Limestone	6	30
Weight (T_w)	1		Sandstone	6	30
Hydraulic conductivity			Bedded limestone, sandstone, shale	6	30
Range (m/d)	Ratings (C_r)	Sub-index ($C_w \times C_r$)	Sand and gravel with significant silt and clay	6	30
0.04–4.07	1	3	Metamorphic/igneous	4	20
4.07–12.22	2	06	Sand and gravel	8	40
12.22–28.52	4	12	Basalt	9	45
28.52–40.74	6	18	Karst limestone	10	50
40.74–81.49	8	24			
81.49+	10	30			
Weight (C_w)	3		Weight (I_w)	5	

resources; demonstrating the implications and consequences of their decisions; providing direction for using groundwater resources; highlighting about proper land use practices and activities; and educating the general public regarding the consequences of groundwater contamination (NRC 1993).

Table 2 Data types and its sources for creation of output layers

Sl. No.	Data types	Sources	Output layer
1	Well data	Real time observation using GPS and tape	Depth to water
2	Average annual rainfall	Indian Meteorological Department, India	Net recharge
3	Geologic map	Central Ground Water Board, PATNA	Aquifer media
4	Soil map	Birsa Agricultural University, RANCHI	Soil media
5	SRTM data	USGS GLOVIS visualization viewer	Topography
6	Soil map	Birsa Agricultural University, RANCHI	Impact of vadose zone
7	Geologic map	Central Ground Water Board, PATNA	Hydraulic conductivity

The concept of aquifer vulnerability to external pollution was introduced in 1960s by (Margat 1968), with several systems of aquifer vulnerability assessment developed in the following years (Aller et al. 1987; Civita 1994; Vrba and Zaporozec 1994; Sinan and Razack 2009; Polemio et al. 2009; Foster 1987). They found that the reason behind the different vulnerabilities is the different hydrogeological settings. Many approaches have been developed to evaluate aquifer vulnerability. These are overlay/index methods, process-based methods and statistical methods (Zhang et al. 1996; Tesoriero et al. 1998). The overlay/index methods use location-specific vulnerability indices based on the hydrogeological factors controlling movement of pollutants from the land surface to the water bearing strata. The process-based methods use contaminants transport models to estimate the contaminant migration (Barbash and Resek 1996). Statistical methods estimate the associations between the spatial variables and the occurrence of pollutants in the groundwater using various statistics.

Among all the approaches mentioned above, the overlay and index method has been the most widely adopted approach for wide-scale groundwater vulnerability assessments. Scientist started giving predictions of groundwater pollution potential based on hydrogeological settings (Polemio et al. 2009; Almasri Mohammad 2008; Berkhoff 2008; Rahman 2008; Massone et al. 2010; Kwansiririkul et al. 2004; Kim and Hamm 1999; Secunda et al. 1998; Vias et al. 2005; Brosig Karolin et al. 2008; Ferreira Lobo and Oliveira Manuel 2004; Nouri and Malmasi 2005; Herlinger and Viero 2007; Shirazi et al. 2012, 2013; Neshat Aminreza et al. 2014). This paper aims to demonstrate a GIS-based DRASTIC model for groundwater vulnerability assessment of Ranchi district. The validation of the model prediction was done on the basis of observed nitrate concentration in groundwater in the study area. Sensitivity analysis of the model was also carried out to understand the influence of the individual input variables on groundwater vulnerability to pollution index.

Study area

The area selected for the proposed study is Ranchi district. Ranchi district lies in the southern part of Jharkhand state and bounded by other district of Jharkhand, viz., Hazaribagh, West Singhbhum, Gumla, Lohardaga, and East Singhbhum. This is also bounded by Purulia district of West Bengal. The district has a total area of 4,912 km^2 and is located between 22°45′–23°45′ North latitude to 84°45′–84°50′ East longitude. The district comprises of 14 blocks namely Ormanjhi, Kanke, Ratu, Bero, Burmu, Lapung, Chanho, Mandar, Bundu, Tamar, Angara, Sonahatu, Silli, Namkum as shown in Fig. 1.

The climate of Ranchi district is a subtropical climate. This is characterized by hot summer season from March to May and well-distributed rainfall during southwest monsoon season from June to October. Ranchi district has varied hydrogeological characteristics and hence the groundwater potential differs from one location to another. The three-fourth of the district area is underlain by Chotanagpur granite gneiss of pre-Cambrian age (CGWB 2009). In two blocks (Ratu and Bero) thick lateritic capping is placed above granite gneiss. A big patch of older alluvium exists in Mandar block and limestone rock exists in northernmost portion of Burmu block. The northernmost and southernmost parts of the district are mainly covered with hillocks and forests. In general, the altitude of the area varies from 500 to 700 m above mean sea level, but there are many hillocks through the district having altitude more than 700 m above mean sea level. Two types of aquifers (Weathered aquifer and fractured aquifers) exist in the study area. Thickness of weathered aquifers varies from 10 to 25 m in granite terrain and 30 to 60 m in lateritic terrain. In weathered aquifers groundwater occurs in unconfined condition, while in fractured aquifer groundwater occurs in semi-confined to confined conditions.

Materials and methods

Groundwater vulnerability was evaluated using hydrogeological factors that can influence the pollutant transport through the vadose zone to the water bearing strata using GIS-based DRASTIC (Aller et al. 1987) method. The flowchart (shown in Fig. 2) represents the general overview of the research methodology. In the present study, seven hydrogeological parameters [Depth to water table (D), net recharge (R), aquifer media (A), soil media (S),

Table 3 Ranges and ratings for various hydrogeological settings using DRASTIC data for the Study area

Sl. No.	Depth to groundwater (ft)	Ratings (D_r)	Net recharge (cm)	Ratings (R_r)	Aquifer media	Rating (A_r)	Soil media	Soil Rating (S_r)	Topography/ slope (in %)	Ratings (T_r)	Impact of vadose zone	Ratings (I_r)	Conductivity (m/d)	Ratings (C_r)
1	6	9	92	9	Weathered metamorphic/igneous	4	Silty loam	4	12	3	Silty loam	4	12.22	4
2	10	9	92	9		4		4	12	3		4	12.22	4
3	9	9	6	3		4		4	12	3		4	12.22	4
4	10	9		9		4		4	12	3		4	12.22	4
5	5	9	91	9		4	Clay loam	3	12	3		4	12.22	4
6	6	9	92	9		4		3	12	3	Clay loam	3	12.22	4
7	5	9	88	9		4		3	12	3		3	12.22	4
8	12	9	90	9		4	Silty loam	4	12	3	Silty loam	4	12.22	4
9	8	9	91	9		4		4	12	3		4	12.22	4
10	9	9	91	9		4		4	12	3		4	12.22	4
11	6	9	90	9		4	Clay loam	3	12	3	Clay loam	3	12.22	4
12	5	9	90	9		4		3	6	5		3	12.22	4
13	5	9	92	9	Bedded sandstone, limestone and shale sequences	5		3	12	3		3	0.04	1
14	4	9	90	9		5		3	19	1		3	0.04	1
15	2	10	88	9		5	Silty loam	4	12	3	Silty loam	4	0.04	1
16	5	10	89	9	Weathered metamorphic/ igneous	4	Clay loam	3	12	3	Clay loam	3	12.22	4
17	2	10	90	9		4		3	12	3		3	12.22	4
18	5	9	91	9		4	Silty loam	4	12	3	Silty loam	4	12.22	4
19	2	9	92	9		4		4	12	3		4	12.22	4
20	2	9	92	9		4		4	6	5		4	12.22	4
21	4	10	92	9	Bedded sandstone, limestone and shale sequences	5	Clay loam	3	6	5	Clay loam	3	0.04	1
22	15	9	91	9	Weathered metamorphic/ igneous	4	Silty loam	4	12	3	Silty loam	4	12.22	4
23	5	9	90	9		4		4	12	3		4	12.22	4
24	2	10	6	3		4		4	12	3		4	12.22	4
25	5	10	90	9	Basalt	9		4	12	3		4	82	10
26	8	10	6	3	Weathered metamorphic/ igneous	4		4	12	3		4	12.22	4
27	5	9	92	9		4	Clay loam	3	12	3	Clay loam	3	12.22	4
28	6	10	6	3		4	Silty loam	4	12	3	Silty loam	4	12.22	4
29	2	9	90	9		4		4	12	3		4	12.22	4
30	5	10	92	9		4		4	12	3		4	12.22	4

topography (T), impact of vadose zone (I) and hydraulic Conductivity (C)] were considered for assessing the groundwater vulnerability.

Thematic maps of seven factors (D, R, A, S, T, I, and C) were generated and used for producing the final groundwater vulnerability to pollution index map. The thematic values in each of the seven hydrogeological maps were classified into corresponding ranges as per the DRASTIC model. Each range was assigned their corresponding ratings as per the DRASTIC model. Weight multipliers were then used for each factor to balance and enhance its importance. The final vulnerability map was computed as the weighted sum overlay of the seven layers using Eq. (1) and was termed as DRASTIC INDEX (DI).

$$\text{DRASTIC.INDEX (DI)} = D_r D_w + R_r R_w + A_r A_w + S_r S_w + T_r T_w + I_r I_w + C_r C_w \qquad (1)$$

where, D_r, R_r, A_r, S_r, T_r, I_r, and C_r are ratings assigned to depth to water table, net recharge, aquifer media, soil media, topography or slope, impact of vadose zone, and hydraulic conductivity, respectively.

D_w, R_w, A_w, S_w, T_w, I_w, and C_w are weights assigned to depth to water table, net recharge, aquifer media, soil media, topography or slope, impact of vadose zone, and hydraulic conductivity, respectively.

Every parameter in the model assigned a fixed weight (listed in Table 1) indicating the relative influence of the parameter in transporting contaminants to the groundwater. Each input factor has been divided into either ranges or significant media types that affect groundwater vulnerability. The media types such as aquifer material, soil type and impact of vadose zone, cannot be measured numerically and hence ratings were assigned to each type of media. Each range of each DRASTIC parameter has been evaluated with respect to the others to determine its relative significance to pollution potential, and has been assigned a rating of 1–10. The "easiest to be polluted" was assigned a rating ten, except net recharge (which is 9) and the "most difficult to pollute" was assigned a rating of one. The numerical ratings, which were established using the Delphi technique (Aller et al. 1987), are well defined and have been used worldwide (Al-Adamat et al. 2003; Anwar et al. 2003; Chandrashekhar et al. 1999; Dixon 2005). The ratings for each parameter are listed in Table 1 for all the ranges and types.

Data sources and generation of thematic layers

The raw data were collected or derived from various published reports/maps for the generation of the thematic layers and are listed in Table 2. Thematic layers for each hydrogeological parameter were generated using ArcGIS software version 9.3.

Fig. 3 **a** Depth to groundwater, **b1** rainfall map, **b2** land use map, **b** net recharge, **c** aquifer media, **d** soil media, **e** topography, **f** impact of vadose zone, **g** hydraulic Conductivity ▶

Depth to groundwater

The depth to groundwater table parameter was derived from water level data collected from Central Ground Water Board (CGWB), Ranchi. The depth to groundwater table is shallow and has a range of 2–15 ft below ground level. The well data were then used to generate the map for depth to water table contoured by interpolating using inverse distance weighted (IDW) method. The study area was extracted using the district boundary as a mask. The thematic map was reclassified into two classes, corresponds to the DRASTIC model range value (listed in Table 1). Though, the ranges defined for different classes (in Table 1) are in meter, these values were converted into feet during rating assignment. The depth to water table values and their corresponding ratings are shown in Table 3. The map generated for depth to water is shown in Fig. 3a.

Net recharge

The thematic map of precipitation was generated using the rainfall data collected from Indian Meteorological Division (IMD), India as shown in Fig. 3b1. The evapotranspiration map was derived from precipitation map by assuming the rate of evapotranspiration to be 5 % of the precipitation (value taken from a report of IMD, Ranchi). The land use map of the study area was prepared and reclassified into five categories as agricultural land, built-up area, forest area, waste land and water bodies. The runoff coefficient assigned to different categories ranges from 0 to 1 depending on the land use type as shown in Fig. 3b2. The values were selected on the basis of rational formula for runoff coefficient (Source: http://water.me.vccs.edu/courses/CIV246/table2b.htm).

The net recharge map was derived in GIS using the formula as

$$\text{net recharge} = \text{precipitation (rainfall)} - 0.05 \times \text{precipitation (rainfall)} - \text{precipitation (rainfall)} \times \text{run off coefficients}$$

The net recharge in the thematic map was reclassified into two types and assigned their corresponding ratings. The map layer for net recharge is shown in Fig. 3(b).

Aquifer media

Aquifer media map was prepared from the geologic map of Ranchi district. Aquifer media in the study area

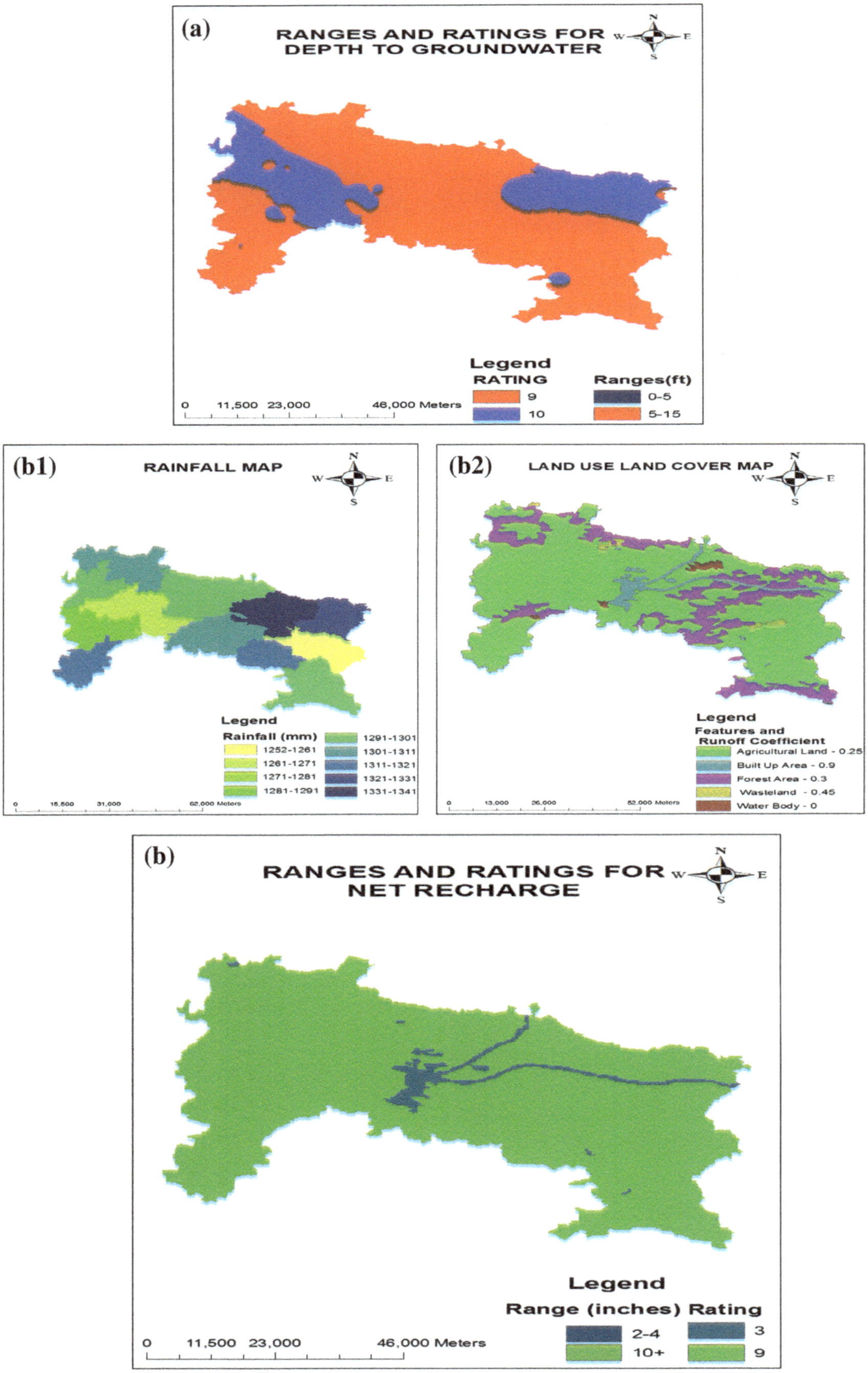
(a)
RANGES AND RATINGS FOR
DEPTH TO GROUNDWATER
N
W
E
S
Legend
RATING
Ranges(ft)
9
10
0-5
5-15
0
11,500
23,000
46,000 Meters
(b1)
RAINFALL MAP
N
W
E
S
Legend
Rainfall (mm)
1252-1261
1261-1271
1271-1281
1281-1291
1291-1301
1301-1311
1311-1321
1321-1331
1331-1341
0
15,500
31,000
62,000 Meters
(b2)
LAND USE LAND COVER MAP
N
W
E
S
Legend
Features and
Runoff Coefficient
Agricultural Land - 0.25
Built Up Area - 0.9
Forest Area - 0.3
Wasteland - 0.45
Water Body - 0
0
13,000
26,000
52,000 Meters
(b)
RANGES AND RATINGS FOR
NET RECHARGE
N
W
E
S
Legend
Range (inches) Rating
2-4
10+
3
9
0
11,500
23,000
46,000 Meters

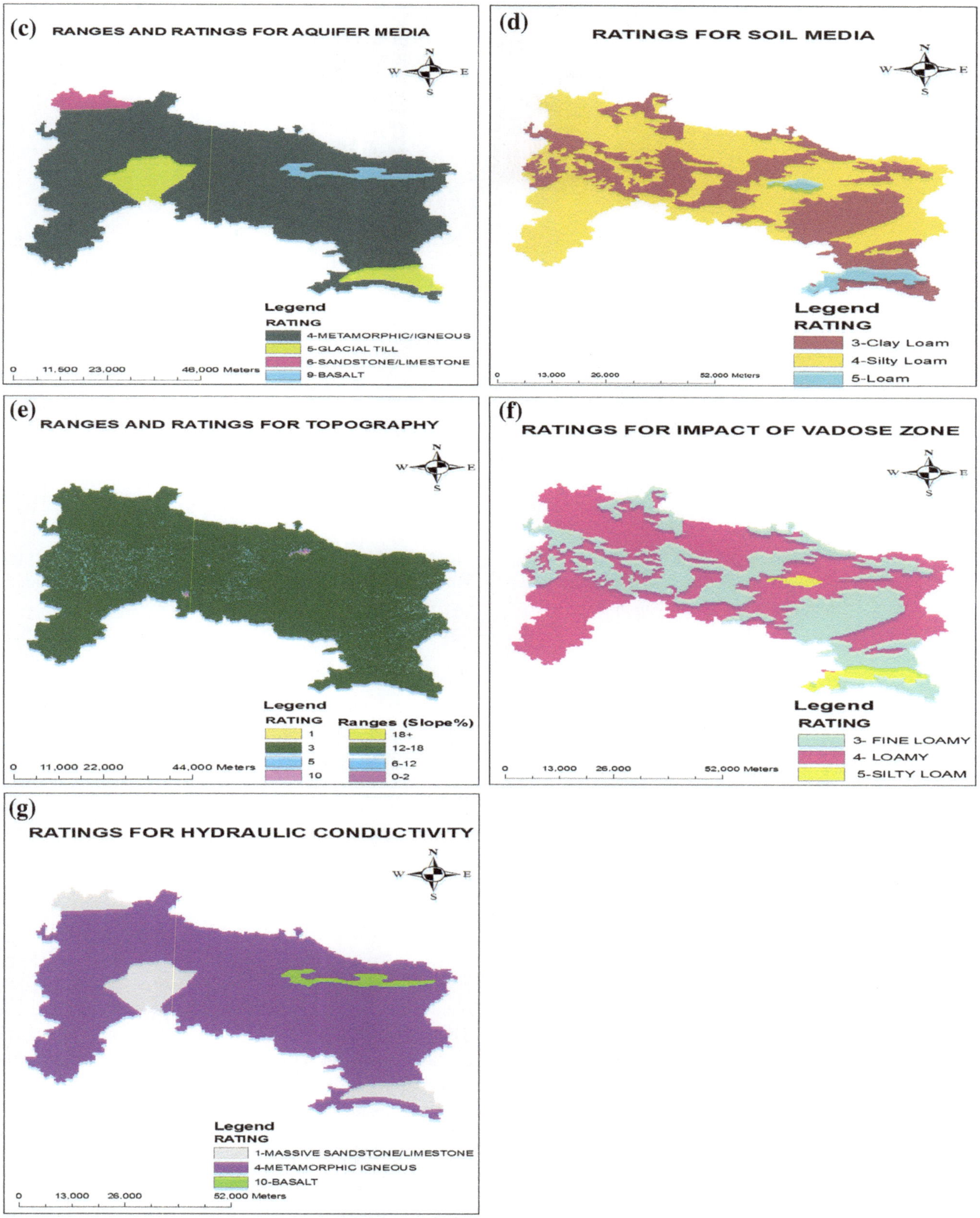

Fig. 3 continued

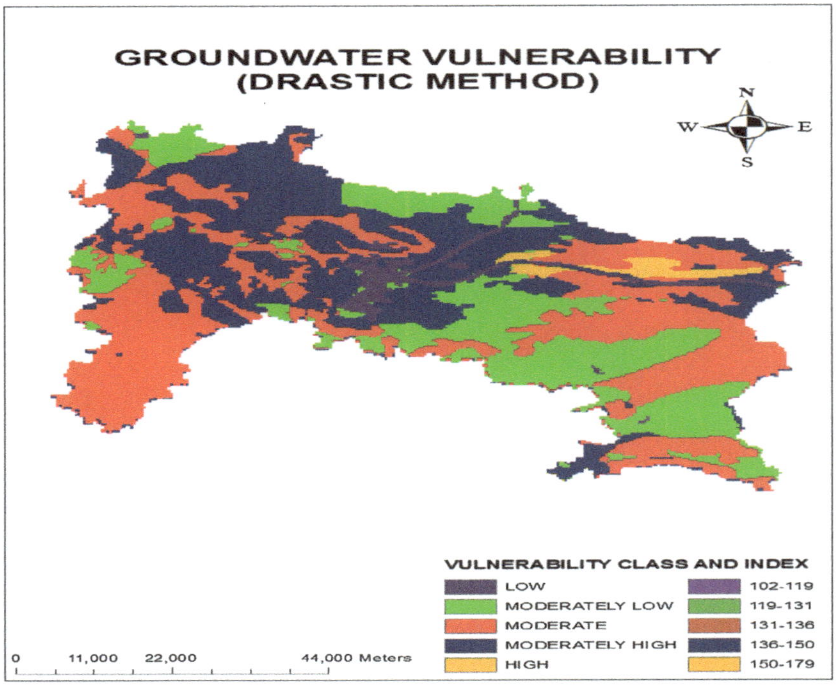

Fig. 4 Relative potential of groundwater vulnerability to pollution map

were reclassified into four types and their corresponding ratings were assigned for each aquifer media as given in Table 3. The thematic map is shown in Fig. 3c.

Soil media

Soil media map was prepared from the soil map of Ranchi district. The soil profile was collected from Birsa Agriculture University (BAU), Ranchi. It was digitized in ArcGIS for generating the thematic map of soil media. The study area consists of fine to coarse loamy-type soil. The soil type was classified into three types and their corresponding ratings were assigned for each type of soil media. The map generated for soil media is shown in Fig. 3d.

Topography

The topography map was prepared using the shuttle radar topography mission (SRTM) data. The percentage slope raster file was created from Digital Elevation Model (DEM) using spatial analyst. The slope percentage in the study area was reclassified into four classes and assigned their corresponding ratings as given in Table 3. The thematic map layer of topography is shown in Fig. 3e.

Impact of vadose zone

Due to unavailability of vadose zone data in the study area, information of the soil media was used to derive the approximate ratings for Vadose zone. The map was converted to a raster data by defining ratings for the vadose zone media (using soil media data) (Table 3; Fig. 3d). The thematic map of the impact of vadose zone is shown in Fig. 3f.

Hydraulic conductivity

Due to unavailability of hydraulic conductivity data in the study area, information of the aquifer media was used to derive the approximate ratings for hydraulic conductivity. It was converted to raster data according to the defined ratings. The ratings of the hydraulic conductivity were assigned (using aquifer media data instead here) as per Table 3. The map of hydraulic conductivity is shown in Fig. 3g.

Results and discussion

The GIS-based DRASTIC model was developed for generating the aquifer vulnerability map of Ranchi District. This will reflect the aquifer's inherent capacity to become contaminated. The final map represents the range of the vulnerability indices. The higher the vulnerability index, the higher is the capacity of the hydrogeologic condition to readily move contaminants from surface to the groundwater. On the other hand, low indices represent groundwater is better protected from contaminant leaching by the natural environment. The final vulnerability map was obtained by overlaying the seven hydrogeological thematic layers in ArcGIS software version 9.3. The final groundwater vulnerability map is shown in Fig. 4. The range of the vulnerability indices was reclassified into five classes (low, moderately low, moderate, moderately high, and high) on the basis of Jenks natural breaks that describe the relative probability of contamination of the groundwater resources. A regional scale has been used for comparing the relative vulnerability of groundwater resources.

The result of groundwater vulnerability to pollution assessment indicates that the index value ranged from 102 to 179. The maximum and minimum vulnerability indices were calculated by the sum of the product of maximum and minimum ratings for all the parameters with its corresponding weightage, respectively. The study area was divided into five zones of relative vulnerability: low groundwater vulnerability risk zone (index: 102–119); moderately low vulnerability risk zone (index: 119–131), moderate vulnerability zone (index: 131–136), moderately high vulnerability zone (index: 136–150), and high vulnerability zone (index: 150–179).

The results reveal that the percentage of total area under different vulnerability classes is 3.45 % (168.13 km^2), 22.12 % (1,075.45 km^2), 38.85 % (1,890.99 km^2), 33.63 % (1,636.96 km^2), and 1.85 % (94.97 km^2) for low, moderately low, moderate, moderately high and high, respectively. The high vulnerability zones are mainly lie in the blocks of Sonahatu, Angara, and Silli.

$$\text{maximum vulnerability index} = \sum_{i=1}^{7} \text{rating}_i \times \text{weightage}_i$$
$$= 10 \times 5 + 9 \times 4 + 10 \times 3 + 10 \times 1 + 10 \times 5 + 10 \times 3 = 226$$

similarly,

$$\text{minimum vulnerability index} = \sum_{i=1}^{7} \text{rating}_i \times \text{weightage}_i$$
$$= 1 * 5 + 1 * 4 + 1 * 3 + 1 * 2 + 1 * 1 + 1 * 5 + 1 * 3 = 23$$

Though, it is very difficult to say the role of a particular parameter on the spatial changes in the vulnerability index without sensitivity analysis. This is because variation in depth to groundwater table in the study area was found to be low. But the vulnerability map clearly reveals that the depth to groundwater has an insignificant role in spatial changes in vulnerability index. It is clear from the map that the vulnerability is low in the area having higher depth to groundwater and vice versa. Furthermore, the variation of net recharge was very high in the study area and hence had a high influence on the spatial changes in the vulnerability index. Thus, to understand the influence of each parameter sensitivity analysis was carried out. This is explained in sensitivity analysis section.

Validation of the model

The model was validated by comparing the model output (vulnerability index) with the observed nitrate concentration in groundwater in the study area. The reason behind the selection of nitrate was that the major sources of nitrate in groundwater are various anthropogenic activities like fertilizer used in the agricultural field. The DRASTIC model assumes that the contaminant has the mobility of water. Nitrate being completely soluble in water and hence very nearly satisfies this assumption. Groundwater samples were collected from 30 locations in the study area and analyzed in laboratory for measuring the nitrate concentrations. The spatial locations of the sampling points were recorded by a handheld GPS meter. Nitrate analysis was done as per the standard methods (APHA 1995) using UV–VIS spectrophotometer. In this method generally the absorption was measured twice, i.e., at 220 nm for nitrate concentration and at 275 nm for organic matters which cause hindrance. Then the absorption at 220 nm was subtracted from twice the absorption at 275 nm to obtain the actual nitrate concentration of a given water sample. Nitrate concentrations were found to be in the range of 10.12–51.34 mg/l. The DRASTIC indices of the corresponding points were determined from vulnerability index map (Fig. 4). Figure 5 clearly indicates that the trends of nitrate concentration and vulnerability indices were matched closely in most of the occasions except a few. The correlation between the vulnerability indices and observed nitrate concentration was found to be 0.859.

This clearly indicates that the model can be accepted for vulnerability assessment and predict with 85.9 percent accuracy. The comparative values of observed nitrate concentration at 30 sampling locations were superimposed on the vulnerability index map as shown in Fig. 6. The range of observed nitrate concentration was classified into three levels (<30 mg/l; 30–45 mg/l; and >45 mg/l) and compared with the corresponding vulnerability indices. Figure 6 clearly indicates that nitrate concentrations of the samples lie down in level 1 (<30 mg/l) were mainly stretched out in the class of moderately low or moderate vulnerability class. Similarly, the nitrate concentration of the samples which lies down in level 2 (30–45 mg/l) was mainly stretched out in the class of moderate or moderately high vulnerability class. It was observed that only one sample had the concentration level greater than 45 mg/l that is in level 3 and this location was stretched out in high vulnerability class. There were few samples that contradicted the law (high vulnerability class zone has high nitrate concentration) due to some other factors which were not defined in the model.

Sensitivity analysis

The ideas or views of scientists' conflict in regard to DRASTIC model for groundwater vulnerability to pollution assessment. Some scientists agreed that groundwater

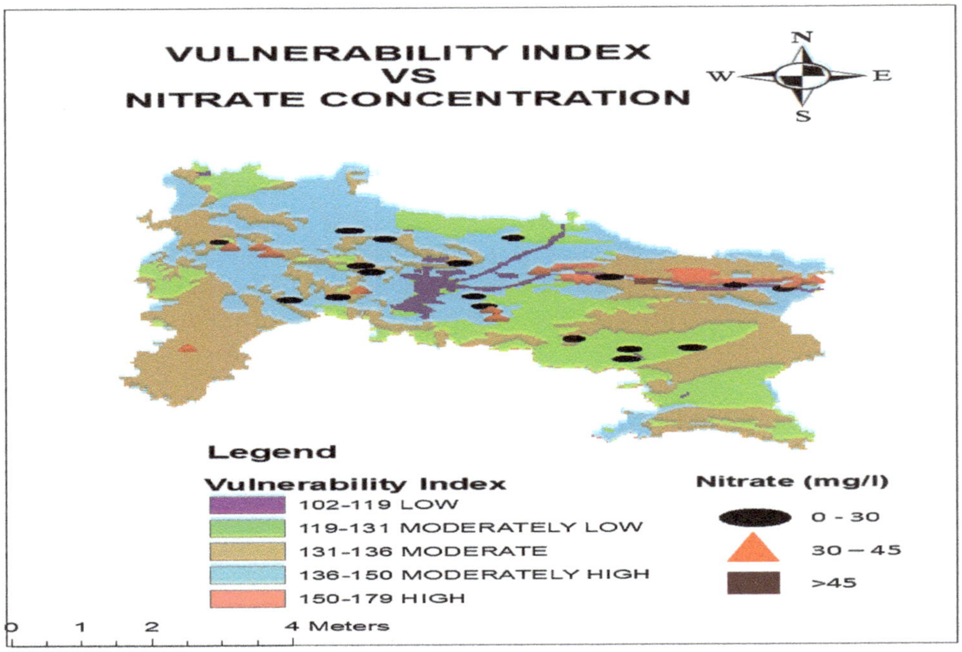

Fig. 5 Vulnerability index vs. nitrate concentration

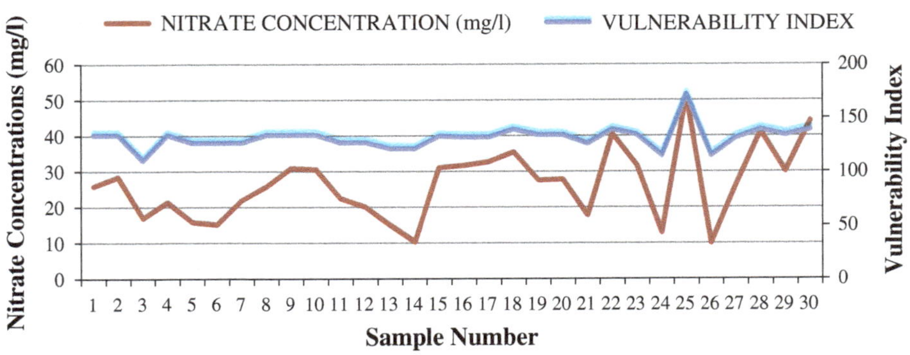

Fig. 6 Vulnerability index and corresponding nitrate concentration

vulnerability assessment can be studied without considering all the factors of the DRASTIC model (Merchant 1994), while some others not agreed with the ideas (Napolitano and Fabbri 1996). To make a common consensus sensitivity analysis of the model and groundwater contamination analysis are carried out.

Sensitivity analysis provides information on the influence of rating and weights assigned to each of the factors considered in the model (Gogu and Dassargues 2000b). Lodwik et al. (1990) defined the measure of map removal sensitivity. This explains the degree of sensitivity associated with removing one or more map layers. The sensitivity analysis mentioned above can be measured by removing one or more layer maps using the following equation:

$$S_i = \left| \frac{V_i}{N} - \frac{V_{xi}}{n} \right|. \tag{2}$$

where, S_i represents sensitivity for ith sub-area associated with the removal of one map (x-factor), V_i is vulnerability index computed using Eq. (1) for the ith sub-area, V_{xi}, vulnerability index computed for ith sub-area excluding o map layer (x), N is the number of map layers used to compute vulnerability index in Eq. (1) and n is number of map layers used for sensitivity analysis. To assess the magnitude of the variation created by removal of one parameter, the variation index was computed as:

$$\mathrm{Var}_i = \left(\frac{V_{i-Vx_i}}{V_\mathrm{i}} \right) \times 100 \tag{3}$$

where, Var_i is variation index of the map removal parameter; and V_i and V_{xi} represent vulnerability index for the ith sub-area in two different conditions as mentioned above. Variation index estimates the effect on vulnerability indices due to removal of each parameter. Its value can be either positive or negative, depending on vulnerability index. Variation index directly depends upon the weighting system.

The single parameter sensitivity test was carried out to estimate the role of each parameters considered in the model on the vulnerability measure. The objective of this analysis is to compare the real or ''effective'' weight of each parameter with that of the corresponding assigned or ''theoretical'' weight. The effective weight of a parameter in ith sub-area can be determined using the following equation:

$$W_{xi} = \left(\frac{X_{ri} \times X_{wi}}{V_i}\right) \times 100 \quad (4)$$

where, X_{ri} and X_{wi} represent the rating and the weight assigned to a parameter X, respectively, in ith sub-area and V_i is the vulnerability index as mentioned above. The sensitivity analysis helps to validate and evaluate the consistency of the analytical results and is the basis for proper evaluation of vulnerability maps. A more efficient interpretation of the vulnerability index can be achieved through sensitivity analysis. The summary of the results of sensitivity analysis that was performed by removing one or more data layer is represented in Tables 4 and 5. Statistical analysis results (shown in Table 4) indicate that the most sensitive to groundwater pollution is the parameter D, followed in importance by factors R, I, A, C, S and T. The highest mean value was associated with the depth to groundwater table (4.52) whereas the impact of vadose zone shows the lowest sensitive value (0.32). The results of variation index (shown in Table 5) clearly indicate that the parameter R has the highest variation index (0.274) followed by parameter I of variation index (0.234). This variation index explains the effect on vulnerability index on removal of any parameter.

Variation index is directly associated with the weighting system of the model. New or effective weights for each input parameters were computed using the Eqs. (3) and (4) and reported in Table 5. The effective weight factor results clearly indicate that the parameter D dominates the vulnerability index with an average weight of 23.84 % against the theoretical weight of 21.74 %. The actual weight of parameter I (16.77 %) is smaller than the theoretical weight (21.74). The calculated weight of parameter T (7.07 %) is greater than theoretical weight (4.35 %). The highest effective weight of parameter D clearly indicates the presence of shallow groundwater table in the most part of the study area and the calculated effective weight of parameter T is more than theoretical weight due to the fact that the slope in most of the part of the study area is <6 %.

It is clearly observed in the study that the calculated effective weights for each parameter are not equal to the theoretical weight assigned in DRASTIC method. This is due to the fact that weight factors are strongly related to the value of a single parameter in the context of value chosen for the other parameters. Therefore, the determination of effective weights is very useful to revise the weight factors assigned in DRASTIC method and may be applied more scientifically to address the local issues.

Table 4 Statistics of single parameter sensitivity analysis

Parameter	Minimum value	Maximum value	Mean	Standard deviation
D	3.47	5.71	4.52	0.35
R	0.45	3.11	2.71	0.42
A	0	1.57	1.09	0.27
S	1.21	3.02	1.98	0.19
T	1	3.92	2.67	0.23
I	0	1.52	0.32	0.28
C	0.45	2.92	1.33	0.47

Conclusions

A GIS-based DRASTIC model was used for computing the groundwater vulnerability to pollution index map of Ranchi district. The study area was divided into five zones (low, moderately low moderate, moderately high and high) on the basis of relative groundwater vulnerability to pollution index. Higher the value of the vulnerability index, higher is the risk of groundwater contamination. The results reveal that moderate vulnerable class covers the maximum percentage of the area (38.85 % of the total area). Moderately high vulnerability class and moderately

Table 5 Assigned weights and effective weights

Parameter	Assigned weight	Assigned weight (%)	Variation index (W_{xi})	Calculated effective weight after rescaling (X_{wi})	Calculated effective weight (%)
D	5	21.74	0.26–0.46	6.09–10.65	26.47–46.3
R	4	17.39	0.08–0.29	1.89–6.73	8.21–29.27
A	3	13.04	0.08–0.19	1.84–4.5	8–19.57
S	2	8.69	0.03–0.08	0.79–1.8	3.42–7.84
T	1	4.35	0.03–0.08	0.72–1.85	−3.14–8.06
I	5	21.74	0.09–0.22	2.06–5.13	8.98–22.32
C	3	13.04	0.02–0.22	0.47–5.0	2.06–21.74

low vulnerability class also cover significant share of the area.

Sensitivity analysis results indicate that the new effective weights for each parameter are not equal to the theoretical weight assigned in DRASTIC method. Thus, the computation of effective weights is very useful to revise the weight factors assigned in DRASTIC method and may be applied more scientifically to address the local issues.

Groundwater has an important role in drinking water supply in Ranchi district. The study suggests that the GIS-based DRASTIC model can be used for identification of the vulnerable areas for groundwater quality management. In the vulnerable areas, detailed and frequent monitoring of groundwater should be carried out for observing the changing level of pollutants. Furthermore, the present study also helps for screening the site selection for waste dumping.

Acknowledgments The authors are thankful to the University Grants Commission (UGC), New Delhi for providing financial support [F.N.39-965/2010 (SR)] which made this study possible. The support of the JSAC, Ranchi, CGWB, New Delhi, and BAU, Ranchi is acknowledged for providing some data. The authors are also thankful to the anonymous reviewers and editors to make the paper more presentable and good.

References

Al-Adamat RAN, Foster IDL, Baban SMJ (2003) Groundwater vulnerability and risk mapping for the basaltic aquifer of the Azraq basin of Jordan using GIS, remote sensing and DRASTIC. Appl Geogr 23:303–324

Aller L, Bennett T, Lehr JH, Petty RJ (1987) DRASTIC: a standardized system for evaluating groundwater pollution potential using hydrogeologic settings. USEPA, Robert S. Kerr Environmental Research Laboratory, Ada, OK. EPA/600/2 85/0108

Almasri Mohammad N (2008) Assessment of intrinsic vulnerability to contamination for Gaza coastal aquifer, Palestine. J Environ Manag 88:577–593

Aminreza Neshat, Pradhan B, Pirasteh S, Shafri HZM (2014) Estimating groundwater vulnerability to pollution using a modified DRASTIC model in the Kerman agricultural area, Iran. Environ Earth Sci 71:3119–3131

Anwar MC, Prem, Rao VB (2003) Evaluation of groundwater potential of Musi River catchment using DRASTIC index model. In: Venkateshwar B R, Ram MK, Sarala C S, Raju C (eds) Hydrology and watershed management. Proceedings of the International Conference 18–20, 2002, B S Publishers, Hyderabad

APHA Standard Methods (1995) Standard methods for the examination of water and wastewater. In: Andrew D Eaton, Lenore S Clesceri, Arnold E Greenberg (eds) American Public Health Association (APHA), 19th edition, Washington

Babiker IS, Mohamed AAM, Hiyama T, Kato K (2005) A GIS-based DRASTIC model for assessing aquifer vulnerability in Kakamigahara Heights, Gifu Prefecture, Central Japan. Sci Total Environ 345(1–3):127–140

Barbash JE, Resek EA (1996) Pesticides in groundwater: distribution, trends, and governing factors. Pesticides in the hydrologic system series, vol 2. Ann Arbor Press, Chelsea, Michigan, p 590

Berkhoff K (2008) Spatially explicit groundwater vulnerability assessment to support the implementation of the Water Framework Directive—a practical approach with stakeholders. Hydrol Earth Syst Sci 12:111–122

Burkart MR, Feher J (1996) Regional estimation of groundwater vulnerability to non-point sources of agricultural chemicals. Water Sci Technol 33:241–247

Central Ground Water Board (CGWB) (2009) Groundwater information booklet. Available on http://cgwb.gov.in/District_Profile/Jharkhand/RANCHI.pdf. Accessed 15th Dec 2012

Chae G, Kim K, Yun S, Kim K, Kim S, Choi B, Kim H, Rhee CW (2004) Hydrogeochemistry of alluvial groundwater in an agricultural area: an implication for groundwater contamination susceptibility. Chemosphere 55:369–378

Chandrashekhar H, Adiga S, Lakshminarayana V, Jagdeesha CJ, Nataraju C (1999) A case study using the model 'DRASTIC' for assessment of groundwater pollution potential. In Proceedings of the ISRS national symposium on remote sensing applications for natural resources, June 19–21, Bangalore

Civita MV (1994) "Le carte della vulnerabilità degli acquiferi all'inquinamento: Teoria & pratica," (Groundwater vulnerability maps to contamination: Theory and practice) Pitagora Editrice, Bologna, pp. 325 (with bibliography)

Connell LD, Daele G (2003) A quantitative approach to aquifer vulnerability mapping. J Hydrol 276:71–88

Dixon B (2005) Groundwater vulnerability mapping: a GIS and fuzzy rule based integrated tool. Appl Geogr 25:327–347

Ferreira Lobo JP, Oliveira Manuel M (2004) Groundwater vulnerability assessment in Portugal. Geofis Int 43(4):541–550

Fobe B, Goossens M (1990) The groundwater vulnerability map for the Flemish region: its principles and uses. Eng Geol 29:355–363

Foster SSD (1987) Fundamental concepts in aquifer vulnerability, pollution risk and protection strategy. In: van Duijvenbooden, W, van Waegeningh GH (eds) TNO Committee on Hydrological Research, the Hague, Proceedings and Information, 38:69–86

Giambelluca TW, Loague K, Green RE, Nullet MA (1996) Uncertainty in recharge estimation: impact on groundwater vulnerability assessments for the Pearl Harbor Basin, O'ahu, Hawai'i, USA. J Contam Hydrol 23:85–112

Gogu RC, Dassargues A (2000a) Sensitivity analysis for the EPIK method of vulnerability assessment in a small karstic aquifer, Southern Belgium. Hydrogeol J 8(3):337–345

Gogu RC, Dassargues A (2000b) Current trends and future challenges in groundwater vulnerability assessment using overlay and index methods. Environ Geol 39:549–559

Herlinger R Jr, Viero AP (2007) Groundwater vulnerability assessment in coastal plain of Rio Grande do Sul State, Brazil, using drastic and adsorption capacity of soils. Environ Geol 52:819–829

Karolin Brosig, Geyer T, Subah A, Sauter M (2008) Travel time based approach for the assessment of vulnerability of karst groundwater: the Transit Time Method. Environ Geol 54:905–911

Kim YJ, Hamm S (1999) Assessment of the potential for groundwater contamination using the DRASTIC/EGIS technique, Cheongju area, South Korea. Hydrogeol J 7:227–235

Kwansiririkul K, Singharajwarapan FS, Mackay R, Ramingwong T, Wongpornchai P (2004) Vulnerability assessment of groundwater resources in the Lampang Basin of Northern Thailand. J Environ Hydrol 12(23):1–15

Lake IR, Lovett AA, Hiscock KM, Betson M, Foley A, Sunnenberg G, Evers S, Fletcher S (2003) Evaluating factors influencing groundwater vulnerability to nitrate pollution: developing the potential of GIS. J Environ Manag 68:315–328

Lodwik WA, Monson W, Svoboda L (1990) Attribute error and sensitivity analysis of maps operation in geographical information systems: suitability analysis. Int J Geogr Inf Syst 4:413–428

Margat J (1968) Groundwater vulnerability to contamination. 68, BRGM, Orleans, France. In: Massone et. al. 2010, Enhanced groundwater vulnerability assessment in geological homogeneous areas: a case study from the Argentine Pampas. Hydrogeol J 18: 371–379

Massone H, Londoño MQ, Martínez D (2010) Enhanced groundwater vulnerability assessment in geological homogeneous areas: a case study from the Argentine Pampas. Hydrogeol J 18:371–379

Merchant JW (1994) GIS-based groundwater pollution hazard assessment: a critical review of the DRASTIC model. Photogramm Eng Rem S 60(9):1117–1127

Napolitano P, Fabbri AG (1996) Single parameter sensitivity analysis for aquifer vulnerability assessment using DRASTIC and SINTACS. In: Proceedings of the 2nd HydroGIS Conference, vol. 235. IAHS Publication, Wallingford, pp 559–566

National Research Council (NRC) (1993) Groundwater vulnerability assessment: predicting relative contamination potential under conditions of uncertainty. National Academy Press, Washington

Nouri J, Malmasi S (2005) The role of groundwater vulnerability in urban development planning. Am J Environ Sci 1:16–21

Polemio M, Casarano D, Limoni PP (2009) Karstic aquifer vulnerability assessment methods and results at a test site (Apulia, southern Italy). Nat Hazards Earth Syst Sci 9:1461–1470

Rahman A (2008) A GIS based DRASTIC model for assessing groundwater vulnerability in shallow aquifer in Aligarh, India. Appl Geogr 28:32–53

Rational formula for runoff coefficient. Available on http://water.me.vccs.edu/courses/CIV246/table2b.htm. Accessed 02nd March 2014

Rupert MG (2001) Calibration of the DRASTIC groundwater vulnerability mapping method. Ground Water 39:630–635

Secunda S, Collin M, Mellou AJ (1998) Groundwater vulnerability assessment using a composite model combining DRASTIC with extensive land use in Israel. J Environ Manag 54:39–57

Shirazi SM, Imran MH, Akib S (2012) GIS based DRASTIC method for groundwater vulnerability assessment: a review. J Risk Res 15:991–1011

Shirazi SM, Imran MH, Akib S, Yusop Z, Harun ZB (2013) Groundwater vulnerability assessment in Melaka state of Malaysia using DRASTIC and GIS techniques. Environ Earth Sci 70:2293–2304

Sinan M, Razack M (2009) An extension to the DRASTIC model to assess groundwater vulnerability to pollution: application to the Haouz aquifer of Marrakech (Morocco). Environ Geol 57:349–363

Soutter M, Musy A (1998) Coupling 1D Monte-Carlo simulations and geostatistics to assess groundwater vulnerability to pesticide contamination on a regional scale. J Contam Hydrol 32:25–39

Tesoriero AJ, Inkpen EL, Voss FD (1998) Assessing groundwater vulnerability using logistic regression. Proceedings for the Source Water Assessment and Protection 98 Conference, Dallas, TX, 157–165

Thapinta A, Hudak P (2003) Use of geographic information systems for assessing groundwater pollution potential by pesticides in Central Thailand. Environ Int 29:87–93

Vias JM, Andreo B, Perles MJ, Carrasco F (2005) A comparative study of four schemes for groundwater vulnerability mapping in a diffuse flow carbonate aquifer under Mediterranean climatic conditions. Environ Geol 47:586–595

Villeneuve JP, Banton O, Lafrance P (1990) A probabilistic approach for the groundwater vulnerability to contamination by pesticides: the VULPEST model. Ecol Model 51:47–58

Vrba J, Zaporozec A (1994) Guidebook on mapping groundwater vulnerability. Int. Association of Hydrogeologists. Int. Contributions to Hydrogeology; 16. Verlag Heinz Heise, Hannover

Worrall F, Besien T (2004) The vulnerability of groundwater to pesticide contamination estimated directly from observations of presence or absence in wells. J Hydrol 303:92–107

Worrall F, Besien T, Kolpin DD (2002) Groundwater vulnerability: interactions of chemical and site properties. Sci Total Environ 299:131–143

Zektser IS, Karimova OA, Bujuoli J, Bucci M (2004) Regional estimation of fresh groundwater vulnerability: methodological aspects and practical applications. Water Resour 31:645–650

Zhang R, Hamerlinck JD, Gloss SP, Munn L (1996) Determination of nonpoint-source pollution using GIS and numerical models. J Environ Qual 25(3):411–418

4

Hydrochemical characteristics and quality assessment of groundwater along the Manavalakurichi coast, Tamil Nadu, India

Y. Srinivas[1] · T. B. Aghil[1] · D. Hudson Oliver[2] · C. Nithya Nair[1] · N. Chandrasekar[1]

Abstract The present study was carried out to find the groundwater quality of coastal aquifer along Manavalakurichi coast. For this study, a total of 30 groundwater samples were collected randomly from open wells and borewells. The concentration of major ions and other geochemical parameters in the groundwater were analyzed in the laboratory by adopting standard procedures suggested by the American Public Health Association. The order of the dominant cations in the study area was found to be $Na^+ > Ca^{2+} > Mg^{2+} > K^+$, whereas the sequence of dominant anions was $Cl^- > HCO_3^- > SO_4^{2-}$. The hydrogeochemical facies of the groundwater samples were studied by constructing piper trilinear diagram which revealed the evidence of saltwater intrusion into the study area. The obtained geochemical parameters were compared with the standard permissible limits suggested by the World Health Organization and Indian Standard Institution to determine the drinking water quality in the study area. The analysis suggests that the groundwater from the wells W25 and W26 is unsuitable for drinking. The suitability of groundwater for irrigation was studied by calculating percent sodium, sodium absorption ratio and residual sodium carbonate values. The Wilcox and USSL plots were also prepared. It was found that the groundwater from the stations W1, W25 and W26 is unfit for irrigation. The Gibbs plots were also sketched to study the mechanisms controlling the geochemical composition of groundwater in the study area.

Keywords Groundwater · Quality · Assessment · Tamil Nadu · India

✉ Y. Srinivas
drysv@yahoo.co.in

1 Centre for GeoTechnology, Manonmaniam Sundaranar University, Tirunelveli, Tamil Nadu 627012, India

2 Senthamarai College of Arts and Science, Vadapalanji, Madurai, Tamil Nadu 625021, India

Introduction

Groundwater is an essential source of drinking for numerous people around the world. Saline water intrusion in the coastal aquifers due to overexploitation of groundwater and other anthropogenic activities is a serious environmental issue nowadays. To protect the aquifers from quality degradation, geochemical assessment and monitoring is necessary. It is estimated that approximately one-third of the world's population use groundwater for drinking purpose (Nickson et al. 2005). In India, more than 90 % of rural and nearly 30 % of urban population depends on groundwater for their drinking and domestic requirements (Jaiswal et al. 2003). According to Babiker et al. (2007), the chemistry of groundwater is not only related to the lithology of the area and the residence time the water is in contact with rock material, but also reflects inputs from the atmosphere, soil and weathering as well as pollutant sources such as saline intrusion, mining, and industrial and domestic wastes. Excessive irrigation activities also resulted in groundwater pollution in India (Pawar and Shaikh 1995; Sujatha and Reddy 2003).

The overexploitation of groundwater along the coastal area has become a major environmental issue in the entire world. In many coastal areas, human settlements together with the day-by-day development of industrial, agricultural and tourist activities have led to overexploitation of aquifers. Because of this exploitation, the chances of intrusion of seawater into the inland aquifer are very high and also affects the quality of groundwater. In the present study, the geochemical characteristics of groundwater is determined to know the groundwater quality along the coastal aquifer of Manavalakurichi, located in Kanyakumari District of Tamil Nadu, India.

The study area Manavalakurichi is a town panchayat located in the southern end of Tamil Nadu, India. The study area extends from 8°7′ to 8°10′N longitudes and 77°19′ to 77°19′E latitudes. It is connected by road network to Trivandrum, Tirunelveli and Kanyakumari. The study area is situated in Kanyakumari District and the southern side of the study area is bounded by the Indian Ocean and in the west by the Arabian Sea. Manavalakurichi and its surroundings are well known for their heavy mineral deposits. Several mining areas under the control of Indian Rare Earth Limited (IREL) are situated here. The average temperature in the study area varies from 22.8 to 33.6 °C with an average rainfall of about 846–1456 mm. The present study area is shown in Fig. 1. A major part of the study area is geologically covered by Quaternary fluvio-marine deposits. The north and northwestern regions of the district are completely occupied by the Western Ghats Mountain with a maximum elevation of 1658 m (CGWB 2008). The coastal tract in the south of the study area is a thin strip of plain region that has a width of 1–2.5 km (Srinivas et al. 2013) and is mostly covered by marshy swamps and a number of sand dunes (Teri sands). The soil type of the present area is classified into red soil, red lateritic soil, brown soil and coastal sand. The soils of this district are in situ, in nature, pale reddish, lateritic and earthy in color (CGWB 2008).

Methodology

The study area base map was scanned and digitized from the Survey of India (SOI) toposheet No. 58 H/8 (1:50,000) (GSI 1995). ArcGis 9.3 software was used for digitization and to analyze the data for groundwater quality evaluation.

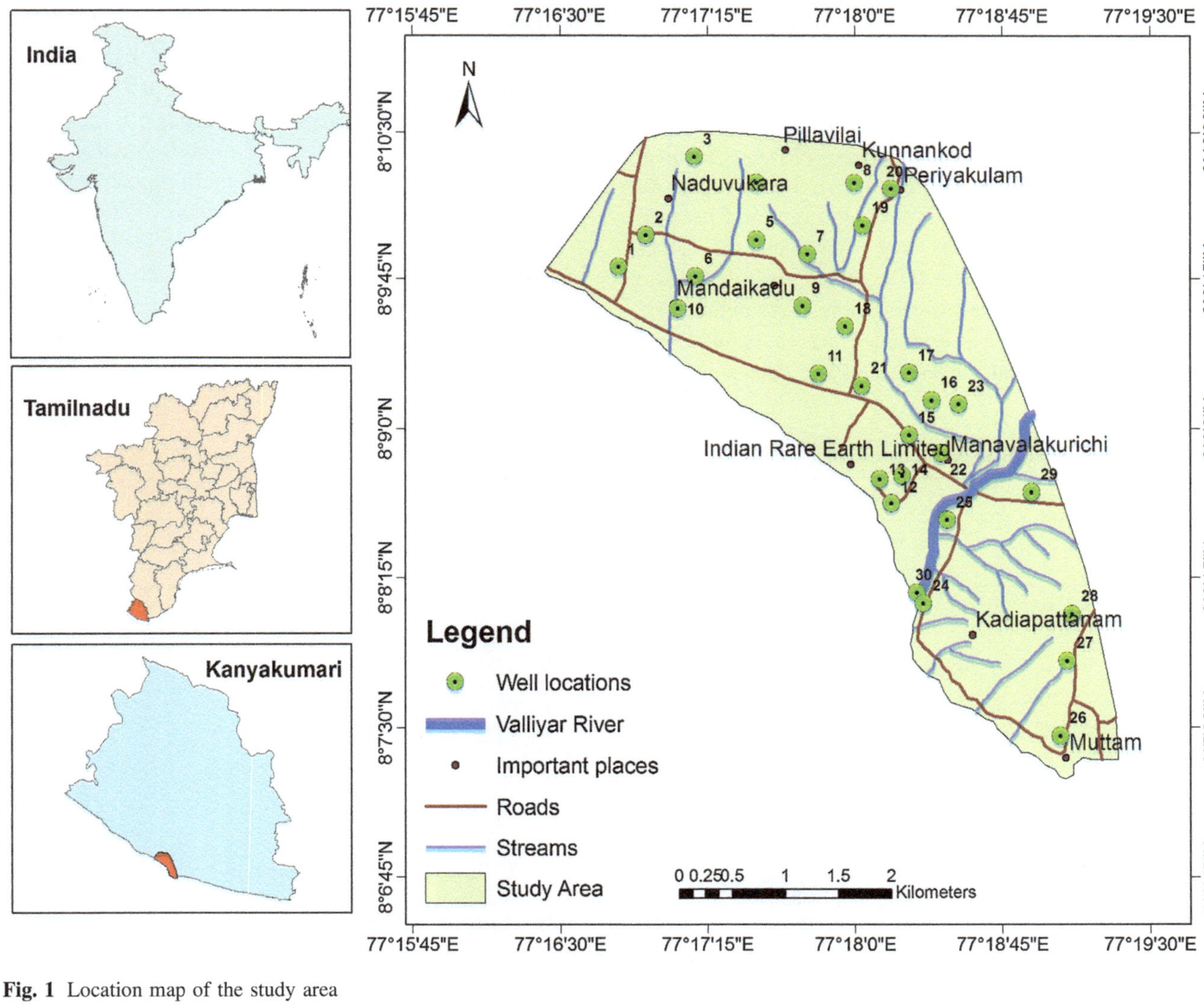

Fig. 1 Location map of the study area

Groundwater samples were randomly collected from 30 open wells and borewells during May 2013, representing the premonsoon season (Fig. 1). For sample collection, high-density polyethylene bottles were used. The bottles are immediately sealed after collecting the sample to avoid reaction with the atmosphere. The sample bottles were labeled systematically. The collected samples were analyzed in the laboratory for various physicochemical parameters. During sample collection, handling, preservation and analysis, standard procedures recommended by the American Public Health Association (APHA 1995) were followed to ensure data quality and consistency.

The pH and electrical conductivity (EC) were measured in situ by using Hanna (HI9828 USA) multi-parameter probe, and the major ions (Ca^{2+}, Mg^{2+}, Na^{+}, K^{+}, HCO_3^-, SO_4^{2-}, Cl^-) were analyzed in the laboratory using the standard methods suggested by the American Public Health Association (APHA 1995). Among the analyzed ions, sodium (Na^+) and potassium (K^+) were determined using flame photometer. Calcium (Ca^{2+}), magnesium (Mg^{2+}), bicarbonate (HCO_3^-) and chloride (Cl^-) were analyzed by volumetric methods and sulfate (SO_4^{2-}) was estimated using spectrophotometer. The piper diagram, Gibbs plot, Wilcox diagram and the USSL plot were used for studying the quality of the groundwater in detail.

The spatial analysis of various physico-chemical parameters was carried out using ArcGIS 9.3 software. An inverse distance weighed (IDW) algorithm was used to interpolate the data spatially and to estimate the values between measurements. The IDW technique calculates a value for each grid node by examining surrounding data points that lie within a user-defined search radius (Burroughs and McDonnell 1998). All of the data points are used in the interpolation process and the node value is calculated by averaging the weighted sum of all the points.

Results and discussions

The analytical results have been evaluated to find out the suitability of groundwater in the study area for drinking and agricultural uses. By comparing the obtained values of different water quality parameters with the guidelines of the World Health Organization (WHO 2011) and Indian Standard Institute (ISI 1983), we obtain the suitability of groundwater for drinking and agricultural purposes (Table 1).

The concentration and behavior of major ions such as Ca^{2+}, Mg^{2+}, Na^{+}, K^{+}, HCO_3^-, $SO_4^{2-}$and Cl^- and some important physico-chemical parameters such as pH, EC, total dissolved solids (TDS) and the suitability of groundwater in the study area are discussed below.

pH

The balance between the concentration of hydrogen ion and hydroxyl ions in the water is termed as pH. The limit of pH value for drinking water is specified as 6.5–8.5 (WHO 2011; ISI 1983). The pH of water provides vital information in many types of geochemical equilibrium or solubility calculations (Hem 1985). The pH value of most of the groundwater samples in the study area varies from 5.3 to 7.4, which clearly shows that the groundwater in the study area is slightly acidic in nature. This may be attributed to anthropogenic activities such as sewage disposal and use of fertilizers in agricultural lands (paddy fields and coconut

Table 1 Ranges of chemical parameters and their comparison with the WHO and the Indian standards for drinking water

Parameter	Range	Mean	Permissible limits		Samples exceeding the permissible limits	
			WHO (2011)	ISI (1983)	WHO (2011)	ISI (1983)
pH	5.3–7.4	6.3	6.5–8.5	6.5–9.2	W2–W11, W16–W19, W21, W25, W29, W30	W2–W11, W16–W19, W21, W25, W29, W30
TDS (mg/l)	41.6–1775	523.5	1000	1500	W1, W12, W16, W26, W27	W26
Na^+ (mg/l)	18–560	121.5	200	–	W1, W25, W26	–
K^+ (mg/l)	0.5–83	18.9	12	–	W1, W12–W17, W20, W22, W24, W26	–
Ca^{2+} (mg/l)	6.4–124.2	34.3	200	200	–	–
Mg^{2+} (mg/l)	0.5–118.2	39.1	150	150	–	–
HCO_3^- (mg/l)	110–549	224.7	500	–	W26	–
SO_4^{2-} (mg/l)	2–78.5	34.3	250	400	–	–
Cl^- (mg/l)	21.3–756.2	169.7	600	1000	W25	–

plantation) of the study area. It may also be due to natural phenomena like intrusion of brackish water into the sandy aquifer, which initiates the weathering process of the underlying geology (Sarat Prasanth et al. 2012).

Electrical conductivity (EC)

EC is a measure of capacity of water to convey electric current. The most desirable limit of EC in drinking water is prescribed as 1500 μS/cm (WHO 2011). The EC of the groundwater in the study area varies from 67.2 and 2771.8 μS/cm with an average value of 818 μS/cm. Higher EC in few groundwater samples indicates the enrichment of salts in the groundwater. The value of EC may be an approximate index of the total content of dissolved substance in water. It depends upon temperature, concentration and types of ions present (Hem 1985). The effect of saline intrusion may be the reason for medium enrichment of EC in the study area. The effect of pH may also increase the dissolution process, which eventually increases the EC value (Sarat Prasanth et al. 2012).

Calcium and magnesium

The concentration of calcium ions (Ca^{2+}) in the groundwater samples ranges from 6.4 to 124.2 mg/l with an average value of 34.3 mg/l and that of magnesium ions (Mg^{2+}) ranges from 0.5 to 118.2 mg/l (average value of 39.1 mg/l) (Table 1). The principal sources of calcium and magnesium in most of the groundwater samples are detrital minerals such as plagioclase feldspar, pyroxene, amphibole and garnet (Hounslow 1995). The limestone, dolomite, gypsum, anhydrates and clay minerals among the sedimentary rocks of the coastal region also enhance the calcium and magnesium content in the groundwater (Chandrasekar et al. 2013). Reverse cationic exchange, i.e., the replacement of calcium and magnesium ions by sodium ion in the groundwater may be the reason for lower concentration of Ca^{2+} and Mg^{2+} in some areas (Jacob et al. 1999). Figure 2a, b shows the spatial distribution of calcium and magnesium in the study area.

Sodium and potassium

The concentration of sodium ions (Na^{+}) in the collected groundwater samples varies from 18 to 560 mg/l (Table 1). The maximum permissible limit of sodium is 200 mg/l and the present study reveals that few samples exceed the permissible limit of WHO and ISI. Groundwater with high Na^{+} content is not suitable for agricultural use, as it tends to deteriorate the soil. Potassium is a naturally occurring element; however, its concentration remains quite lower compared with Ca^{2+}, Mg^{2+} and Na^{+}. Its concentration in drinking water seldom reaches 20 mg/l (Hounslow 1995). The concentration of K^{+} is between 0.5 and 83 mg/l in the groundwater of the study area. The maximum permissible limit of potassium in drinking water is 12 mg/l, and it was found that few samples exceeded the permissible limit of WHO. The higher concentration of potassium in groundwater is due to the anthropogenic sources and saline water intrusion (Krishnakumar et al. 2009). Figure 2c, d shows the spatial variation of sodium and potassium in the study area.

Chloride

Chloride ion (Cl^{-}) concentration in the collected groundwater samples varies from 21.3 to 756.2 mg/l with a mean value of 169.7 mg/l (Table 1). According to the WHO (2011) standards, the maximum permissible limit of chloride in the groundwater is 600 mg/l. The increased chloride concentration in a freshwater aquifer is one of the indicators of seawater intrusion. Appelo and postma (1993) and Raju et al. (2011) have suggested that the high concentration of chloride may also result from pollution by domestic sewage waste, sand leaching and saline residues in soil.

Based on the ionic concentration, sodium and chloride are found to be the dominant cation and anion, respectively. Mostly, sodium and chloride dominate the seawater ionic content, while calcium and bicarbonate are generally the major ions of freshwater (Hem 1985). Therefore, the higher concentration of sodium and chloride ions in the groundwater from the study area indicates the significant effect of saltwater intrusion in few regions. The spatial distribution of chloride in the study area is shown in Fig. 3a.

Bicarbonate

The observed HCO_3^- values in the groundwater samples range from 110 to 549 mg/l (Table 1) Bicarbonate is the dominant anion in the majority of the samples, except in few groundwater samples collected near the coast. The higher concentration of HCO_3^- in the water shows the dominance of mineral dissolution (Stumm and Morgan 1996). The spatial distribution of the bicarbonate ion in the study area is shown in Fig. 3b.

Sulfate

The concentration of sulfate (SO_4^{2-}) ranges from 2 to 78.5 mg/l with an average value of 34.3 mg/l (Table 1). High value of sulfate content in the groundwater is due to

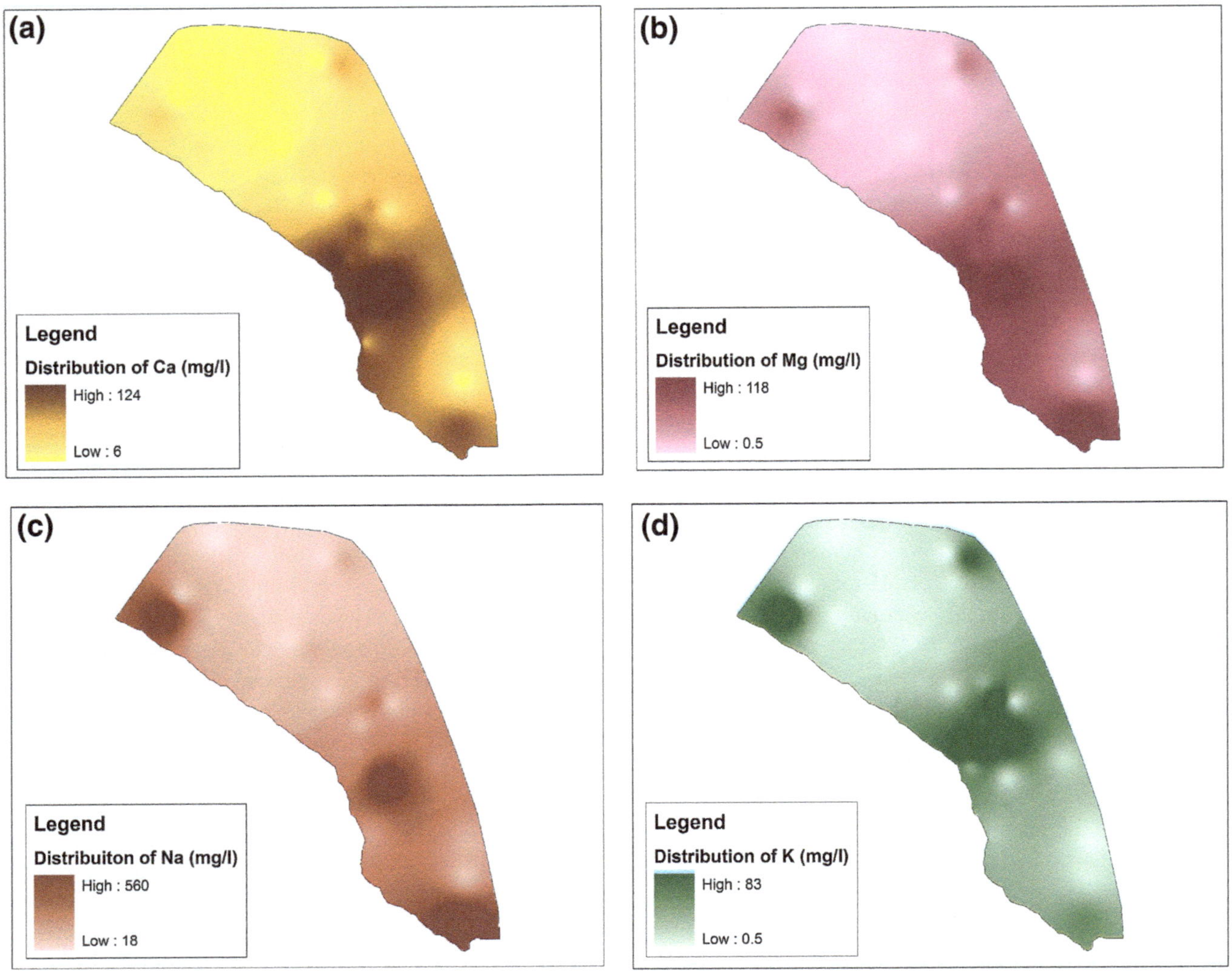

Fig. 2 Spatial distribution of calcium (Ca^{2+}) in the study area. **a** Spatial distribution of magnesium (Mg^{2+})in the study area. **b** Spatial distribution of sodium (Na^{+}) in the study area. **c** Spatial distribution of potassium (K^{+}) in the study area

reduction, precipitation, solution and concentration of the traverse through the sedimentary rock such as gypsum and anhydrite (Hounslow 1995). In the study area, the sulfate concentration is within the limit of WHO and ISI standards. The spatial distribution of sulfate ions in the study area is given in Fig. 3c.

Total dissolved solids (TDS)

According to WHO specifications, TDS in groundwater up to 500 mg/l is the highest desirable and up to 1000 mg/l is maximum permissible. In the study area, the TDS values vary between a minimum of 41.6 mg/l and a maximum of 1775 mg/l (Table 1). According to the Davis and De wiest (1966) classification of groundwater samples based on TDS, 60 % of the total groundwater samples in the study area are desirable for drinking (TDS < 500 mg/l), 23.3 % of samples are permissible for drinking (500–1000 mg/l) and 16.6 % are suitable for irrigation purposes. The spatial distribution of TDS in the study area is given in the Fig. 3d.

Hydrogeochemical facies

The geochemical evolution of groundwater can be understood by plotting the concentrations of major cations and anions in the piper trilinear diagram (Piper 1944). The nature and distribution of hydrochemical facies can be determined by providing insights into how groundwater quality changes within and between aquifers. Trilinear diagrams can be used to delineate the hydrogeochemical facies, because they graphically demonstrate the relationships between the most important dissolved constituents in a set of groundwater samples. The hydrogeochemistry of groundwater in the study area was evaluated using the concentrations of major cations (Ca^{2+}, Mg^{2+}, Na^{+} and K^{+}) and anions (HCO_3^-, SO_4^{2-} and Cl^-) in meq/l. The

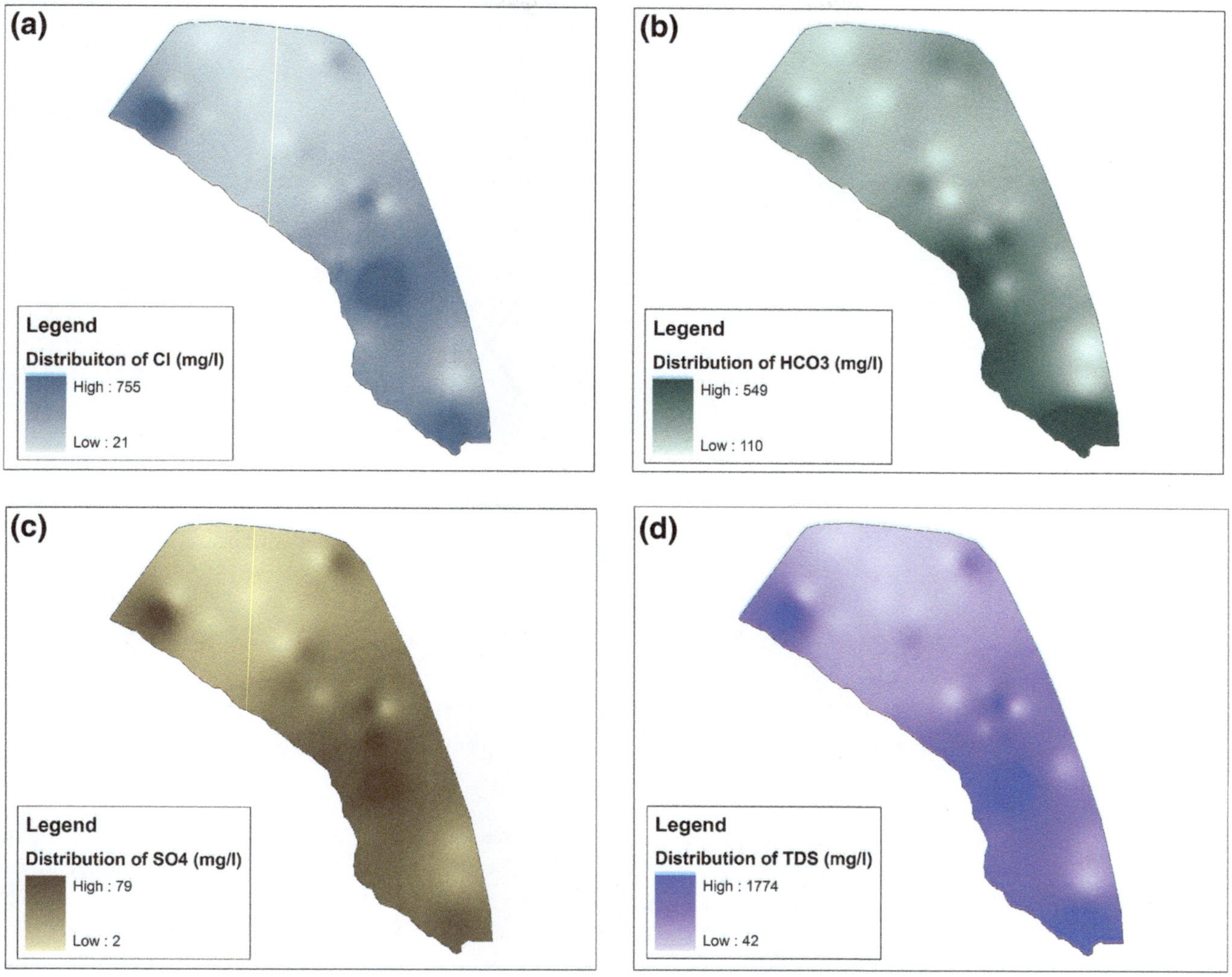

Fig. 3 **a** Spatial distribution of chlorine (Cl^-) in the study area. **b** Spatial distribution of bicarbonate (HCO_3^-) in the study area. **c** Spatial distribution of sulfate (SO_4^{2-}) in the study area. **d** Spatial distribution of total dissolved solids (TDS) in the study area

hydrogeochemical facies for groundwater in the study area is shown in a piper diagram (Fig. 4). The piper diagram indicates that sodium is the major cation and chloride is the major anion.

Pulido-Leboeuf (2004) pointed out that the groundwater of Na–Cl type will generally influence a strong seawater intrusion; anthropogenic contamination is caused by excessive use of fertilizers in the coastal land. In some regions, calcium- and magnesium-based water type is seen. The sources of calcium and magnesium in these regions are probably the shell deposits of fluvio-marine origin (Hounslow 1995).

Sodium percentage

The sodium in irrigation waters is usually denoted as percent sodium or sodium percentage. According to Wilcox (1955), for all natural waters Na % is a common parameter to assess its suitability for irrigational purposes. The sodium percent (Na %) values are obtained using the following equation:

$$\text{Na}\ \% = \text{Na}^+ \times 100/(\text{Ca}^{2+} + \text{Mg}^{2+} + \text{Na}^+ + \text{K}^+),$$

where all ionic concentrations are expressed in meq/l.

The groundwater samples according to Wilcox classification (Table 2) on the basis of sodium percent shows that 53.33 % of the groundwater samples are of uncertain quality for irrigation, 16 % are within the permissible limit, 30 % are under good quality limit and no samples are in the unsuitable category.

Sodium absorption ratio

Sodium absorption ratio (SAR) is a measure of the suitability of groundwater for irrigation usage, because sodium concentration can reduce the soil permeability and soil

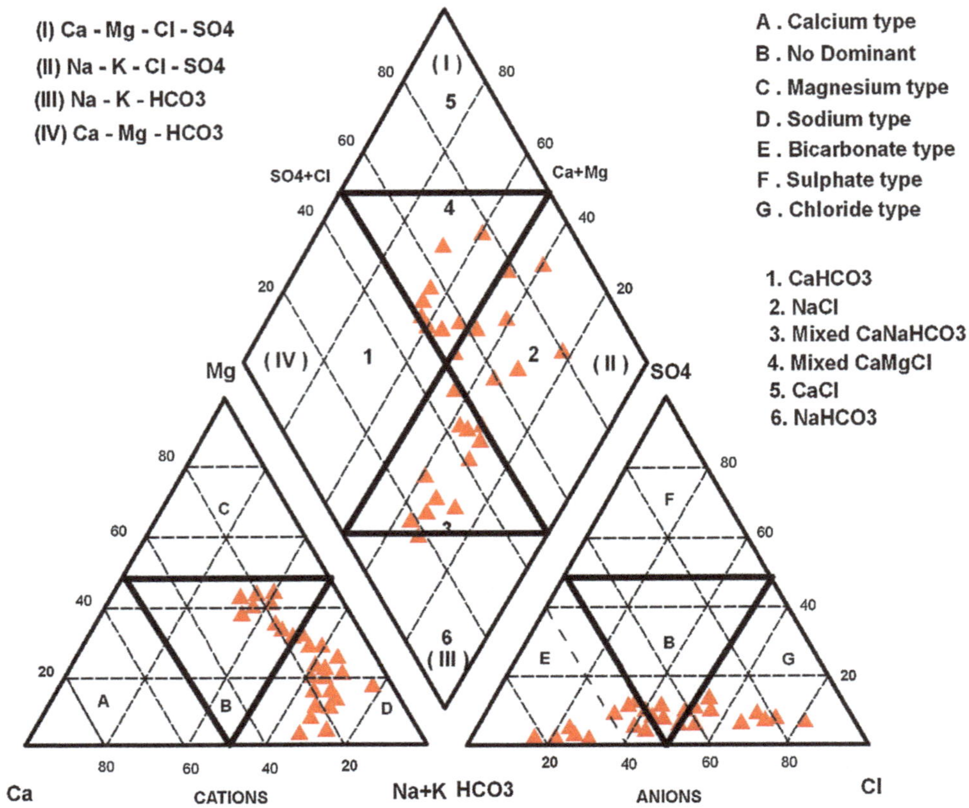

Fig. 4 Piper trilinear diagram representing the geochemical evolution of groundwater

Table 2 Classification of groundwater for irrigation purpose on the basis of Na % and SAR

Parameters	Range	Groundwater class	No. of samples
Na % (Wilcox 1955)	<20	Excellent	–
	20–40	Good	9
	40–60	Permissible	5
	60–80	Doubtful	16
	>80	Unsuitable	–
SAR (Bouwer 1978)	<6	No problem	27
	6–9	Increasing problem	2
	>9	Severe problem	1

structure (Todd 2007). SAR is a measure of alkali/sodium hazard to crops and is estimated by the following formula:

$$SAR = Na^{+} \Big/ [(Ca^{2+} + Mg^{2+})/2]^{0.5},$$

where the concentrations of sodium, calcium and magnesium are in meq/l.

The SAR value of water for irrigation purposes has a significant relationship with the extent to which sodium is absorbed by the soils. Irrigation using water with high SAR values may require soil amendments to prevent long-term damage to the soil, because the sodium in the water can displace the calcium and magnesium in the soil. This will cause a decrease in the ability of the soil to form stable aggregates and loss of soil structure. This will also lead to a decrease in infiltration and permeability of the soil to water, leading to problems with crop production (Chandrasekar et al. 2013). The calculated values of SAR in the study area vary between 1.55 and 11.18 meq/l. The classification of groundwater samples based on SAR values are given in Table 2. The SAR values of the majority of the samples are found within the range of excellent to good category, except sample nos 1, 25 and 26 which were found to be unsuitable for irrigation purpose.

USSL plot

The United States Salinity Laboratory (USSL) has constructed a diagram for the classification of irrigation waters (Wilcox 1955) describing 16 classes with reference to SAR as index for sodium hazard and EC as an index for salinity hazard. The USSL diagram highlights that 33 % of the samples fall under the field of C3S1, which indicates water having high salinity and low sodium alkali hazard (Fig. 5). Three of the samples fall under the C4S3, C3S3 and C3S2 category, respectively, indicating high to very high salinity hazard and medium to high sodium hazard, while the remaining samples fall under the C2S1 and C1S1 regions indicating low and medium saline and sodium hazards.

Fig. 5 USSL diagram representing the salinity and sodium hazard

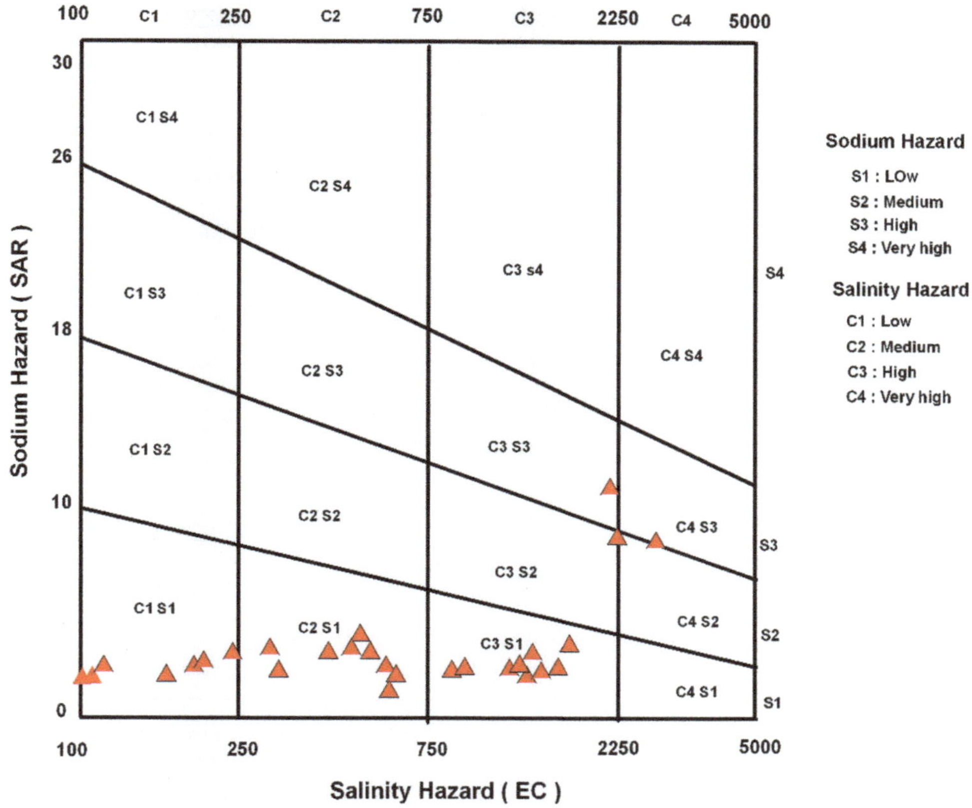

Wilcox diagram

Wilcox (1955) classified groundwater for irrigation purposes by correlating percent sodium and EC, which suggests that 60 % of the total of 30 samples fall under excellent to good limit and 30 % fall under good to permissible limit (Fig. 6). Only 10 % of the samples (W1, W25, W26) fall under the doubtful to unsuitable limit, which illustrates that the groundwater from these wells are not fit for agricultural usages.

Mechanisms controlling the groundwater chemistry

Reactions between aquifer minerals and groundwater have a significant role on water quality, which are also useful to understand the genesis of groundwater (Cederstorm 1946). Commonly, groundwater chemistry in the study region is controlled by different processes and mechanisms. Gibbs plot is used here to understand and differentiate the influences of rock–water interaction, evaporation and precipitation on water chemistry (Gibbs 1970). The Gibbs diagram (Fig. 7a, b) illustrates that most of the groundwater samples from the study area falls under the rock-dominant region. The cation versus TDS curves denote that the cations in some of the wells may be derived from evaporation or crystallization processes. The Gibbs diagrams suggests that the weathering of rocks primarily controls the major ion chemistry of groundwater in this region.

Conclusions

Contamination of groundwater generally results in poor drinking water quality, loss of water supply, high cleanup costs, high-cost alternative water supplies and potential health problem. In the present study, the interpretation of hydrochemical analysis reveals that the groundwater in the study area is slightly acidic in nature. The sequence of the abundance of the major cations is $Na^+ > Ca^{2+} > Mg^{2+} > K^+$ and anions is $Cl^- > HCO_3^- > SO_4^{2-}$. The dominant cation and anion are sodium and chloride respectively, which shows the saline nature of the groundwater in some regions. In the study area, rock weathering and evaporation processes are the major hydrogeochemical processes responsible for the concentration of major ions in groundwater. The drinking water quality analysis shows that groundwater from the wells W25 and W26 are not fit for drinking, as they have higher concentration of ions than the standard permissible limits. The groundwater classification based on their suitability for irrigation also reveals that the wells W1, W25 and W26 are

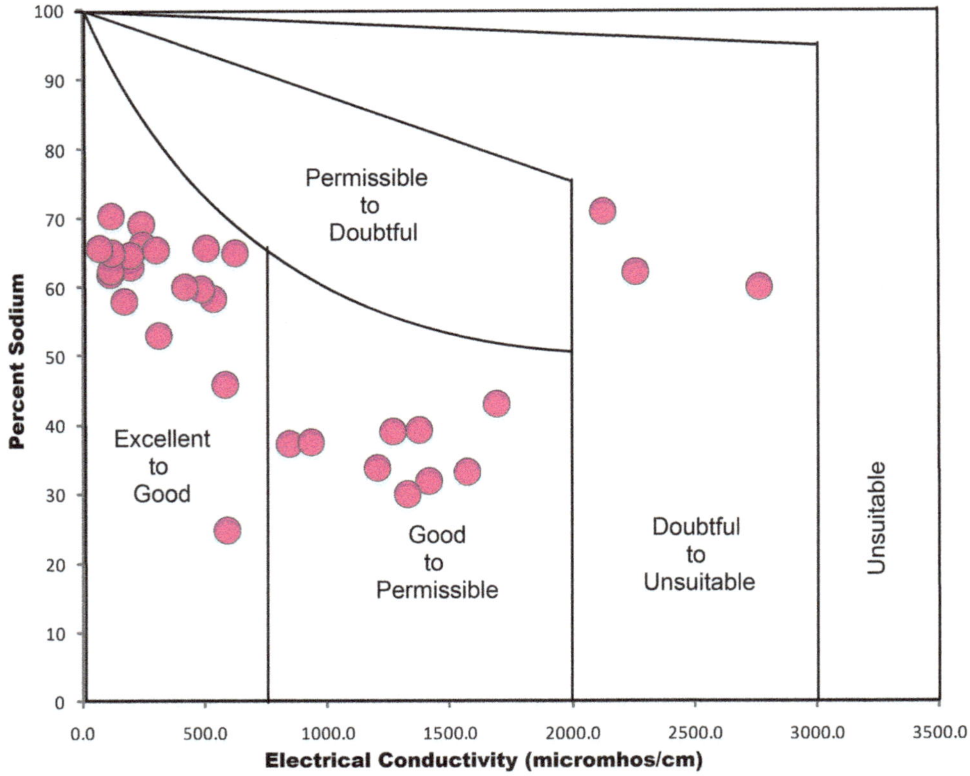

Fig. 6 Suitability of groundwater for irrigation in Wilcox diagram

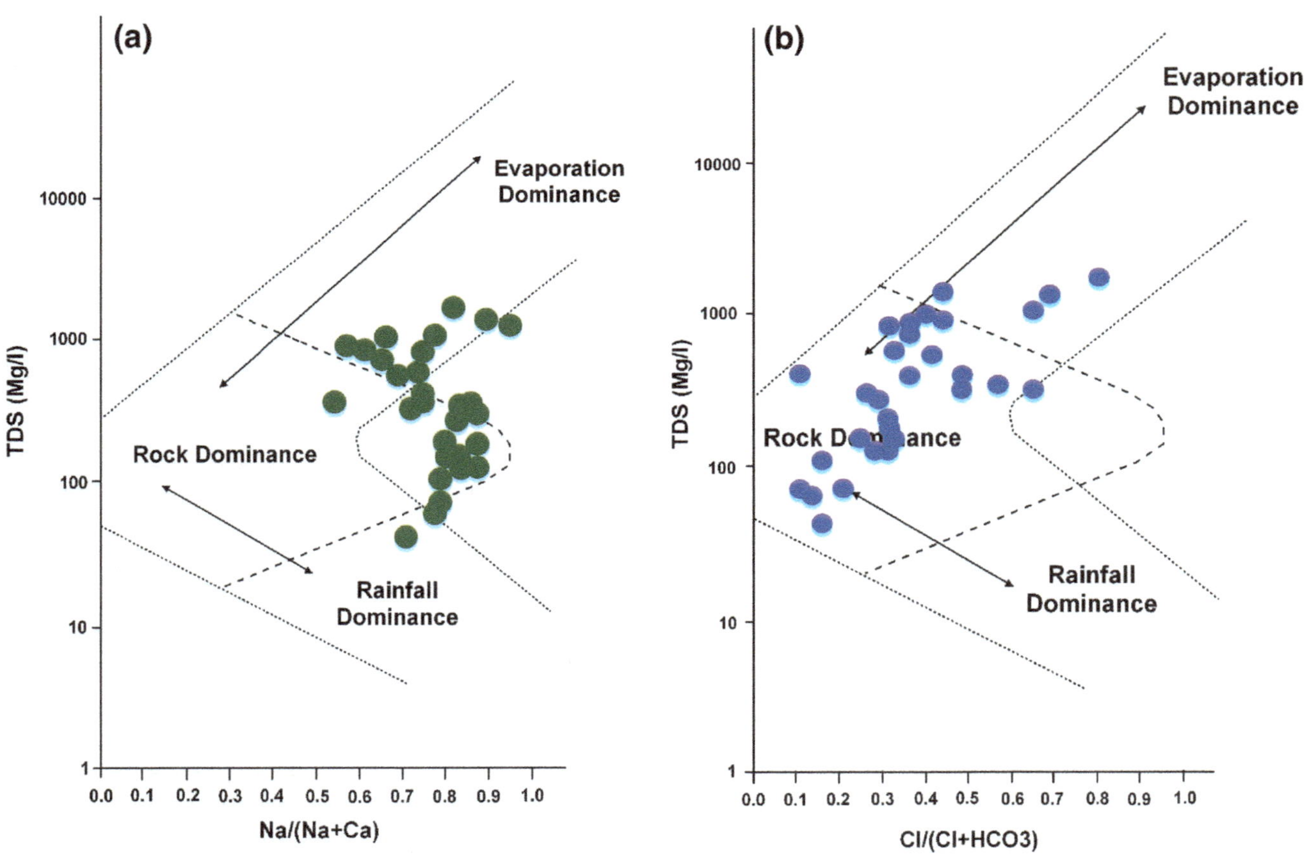

Fig. 7 Gibbs diagrams representing the mechanism controlling the chemistry of groundwater. **a** Major anions versus TDS. **b** Major cations versus TDS

unsuitable for irrigation. So, among the 30 well samples analyzed, the groundwater from the wells W1, W25, W26 were found to be more hazardous in the study area. The harmful nature of groundwater may be due to natural saline water intrusion and also because of anthropogenic activities.

Acknowledgments The present study was carried out under the UGC major research project. The first author is grateful to the UGC for providing the financial support under this major research project No. 41-940/2012 (SR). The authors are very much thankful to the anonymous reviewers for their critical review and comments that helped immensely in improving the quality of the manuscript.

References

APHA (1995) Standard methods for the examination of water and waste water, 19th edn. American Public Health Association, Washington, DC

Appelo CAJ, Postma D (1993) Geochemistry, groundwater and pollution. AA Balkema, Rotterdam

Babiker IS, Mohamed MAA, Hiyama T (2007) Assessing groundwater quality using GIS. Water Resour Manag 21:699–715

Bouwer H (1978) Ground water hydrology. McGraw-Hill, New York

Burroughs PA, McDonnell RA (1998) Principles of geographical information systems. Oxford University Press, Oxford, p 333

Cederstorm DJ (1946) Genesis of groundwater in the coastal plain of Virginia. Environ Geol 41:218–245

CGWB (2008) District groundwater Brochure, Kanyakumari district, Tamilnadu. Government of India, Ministry of Resources

Chandrasekar N, Selvakumar S, Srinivas Y, John Wilson JS, Simon Peter T, Magesh NS (2013) Hydrogeochemical assessment of groundwater quality along the coastal aquifers of southern Tamil Nadu, India. J Environ Earth Sci 71(11):4739–4750. doi:10.1007/s12665-013-2864-3

Davis SN, De Wiest RJM (1966) Hydrogeology, vol 463. Wiley, New York

Gibbs RJ (1970) Mechanisms controlling world water chemistry. Sci J 170:795–840

GSI (1995) Geological survey of India, geological and mineral map of Tamil Nadu and Pondicherry (scale 1:500000)

Hem JD (1985) Study and interpretation of the chemical characteristics of natural waters, 3rd edn. USGS Water Supply Paper. 2254:117–120

Hounslow AW (1995) Water quality data—analysis and interpretation. Lewis Publishers, Boca Raton

Indian Standard Institution (ISI) (1983) Drinking water standard substances or characteristic affecting the acceptability of water for domestic use. ISI0500, pp 1–22

Jacob CT, Azariah J, Viji Roy AG (1999) Impact of textile industries on river Noyyal and riverine groundwater quality of Tirupur, India. Pollut Res 18(4):359–368

Jaiswal RK, Mukherjee S, Krishnamurthy J, Saxena R (2003) Role of remote sensing and GIS—techniques for generation of groundwater prospect zones towards rural development-an approach. Int J Remote Sens 24(5):993–1008

Krishnakumar S, Ram Mohan V, Dajkumar SJ, Jeevanandam M (2009) Assessment of groundwater quality and hydrogeochemistry of Manimuktha River basin, Tamil Nadu, India. Environ Monit Assess 159(1–4):341–351

Nickson R, McArthur JM, Shrestha B, Kyaw-Myint TO, Lowry D (2005) Arsenic and other drinking water quality issues, Muzaffargarh district, Pakistan. Appl Geochem 20:55–68

Pawar NJ, Shaikh IJ (1995) Nitrate pollution of ground waters from shallow basaltic aquifers, Deccan trap Hydrologic Province, India. Environ Geol 25:197–204

Piper AM (1944) A graphic procedure in geochemical interpretation of water analysis. Trans Am Geophys Union 25(6):914–928

Pulido-Leboeuf P (2004) Seawater intrusion and associated processes in a small coastal complex aquifer (Castell deFerro, Spain). Appl Geochem 19:1517–1527

Raju NJ, Shukla UK, Ram P (2011) Hydrogeochemistry for the assessment of groundwater quality in Varanasi: a fast-urbanizing center in Uttar Pradesh, India. Environ Monit Assess 173:279–300

Sarath Prasanth SV, Magesh NS, Jitheshlal KV, Chandrasekar N, Gangadhar K (2012) Evaluation of groundwater quality and its suitability for drinking and agricultural use in the coastal stretch of Alappuzha district. Appl Water Sci, Kerala, India. doi:10.1007/s13201-012-0042-5

Srinivas Y, Hudson Oliver D, Stanley Raj A, Chandrasekar N (2013) Evaluation of groundwater quality in and around Nagercoil town, Tamilnadu, India: an integrated geochemical and GIS approach. J Appl Water Sci 3:631–651

Stumm W, Morgan JJ (1996) Aquatic chemistry. Wiley, New York, p 1022

Sujatha D, Reddy RB (2003) Quality characterization of groundwater in the south eastern parts of the Ranga Reddy district, Andhra Pradesh, India. Environ Geol 44(5):570–576

Todd DK (2007) Groundwater hydrology, 3rd edn. Wiley, New York, p 347

Wilcox LV (1955) Classification and use of irrigation waters, USDA Circular No. 969, p 19

World Health Organization (WHO) (2011) Guidelines for drinking water quality, 4th edn. World Health Organization, Geneva

5

Coalbed methane-produced water quality and its management options in Raniganj Basin, West Bengal, India

Vinod Atmaram Mendhe[1] · Subhashree Mishra[1] · Atul Kumar Varma[1] · Awanindra Pratap Singh[1]

Abstract Coalbed methane (CBM) recovery is associated with production of large quantity of groundwater. The coal seams are depressurized by pumping of water for regular and consistent gas production. Usually, CBM operators need to pump >10 m^3 of water per day from one well, which depends on the aquifer characteristics, drainage and recharge pattern. In India, 32 CBM blocks have been awarded for exploration and production, out of which six blocks are commercially producing methane gas at 0.5 million metric standard cubic feet per day. Large amount of water is being produced from CBM producing blocks, but no specific information or data are available for geochemical properties of CBM-produced water and its suitable disposal or utilization options for better management. CBM operators are in infancy and searching for the suitable solutions for optimal management of produced water. CBM- and mine-produced water needs to be handled considering its physical and geochemical assessment, because it may have environmental as well as long-term impact on aquifer. Investigations were carried out to evaluate geochemical and hydrogeological conditions of CBM blocks in Raniganj Basin. Totally, 15 water samples from CBM well head and nine water samples from mine disposal head were collected from Raniganj Basin. The chemical signature of produced water reveals high sodium and bicarbonate concentrations with low calcium and magnesium, and very low sulphate in CBM water. It is comprehend that CBM water is mainly of Na–HCO_3 type and coal mine water is of Ca–Mg–SO_4 and HCO_3–Cl–SO_4 type. The comparative studies are also carried out for CBM- and mine-produced water considering the geochemical properties, aquifer type, depth of occurrence and lithological formations. Suitable options like impounding, reverse osmosis, irrigation and industrial use after prerequisite treatments are suggested. However, use of this huge volume of CBM- and mine-produced water for irrigation or other beneficial purposes may require careful management based on water pH, EC, TDS, alkalinity, bicarbonate, sodium, fluoride, metals content and SAR values.

Keywords CBM and coal mine water · Quality · Geochemical · Utilization and disposal options

Abbreviations

TDS	Total dissolved solids
EC	Electrical conductivity
SAR	Sodium adsorption ratio
NTU	Nephelometric turbidity units

Introduction

In India, CBM recovery is increasing day-by-day and expected to rise from current 0.5 to 7 mmscmd by 2020. The commercial methane production in India has been started since 2007, first in Raniganj Coalfield by GEECL and consequently by Essar and ONGC. Production of CBM is associated with pumping of large quantity of aquifer

✉ Vinod Atmaram Mendhe
vamendhe@gmail.com

[1] Central Institute of Mining and Fuel Research, Dhanbad, Jharkhand, India

water to reduce hydrostatic pressure existing on coal seams. Produced water quality and quantity vary widely, and it is necessary to manage through some combinations of treatment, storage, disposal and use. Unlike conventional gas reservoirs, coal is both the reservoir rock and the source rock for methane. CBM wells, in comparison with conventional oil and gas wells, produce large volume of water early in their life, and the water volume declines over time (Khatib and Verbeek 2003). Usually, CBM-produced water is discharged into associated unlined holding ponds (Reddy et al. 2003). Management of CBM-produced water is associated with challenges, and it is also very expensive for operators. Understanding about produced water characteristics can help in increasing the production and also knowledge of its chemical constituents; operators can determine the proper application of scale inhibitors and well treatment chemicals as well as identify potential wellbore or reservoir problem areas (Breit et al. 1998). The geochemical properties of CBM-produced water vary with the original depositional environment, depth of burial and coal type, and it vary significantly across production areas (Jackson and Myers 2002). CBM-produced water can be beneficially used, but the presence of some of the chemical parameters and their concentrations may limit the use of these waters in certain areas (Shramko et al. 2009). The suitability of CBM-produced water for agricultural purposes generally irrigation or stock watering, will depend not only on the quality of the produced water but also on the conditions of the receiving areas (ALL 2003).

This paper presents the basic information on various physical and geochemical aspects of CBM- and coal mine-produced water. It also focuses on, how it is to be managed and regulated using suitable suggested options under environmental settings at Raniganj Coalfield.

Process of CBM production

Methane occurs in adsorbed state within the micropores of coal; in order to recover it, the CBM reservoirs are depressurized by pumping of water (Mendhe et al. 2010). Typically, water must be produced continuously from coal seams to decrease the reservoir pressure and release the gas (Dart Energy International 2013). Once the pressure in the cleat/fracture system is lowered by water production to the "critical desorption pressure", gas gets desorbs from the coal matrix. The CBM reservoirs are of low pressure and initially produce large quantity of water to reach desired

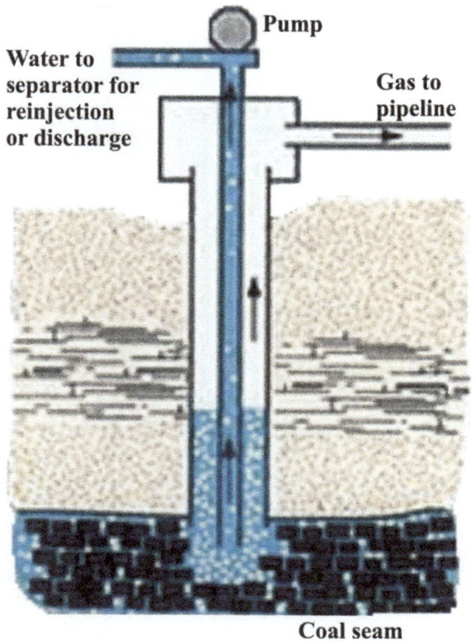

Fig. 1 CBM extraction process

rate of gas production. The produced water needs to be managed considering its geochemical properties, surface drainage pattern and low-cost methods for its treatment and use, because the cost of treatment and disposal of the produced water may be a critical factor in the economics of a coalbed methane project. The schematic of CBM production process and curve is given in Figs. 1 and 2, respectively.

Study area

The Raniganj Coalfield is the easternmost depository within the Damodar Valley of Gondwana Basin (Ghosh 2002). It is bounded by latitudes 23°03′ and 23°51′N and longitudes 86°42′ and 87°28′E (Murthy et al. 2010). Raniganj Formation of the Upper Permian age bearing thick coal seams is the most prolific for CBM reserve (Datta 2003; DGH 2006). The gas-bearing coal seams laterally varying in thickness and depth range from 1.6 to 22 and 26 to 1250 m, respectively. There are three CBM blocks that have been awarded for exploration and production development: GEECL—south block, ONGC—north-central block and Essar—northeast block.

The generalized stratigraphic succession of Raniganj Coalfield is given below (after Gee 1932).

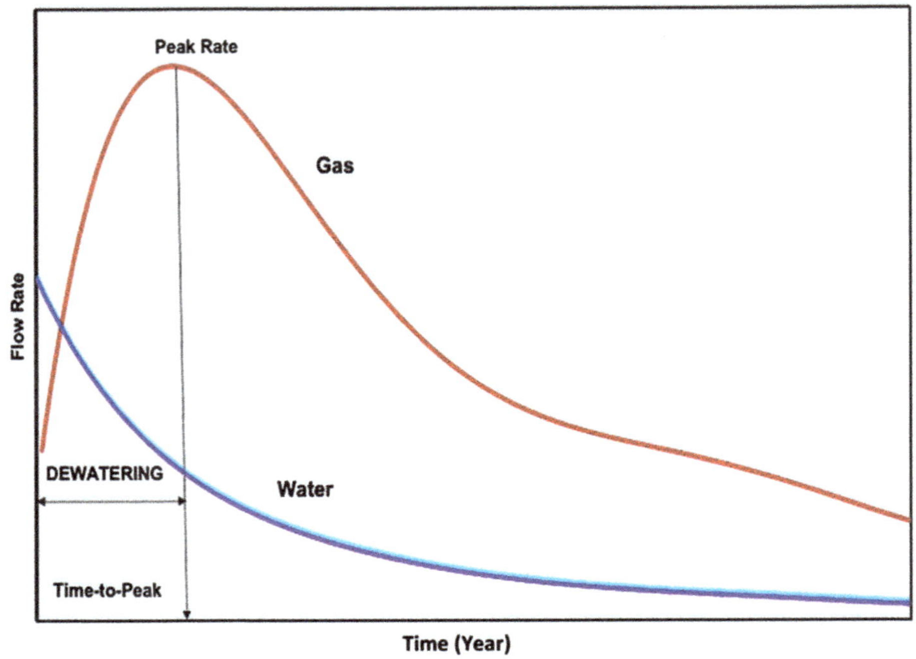

Fig. 2 Typical production curve of a CBM reservoir

Age	Formation	Thickness (m)	Lithology
Recent and quaternary			Alluvium and sandy soil, lateritic gravel and clay
Unconformity			
Jurassic	Igneous intru		Dolerite dykes, mica peridotite dykes and sills
Upper Triassic	Supra Panchet	300	Coarse red-yellow-grey sst, quartzite, conglomerate and shale bands
Lower Triassic	Panchet	600	Coarse red-yellow-grey soft mica false bedded sst with thick clay
Upper Permian	Raniganj	1050	Fine to medium grained grey and greenish sst, shales and coal seams
Middle Permian	Barren measures	550	Carb. Shale with bands of sandy mica shales and clay iron stone
Lower Permian	Barakar	650	Coarse white and grey sst, conglomerate shales and coal seams
Upper Carboniferous	Talchir	300	Coarse sst, white-variegated green shales and fine grained sst with undecomposed feldspar and boulder beds at the base
Unconformity			
Archaeans			Granites, gneisses and schists

Essar is producing >1 lakh m^3 of gas from 25 wells, and GEECL is producing >2.5 lakh m^3 of gas from 40 wells along with large quantity of water at 10 m^3 per well per day. The location of produced water samples from CBM wells and mines is marked in Fig. 3.

Methods and experiments

Produced water samples were collected from five CBM production wells and five coal mine water disposal heads in Raniganj Coalfield. The standard methods for examination of water and wastewater suggested by APHA.AW-WA.WPCF (1992) were used for analysis of water samples drawn from CBM wells and coal mine heads. The water samples were kept in dry place under normal atmospheric temperature and then analysed for pH, electrical conductivity and turbidity. The water samples were filtered and divided into two halves. Half samples were acidified to pH 2.0 with concentric nitric acid (HNO_3), and other half left as un-acidified. The un-acidified samples were analysed for anions such as SO_4^{2-}, Cl^-, F^- and NO_3^- using ion chromatography (IC) and for cations (Ca^{2+}, Na^+, Mg^{2+} and K^+) by atomic adsorption spectrophotometry (AAS). Acidified water samples were analysed for metals such as Fe, Al, Cr, Mn, Pb, Cu, As, Zn, Se, Mo, Cd, Ba and B by inductively coupled plasma mass spectrophotometry (ICP-MS). Bicarbonate and total alkalinity [phenolphthalein alkalinity (calcium carbonate ($CaCO_3^-$)) + methyl orange alkalinity (HCO_3^-)] on un-acidified samples were

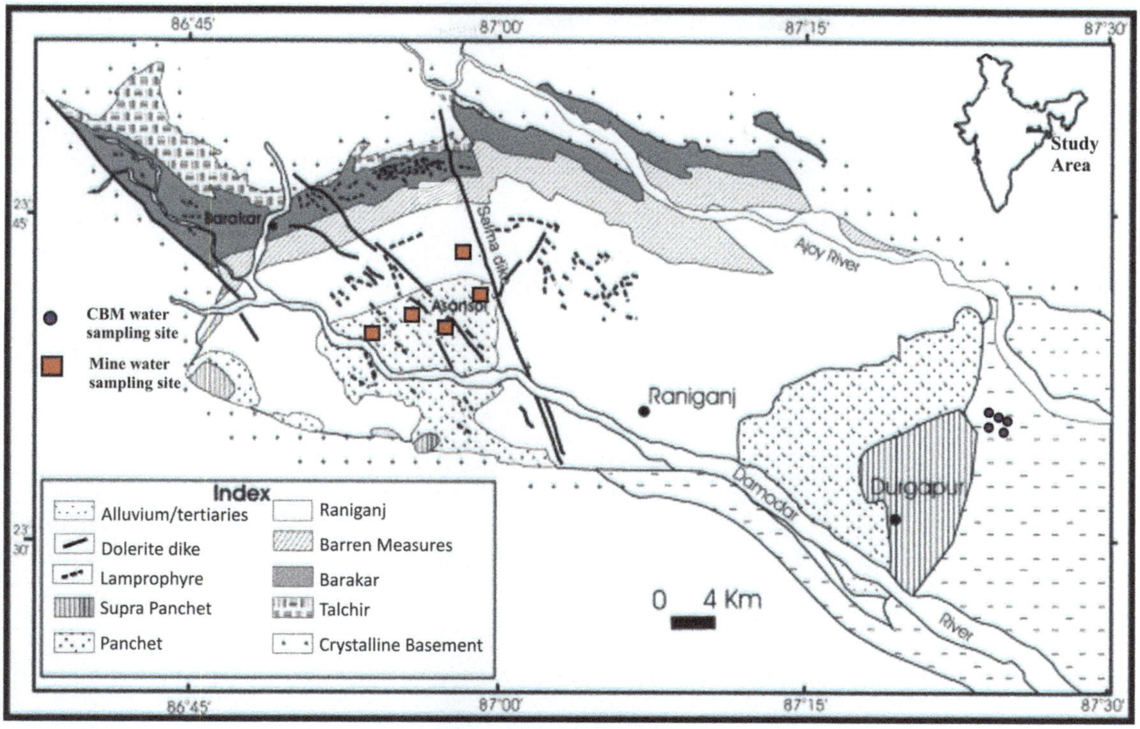

Fig. 3 Location of water samples in Raniganj Coalfield (modified after Gee 1932)

determined by acid titration method. The first is to titrate the water with acid titrant to the phenolphthalein end point. This is called the phenolphthalein alkalinity. Since phenolphthalein changes colour at pH ~8.3, this corresponds to a pH where all the CO_3^{2-} present were protonated. Second, acid titration to a methyl orange end point, pH ~4.3, further converts the bicarbonate to aqueous carbon dioxide. At this end point, some of the weaker conjugate bases are protonated. The sum of phenolphthalein alkalinity and methyl orange alkalinity indicates total alkalinity.

Results and discussion

The results of different analysis of CBM- and mine-produced water samples are given in Table 1. The pH, EC, turbidity and TDS values for CBM water vary from 8.260 to 8.720, 3090 to 4600 µs/cm, 0.600 to 2.360 NTU and 2070.300 to 3082.000 mg/L, respectively. The pH, EC, turbidity and TDS values for mine water range from 6.820 to 8.580, 623 to 1513 µs/cm, 0.740 to 2.300 NTU and 417.410 to 1013.710 mg/L, respectively.

The mine water of the Raniganj Coalfield is mildly acidic to alkaline in nature, and the variation between CBM and mine water is shown in Fig. 4. CBM-produced water in Raniganj Coalfield typically has rich concentrations of total dissolved solids than coal mine water (Fig. 5). The distribution of major ions and SAR is given in Fig. 6, which shows that bicarbonate and sodium concentration in CBM water are relatively high ranging from 2129.400–2771.300 to 349.800–976.100 mg/L, respectively, whereas for mine water it varies within values of 132.450–1023.950 and 0.000–297.300 mg/L. Heavy metals have similar range of distribution in both CBM and coal mine water, except manganese concentration is observed relatively high in mine water as shown in Fig. 7. The relationship between TDS and $HCO3^-$ is presented in Fig. 8. It displays a very good correlation separately for CBM and mine water. SAR and Na^+ concentrations vary proportionately to each other (Fig. 9). Ternary diagram showing cations and anions distribution for CBM and mine water is given in Figs. 10 and 11. The stiff plots of cations and anions of CBM and mine water are presented in Figs. 12 and 13. It is observed that CBM water contains wide distribution of Na^+ and HCO_3^-, while mine water contains SO_4^{2-} and HCO_3^-.

Water that is produced from deeper coal formations can contain NO_3^-, Cl^-, metals and high levels of total dissolved solids, which makes it unsafe for drinking purposes (Jamshidi and Jessen 2012). The mine water can be used for domestic uses after proper treatment and disinfection. Sulphate is usually derived from the weathering of sulphide-bearing minerals like pyrite (FeS_2), or dissolution of gypsum ($CaSO_4 \cdot 2H_2O$) or anhydrite ($CaSO_4$). Pyrite (FeS_2) occurs as a secondary mineral in the Gondwana

Table 1 Physical and chemical properties of CBM and mine water samples

Sample no.	pH	E.C (µs/cm)	TDS (mg/L)	Turbidity (NTU)	Total alkalinity (mg/L)	Anions				HCO_3^- (mg/L)
						SO_4^{2-} (mg/L)	F^- (mg/L)	Cl^- (mg/L)	NO_3^{2-} (mg/L)	
CBM water										
C-1	8.560	4600	3082.000	1.010	2150	0.000	3.925	407.800	47.050	2771.300
C-2	8.700	3910	2619.700	0.600	2100	0.000	5.760	231.700	13.910	2521.700
C-3	8.720	3090	2070.300	0.750	1650	0.449	5.380	104.200	47.489	2129.400
C-4	8.260	4120	2760.400	2.360	1850	1.161	5.080	353.250	43.400	2343.400
C-5	8.480	3700	2479.000	0.800	1900	0.000	7.025	225.700	50.380	2307.700
Coal mine water										
M-1	8.580	623	417.410	0.740	250	35.900	0.284	27.500	9.765	417.730
M-2	7.930	1349	903.830	1.330	800	42.450	0.250	55.000	5.151	1023.950
M-3	6.820	1513	1013.710	2.300	100	554.450	0.168	15.000	20.162	168.110
M-4	8.230	1158	775.860	0.800	550	147.000	0.473	40.000	1.849	881.310
M-5	6.860	741	496.470	1.010	150	310.650	0.244	45.000	5.372	132.450

Sample no.	Cations				SAR (meq/L)	Metals				
	Na^+ (mg/L)	K^+ (mg/L)	Mg^{2+} (mg/L)	Ca^{2+} (mg/L)		Fe (mg/L)	Mn (mg/L)	Zn (mg/L)	Sr (mg/L)	Al (mg/L)
CBM water										
C-1	832.500	0.000	1.870	19.340	48.910	0.328	0.002	0.025	0.247	0.013
C-2	486.000	0.000	0.530	15.130	33.540	0.039	0.008	0.034	0.208	0.026
C-3	349.800	0.000	0.480	16.820	23.040	0.184	0.001	0.020	0.218	0.011
C-4	976.100	0.000	0.460	17.660	63.340	0.101	0.005	0.027	0.464	0.026
C-5	974.000	0.000	0.650	10.930	77.000	0.080	0.059	0.029	0.347	0.004
Coal mine water										
M-1	0.000	0.000	0.000	0.000	0.000	0.218	0.010	0.004	0.921	0.021
M-2	297.300	0.000	6.120	14.670	19.290	0.166	0.006	0.013	0.333	0.026
M-3	26.300	9.250	102.100	192.970	0.380	0.923	1.693	0.058	0.617	0.007
M-4	218.200	0.000	22.620	24.820	7.590	0.301	0.015	0.003	0.785	0.018
M-5	12.660	6.110	41.570	85.050	0.280	0.426	0.800	0.055	0.284	0.031

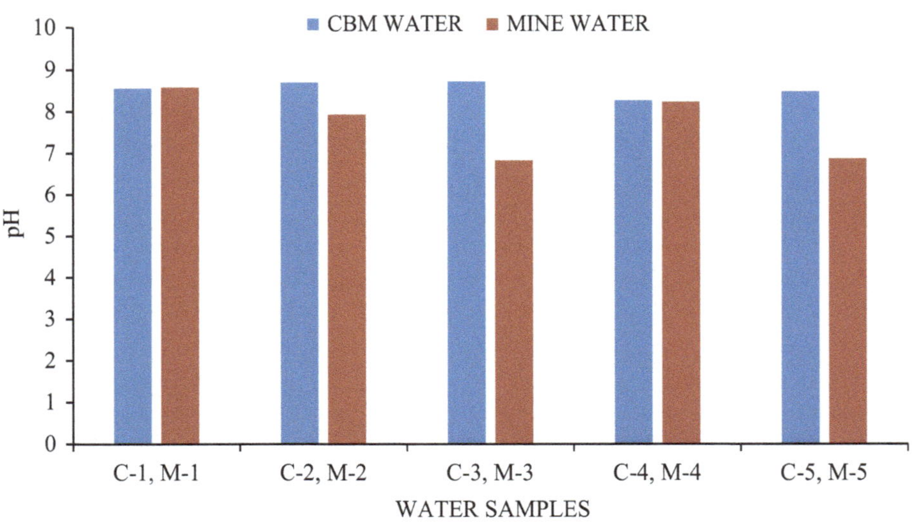

Fig. 4 Variation in pH of CBM and mine water

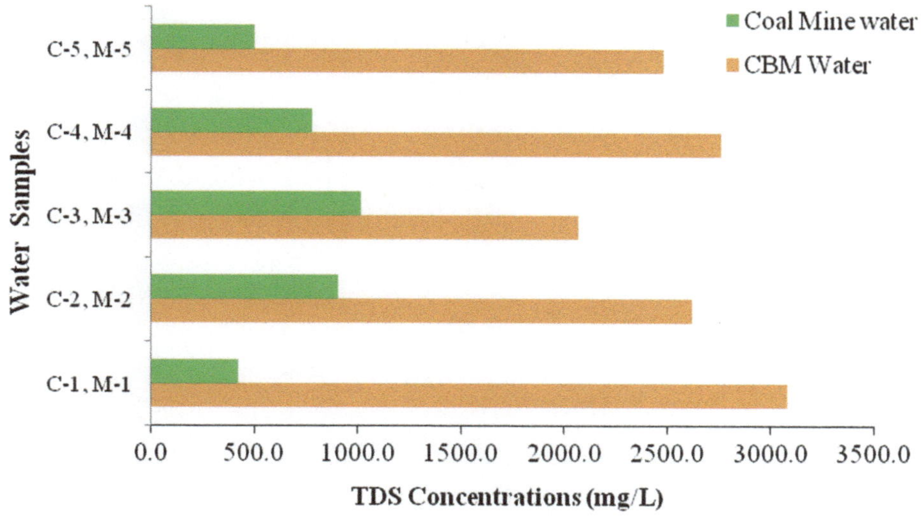

Fig. 5 Variation in TDS concentration of CBM and mine water

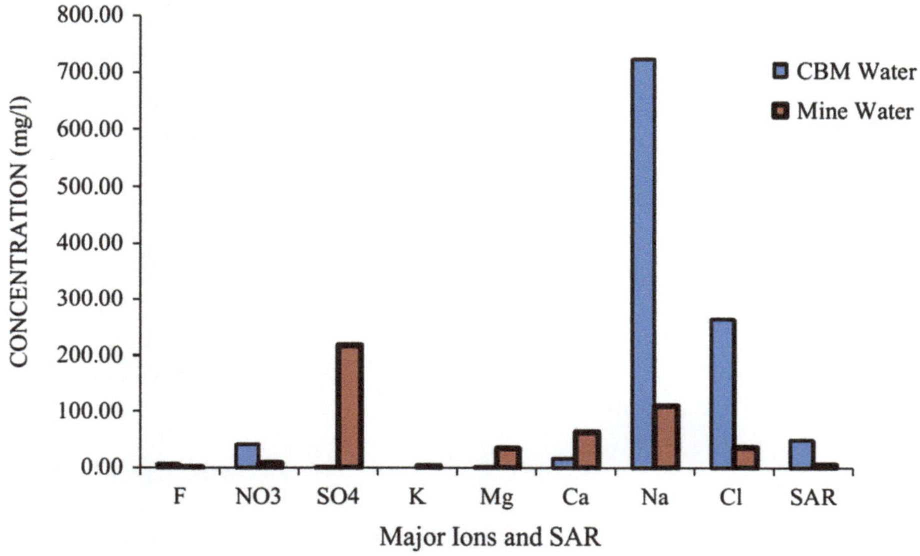

Fig. 6 Variation of major ions and SAR in CBM and mine water

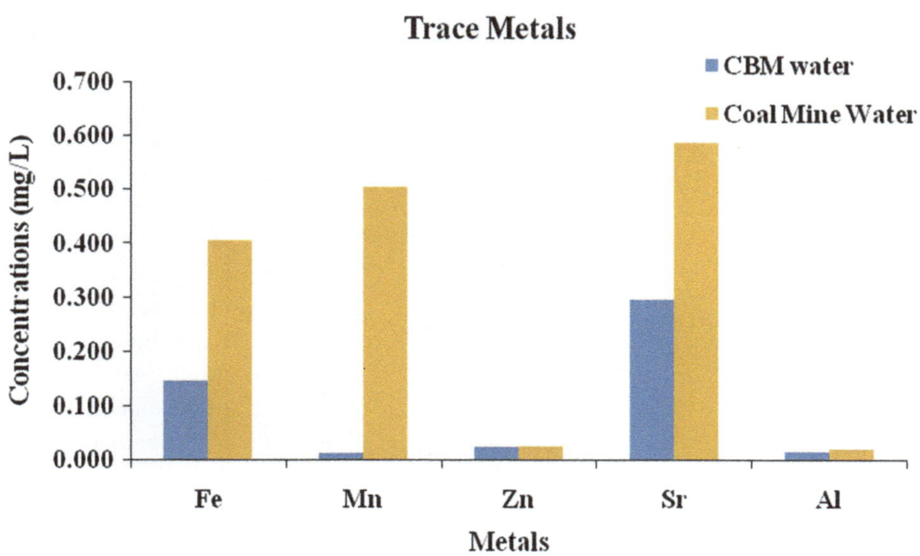

Fig. 7 Variation of trace metals in CBM and mine water

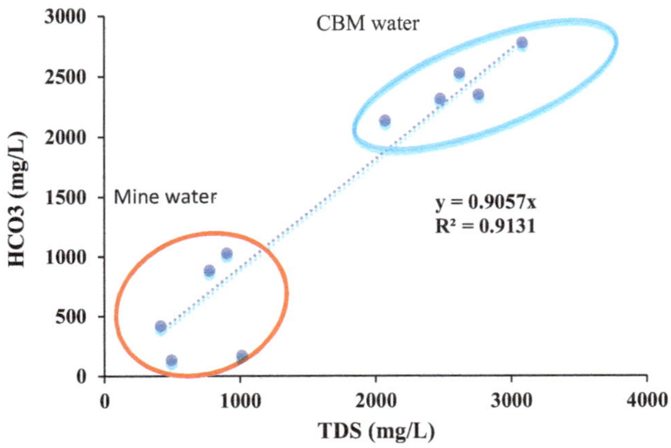

Fig. 8 TDS versus HCO_3 of CBM and mine water

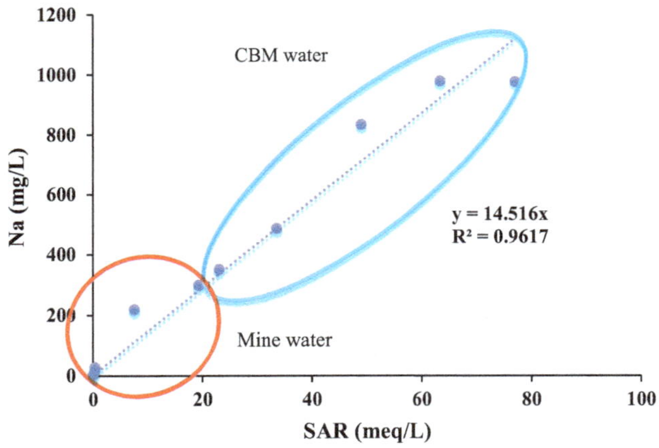

Fig. 9 SAR versus Na of CBM and mine water

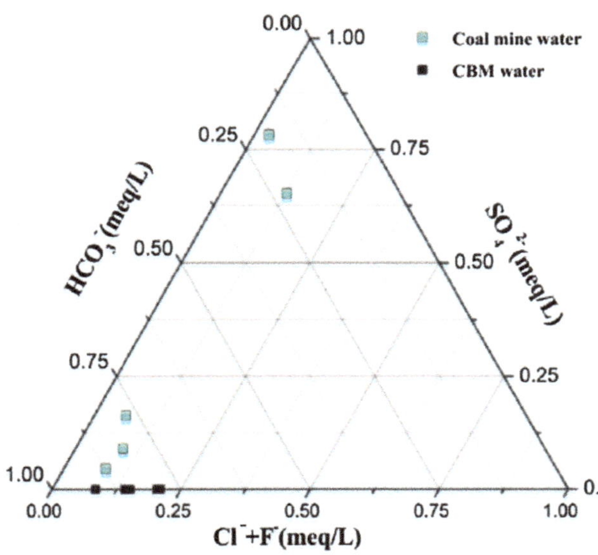

Fig. 10 Ternary diagram for anions of CBM and mine water

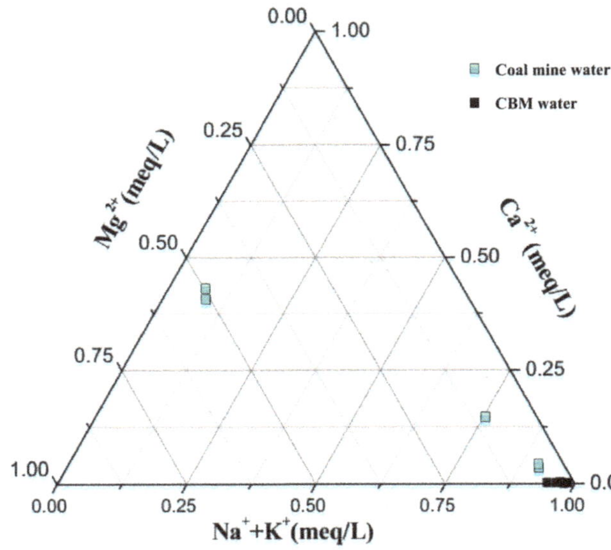

Fig. 11 Ternary diagram for cations of CBM and mine water

coals and associated sediments. The surface disposal and agriculture use of CBM-produced water are restricted due to high values of SAR

$$\left(\mathrm{SAR} = \frac{\mathrm{Na}^{+}}{\sqrt{\frac{1}{2}\left(\mathrm{Ca}^{2+} + \mathrm{Mg}^{2+}\right)}}\right),$$

which may cause infiltration, surface crusting and also reduces the permeability of soil (Van Voast 2003).

CBM-produced water disposal options

Management of large volumes of associated water with CBM production is a potential concern due to the presence of elevated water salinity and sodicity. The produced water is managed in different ways in different areas of the USA and other countries. Existing production in the Powder River Basin utilizes a variety of options to manage CBM-produced water. Deep injection, aquifer storage, surface water discharge, land application (irrigation with amendments), livestock watering and impoundment are all being used to manage produced water. Land application of the CBM-associated high saline–sodic water is a common management method that has been practiced in the Powder River Basin of Wyoming and Montana. The agricultural use of the co-produced waters from CBM is another management option. However, the use of produced water for irrigation can result in deterioration in soil quality and

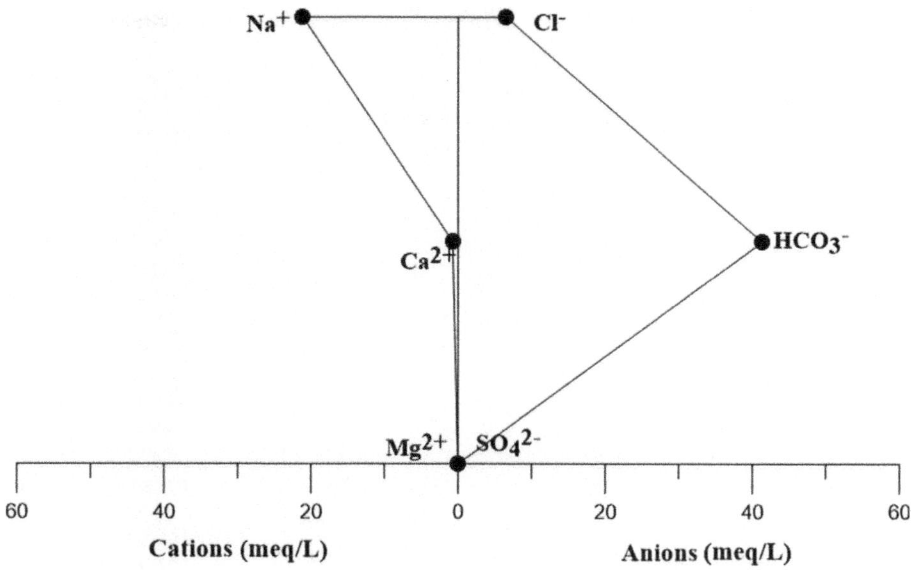

Fig. 12 Stiff diagram for CBM water (sample C-2)

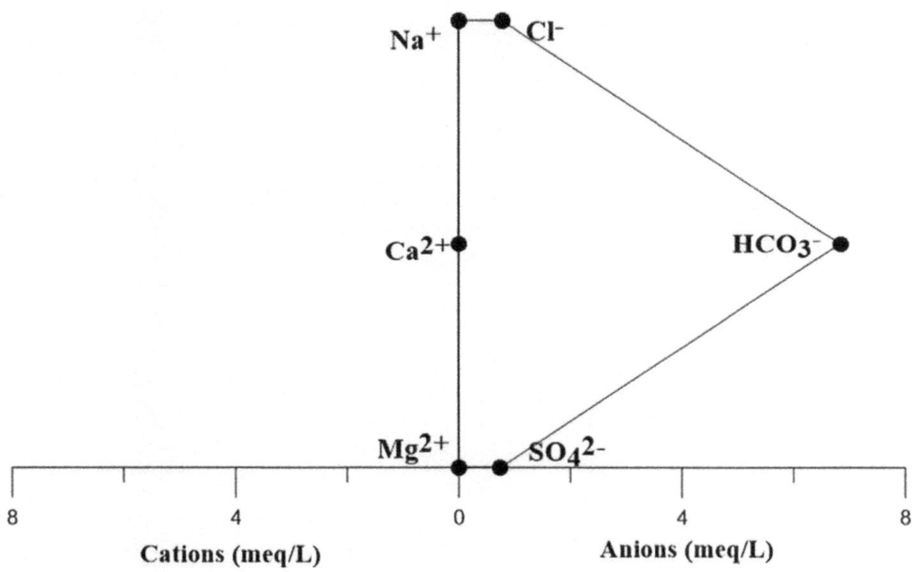

Fig. 13 Stiff diagram for coal mine water (sample M-1)

changes in physical and chemical parameters of the soil (Veil and Clark 2011).

Considering the quality and quantity of produced water, following options may be useful for appropriate use and disposal of CBM water in Raniganj Coalfield. Irrigation may be a suitable option for CBM-produced water only after desalinization and proper treatment. Irrigation has several critical aspects which need to be taken care for proper balance of soil quality and crops grown in the area. Impounding CBM water by pumping it into storage facilities, reservoirs and ponds has traditionally been a preferred water management option for CBM operators and may be one of the effective methods in Raniganj Coalfield. These impoundments are well known as infiltration ponds, evaporation ponds, or zero-discharge ponds. Drinking water availability is the major issue in Raniganj Coalfield. The large quantity of water generated from CBM production wells can be potential freshwater sources for various applications, including potable consumption. These challenges include high treatment cost, potential chronic toxicity of the treated produced water and public acceptance. Because of the need of desalination and removal of a large number of chemical compounds, RO will most likely be used for potable reuse applications. It is emphasized that the main challenges present in produced water are desalination, degassing, suspended solids removal, organic compounds removal, heavy metal and others. Achieving the various treatment goals requires the use of multiple treatment technologies, including physical, chemical, and biological treatment processes (Ahmadun et al. 2009).

Some of the technologies are removal of TDS by precipitation, electrochemical or photocatalytic oxidation, nanofiltration or reverse osmosis, removal of metal through aeration, settling, sand filtration with suspended solids removal, coagulation/flocculation, sedimentation and filtration. The surface discharge and sub-surface injection of the produced water should be treated up to the requirements of the locals and state regulatory limitations for discharge and injection.

Conclusions

The appraisal of CBM- and mine-produced water is useful for evaluating water quality from different geological formations, which normally have distinctly different geochemical signatures. Coal mine water is relatively higher in dissolved calcium (Ca^{2+}), magnesium (Mg^{2+}), chloride (Cl^-) and sulphate (SO_4^{2-}), whereas water from the deep coalbeds associated with adsorbed methane gas is comparatively higher in dissolved sodium (Na^+) and bicarbonate (HCO_3^-). The CBM water is categorized as Na–K type, Na–HCO_3 type and HCO_3 type, whereas the coal mine water may be categorized as the Ca–Mg–HCO_3, HCO_3–Cl–SO_4 and Na–HCO_3 type in Raniganj Coalfield. The relevant options for management and surface/subsurface disposal of large volume of produced water from CBM wells are impounding, irrigation and drinking water on the basis of water pH, EC, TDS, alkalinity, bicarbonate, sodium, fluoride, metals content and SAR values. The effective management of CBM and coal mine water in Raniganj Coalfield required more specific scientific investigation before adoption of any disposal method.

Acknowledgments The authors are grateful to the Director CIMFR for granting permission to carry out different analysis and publication of this paper.

References

Ahmadun FL, Pendashteh A, Abdullah LC, Biak DRA, Madaeni SS, Abidin ZZ (2009) Review of technologies for oil and gas produced water treatment. J Hazard Mater 170:530–551

ALL (2003) Handbook on coal bed methane produced water: management and beneficial use alternatives. Prepared by ALL Consulting for the Ground Water Protection Research Foundation, U.S. Department of Energy, and U.S. Bureau of Land Management

APHA.AWWA.WPCF (1992) Standard methods for the examination of water and waste water, 16th edn. APHA, Washington

Breit G, Klett TR, Rice CA, Ferderer DA, Kharaka Y (1998) National compilation of information about water co-produced with oil and gas. In: 5th international petroleum environmental conference, Albuquerque, NM, Oct 20–23

Coal bed Methane (2013) Dart Energy International. http://www.dartgas.com/

Datta D (2003) Coal resources of West Bengal. Bulletin of the Geological Survey of India, Series 'A', No. 45, Coalfields of India, Volume V

DGH (2006) Directorate General of Hydrocarbons Annual Report "Exploration and Production Activities in India" Ministry of Petroleum, Govt. of India, pp 1–76

Ghosh SC (2002) The raniganj coalfield: an example of an Indian Gondwana rift. Sed Geol 147:155–176

Jackson L, Myers J (2002) Alternative use of produced water in aquaculture and hydroponic systems at Naval Petroleum Reserve No. 3. In: Ground Water Protection Council produced water conference, Colorado Springs, CO, Oct 16–17

Jamshidi M, Jessen K (2012) Water production in enhanced coalbed methane operations. J Petrol Sci Eng 92–93:56–64

Khatib Z, Verbeek P (2003) Water to value—produced water management for sustainable field development of mature and green fields. J Pet Technol 55(1):26–28

Mendhe VA, Mishra P, Varade AM (2010) Coal seam reservoir characteristics for Coalbed Methane in North and South Karanpura Coalfields, Jharkhand, sedimentary basins of India: recent developments. Gondwana Geol Mag 12(2010):141–152

Murthy S, Chakraborti B, Roy MD (2010) Palynodating of subsurface sediments, Raniganj Coalfield, Damodar Basin, West Bengal. J Earth Syst Sci 119(5):701–710

Reddy KJ, McBeth I, Skinner QD (2003) Chemistry of trace elements in coalbed methane product water. Water Res 37:884–890

Shramko A, Palmgren T, Gallo D, Dixit R, Swaco M-I (2009) Analytical characterization of flow-back waters in the field. In: 16th annual petroleum & biofuels environmental conference (IPEC), Houston, November 3–5 2009

Van Voast WA (2003) Geochemical signature of formation waters associated with coalbed methane. AAPG Bull 87:667–676

Veil J, Clark CE (2011) Produced-water-volume estimates and management practices. SPE Prod Oper 26(3):234–239

6

An overview of impact of subsurface drainage project studies on salinity management in developing countries

Priyanka Tiwari[1] · Arun Goel[1]

Abstract Subsurface drainage has been used for more than a century to keep water table at a desired level of salinity and waterlogging control. This paper has been focused on the impact assessment of pilot studies in India and some other countries from 1969 to 2014. This review article may prove quite useful in deciding the installation of subsurface drainage project depending on main design parameters, such as drain depth and drain spacing, installation area and type of used outlet. A number of pilot studies have been taken up in past to solve the problems of soil salinity and waterlogging in India. The general guidelines that arise on the behalf of this review paper are to adapt drain depth >1.2 m and spacing depending on soil texture classification, i.e., 100–150 m for light-textured soils, 50–100 m for medium-textured soils and 30–50 m heavy-textured soils, for better result obtained from the problem areas in Indian soil and climatic conditions. An attempt has been made in the manner of literature survey to highlight the salient features of these studies, and it is hopeful to go a long way in selecting design parameters for subsurface drainage problems in the future with similar soil, water table and climatic conditions.

Keywords Subsurface drainage · Salinity · Waterlogging and groundwater table

✉ Priyanka Tiwari
tiwari.priyankatiwari.p2@gmail.com

Arun Goel
drarun_goel@yahoo.co.in

[1] Department of Civil Engineering, National Institute of Technology (Institute of National Importance), Kurukshetra, Haryana 136119, India

Introduction

Land and water are two basic natural resources. Due to rapid population growth and fast industrialization, these resources are facing immense pressure and are depleting day by day. According to the FAO land and plant nutrition management service (1994), over 6 % of the world's land is affected by either salinity or sodicity. Much of the world's land is not cultivated; however, a significant proportion of cultivated land is salt-affected, and of the current 230 million ha of irrigated land, 45 million ha is salt-affected (19.5 %), and of the 1500 million ha under dry land agriculture, 32 million ha is salt-affected to varying degrees, i.e., 2.1 % (Hefny et al. 2013). World's large irrigated regions with serious salinity problems are Yellow River Plain in China, San Joaquin Valley in California, KaraKum Canal project in Turkmenistan, Indus Plain in Pakistan, Tigris–Euphrates Plain in Iraq, Murray–Murrumbidgee Area in Australia and lower Nile Valley in Egypt, which need a serious attention of researchers (Ghassemi et al. 1995).

Subsurface drainage is considered as a most suitable approach for groundwater balance and land and water management practices containing the groundwater table at a suitable level (Luthin 1978; Gates and Grismer 1989). Agricultural subsurface drainage is a process of removal of excess groundwater from the crop root zone system which promotes safe environment for efficient crop growth and for better health in rural and urban areas. Subsurface drainage lowers the high water tables, and the main causes of the rise in water table are precipitation, excess irrigation, leaching water, seeps from higher land or irrigation canal and ditches and groundwater under artesian pressure. This technique has gained international acceptance. Subsurface agricultural drainage provides agronomical and environmental benefits in terms of improved crop yield, improved soil trafficability, field operations and reduction in sediment and phosphorus losses from agricultural fields (Kornecki et al. 2001).

Subsurface drainage has been found to be the only solution for providing land reclamation on a long-term basis when salts are present in the soil and groundwater. Subsurface drainage has been provided in 75–80 % of irrigated area in Egypt and 25–30 % irrigated area in western USA Goel and Tiwari (2013). Comprehensive reclamation programs involving provision of subsurface drainage in irrigated areas have been embarked in a big way in Pakistan over the past 50 years, while these started much earlier in India during late 1920s (Thatte and Kulkarni 2000), and FAO (1999) reported that India has 2.5 million ha waterlogged land and 3.1 million ha salinity-affected area, while this hazard on the state level in India is extended, i.e., Uttar Pradesh has salinity-affected area over 1 million hectares, 1 million hectares in Gujarat, 0.5 million ha in Punjab, 0.2 million ha in Haryana and a smaller area in Rajasthan, and significant less proportionate area in Tamil Nadu in the south is also affected. At the time of nineties, subsurface drainage installed <10 % of the total irrigated area Zimmer et al. (2002). But currently <0.02 % irrigated area in India is provided with subsurface drainage Tiwari (2011), and Maharashtra state has minimum salt-affected area to the extent of 0.6 M ha (Sethi et al. 2010).Still India is facing varying degree of salinity problems such as saline soil, costal saline and alkalinity (Patel et al. 2002; Mandal and Sharma 2011).

Some past review papers and past remarkable studies on subsurface drainage and salinity management were discussed in this session. The purpose of this literature survey is that the findings and recommendations of these studies would be useful to take as guidelines to plan a new subsurface drainage system in an efficient, effective, economic and ecofriendly manner for future in India and other countries.

Gupta (2002, 2003) has summarized past 100-year Indian efforts to control salinity and waterlogging problems, which were conducted under the supervision of Central Soil Salinity Research Institute (CSSRI) from 1972 to 2002. He has covered several pre- and post-independence pilot studies and their salient findings from 1873 to 1975. He has provided general drainage design guidelines and operationalizes the subsurface drainage system for Indian conditions.

Kaledhonkar et al. (2009) have summarized preliminary studies conducted in India from 1980 to 2008. He has been enlightening on the 18 Haryana, 3 Gujarat, 2 Maharashtra, 2 Andhra Pradesh, 1 Karnataka and 1 Rajasthan pilot studies of their saline areas in the listed states, and he made efforts to focus the subsurface drainage operational and subsurface drainage effluent management problems. They discussed the drainage design parameters, crop production, cropping intensity and cost investment for the subsurface drainage installations but did not considered the maintenance of SSD.

Several researchers have suggested different strategies for salinity management in developing countries from 1969 to 2014. Special issues in SSD are discussed/highlighted, and specific methodologies for design of SSD are summarized in this paper. In the end, some of the general guidelines have been proposed which can be kept in mind while designing any SSD project in future.

Subsurface drainage pilot or project studies

India

The requirement of salinity management was firstly reported by Punjab Govt. in 1865 to Governor General. Mr. Robertson and started India's first drainage project in 1873 to reclaim the land from waterlogging, and people cure from malaria disease. History of northwest India's salinity has to reclaim through subsurface drainage installation and keep functioning it till achieving the aim of installation (Hanumanthaiah and Satyanarayana 2003). But they always worried about drain discharge disposal. Till few years ago, the drain discharge was disposed in canals. And also they tried to reuse the saline drain water in crop production, but the problem in those days was the pumping out of large quantity of drain discharge. So nowadays, the drainage installation is costly and seen more problematic in disposal point of view. A number of pilot scales subsurface drainage projects undertaken by CSSRI Karnal (Haryana), India, during 1980s have slowly paved the way for communication of large-scale mechanically installed drainage projects in Haryana, Rajasthan, Gujarat, Punjab, Andhra Pradesh, Maharashtra and Karnataka. India is divided into ten climatic zones which played very important role in SSD installation because any region is more affected by their weather conditions. On the basis of the climatic zones, experimental sites are through the India map (Fig. 1).

About 35,000 ha waterlogged saline soils have been reclaimed in these states with significant improvement in crop intensity, yield, land value and farmers income. India's major pilot studies literature survey has been discussed in Table 1. Here we are discussing the important factor of the drainage design criteria with their major outcomes. The following table (Table 1) have described about India's 56 successful pilot studies with their study area, installation year, operational duration, necessary drainage design parameters and their major conclusions.

On the basis of the Table 1, the pilot studies following guidelines are framed which would be helpful in drainage

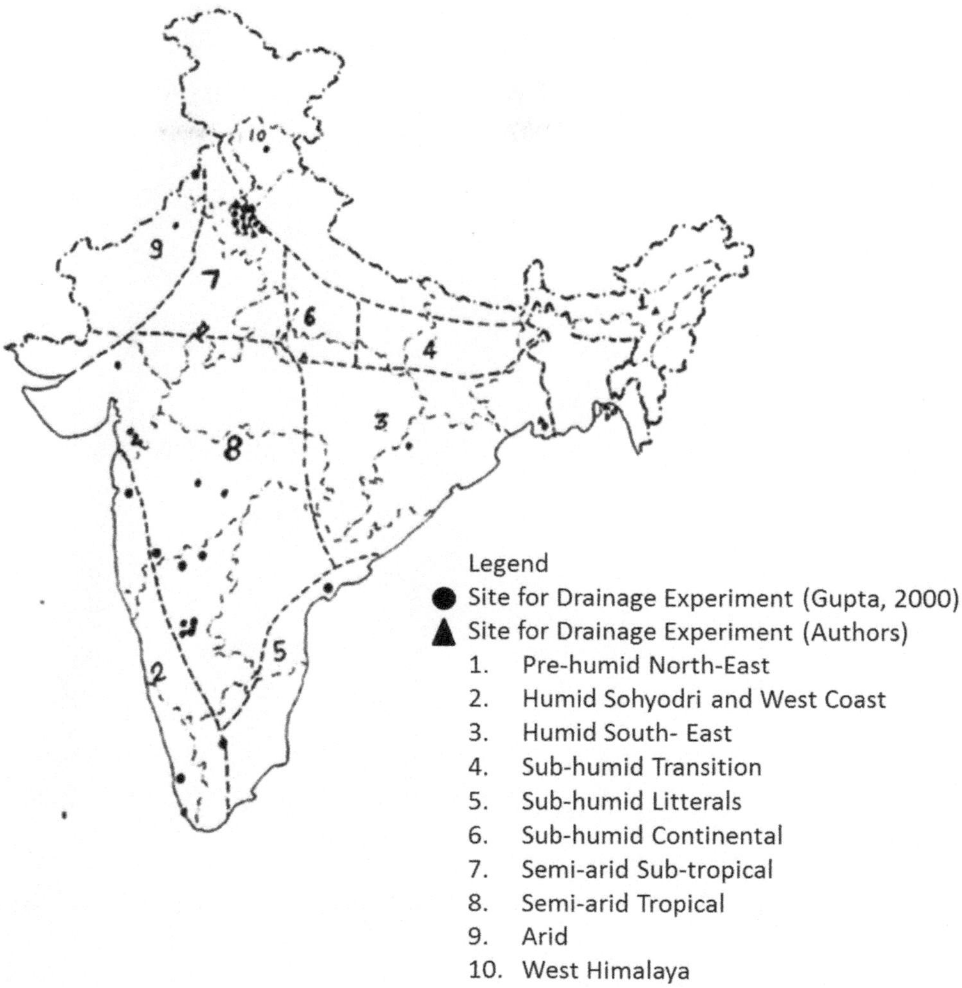

Fig. 1 Climatic zone wise distribution of SSD studies in India

design subsurface drainage system in India. It is expected that these salient points would be useful in new wherein similar soil, water table and climatic conditions existing in India.

Basic guidelines for Indian subsurface drainage design

Although Drainage manual (1978) and Drainage design factors (1986) have recommended general guidelines for installation procedure and drainage design parameters for more effective performance yet on the basis of pilot studies undertaken in the past 50 years in India, few more guidelines have emerged which are mentioned below.

1. Before subjecting any drainage project into a problematic area, the primary data (technical, socioeconomic, geo-hydrological climate) and secondary (water quality) have to be collected, which are relevant to the particular area.
2. In past, cement concrete pipes were very popular in alkaline and saline areas because these pipes have not given any complaints up to 20 years in alkaline areas and 8 years in saline zones after installation. But nowadays PVC pipes are more popular than others due to its portability and light in weight.
3. Drainage depth is an issue of increase or decrease in cost of installation and also availability of machineries and labor. But it is depending on position of groundwater table, soil type and hydraulic characteristics. For agricultural drainage, recommended depth of the lateral is >1.2 m because agricultural activities may damage the laterals.
4. Drain spacing is classified on the basis of the texture of the soil. The lateral spacing, size and grade (in %) for light-textured soils are ranging between 100 and 150 m, 100 mm and 0.10 %, respectively, 50 and 100 m, 125 mm and 0.075 % for medium-textured soil and 30 and 50 m, 150 mm and 0.05 % for heavy-textured soils.
5. On the basis of climatic conditions, drainage coefficient has an optimum value which lies between 1 and 3 mm/day (i.e., arid region is a 1 mm/day, semiarid region is a 2 mm/day, and subhumid region is 3 mm/day).

Table 1 Pilot studies on subsurface drainage in India

S.no.	References	Study area description for SSD Installation		Drainage design parameters				Major conclusion
		Place and area (ha)	Year of installation	Drain depth (m)	Drain spacing (m)	Type of drain	Type of outlet and envelope material	
Andhra Pradesh								
1	Channabasiah (1972)	Siruguppa	1972	1.2	12	Open drain	Gravity	Significant increase in crop yield is observed
2	Devadattam and Chandra (1995)	Machilipatnam	1995	–	–	Tile drain	–	Increased rice crop yield
3	Joint completion report on IDNP (2007)	Nagarjuna Sagar Right Command and Krishna Western Delta	–	1 1.0–1.2	60 75	Pipe drain Open drain	Pumped Geo-textile	Paddy crop yields have increased around 50 % in 2-year duration
4	Srinivasulu et al. (2005)	Konanki village in Prakasam district (21.63 ha)	1998–1999	0.9 and 1.1, 1	100 30, 60	Open Pipe Pipe	Gravity Geo-textile Nylon mesh	Late sowing kharif crop rice yield increase by 51 %
5	Ritzema et al. (2008)	Konanki and Uppugunduru (8 and 5 ha)	1999	0.90–1.75	45–150	Open drain Pipe drains	Pumped Geo-textile Nylon mesh	Water table has controlled and cropping intensity increased
Assam								
6	Bhattacharya et al. (1996)	Mornai	1996	–	–	Tile drain	–	Controlled water table and increased yield of tea leaves are observed
Gujarat								
7	Michael and Ojha (2013)	Saurashtra (22 km)	1972	1.2–1.8	–	Open drains	–	Land has reclaimed from salinity and waterlogging problems
8	Anonymous (1988)	Bidaj	1981	–	–	Open drains	–	Controlled salinity and waterlogging.
9	Ritzema (2009)	Segwa, Surat (5.6 ha, 2.6 ha, 5.7 ha and 4.6 ha)	1998–1999	0.9 1.0–1.2 1.2 1.2	45 30 45 60	Singular and Composite open pipe drains	Gravity	Water quality (EC reduced from 18.2 to 1.4 dS/m during 2001) has improved after one year and has used in irrigation
10		Sisodra, Surat (4.2 ha, 5.4 ha, 2 ha and 14 ha	2000	0.9 0.9 0.7 0.8	30 60 30 60	Singular and open pipe drains	Gravity	Cropping intensity has increased to 102 % after SSD installation
11	Kaledhonkar et al. (2009)	DabuoKheda (50 ha)	1987–1988	–	–	–	–	Water table has controlled, and this has given positive effect on crop production.
12		Muraj Kheda (52 ha)	1990–1991	–	–	–	–	Water table has controlled
Haryana								

Table 1 continued

S.no.	References	Study area description for SSD Installation		Drainage design parameters				Major conclusion
		Place and area (ha)	Year of installation	Drain depth (m)	Drain spacing (m)	Type of drain	Type of outlet and envelope material	
13	Jaiswal and Dhruvanarayana (1972)	Karnal, Haryana	1972	1.5	10–30	Tile and open drains	Gravity Pumped	Open drain has not been found suitable in alkaline soils
14	Kaledhonkar et al. (2009), Anonymous (1988)	Kailana khas (1.6 ha)	1980–1982	–	–	Open and Tile drains	Gravity	Groundwater table has been controlled
15	Rao et al. (1986), Kamra et al. (1991), Kaledonkar et al. (1998), Gupta (2002)	Sampla, Rohtak (1 ha, 10 ha, 27 ha, 13 ha)	1980–1981 1983–1984 1986–1987 1991–1992	1.5 1.5 1.75	20 50 25, 50, 75	Open drains Tile drains	Pumped and gravel for deep drain system Synthetic for shallow drain system	Wheat and barley crops yield was obtained 2.5–4.9 (t/ha) and 2.1–4.2(t/ha) at the zero production area Cropping intensity has been increased up to 200 % after drainage installation
16	Hirekhanet al. (2007), Pali et al. (2014)	Mudlana, Sonepat (50 ha)	1984–1985	1.75 and 1.6	50, 67, and 84	PVC corrugated pipes	Pumped	Salinity has been controlled
17	Kelleners et al. (2000)	CIRB, HAU Hisar (75 ha)	1990–1991	1.5	24	Tile drain	Pumped	Water table has controlled and crop yield increased
18	Kaledhonkar et al. (2009)	HLRDC farm, Hisar	1989–1990	–	–	Tile drain	Pumped	Groundwater table has controlled and increased crop yield
19		Ujana, Sonepat (32 ha)	1986–1987	–	–	Tile drain	Pumped	
20		Bhan Brahmana (105 ha)	1988–1990	–	–	Tile drain	Pumped	
21		Kolekhan, Jind (41 ha)	1989–1990	–	–	Tile drain	Pumped	
22		Ishapur Kher (58 ha)	1989–1990	1.0–1.2	–	Tile drain	Pumped	
23		Rori, Sirsa (1500 ha)	2004–2005	–	–	Poly-pipes	Pumped	
24	Joint Completion Report on IDNP (2007)	Gohana (1236 ha)	1979–1999	1.6	60–67	PVC corrugated pipes	Pumped Synthetic	Salinity has decreased 35 % from the year of installation to 2000
25	Lal (2011)	Jhajjar (Beri, 800 ha)	2004–2005	1.75	60–67	Poly-pipes	Pumped	Land has been reclaimed from salinity hazard
26		Dadri, Bibhani (800 ha)	2004–2005	–	–	Poly-pipes	Pumped	Agricultural land removes salinity at a desired level
27		Kalyant (1000 ha)	2004–2005	–	–	Poly-pipes	Pumped	Cropping intensity has increased
28		Sonipat (1500 ha)	2004–2005	–	–	Poly-pipes	Pumped	Water table has controlled
Karnataka								
29	Joint Completion Report on IDNP (2007)	Islampur, Karnataka (Upper Krishna Canal Command, 29.3 ha, 14.4 ha)	1998–1999 1998–1999-	1.0–1.1 1.0–1.1	50 50	PVC perforated corrugated Pipe Open Drain	Pumped Geo-textile	Cropping intensities have increased 27, 35 and 100 % in first three years
30				1.1–1.2 1.1–1.2	50 30	Pipe drain Pipe drain	Pumped	Cropping intensities have increased 114 and 156 % in first two years
31	Datta et al. (2004)	Mudhol Taluk Karnataka Bagal kot (19.35 ha)	1995–1996	–	1	–	Gravity	Water logged area has reclaimed

Table 1 continued

S.no.	References	Study area description for SSD Installation		Drainage design parameters				Major conclusion
		Place and area (ha)	Year of installation	Drain depth (m)	Drain spacing (m)	Type of drain	Type of outlet and envelope material	
32	Satyanarayana et al. (2002) Srinivasulu et al. (2014)	Uppugunduru	1999	0.85–1.1	50	Open	Pumped	Water table has controlled and cropping intensities increased to 170 % after drainage installation
		In Krishna Western Delta(7 and 5 ha)			30	Pipe	Geo-textile	
					45	Pipe	Nylon mesh	
					60	Pipe		
Kerala								
33	Mathew et al. (1993)	karumady	1993	–	–	Tile drain	–	Increased rice yield
Madhya Pradesh								
34	Yadav (1975), Anonymous (1988)	IARI, New Delhi	1975	1.0	32	Open drains	Gravity	Significant increase in sorghum and wheat yield
35	Yadav (1975), Anonymous (1988)	Indore, M.P.	1975	0.6–0.9	6.5–40	Tile drains	Gravity	Significant increase in crop yield and coarse sand filters are not suitable for alkaline vertisols
36	Sharma et al. (1995)	Barna, M.P.	1995	–	–	Open drain	–	Increased crop yield
37	Ramana et al. (2009)	CIAE Bhopal (M.P.)	–	0.60	2, 4 and 6	Tile drain and mole drain	Gravity	Water table has controlled
Maharashtra								
38	Gupta (2002), Kaledhonkar et al. (2009)	Manjri	1920	1.5	–	combination of open and tile drains	Gravity	Crop yield has increased
39	Anonymous (1988)	Manjri	1941	–	–	Tile drains and open drain	Gravity	Crop yield has increased
40		Parbhani	1982			Tile drain	Gravity	Increased net profit
41	Kamble et al. (2006)	Dhudgan Sangli (1100 ha)	–	2.5		Corrugated perforated PVC pipes	Gravity	Water table has controlled
42	Kaledhonkar et al. (2009)	Kasabe Digraj Sangli (1065 ha)	2004–2006	2.5	10.75	Tile drains	Gravity	Water table has controlled
43	Kolekar et al. (2014)	College of Agril Engg and Tech., Parbhani	2006–2008	2, 4 and 6	–	Mole drain	–	The total yield of groundnut has highest in 4 m mole spacing, i.e., 69.20 %
Punjab								
44	Gupta (2002)Kaledhonkar et al. (2009)	Amritsar	1873	–	–	Stone, tile drains and 16-tube well (0.0425 m^3/sec)	Gravity	Malaria disease has mitigated and land free from waterlogging problem
45		Chakanwal	1920	0.9	33–66	Open drain	Gravity	This has played a vital role in crop production
46	Anonymous (1988)	Ludhiana	1967	–	–	Tile drains and open drain	Gravity	Maize yield increased

Table 1 continued

S.no.	References	Study area description for SSD Installation		Drainage design parameters				Major conclusion
		Place and area (ha)	Year of installation	Drain depth (m)	Drain spacing (m)	Type of drain	Type of outlet and envelope material	
47	Michael and Ojha (2013)	Ranbirpura, Patiala (800 h)	1959–1960, 1971	–	–	Open drain	–	This area has significantly free from salinity
48	Michael (1967), Michael and Ojha (2013)	Ludhiana	1966	1.2	–	Interceptor and combination of open and tile drains	Pumped	System was not much effective in rabi season
49	Shakya et al. (1995)	Golewala	1995	–	–	Multiple well one point system	Pumped	Permitted conjunctive use of groundwater
Rajasthan								
50	Lovas (1972), Anonymous (1988)	Digod	1972	0.6–1.0 1.0–1.3	2–2.5 8–12 15–20	Mole, plastic asbestos pipes	Gravity	Mole drains have found to be cheaper and less time-consuming
51	Datta et al. (2004)	Chambal command area RAJAD (1420 ha, 12580 ha)	1992–1993 1993–1997	1–1.2	30, 60	Tile and open drains	Gravity	Cropping intensity has increased
52	Anonymous (1995)	Kota	1995	–	–	Tile and open drains	Gravity	Project benefit cost ratio 2.5
53	Shekhawat (2007)	Lakhuwali (15 ha, 60 ha)	2000 2001	1.2	100, 150, 200	Pipe Drain	Pumped 300–350 Microns geo-textile nonwoven	Kharif (paddy, cotton,) and Rabi (wheat barley, mustard) crops has obtained just double yield and cropping intensity has increased 160 %
Uttar Pradesh								
54	Chauhan and Ram (1972)	Pant Nagar, U.P.	1972	1.0	15–35	Tile drains	Gravity	More water storage has taken place in the profile and less runoff from draining lands
West Bengal								
55	Maity et al. (1996)	Midnapur	1996	–	–	Open drain	–	Increase in observed crop yield
56	Bhattacharya (1999)	Dankuni Basin	1999	–	–	Raised bed pond	–	Salvaged waterlogged land for agriculture activities

Table 2 Major four pilot studies of Pakistan

S. no.	References	Site information	Main design parameters						Shortcomings
			Drain depth (m)	Drain spacing (m)	Design discharge (mm/day)	Design $D_{W.T.}$ (m)	Type of drainage	Outlet and sump	
1	Ghumman et al. (2012)	Khushab SCARP, Pakistan (42,000,000 ha)	1.68–1.98, 2.1	150–300	1.8	1.2	Pipe drainage	Pumped	In Khushab, the water salinity measured in terms of dS/m was reduced by 33 %, respectively, in 12 years
2		FDP (Fourth Drainage Project, Faisalabad), Pakistan (30,351 ha)	1.68–1.98	500–600	2.44	1.2	Pipe drainage	Pumped	In Faisalabad, the water salinity measured in terms of dS/m was reduced by 25 %, respectively, in 12 years
3	Azhar and Latif (2011), Ghumman et al. (2012)	Mardan SCARP Project (MSP) Pakistan (29,542 ha)	2.4	150–200	3	1.05	Pipe drainage	Pumped	Increase in rice, sugarcane, kharif fodder, wheat, sugar beet and tobacco yield was 46, 49, 50, 56,43 and 23 %, respectively
4		Mirpukhas Tile Drainage Project Pakistan (24,281 ha)	1.8–2.4	–	0.95		Pipe drainage	Pumped	Crop yield had increased by the use of saline drain water
5	Ritzema (2007)	East Khaipur Tile Drainage	1.95	–	2.5–3.5	1.0	Pipe drain	Pumped	Crop yield had increased
6	Sarwar and Muhammad (2008)	Swabi SCARP	1.8–2	100–150	2	1.0	Pipe Drain	Pumped	Yields of sugarcane, rice and wheat were increased by 16, 24 and 37 %, respectively
7	Asian Development Bank (1984)	Chashma Command Area Development (18160 ha)	2.10 1.8–2.7	–	1.2–2.6	1.4	Interceptor drain, PVC pipe drain	Pumped	Cropping intensities increased 105 to 135 % after installation of SSD
8	Ritzema (2007)	DC Khan SCARP	–	–	1.88	–	Pipe drain	Pumped	Crop yield has increased and water table is controlled
9	Ritzema (2007)	Fordwah Eastern Sadiqia (South)	2.1		1.5	1.2	Pipe drain	Pumped	Crop yield has increased

Table 3 Pilot studies on desalinization and waterlogging

S. no.	References	Location states and countries	Spacing(m) and	depth (m)	Type of drains and	type of outlet	Major conclusions
1	Kale 2011	Konya–C¸ umra Plain, Turkey	–	–	–	Pumped	Water table, soil salinity have been controlled, and wheat crop yield also increased
2.	Ozturk (2004)	Anatolia, Turkey	21.3	1.8–2.0	Tile drains	Gravity	Crop yield increased, and water table has been controlled
3.	El-Mowelhi et al. (1988)	Nile Delta Egypt	20, 75	1.5	Tile drains	Gravity	Crop yield increased

6. Continuous movement of water in the pipe and aquifer system collects sediment in the pipes which may affect the performance of the drainage system. For resolving the problem, the provision of filters and envelopes on the drain pipes has to be adopted. Filter/envelope material is used to filter that surrounds drain pipe, and these are commonly used along with drain pipes (geotextile, polypropylene, coconut fiber, polystyrene and foam plastic). The traditional filter material is a combination of gravel and coarse sand.
7. Normally the drain effluent is disposed in the canal, salt-making ponds, fishpond, or it can be reused in crop production. The several methods have been suggested by researchers for effluent reuse in irrigation such as blending and mixing.

Subsurface drainage pilot studies in Pakistan

As a consequence, waterlogging and salinity now are serious threats to irrigated agriculture of the 16.7 million hectare in the Indus Basin; about 2 million hectare is waterlogged, and 6 million hectare is salt-affected (Nijland et al. 2005).

Pakistan made an effort for salinity control by starting a series of salinity control and reclamation projects in 1959, and results show a very good agreement to reduce salinity level in the surface and soil profile salinity. The salinity in percentage has been compared for 17-year variation (between the initial salinity year 1960 and 1977–1979) after installation of drainage, and this is also indicated that the salt-free areas increased up to 20 % in surface and profile salinity from past decades (Bhutta 2007).

The subsurface drainage installations in Pakistan are still <1 % of the total cultivable commanded areas so it needs to be increased (Ghumman et al. 2012; Azhar et al. 2004). Bhutta (2007) has discussed salient findings of drainage research and its benefits in Pakistan. Pakistan has world's largest irrigation system, namely the Indus Basin Irrigation System that commands about 14.2 M ha canal irrigated area (Niazi 2008). Due to poor maintenance and neglecting behavior of academician of Pakistan, most of the parts of canal are still unlined and result in the basic cause of waterlogging problem, which directly covered major part of agricultural lands of Pakistan. Niazi (2008) and Ghumman et al. (2010) reported that Government of Pakistan has made effort to save major fertile land of agriculture by the installation of major subsurface drainage projects from past few decades. Various methodologies have been used to protect agricultural lands of Pakistan from the salinity, but most of the subsurface drainage project had not proved informative and beneficial to solve the problem (Smedema 1990; Sarwar 2000; Kahlown and Khan 2004). So many approaches like surface drainage, subsurface pipe drainage and tube wells had been installed for reclamations of fertile lands of Pakistan, but subsurface pipe drainage was proved more beneficial than other two methods (Azhar et al. 2010). Four decades ago, Pakistan government was more concerned about the protection of the productive agricultural lands, and they passed eight subsurface pipe drain projects, but Sarwar and Feddes (2000) and Azhar et al. (2004) have reported that the installed drainage projects had not proved beneficial as per the expectation because these projects have not been technically sound and designed in that particular decade. After that Pakistan's irrigation and drainage engineers and scientists continuously worked on the drainage design parameters such as drainage coefficient, drain spacing Naz et al.2009) and drain depth. Kahlown et al. (2007) and Azhar and Latif (2011)have been worked on three experimental sites NIA, Bughio and Nawazabad farms in Pakistan in order to find the drainage coefficient for the efficient performance of futuristic projects and problems. History of Pakistan's irrigation and drainage researches are mainly explained by existence of nine different subsurface drainage projects (Azhar et al. 2005) as presented in Table 2.

Pakistan drainage design guideline

In Pakistan, in drained areas where a deep water table is maintained, farmers sometimes complain about the increased need of irrigation water (Qureshi et al. 1997). A shallow water table, especially in the fine soils of the Indus plains, is capable of water delivery to the crops through capillary rise. In areas with an 'acceptable' groundwater quality, there is no need to maintain a deep water table. The following guideline arises from the studies of the different research works. On the bases of the pilot studies, following steps are recommended.

1. In Pakistan, drainage design depth of the subsurface drainage systems should be in between 1.8 and 2.4 m while collector depth >3 m.
2. SSD drain discharge has to be designed as 0.95–3.5 mm/day.
3. PVC pipes are generally popular in field drains with fixed dimensions, i.e., diameters 100–200 mm and length up to 800 m.
4. In collector drains, PVC and PE pipes are used with fixed dimensions of 200–380 mm in diameter and length >4 km.
5. Gravel filters are more popular in Pakistan subsurface drainage installation.

© Iraq and Egypt

In Iraq, for example, more than 50 % of the lower Rafidain Plains faces a stern salinity and waterlogging problems (El-Hinnawi 1993). Similarly, salinity and waterlogging have been an inevitable problem in Egypt. These problems have existed during the pre- and post-Aswan High Dam periods (Ritzema 2009; IPTRID Secretariat 2007). To overcome this twin problem, subsurface drainage projects were commissioned in 1942. About 55 % of agricultural land was reported as saline in Iran (FAO 1994). Waterlogging and drainage problems occur in the central and southern parts of the Saudi Arabia. In some projects, like Al-Hassa irrigation project, the agricultural drainage water is mixed with fresh groundwater and reused for irrigation. So, due to the poor quality of irrigation water, soil salinity problem is increasing (AQUASTAT 2008). Multi-level subsurface drainage has also proved beneficial (Hornbuckle et al. 2012). Hirekhan et al. (2007) have suggested that field observation must be taken before and after the installations of SSD in a semiarid climate area and have stressed on rigorous analysis before adopting a subsurface drainage technique. Chahar and Vadodaria (2010) and Chahar and Vadodaria (2012) investigated the optimal spacing in an array of fully penetrating ditches for subsurface drainage. They developed an explicit equation for computing the optimal spacing between the ditches. Eldeiry and Garcia (2010) have compared ordinary kriging, regression kriging and co-kriging techniques to estimate the soil salinity using various images. The best combinations have evaluated to estimate soil salinity with different crop types. Gammal El and Ali (2010) reported that subsurface drainage water has proved very beneficial for crop production in Egypt. Salinity and waterlogging problems affected countries are trying to reclaim their waste land, which are severely affected from these hazardous reasons presented in Table 3.

Conclusions

This paper briefly discussed the old to current state of art in subsurface drainage system, positive effects of drainage on crop production and also what the future hold for this technical approach. Pilot studies reviewed subsurface drainage investigations from 1969 to 2014 (India and abroad) with their outcomes. This study is to suggest the numeric value of drain depth which is kept greater than 1.2 m, i.e., >1.2–1.8 m and drain spacing depending on the soil texture classification, i.e., 100–150 m for light-textured soils, 50–100 m for medium-textured soils and 30–50 m for heavy-textured soils for Indian subsurface drainage installation in the problem area, and Pakistan subsurface drainage project studies show drain depth kept more than 1.8 m. With the help of this paper, researchers will be able to decide the drain depth, spacing, types of subsurface drainage and type of outlets.

References

Anonymous (1988) Handbook for drainage of irrigated areas in India. Irrigation Management and Training Project: technical report no. 5. LBII/WAPCOS, Room 213, Ansal Chambers, Bhikaji Cama Place, New Delhi, pp 12.3 & 12.5

Anonymous (1995) Analysis of subsurface drainage design criteria. Rajasthan Agricultural Drainage Research Project, Chambal Command Area Development, Kota, Rajasthan, pp 7.3

AQUASTAT (2008). Irrigation in the Middle East Regionc in figures. Water Report No 34, FAO. Rome, Italy

Azhar AH, Latif M (2011) Subsurface drainage impact assessment on crop yield. J Anim Plant Sci 21(2):215–219

Azhar AH, Alam MM, Rafiq M (2004). Impact of subsurface drainage systems on land and water conditions. IWASRI Publication 261, WAPDA, Lahore

Azhar AH, Alam MM, Rafiq M (2005) Agricultural impact assessment of sub-surface drainage projects in Pakistan—crop yield analysis. Pak J Water Resour 9(1):1–8

Azhar AH, Bhutt MN, Latif M (2010) Reclamation irrigated agriculture through tile drainage at Fourth Drainage Project, Faisalabad. J Anim Plant Sci 20(3):211–216

Bhattacharya AK (1999) Drainage of agricultural lands. In: Singh GB, Sharma BR (eds) 50 Years of natural resource management research. Division of Natural Resource Management, ICAR, New Delhi, pp 347–362

Bhattacharya AK, Ray SB, Sinha AK, Ghosh G (1996) Drainage in tea plantation. AICRP on Agricultural Drainage, Water Technology Centre, IARI, New Delhi, vol 23, p 4

Bhutta MN (2007) Salient findings of drainage research and its benefits in Pakistan. Pak J Water Resour 11(1):31–38

Chahar BR, Vadodaria GP (2010) Optimal spacing in an array of fully penetrating ditches for subsurface. J Irrig Drain Eng ASCE 136(1):63–67

Chahar BR, Vadodaria GP (2012) Steady subsurface drainage of ponded surface by an array of parallel ditches. J Hydrol Eng 17:895–908

Channabasiah HSM (1972) Sub-surface drainage in salt affected medium to deep black soils in India. Lecture delivered at the Summer Institute on Saline and Alkali Soils and their Management at C.S.S.R.I., Karnal, India

Chauhan HS, Ram S (1972) Surface and sub-surface drainage hydrology. In: Proceedings symposium on waterlogging causes and measures for its prevention. CBIP. Publication, vol 118(1), pp 19–20

Datta KK, Tewari L, Joshi PK (2004) Impact of subsurface drainage on improvement of crop production and farm income in northwest India. Irrig Drain Syst 18:43–55

Devadattam DSK, Chandra RS (1995) Agricultural land drainage in coastal saline soils. AICRP on agricultural drainage, A.N.G. Ranga Agricultural University, Hyderabad, 40 p

Drainage Design Factors (1986) Irrigation and drainage. Paper no. 38. Food and Agriculture Organization of the United Nations, Rome, Italy

Drainage manual (1978) U.S. Department of the Interior, Bureau of Reclamation. U.S. Government Printing Office, Washington, DC

Eldeiry AA, Garcia LA (2010) Comparison of ordinary kriging, regression kriging, and cokriging techniques to estimate soil salinity using LANDSAT images. J Irrig Drain Eng ASCE 136(6):355–364

El-Hinnawi E (1993) Population, environment, and development in the Arab World. Paper presented to the Arab Population Conference, Amman, United Nations Economic and Social Commission for Western Asia (ESCWA)

El-Mowelhi N, El-Bershamgy A, Hoffman GJ, Chang AC (1988) Enhancement of crop yields from subsurface drains with various envelopes. Agric Water Manag 15(2):131–140

FAO (1994) Land degradation in south Asia: its severity, causes and effects upon the people. Food and Agriculture Organization of The United Nations, World Soil Resources Reports, Rome

Gammal EL HAA, Ali HMM (2010) Commissioning of abandoned drainage water reuse systems in Egypt: a case study of upgrading the Umoum project, Nile Delta. Irrig Drain 60(1)115–122

Gates TK, Grismer ME (1989) Irrigation and drainage strategies in a salinity affected region. J Irrig Drain Eng 115(2):87–99 **128(2)**

Ghassemi F, Jakeman AJ, Nix HA (1995) Salinization of land and water resources. University of New South Wales Press, Sydney

Ghumman AR, Ghazaw YM, Niazi MF, Hashmi HN (2010) Impact assessment of subsurface drainage on waterlogged and saline lands. J Environ Monit Assess 172:189–197

Ghumman AR, Ghazaw YM, Hashmi HN, Kamal MA, Niazi MF (2012) Environmental and socio-economic impacts of pipe drainage in Pakistan. Environ Monit Assess 184:1671–1681

Goel A, Tiwari P (2013) Review of computer based software tools for salinity management in agricultural lands. J Indian Water Resour Soc 33(4):25–32

Gupta SK (2002) A century of subsurface drainage research in India. Irrig Drain Syst 16:69–84

Gupta SK (2003) A historical perspective of subsurface drainage research in India. Asian Agric Hist 7(1):45–54

Hanumanthaiah CV, Satyanarayana TV (2003) Cost effectiveness of subsurface drainage technology under Indian conditions—a boon to farming community in the areas of saline soils, paper no 118. Presented at the 9th international drainage workshop, Utrecht, The Netherlands

Hefny MM, Mohamed E, Metwali R, Mohamed AI (2013) Assessment of genetic diversity of sorghum (Sorghum bicolor L. Moench) genotypes under saline irrigation water based on some selection indices. Australian J Crop Sci (AJCS) 7(12):1935–1945

Hirekhan M, Gupta SK, Mishra KL (2007) Application of Wasim to assess performance of a subsurface drainage system under semiarid monsoon climate. J Agric Water Manag 88:224–234

Hornbuckle JW, Christen EW, Faulkner RD (2012) Analytical solution for drainflows from bilevel multiple-drain subsurface drainage systems. J Irrig Drain Eng 138:642–650

IPTRID SECRETARIAT (2007) Egypt's experience in irrigation and drainage research uptake, final report. Electronic Publishing Policy and Support Branch, Information Division, Food and Agriculture Organization, FAO, Viale delle Terme di Caracalla, 00153, Rome, Italy

Jaiswal SR, Dhruva Narayana VV (1972) Design of a sub-surface drainage system. In: Proceedings symposium on waterlogging causes and its prevention CBIP, vol 110(1), pp 91–100

Joint Completion Report on IDNP Result (2007) Recommendation on waterlogging and salinity control based on pilot area drainage research, p 90

Kahlown MA, Khan AD (2004) Tile drainage manual, Pakistan Council of Engineering in Water Resources (PCRWR), Islamabad, Pakistan, ISSB 969-8469-13-3, PCRWR Press, Islamabad

Kahlown MA, Marri MK, Azam M (2007) Design, construction and performance evaluation of small tile drainage systems in the Indus Basin. J Irrig Drain ICID 56(1):217–225

Kale S (2011) Field-evaluation of DRAINMOD-S for predicting soil and drainage water salinity under semiarid conditions in Turkey. Spanish J Agric Res 9(4):1142–1155

Kaledhonkar MJ, Gupta SK, Singh G.(2009) An overview of research, development, opportunities and challenges for subsurface drainage in India. In: 60th international executive council meeting & 5th Asian regional conference, New Delhi, India

Kaledhonkar MJ, Kamra SK, Sharma DP (1998) Drainage measures for control of waterlogging in semi-arid regions. In: Proceedings 9th National Symposium on Hydrology, Amritsar, 26-27th Nov. Amritsar, pp 97–104

Kamble BM, Chougule BA, Rathod SD, Rathod PK (2006) Improvement of salt affected-water logged soils in western Maharashtra through subsurface drainage (Ssd) system. Asian J Biol Sci 1(2):64–67

Kamra SK, Singh SR, Rao KVGK, van Genuchten MTh (1991) A semi-discrete model for water and solute movement in tile-drained soils 1. Governing equations and solution. Water Resour Res 27:2439–2447

Kelleners TJ, Kamra SK, Jhorar RK (2000) Modeling of drainage water salinity of pipe drains. J Hydrol 234:249–263

Kolekar OL, Patil SB, Rathod SD (2014) Effects of different mole drain spacings on the yield of summer groundnut. Int J Res Eng Technol 3(4):2321–7308

Kornecki TS, Fouss JL, Grigg BC, Southwick LM (2001) Drainage research to improve runoff water quality and soil trafficability. Land Water Mag 45(3):10–13

Lal M (2011) Impact of subsurface drainage on land quality, ground water quality and crops yields in Haryana. PhD, Thesis, National Institute of Technology Kurukshetra, Haryana, India

Lovas L (1972) Salinity control trials by sub-soil drainage at Digod. In: Proceedings All India symposium on soil salinity held at Kanpur, pp 156–160

Luthin JN (1978) Drainage Engineering. R. E. Krieger Publishing Co., Huntington

Maity SP, Ghosh RK, Das NC, Patra SK (1996) Drainage and Crop Productivity in Saline Waterlogged Soil. AICRP on agricultural drainage. Bidhan Chandra Krishi Vishwavidyalaya. Kalyani, West Bengal, vol. 55, p 11

Mandal AK, Sharma RC (2011) Delineation and characterization of waterlogged salt affected soils in IGNP using remote sensing and GIS. J Indian Soc Remote Sens 39(1):39–50

Mathew EK, Nair M, Raju TD (1993) Drainage and Rice Productivity. AICRP on Agricultural Drainage. Kerala Agricultural University, Karumady, p 27

Michael AM (1967) Drainage research scheme. Annual report. College of Agricultural Engineering Punjab. Agricultural University, Ludhiana. p 108

Michael AM, Ojha TP (2013) Principles of Agricultural Engineering. Jain Brothers (New Delhi) 2(4):383–395

Naz BS, Ale S, Bowling LC (2009) Detecting subsurface drainage systems and estimating drain spacing in intensively managed agricultural landscapes. Agric Water Manag 96(4):627–637

Niazi MFK (2008) Evaluating the impact of subsurface drainage on land and water. Ph.D. thesis, University of Engineering and technology, Taxila, Pakistan

Nijland HJ, Croon FW, Ritzema HP (2005) Subsurface drainage practices: guidelines for the implementation, operation and maintenance of subsurface pipe drainage systems. ILRI Publication 60, Wageningen, p 607

Ozturk A (2004) Multi-objective drainage requirement of Harran Plain. Pak J Biol Sci 7(2):150–154

Pali AK, Katre P, Khalkho D (2014) An unsteady subsurface drainage equation incorporating variability of soil drainage properties. Water Resour Manag 28:2639–2653

Patel RM, Prasher SO, Bonnell RB, Broughton RS (2002) Development of comprehensive soil salinity index. J Irrig Drain Eng 128(3):185–188

Qureshi AS, Iqbal M, Anwar NA, Aslam M, Ramzan MC (1997) Benefits of shallow drainage. In Proceedings of seminar on on-farm salinity, drainage, and reclamation. IWASRI Publication, p 179

Ramana RKV, Kishore R, Singh R (2009) Mole drainage to enhance soybean production in waterlogged Vertisols. J Agric Eng 46(4):54–58

Rao KVGK, Singh OP, Gupta RK, Kamra SK, Padey RS, KumbharePS, Abrol IP (1986) Drainage investigations for salinity control in Haryana. Bull. no. 10. CSSRI, Karnal, India. p 95

Ritzema H (2007) Performance assessment of subsurface drainage systems case studies from Egypt and Pakistan. Alterra-ILRI Wageningen University and Research Centre Wageningen The Netherlands, p 137

Ritzema PH (2009) Drain for gain, making water management worth its salt, subsurface drainage practices in irrigated agriculture in semi-arid and arid regions. Ph.D. Dissertation, Wageningen University and UNESCOIHE Institute for Water Education, Wageningen, The Netherlands

Ritzema HP, Satyanarayana TS, Raman S, Boonstra J (2008) Subsurface drainage to combat waterlogging and salinity in irrigated lands in India: lessons learned in farmers' fields. Agric Water Manag 95:179–189

Sarwar A (2000) A transient model approach to improve on-farm irrigation and drainage in semi-arid zones. Ph.D. Dissertation, Wageningen University & Research Centre, The Netherlands

Sarwar A, Feddes RA (2000) Evaluation drainage design parameters for the fourth drainage project, Pakistan by using SWAP model: Part ii-modeling result. Irrig Drain Syst 14:281–299

Sarwar T, Muhammad N (2008) Effect of enhanced water allowance on the performance of tile drainage system at Shabaz Ghari pilot project area Swabi SCARP. Sarhad J. Agric 24(1):59–68

Satyanarayana TV, Lakshmi GV, Srinivasulu A, Hanumanthaiah CV, Ratnam M, Hemakumar HV (2002) Drainage and water management for salinity control in canal commands. A comprehensive report of research achievements of Bapatla Network Center, Indo-Dutch Network Operational Research Project, Acharya N.G. Ranga Agricultural University, Bapatla, India

Sethi M, Bundela DS, Sethi M, Lal K, Kamra SK (2010) Remote sensing and geographic information system for appraisal of salt-affected soils in India. J Environ Qual 39(1):5–15

Shakya SK, Gupta, PK Kumar D(1995) Innovative drainage techniques. AICRP on drainage. Punjab Agricultural University, Ludhiana, Punjab. p 34

Sharma RK, Yadav RD, Mourya BM, Kushwaha SS, Sharma AK (1995) Research on agricultural drainage in black soils of Barna command. AICRP on agricultural drainage. Jawaharlal Nehru Krishi Vishwavidyalaya, Powerkheda, Madhya Pradesh, India. p 35

Shekhawat RS (2007) Economic analysis of sub-surface drainage under Indira Gandhi Nahar Priyojna command area—A case study. Agric Econ Res Rev 20:361–374

Smedema LK(1990). Comparative study of the East Khairpur, Mardan and drainage-IV pipe drainage project. Project: PAK/83/022. Technical report no. 90/01, IWASRI Pub. No. 8, Lahore, Pakistan

Srinivasulu A, Satyanarayana TV, Kumar HVH (2005) Subsurface drainage in a pilot area in Nagarjuna Sagar right canal command, India. Irrig Drain Syst 19:61–70

Srinivasulu M, Rao AR, Babu MR, Babu GR, Kumar HVH (2014) Performance of subsurface drainage system installed at Appikatla region of Andhra Pradesh. Int J Argric Eng 7(1):180–185

Thatte CD, Kulkarni SA (2000) Status and strategy for prevention and reclamation of waterlogged and salt-affected lands from irrigated areas of India. 8th ICID international drainage workshop, New Delhi

Tiwari P (2011) Simulation of salt and water dynamics in subsurface drained land using hydrus. M.Tech. Dissertation, College of Agricultural Engineering, JNKVV Jabalpur, Madhya Pradesh India

Yadav JSP (1975) Annual Report. All India coordinated scheme for research on water management and salinity (unpublished)

Zimmer D, Bermond M, Melliand ML, Tato-Serrano S (2002) Global drainage status and needs, private communication, report will be published later by Cemagref/ICID

7

Comparative study on the water quality status of Andra reservoir and Denkada anicut constructed on Champavati River, Vizianagaram, India

G V. S. R. Pavan Kumar[1] · K. Rama Krishna[2]

Abstract The author's present study was carried out for a period of 3 years from 2010 to 2013 to itemize the various physico-chemical parameters, irrigation water quality parameters and heavy metals in Champavathi River waters at Andra reservoir and Denkada anicut. Water samples were collected from the chosen sampling stations of the two reservoirs for every 4 months and analyzed as per APHA standard methods. The results obtained were compared with IS 10500 standards and found to be well within the prescribed values. Though the obtained values were well within the prescribed standard values, it was found that the water quality index, concentration of certain parameters such as calcium, magnesium, sodium and potassium of the waters of Andra reservoir are higher than that of the Denkada anicut, and the concentration of nitrite was found to be higher in the water sample analyzed from Denkada anicut. Except silicon, all the other metals were found to be below the detection limits in the two reservoir waters. The reasons for the same were probed by the authors in the presented study. From the analysis reports, it was found that the water analyzed from the two reservoirs was fit for irrigation, agriculture, industrial and domestic purposes.

✉ G V. S. R. Pavan Kumar
prs.ganti@gmail.com

K. Rama Krishna
rk.karipeddi@yahoo.co.in

[1] Department of Chemistry, M V G R College Of Engineering, Vizianagaram, India

[2] Department of Chemistry, GITAM Institute of Sciences, GITAM University, Visakhapatnam, India

Keywords Water quality index · Physico-chemical characterization · Andra reservoir · Denkada anicut · Vizianagaram · Irrigation projects

Introduction

The hydrochemical study of quality of water reveals that it is suitable for drinking, agriculture, industrial and irrigation purposes (Nagaraju et al. 2014). The availability of good quality water is an indispensible feature for preventing diseases and improving the quality of life (Nabila et al. 2014). The healthy aquatic ecosystem is dependent on physico-chemical and biological characteristics (Tirupathaiah et al. 2012). The analysis of freshwater is an important and sensitive issue in water quality monitoring. Its main objective is to control and reduce the incidence of pollutant-related problems (Akoto and Abankua 2014). Water, the matrix of life, is exposed to pollution, unhealthy environment resulting in human affliction and diseases, the reason being rapid industrialization and increase in population (John Mohammad et al. 2015). Water quality monitoring of aquatic systems like reservoir serves as a tool for planning and management of the river basin. The impact of human activities directly influences the lake habitat and can alter the environment of the lake or reservoir (Pavan Kumar et al. 2013). To make the entire ayacut fertile and for feasible distribution of the available water of the river to irrigation and agricultural purposes, reservoirs are constructed. It is important to know the quality of water at each of such reservoirs constructed across the same river to assess whether the water is fit for human consumption for different purposes. In the present study, the author attempted to compare the physical and chemical parameters, irrigation water quality, heavy metals and water quality data of two selected reservoirs, namely Andra and Denkada anicut constructed on the River Champavathi.

River Champavathi originates in the Eastern Ghats of south India at an altitude of 1200 m above MSL near Andra village, flows eastward and joins the Bay of Bengal at Konada village. The basin has a drainage area of 1410 km^2. For the purpose of irrigation and agriculture, two reservoirs are constructed on the river, one at the originating place, Andra reservoir, and the other, Denkada anicut, which is 30 km away from the estuarine point of the river. Andra reservoir (Fig. 21) is located in Andra village near the originating place of the river, at 18.3500°N 83.2000°E. It has an average elevation of 127 m (419 feet). The project utilizes 0.980 TMC of the available water and the reservoir storage capacity is about 0.9 TMC. The total ayacut of 9426 acres has been stabilized in Bondapalli, Gajapathinagaram and Mentada Mandals of Vizianagaram districts of Andhra Pradesh, India. The Denkada aniuct (Fig. 21) was also constructed across Champavathi River near the estuarine point of the river. The project is located near Saripalli Village, Nellimarla Mandal, Vizianagaram District for irrigation of a total ayacut of 5,153 acres in the district. The project utilizes 0.640 TMC of the available water. The Denakda anicut is situated at 18°7′51″N 83°28′36″E. These two reservoirs were constructed for the purpose of irrigation and hence it is necessary to assess the quality of the waters of the two reservoirs. The distance between the two reservoirs is 50 km. Ten semi-urban towns are situated throughout the flow of the river in between. Pollution due to agricultural waste, manmade acts and disposal of animal waste into the river are observed in the lower catchment of the Andra reservoir to Denkada anicut. Hence, a comparative study is presented

Table 1 A comparative report on physico-chemical parameters

Season	Chloride		Phosphate		Nitrite		Fluoride		TA		Carbonate		Bicarbonate	
	A	D	A	D	A	D	A	D	A	D	A	D	A	D
S1	98.5	135	0.5	0.1	0.05	0.3	0.1	0.2	159.5	206	18.1	18.5	111	185
S2	86.2	51.2	0.8	0.2	0.03	0.2	0.4	0.1	159.5	200	18.1	18.5	111	185
S3	95	51.8	1	0.3	0.03	0.2	0.4	0.1	160.5	199	36.1	35.2	222	193
S4	100	55	0.8	0.3	0.03	0.2	0.4	0.1	150.8	195	38.2	35.2	225	190
S5	100	58	0.9	0.3	0.03	0.1	0.5	0.1	165	199	40	37	228	195
S6	98	52.5	1.2	0.3	0.03	0.2	0.6	0.1	160	201	38.5	37	235	195
S7	98	52	1.2	0.3	0.03	0.2	0.6	0.1	165	200	38.5	41	235	200
S8	95.5	52.5	1.1	0.3	0.03	0.2	0.7	0.1	168	198	38.5	39.5	229	199
S9	98.2	52	1.1	0.3	0.03	0.2	0.9	0.1	165	200	40	39.5	229	199
S10	100	52	1	0.3	0.05	0.2	0.8	0.1	165	206	42	40	231	198
S11	150	55	1.2	0.5	0.06	0.1	0.8	0.3	170	200	45	39	230	190

EC is expressed in μS/cm and all others are expressed in mg/L

Table 2 WQI and irrigation water quality parameters: a comparative analysis report

S.No	SAR		RSC		%Na		Mg hazard		WQI	
S1	1.537	1.31	−5	−3.8	27.9	24.3	46.4	46.4	24.9	39.6
S2	1.308	1.3	−1.5	−0.3	31.3	30.8	49.5	49.6	23.5	31.8
S3	0.376	0.79	−1.2	−1.07	9.71	18.9	32.9	49.2	36.4	32.4
S4	1.374	0.9	−0.03	−1.41	29.25	20.6	39.8	47.4	29.1	32.4
S5	1.132	0.97	−2.9	−1.63	27.75	21.3	37.2	45.7	32.1	26.9
S6	1.922	0.95	−3.5	−1.86	30.1	20.8	36.45	45.5	35.1	32.5
S7	1.922	0.87	−3.5	−1.4	30.1	19.5	36.45	46.8	35.1	32.4
S8	1.827	0.9	−2.9	−1.1	30	20.5	37.43	47.4	35.6	32.4
S9	1.759	0.92	−3.5	−1.8	28.5	19.9	28.6	49.5	38.5	32.5
S10	1.841	0.97	−3.8	−1.12	36.15	20.9	36.31	47.4	38.7	32.4
S11	1.93	1.15	−4.41	−1.29	29.2	25.1	25.1	38.1	36.4	30.2

in this article on the seasonal variation in the physico-chemical parameters, water quality index and heavy metal analysis.

In literature, comparative studies on the reservoir water quality was taken up all over the world (Chwowdary et al. 2011; Janeshwar et al. 2012; Swaranlatha and Narsingrao 1998; Pandey et al. 1993; Bhadja et al. 2013; Vaghela 2013). In light of this, the author presented a comparative study on the water quality of the two reservoirs on the River Champavathi.

Materials and methods

The samples were collected during different seasons of August 2010 (S1), December 2010 (S2), April 2011 (S3), August 2011 (S4), December 2011 (S5), April 2012 (S6), August 2012 (S7), December 2012 (S8) April 2013 (S9), August 2013 (S10) and December 2013 (S11). The samples were collected in previously cleaned polyethylene bottle of capacity 1000 ml and mixed with appropriate preservatives for further analysis. EC, temperature, pH and DO of the collected samples were measured on the spot. The concentration of chloride ion in the water sample was determined by Mohr's method using potassium chromate as an indicator. Concentrations of calcium and magnesium were determined by using EDTA with EBT and murexide indicators. All these analyses were performed based on the APHA methods (Clesceri 1993).

Results and discussion

The complete comparative data on the physico-chemical parameters is presented in Table 1, and irrigation water quality parameters and WQI are presented in Table 2. Comparative seasonal variations in the physico-chemical characters, irrigation water quality parameters and water quality index data are presented in Figs. 1, 2, 3, 4, 5, 6, 7, 8, 9, 10, 11, 12, 13, 14, 15, 16, 17, 18, 19 and 20. A comparison of the results obtained with IS 10500 standards and WHO standards was also done and found that all the values obtained were found to be well within the prescribed standard values (IS 10500, BIS 1991) (WHO 1993).

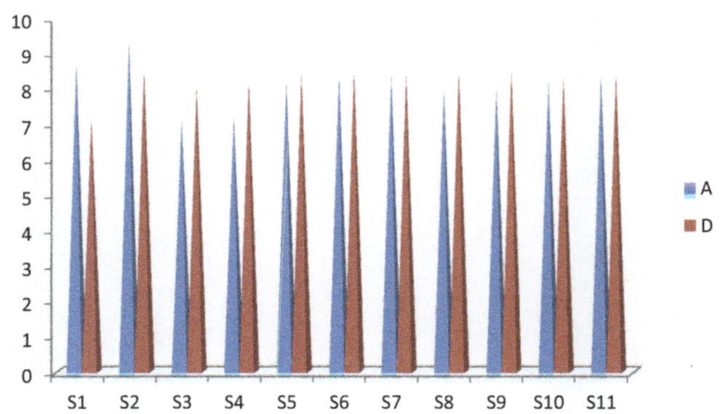

Fig. 1 Seasonal variation in pH for the two reservoir waters—a comparative representation

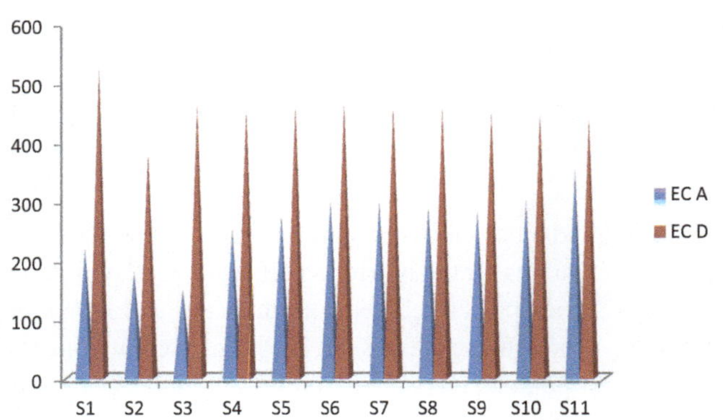

Fig. 2 Seasonal variation in EC for the two reservoir waters—a comparative representation

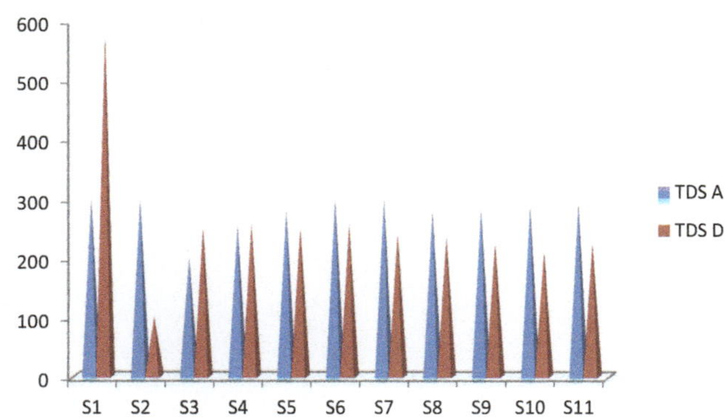

Fig. 3 Seasonal variation in TDS for the two reservoir waters—a comparative representation

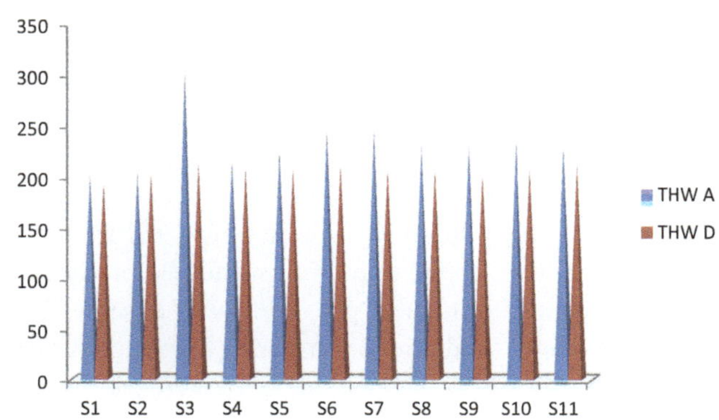

Fig. 4 Seasonal variation in THW for the two reservoir waters—a comparative representation

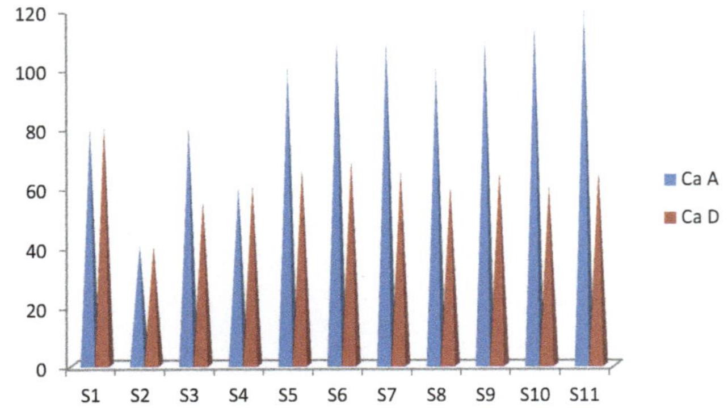

Fig. 5 Seasonal variation in calcium for the two reservoir waters—a comparative representation

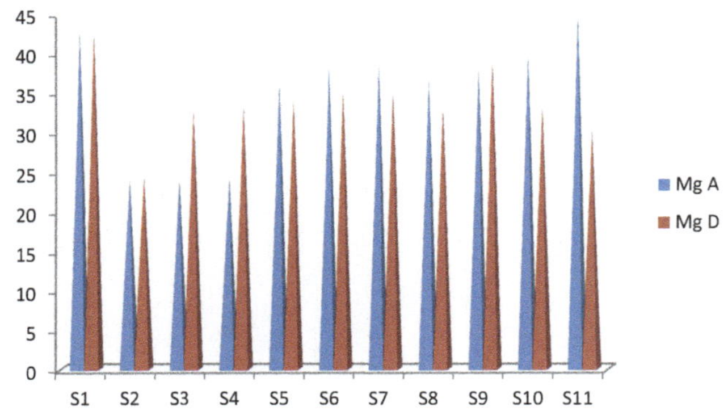

Fig. 6 Seasonal variation in magnesium for the two reservoir waters—a comparative representation

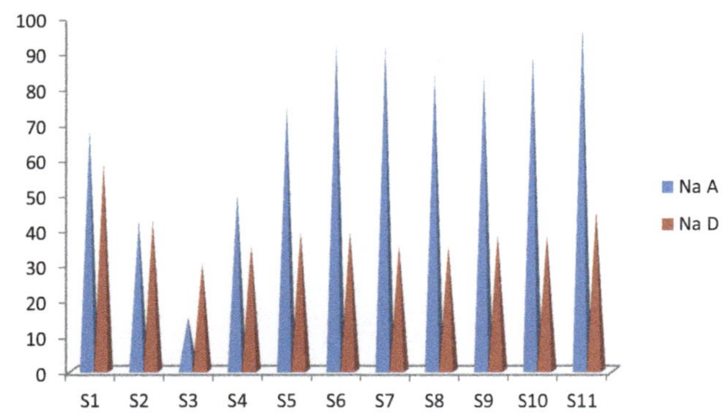

Fig. 7 Seasonal variation in sodium for the two reservoir waters—a comparative representation

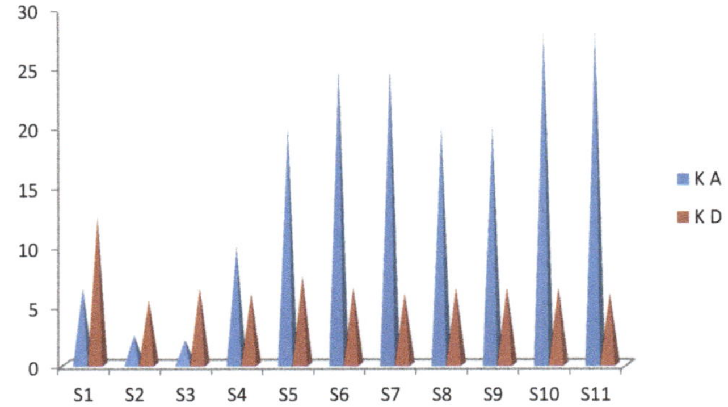

Fig. 8 Seasonal variation in potassium for the two reservoir waters—a comparative representation

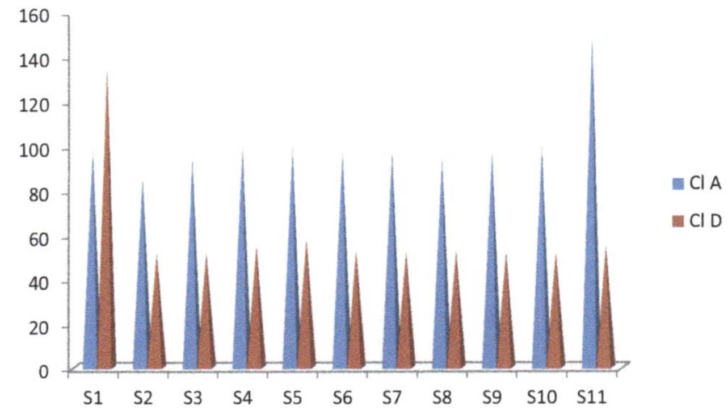

Fig. 9 Seasonal variation in chloride for the two reservoir waters—a comparative representation

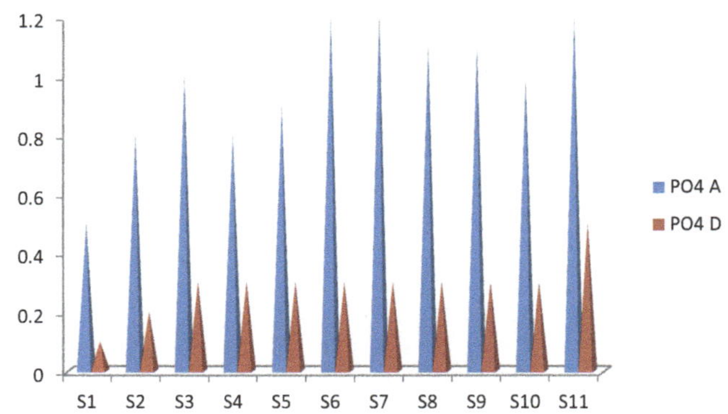

Fig. 10 Seasonal variation in phosphate for the two reservoir waters—a comparative representation

The pH values of the water samples analyzed from the two reservoirs were found to be well within the prescribed standard values. The highest pH of 9.3 was observed for the water from the Andra reservoir during December 2010. From the data, it was concluded that the water was slightly alkaline in both the reservoirs and no appreciable variation was found between them.

From the data of electrical conductivity, it was observed that the EC values for the two reservoirs were well within the standard values prescribed for both drinking and irrigation purposes. Though the values of EC were well within the standards stipulated, on comparison slightly higher values of EC were observed for the water samples collected and analyzed from Andra reservoir. The reservoir is constructed

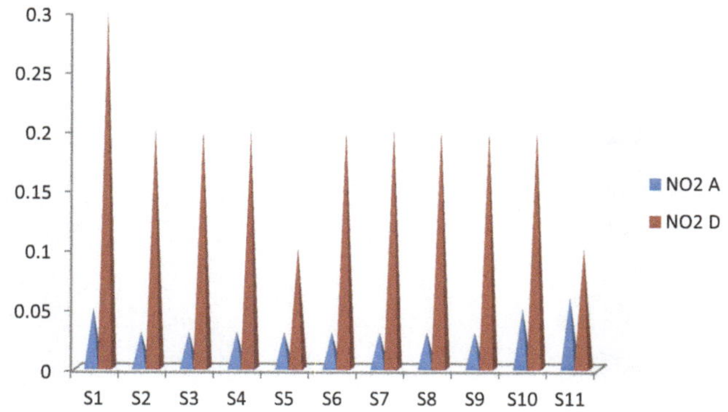

Fig. 11 Seasonal variation in nitrite for the two reservoir waters—a comparative representation

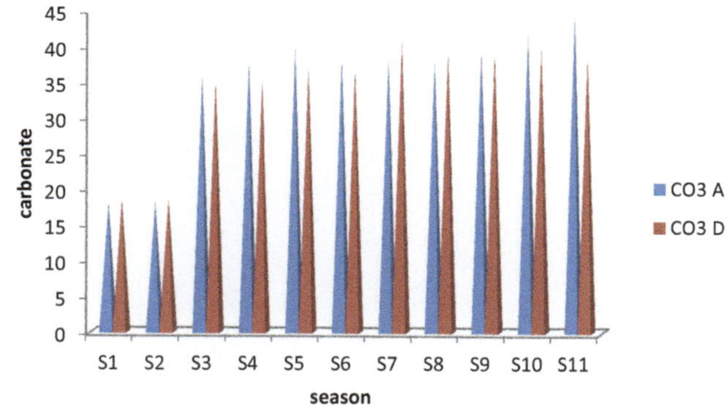

Fig. 14 Seasonal variation in carbonate for the two reservoir waters—a comparative representation

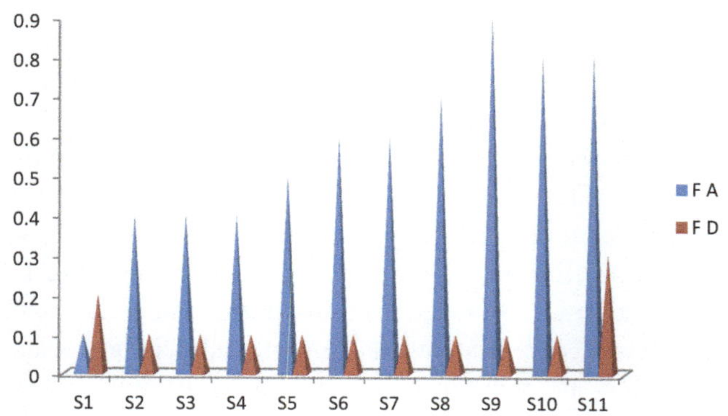

Fig. 12 Seasonal variation in fluoride for the two reservoir waters—a comparative representation

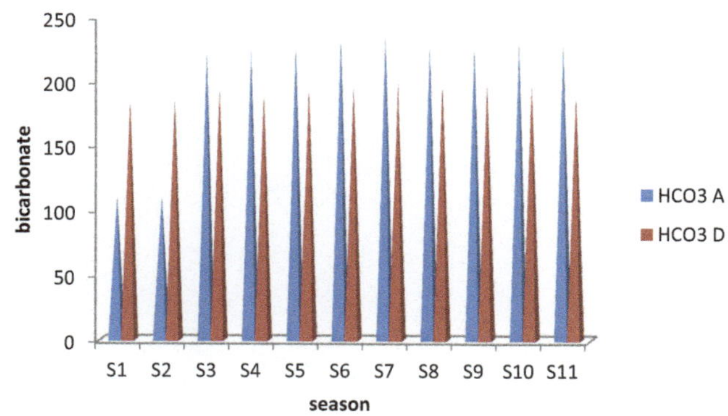

Fig. 15 Seasonal variation in bicarbonate for the two reservoir waters—a comparative representation

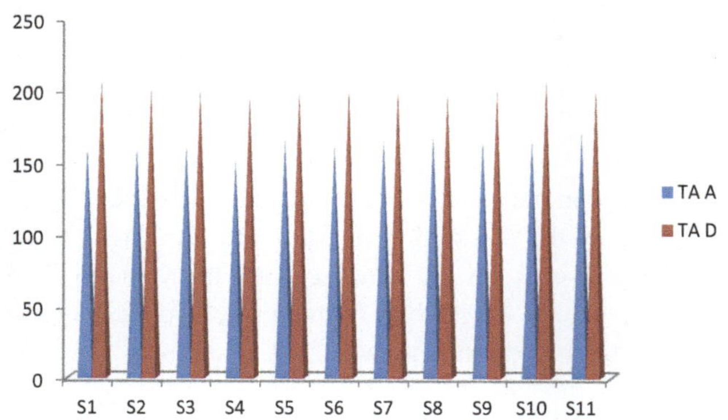

Fig. 13 Seasonal variation in TA for the two reservoir waters—a comparative representation

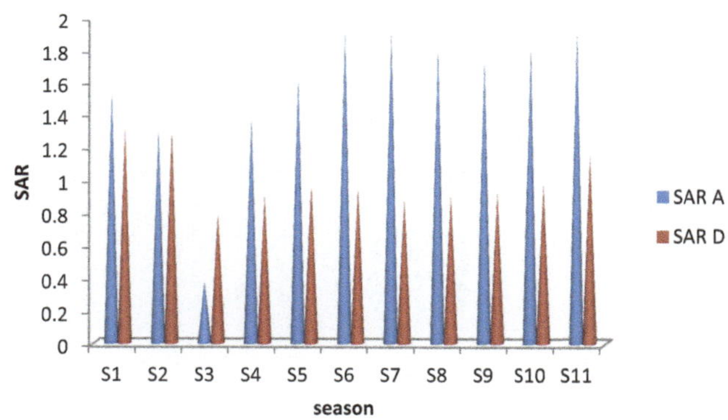

Fig. 16 Seasonal variation in SAR for the two reservoir waters—a comparative representation

close to the origin of the river, water of the reservoir is found not to be influenced by the various physical and chemical pollution or natural or manmade acts.

The concentration levels of total hardness of water, calcium and magnesium were found to be well within the prescribed standard limit values. In all the seasons under study, no variation in the concentration levels of the parameters mentioned was found in the two reservoirs. Sodium, potassium, chloride, fluoride and phosphate were found to be higher in the water analyzed from the Andra reservoir than the Denkada anicut. Though there was variation in the concentration of the parameters between both the reservoirs, all the results obtained were well within the prescribed standard values. The reasons for the

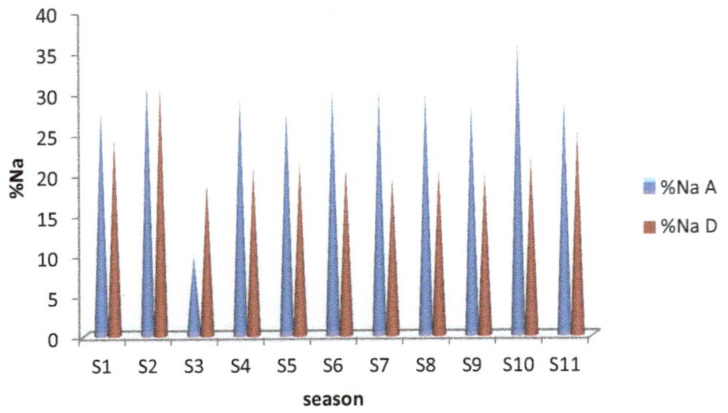

Fig. 17 Seasonal variation in %Na for the two reservoir waters—a comparative representation

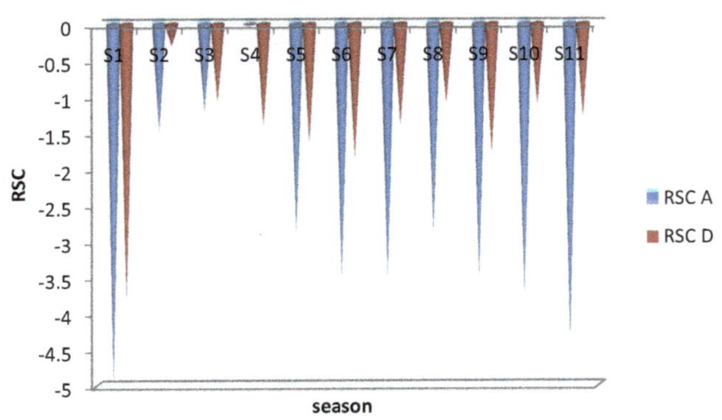

Fig. 18 Seasonal variation in RSC for the two reservoir waters—a comparative representation

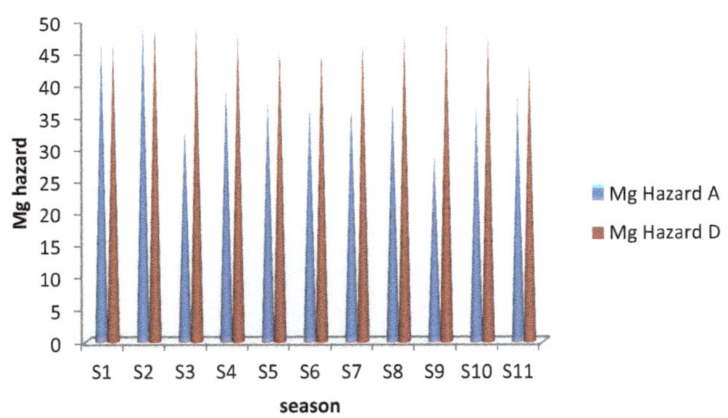

Fig. 19 Seasonal variation in Mg-hazard for the two reservoir waters—a comparative representation

variation in the concentration of different parameters mentioned above are as follows:

(a) It was observed that the Andra reservoir was constructed at the origin of the river. In such conditions, the water of the river is away from the influence of the soil and conditions through which it flows.

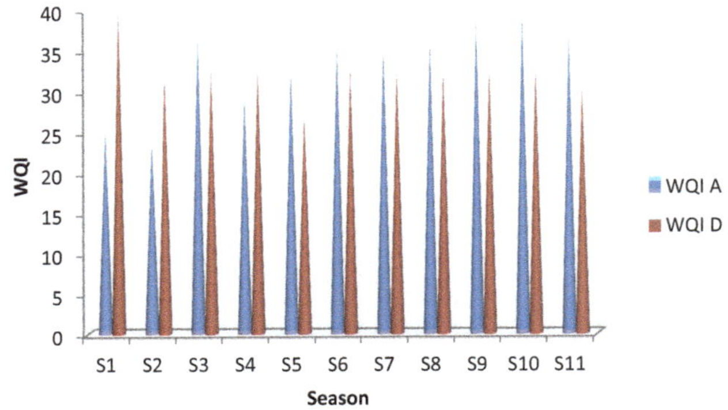

Fig. 20 Seasonal variation in WQI for the two reservoir waters—a comparative representation

(b) As the river flows from the reservoir downstream, during its travel, it comes in contact with a variety of conditionings such as soil (Ray Chaudhary 1941) and contamination with pollutants. It was observed that most of the soils in the district of Vizianagaram were red soils, sandy loams and sandy clay. These soils are porous textured and leaching dominant. As the river flows through this soil, the chemical constituents present in the water may be sedimented and makes the water with lower concentrations of the aforesaid parameters and hence the trend of high levels of the parameters was found in the reservoir at the originating place and lower concentration the same at the other, Denkada anicut.

(c) In most of the seasons under investigation, fluoride ion concentration was found to be lower in the water samples of Denkada anicut to that of Andra reservoir. This indicates that the waters of the Denkada anicut are not contaminated with agricultural runoff (Sharma et al. 2013).

(d) During August 2010, a higher value of TDS was found in the water samples of Denkada anicut. This is due to the dissolution of salts into the water of the reservoir. In all the other seasons, the value of TDS was found to be below the value reported in the seasons mentioned above.

It was found that the concentration of nitrite in the waters of the Denkada anicut was higher than that of Andra reservoir. Nitrite enters water bodies from municipal and industrial wastewater, septic tanks, feed lot discharges, animal wastes (including birds and fish), runoff from fertilized agricultural field and lawns and discharges from automobile exhausts. It was observed that animal waste is fed into the river at a place near the Denkada anicut, which is the reason for the high levels of nitrite in the anicut waters. Wastes and seepage fed into the river under study is presented in Figs. 21, 22 and 23. Though higher levels of nitrite is observed in the Denkada anicut compared to that

Fig. 21 Household seepage fed into Champavathi River

Fig. 22 Solid waste from houses into Champavathi River

Fig. 23 Drainage fed into Champavathi River

of Andra reservoir, the concentration of nitrite in both the reservoir waters was found to be well in accordance with the standard values stipulated.

The water quality index (WQI), calculated from the parameters analyzed was found to be within the range of 23.5–38.7 for the Andra reservoir and 26.9–39.6 for Denkada anicut. From the water quality index ratings, the quality of the water of the two reservoirs was rated as "Good" for human consumption. From the correlation of irrigation water quality data, it was found that all the irrigation water quality parameters analyzed such as SAR, RSC, %Na and magnesium hazard (Pavan Kumar 2014; Sarath Prasanth et al. 2012; Nagaraju et al. 2014) were found to be within the standard values.

In the water samples analyzed from the two reservoirs, silica was present in higher concentrations. Heavy metals such as Al, Zn, Cd, Hg, As, Co, Ni, Ti, Ba and Mn were found to be below detection limits in the water samples analyzed by ICP-OES. From such data, it is inferred that the water analyzed from the two reservoirs is suitable for irrigation, and agricultural and domestic purposes.

Conclusions

From the results obtained, it was found that the water quality index and the concentration of certain parameters such as sodium, potassium, chloride and phosphate of the waters of Andra reservoir are higher than that of Denkada anicut. The water from Andra reservoir is released to reach Denkada anicut, downstream; as most of the soil in the riverbed is red soil and sandy loams, the variation is due to the percolation of the chemicals mentioned above through the soil on which the river flows. The same trend was observed in most of the seasons under investigation. Nitrite ion concentrations were found to be higher in the water samples of Denkada anicut due to the contamination of the water by automobile wastes and household seepage. The lower concentration levels of fluoride in the water samples of Denkada anicut indicate that the water is free from contamination by agricultural runoff. Heavy metals were found to below detection limits in the two reservoir waters. From the analysis reports, it is concluded that the water of the two reservoirs is fit for irrigation, agriculture, and industrial and domestic purposes.

Acknowledgments One of the authors is grateful to the principal and the management of M V G R College of Engineering for their support and encouragement.

References

Akoto O, Abankua E (2014) Evaluation of Oqabi reservoir (Ghana) water quality using factor analysis. Lakes Reserv Res Manag 19:174–182

Bhadja P et al (2013) Assessment of physico-chemical parameters and water quality index of reservoir water. IJPAES 3(3):89–95

BIS (1991) Drinking water specification. India

Chaudhary R (1941) Morphological grouping of red soil. ICAR

Chwowdary R, Rawtani P, Viswakarma M (2011) Comparative study of drinking water quality parameters of three Manmade Reservoirs i.e., Kolar, Kaliasote and Kerwa Dam. Curr World Environ 6(1):145–149

Clesceri LS (1993) Standard methods of examination of water and waste water. APHA

Janeshwar Y et al (2012) Analysis of water quality using physico-chemical parameters, satak reservoir in Khargone district MP, India. Int Res J Environ Sci 2(1):9–11

John Mohammad M et al (2015) Analysis of water quality using limological studies of Wyra reservoir Khammam district, Telangana, India. Int J Curr Microbiol App Sci 4(2):880–895

Nabila B et al (2014) An assessment of the physico-chemical parameters of Oran sebkha basin. Appl Water Sci 4:351–356

Nagaraju A et al (2014) Assessment of ground water quality for irrigation: a case study from Bandalamottu lead mining area, Guntur district, Andhra Pradesh. Appl Water Sci 4:385–396

Pandey AK, Siddiqi SZ, Rao Rama (1993) Physicochemical and biological characteristics of Husain sagar, an industrially polluted lake, Hyderabad. Proc Acad Environ Biol 2(2):161–167

Pavan Kumar GVSR (2014) Water quality status of Andra reservoir in Vizianagaram district of Andhra Pradesh, India. J Appl Chem 3(5):2095–2101

Pavan Kumar GVSR et al (2013) Assessment of Water quality status of Thatipudi reservoir of Vizianagaram district, Andhra Pradesh. INnov J Sci 1(2):20–24

Sarath Prasanth SV et al (2012) Evaluation of ground water quality and its suitability for drinking and agricultural use in the coastal stretch of Alappuzha district, Kerala, India. Appl Water Sci 2:165–175

Sharma KK et al. (2013) Assessment of fluoride in ground and surface water bodies and its seasonal variation at Bishnah Tehsil of Jammu district, J&K: correlation with physic-chemical parameters. Int J Recent Sci Res 4(11):1736–1747

Swaranlatha S, Narsingrao A (1998) Ecological studies of Banjara Lake with reference to water pollution. J Envi Biol 19(2):179–186

Tirupathaiah M et al (2012) Analysis of water quality using physico chemical parameters in lower Maniar reservoir of Karimnagar district of Andhra Pradesh. Int J Environ Sci 3(1):172–180

Vaghela AK (2013) Assessment of physico-chemical parameters and water quality index of reservoir water. IJPAES 3(3):89–95

WHO (1993) Guidelines for drinking water quality-I, recommendations, 2nd edn. World health organization, Geneva

8

Estimate of regional groundwater recharge rate in the Central Haouz Plain, Morocco, using the chloride mass balance method and a geographical information system

Ouassil Ait El Mekki[1] **· Nour-Eddine Laftouhi**[1] **· Lahoucine Hanich**[2]

Abstract Located in the extreme northwest of Africa, the Kingdom of Morocco is increasingly affected by drought. Much of the country is characterised by an arid to semi-arid climate and the demand for water is considerably higher than the supply, particularly on the Haouz Plain in the centre of the country. The expansion of agriculture and tourism, in addition to industrial development and mining, have exacerbated the stress on water supplies resulting in drought. It is therefore necessary to adopt careful management practices to preserve the sustainability of the water resources in this region. The aquifer recharge rate in the piedmont region that links the High Atlas and the Central Haouz Plain was estimated using the chloride mass balance hydrochemical method, which is based on the relationship between the chloride concentrations in groundwater and rainwater. The addition of a geographical information system made it possible to estimate the recharge rate over the whole 400 km^2 of the study area. The results are presented in the form of a map showing the spatialized recharge rate, which ranges from 13 to 100 mm/year and the recharge percentage of the total rainfall varies from 3 to 25 % for the hydrological year 2011–2012. This approach will enable the validation of empirical models covering areas >6200 km^2, such as the Haouz nappe.

✉ Ouassil Ait El Mekki
o.aitelmekki@uca.ma

[1] Department of Earth Sciences, Faculty of Sciences Semlalia, Cadi Ayyad University, BP 2390, Marrakesh, Morocco

[2] Department of Earth Sciences, Faculty of Sciences and Techniques Marrakech, Cadi Ayyad University, BP 549, Marrakesh, Morocco

Keywords Groundwater recharge · Semi-arid regions · Chloride mass-balance method (CMB) · Geographic information system (GIS) · Haouz Plain (Morocco)

Introduction

Water is a limiting factor in semi-arid zones affected by repeated droughts. The evaluation and optimum management of this resource are therefore essential in regions such as the Haouz Plain, Morocco. The water supplies in this region are stressed as a result of a high and unconstrained demand, complicated by problems with water quality associated with the development of industry and mining, together with pollution from agriculture and urbanization.

The Mio-Plio-Quaternary nappe of the Haouz Plain provides more than 400 million m^3 of water and this water is used to irrigate an area covering >200,000 ha. In years of drought, the volume of water required for irrigation is estimated to be >1300 million m^3/year (Abourida et al. 2008). It is therefore very important to have a good understanding of the recharge rate of aquifers. Several different methods can be used to determine this recharge rate, including hydrochemical approaches. Hydrochemisty is becoming increasingly important in solving problems in hydrology and hydrogeology (Eriksson 1985; Adelana and MacDonald 2008) and several researchers have used hydrochemical (Allison et al. 1985; Adar et al. 1988; Adar and Neuman 1988; Al-Bassam and Al-Rumikhani 2003; Stigter et al. 2006; Fernandes et al. 2009; Hagedorn 2015) and geochemical (Holland 1972; Dissanayake 1991; Glynn and Plummer 2005; Xiao et al. 2012) approaches to address issues related to water management. This study used a hydrochemical method to calculate the recharge rate of aquifers in the Haouz Plain, Morocco.

The aquifer recharge rate is a crucial piece of data in hydrogeology and is key to the improved understanding and management of water resources. The elements used are observations of the behaviour, in the groundwater, of hydrochemical tracers, and major and minor ions subject to very sophisticated analyses.

The objective of this work was to estimate the infiltration recharge rate using the chloride mass balance (CMB) method developed by Eriksson and Khunakasem (1969), also called the chloride budget technique. This method has been successfully applied to aquifers in every continent, including: Africa (Edmunds et al. 1988; Takounjou et al. 2010; Diouf et al. 2013), Asia (Ting et al. 1998; Subyani 2004; Liu et al. 2009; Marei et al. 2010; Huang and Pang 2010), Europe (Lo Russo et al. 2003; Alcalá and Custodio 2014), Australia (Allison and Hughes 1978; Guan et al. 2009) and America (Sophocleous 1991; Scanlon 1991; Murphy et al. 1996; Nolan et al. 2007). It can be used in both the saturated and unsaturated zones (Edmunds et al. 2002) and is applicable in arid (Gee and Hillel 1988; Fouty 1989), semi-arid (Wood et al. 1997; Subyani and Şen 2006; Rödiger et al. 2014) and humid climates (Saghravani et al. 2014). However, it has not yet been used in Morocco.

Wood (2014) has highlighted the point that this method in the very specific case of a steady-state, chemically homogeneous and isotropic aquifer with no other sources of chloride other than precipitation. With a constant spatial and temporal concentration of chloride anions present in precipitation, a single sample of groundwater would therefore be sufficient to estimate the overall recharge rate of the whole aquifer. However, with a heterogeneous aquifer, which is the case in almost all natural aquifers, the density of sampling is critical in achieving an unbiased estimate of the average chloride concentration and the average recharge flow-rate. The representativeness of these concentrations then becomes an important issue.

The novelty of the work reported here lies in the use of geomatics through the development of a geographical information system (GIS) that spatializes variables by interpolating data samples. Each point represents a specific data pair, which makes it possible to apply the relation while retaining the reliability of the input data. The study area is a piedmont zone of the Central Haouz Plain. It has a semi-arid climate and contains the sub-basins of the Rheraya, Issil and Ourika rivers.

The principal and major objective of this study was to estimate the recharge of this piedmont region. A broader aim is that the results can be used to validate large-scale empirical models of areas exceeding 6200 km^2, such as the Haouz nappe.

Materials and methods

Study area

The study area is located between 7°46′42″W and 7°57′55″W and 31°19′10″N and 31°31′10″N (Fig. 1b). It is a piedmont zone covering 400 km^2. Most of the region is made up of Quaternary alluvium, Permo-Triassic sandstone formations, marly limestone continental facies Senonian age, with rocky outcrops to the south of the region and rugged areas to the southeast (Fig. 1c). (Piqué et al. 2007). There is no volcanic activity in the region. The distance from the western perimeter of the study area and the Atlantic Ocean is 200 km (Fig. 1a, b). There is no contamination by these waters from this effect.

The altitude of the piedmont ranges from 683 to 1358 m. This zone is crossed by the Ourika and Rheraya rivers and is the source of the Issil. Most of this hydrographic system is dry (Fig. 2), except during snowmelt or following exceptional rainfall. The area is characterised by a semi-arid climate. The average monthly temperatures over a 30-year period have varied between 18.5 and 20.5 °C; the coldest month is January (11–13 °C) and the hottest months are July and August (25–27 °C) (courtesy of ABHT: Tensift Basin Agency, Marrakesh, Morocco).

In the south of the region, near the rivers, the elevation of the potentiometric surface is 3 m; this increases to up to 60 m at points in the north distant from any rivers (Ait El Mekki 2010). According to the piezometric map (Fig. 1c) there is no interaction between surface water and groundwater at the level of the three rivers for the hydrologic year 2011–2012. The generalized water table is important both in terms of volume and function. It is made up of Mio-Plio-Quaternary alluvium in the form of detritus resulting from the erosion of the Atlas chain by the various rivers.

Chloride mass balance method

The CMB method uses chloride anions to calculate the recharge rate. The chloride anions are derived from both surface waters and groundwater, and from precipitation.

The method, developed by Eriksson and Khunakasem (1969), has been described by many researchers (Dettinger 1989; Wood and Sanford 1995; Bazuhair and Wood 1996; Wood 1999; Subyani 2004; Gee et al. 2005; Somaratne and Smettem 2014) and has been implemented in various climates and on all continents. It is based on the assumption that the only source of chloride anions in an unsaturated zone is precipitation. As few plants can use this anion, the concentrations of chloride increase in the soil until the aquifer is reached. The recharge rate therefore decreases as the chloride concentration increases, i.e. the higher the soil concentration, the lower the recharge rate.

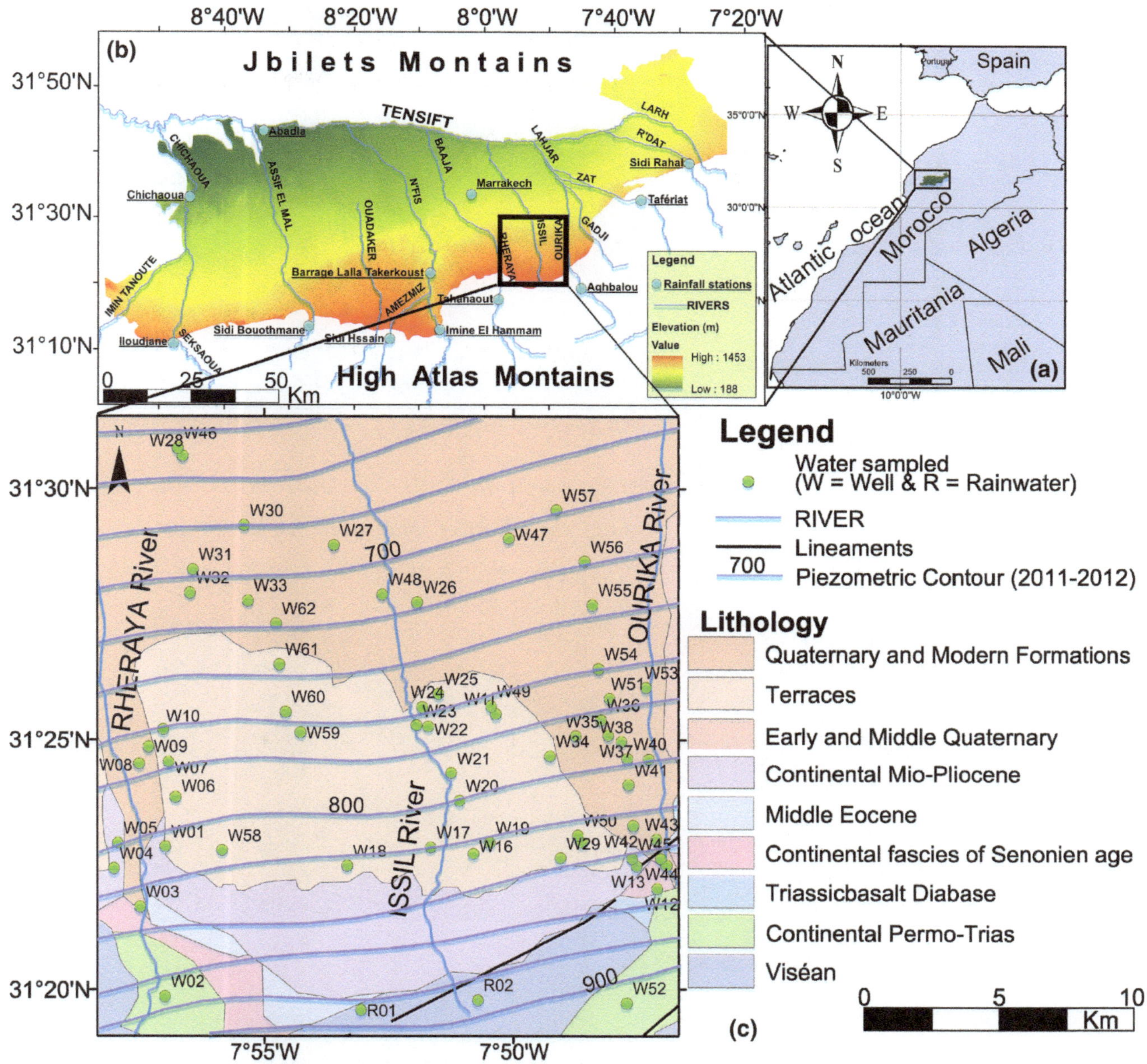

Fig. 1 Map of the study area

The conditions for the use of this method are as follows:

- there are sufficient chloride anions present in precipitation;
- there is no other source of chloride in the ground or in the aquifer (e.g. from urban or industrial waste, mining or agriculture);
- the chloride concentrations in rainwater are stable (i.e. there is no acid rain).

The recharge from rainwater can be calculated using (Eq. 1):

$$R = \frac{P \times C_p}{C_i} - \frac{Q \times C_q}{C_i} \quad (1)$$

where R the effective recharge rate ($L^3\ T^{-1}$), P the annual precipitation ($L^3\ T^{-1}$), C_p the chloride concentration in rainwater (ML^{-3}), C_q the average chloride concentration in runoff (ML^{-3}), C_i the concentration of groundwater chlorides (ML^{-3}) and Q the average runoff ($L^3\ T^{-1}$).

In general, in this type of semi-arid climate, the runoff tends to evaporate or to infiltrate downstream of the mountains. It can therefore be considered to be negligible (Dettinger 1989) and Eq. (1) simplifies to:

$$R = P \times \frac{C_P}{C_i} \quad (2)$$

Fig. 2 Photos of rivers: Issil, Rheraya and Ourika

Determination of variables

Determination of annual effective precipitation

The effective daily rainfall was measured for the hydrological cycle 2011–2012 using data from 12 rainfall stations distributed throughout the Haouz Plain (Fig. 1b).

Determination of concentrations

Rainwater was sampled twice, in November and December 2011. Groundwater was sampled from 62 wells evenly distributed over the study area in a piezometric survey carried out in January (the middle of the hydrological year 2011–2012 and a period that coincides with the start of high water levels) (Fig. 1c).

Physical parameters (pH, conductivity and temperature) were measured on-site using a Yokogawa analyzer. Laboratory samples were collected using a rigorous protocol. Two samples were collected at a time: the first was stored in 500-ml polyethylene bottles for the determination of hydrogen carbonate anions and the second was placed in 60-ml tinted glass vials for analysis by ion-exchange chromatography. All the samples were stored in the field in an ice-box and then in a laboratory refrigerator at 3–4 °C until analysis.

The chloride anions were determined by ion-exchange chromatography using a Dionex ICS-1100 column. These analyses were carried out at the Analysis and Characterization Centre at the Faculty of Sciences Semlalia; Cadi Ayyad University (CAC, FSSM, UCA Marrakesh, Morocco).

Spatialization of rainfall and chloride concentrations

Several researchers (Somaratne and Smettem 2014) have confirmed the validity of the CMB method used punctually or in a chemically stable homogeneous aquifer. However, natural aquifers are heterogeneous, with wide chemical variations. Therefore, the method is not considered to be applicable on a regional scale, which questions the usefulness of Eq. (2). However, if the three parameters of Eq. (2) are spatialized, it is then possible to apply the method over large areas. In the recharge map that is created, each pixel represents a single measurement of rainfall and chloride concentration.

The various data points were transformed into isovalue data layers through inverse distance weighting (IDW) interpolation (Lu and Wong 2008). The data layer for the 2011–2012 rainfall distribution was obtained through IDW interpolation of data from the 12 rainfall stations (courtesy of ABHT). The groundwater chloride layer was constructed using data from 62 wells. ArcGIS software (version 10.2) was used to develop the GIS. Eqn (2) was calculated using the spatial analyst tools module based on spatialized variables in raster format data layers.

Results

Precipitation

The data from the 12 stations show an overall average rainfall of about 317 mm/year (Fig. 1). The lowest value was at the Abadla station (173 mm/year) and the maximum was at the Tahanaout station (519 mm/year). In the overall study area, the minimum and maximum values were 281 and 512 mm/year, respectively. The spatial IDW interpolation makes it possible to map the rainfall distribution (Fig. 3). The map in Fig. 3 corresponds to the variable P in Eq. (2).

Chloride concentration (rainwater and groundwater)

Table 1 gives the results obtained for the measurement of the chloride concentrations. The November and December rainwater concentrations are both very close to 11.5 mg L^{-1} (C_p in Eq. 2). The groundwater concentration varies from 14 to 361 mg L^{-1}. The interpolation can be used to produce the concentration distribution map (Fig. 4). The map in Fig. 4 corresponds to the variable C_i in Eq. (2).

Design of the regional map

Following the preparation of the spatialized variables in data layers in raster format, the application of Eq. (2) led to the development of the regional recharge map. Minimum and maximum values were 12 and 100 mm/year,

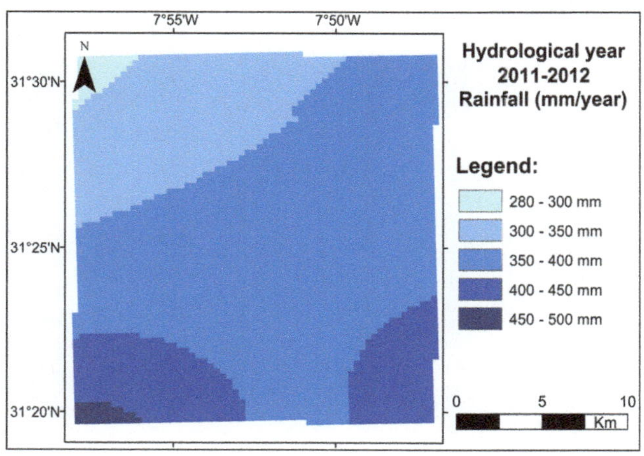

Fig. 3 Rainfall map for the hydrological year 2011–2012 in the Rheraya, Issil and Ourika river sub-basins (in mm/year)

Table 1 Chemical composition of rain and groundwater samples in the study area

No.	Nature	Water type	Date	Longitude[a] (°)	Latitude[a] (°)	Altitude (m)	Cl^- (mg L^{-1})
R01	Rainfall	Rw	11/2011	−7.88400	31.32600	1027	11.5105
R02	Rainfall	Rw	12/2011	−7.84500	31.33200	1009	11.4849
W01	Well	Gw	01/2012	−7.94971	31.38079	880	289.8025
W02	Well	Gw	01/2012	−7.94984	31.33088	980	270.0859
W03	Well	Gw	01/2012	−7.95820	31.36055	906	264.8087
W04	Well	Gw	01/2012	−7.96697	31.37342	904	271.2423
W05	Well	Gw	01/2012	−7.96560	31.38208	851	360.6405
W06	Well	Gw	01/2012	−7.94634	31.39724	841	321.5109
W07	Well	Gw	01/2012	−7.94857	31.40911	855	209.0568
W08	Well	Gw	01/2012	−7.95852	31.40847	796	147.4853
W09	Well	Gw	01/2012	−7.95535	31.41432	805	191.1957
W10	Well	Gw	01/2012	−7.95036	31.42009	810	163.0371
W11	Well	Gw	01/2012	−7.83921	31.42530	734	25.7170
W12	Well	Gw	01/2012	−7.78525	31.36671	906	20.5686
W13	Well	Gw	01/2012	−7.79217	31.37428	882	24.7632
W14	Well	Gw	01/2012	−7.78221	31.37445	871	96.0487
W15	Well	Gw	01/2012	−7.81035	31.38225	875	51.8383
W16	Well	Gw	01/2012	−7.84680	31.37844	852	145.9234
W17	Well	Gw	01/2012	−7.86105	31.38063	836	92.5382
W18	Well	Gw	01/2012	−7.88906	31.37440	882	27.4482
W19	Well	Gw	01/2012	−7.84115	31.38212	837	165.8190
W20	Well	Gw	01/2012	−7.85140	31.39612	786	116.9393
W21	Well	Gw	01/2012	−7.85412	31.40541	798	117.9560
W22	Well	Gw	01/2012	−7.86185	31.42110	750	98.4816
W23	Well	Gw	01/2012	−7.86570	31.42153	732	149.6134
W24	Well	Gw	01/2012	−7.86378	31.42747	788	88.6706
W25	Well	Gw	01/2012	−7.85838	31.43182	722	171.6162
W26	Well	Gw	01/2012	−7.86557	31.46216	677	75.0355
W27	Well	Gw	01/2012	−7.89336	31.48095	665	94.9147
W28	Well	Gw	01/2012	−7.94388	31.51082	622	184.0986
W29	Well	Gw	01/2012	−7.81757	31.37695	872	87.4206
W30	Well	Gw	01/2012	−7.92334	31.48767	646	195.6960
W31	Well	Gw	01/2012	−7.94052	31.47293	656	97.4012
W32	Well	Gw	01/2012	−7.94151	31.46521	673	100.2156
W33	Well	Gw	01/2012	−7.92207	31.46269	670	187.7759
W34	Well	Gw	01/2012	−7.82113	31.41121	772	35.7094
W35	Well	Gw	01/2012	−7.81244	31.41794	764	25.9241
W36	Well	Gw	01/2012	−7.80407	31.42332	761	56.2204
W37	Well	Gw	01/2012	−7.80164	31.41812	777	58.9769
W38	Well	Gw	01/2012	−7.79711	31.41600	784	57.3124
W39	Well	Gw	01/2012	−7.78810	31.41023	787	32.4079
W40	Well	Gw	01/2012	−7.79524	31.41043	792	13.9901
W41	Well	Gw	01/2012	−7.79476	31.40147	817	46.1923
W42	Well	Gw	01/2012	−7.79329	31.38787	847	104.9926
W43	Well	Gw	01/2012	−7.78559	31.38334	852	107.6751
W44	Well	Gw	01/2012	−7.78400	31.37676	868	88.6856
W45	Well	Gw	01/2012	−7.79374	31.37682	874	23.2708
W46	Well	Gw	01/2012	−7.94552	31.51331	600	154.8350

Table 1 continued

No.	Nature	Water type	Date	Longitude[a] (°)	Latitude[a] (°)	Altitude (m)	Cl^- (mg L^{-1})
W47	Well	Gw	01/2012	−7.83488	31.48335	636	81.5666
W48	Well	Gw	01/2012	−7.87715	31.46463	659	47.2504
W49	Well	Gw	01/2012	−7.84076	31.42779	744	18.2450
W50	Well	Gw	01/2012	−7.81193	31.38467	860	38.5559
W51	Well	Gw	01/2012	−7.80113	31.43031	760	71.4605
W52	Well	Gw	01/2012	−7.79545	31.32860	1172	39.3922
W53	Well	Gw	01/2012	−7.78898	31.43396	748	38.4962
W54	Well	Gw	01/2012	−7.80485	31.44003	736	79.0671
W55	Well	Gw	01/2012	−7.80694	31.46126	698	99.9467
W56	Well	Gw	01/2012	−7.80955	31.47558	674	80.0577
W57	Well	Gw	01/2012	−7.81889	31.49285	645	48.1516
W58	Well	Gw	01/2012	−7.93084	31.37957	870	287.6349
W59	Well	Gw	01/2012	−7.90446	31.41916	753	19.0713
W60	Well	Gw	01/2012	−7.90944	31.42594	724	87.4032
W61	Well	Gw	01/2012	−7.91156	31.44152	707	86.4914
W62	Well	Gw	01/2012	−7.91261	31.45505	681	75.8798

Rw rainwater, *Gw* groundwater

[a] Coordinates in decimal degrees (coordinates system: GCS Merchich Degree)

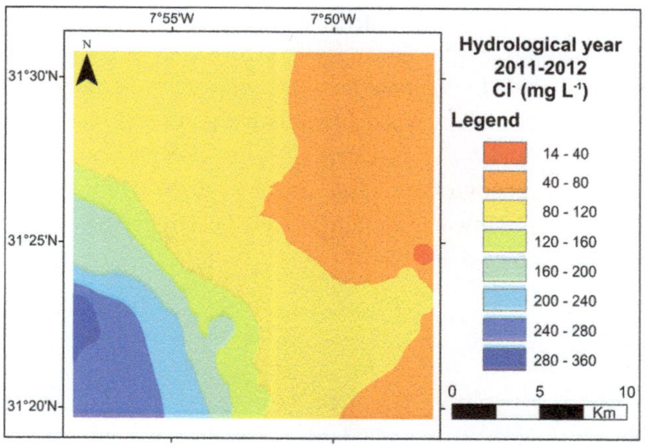

Fig. 4 Distribution of chloride in groundwater (mg L^{-1}) for the hydrological year 2011–2012 in the Rheraya, Issi and Ourika river sub-basins

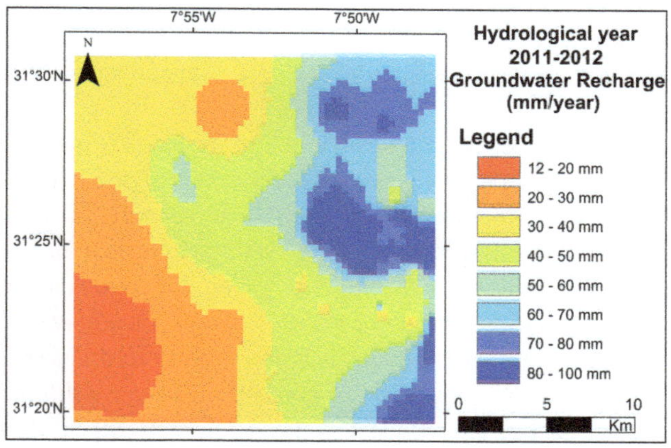

Fig. 5 Spatialized groundwater recharge map for the hydrological year 2011–2012 in the Rheraya, Issil and Ourika river sub-basins (in mm/year)

respectively (Fig. 5), while the percentage of recharge due to rainfall varies from 3 to 25 % (Fig. 6).

Discussion

The choice of study area was based on a number of key criteria. There was no industrial activity, urban discharge, marine intrusion, or volcanic activity, which eliminates any potential contamination from these sources; in addition, there was no agricultural land and therefore no pollution from chemical fertilizers or pesticides. The area was therefore almost entirely free from anthropogenic influences, with the exception of a few, very small, rural populations of shepherds.

The Atlantic coast is at a minimum of 200 km from the sampled wells, which are at an altitude of between 683 and 1358 m. Seawater has no influence.

The climate of the region is semi-arid. Most of this hydrographic system is dry except following exceptional rainfall. Rainfall in the hydrologic year 2011–2012 was not high (<512 mm/year). This intensity is insufficient to create a rise in the water level leading to a concentrated

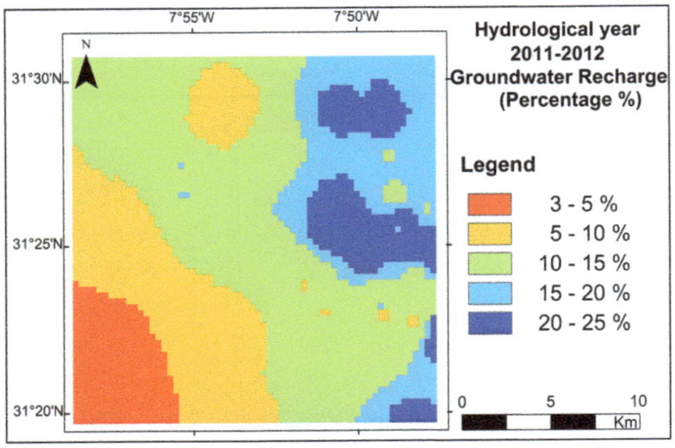

Fig. 6 Spatialized groundwater recharge map for the hydrological year 2011–2012 in the Rheraya, Issil and Ourika river sub-basins (in %)

recharge that would be a second source for the nappe. This observation is confirmed by hydroisohypses lines showing that drainage is generalized throughout the region (Fig. 1c). The latter figure shows that current flows and the drainage axes in the rivers are somewhat parallel. These lines are neither divergent (dismissing the idea of a concentrated recharge at the level of the river), nor convergent (suggesting exfiltration from groundwater). Consequently, there is only diffuse recharge, which is quantized in this study.

97 % of chloride samples are below around 300 mg L^{-1} which is very normal for groundwater in this region. On the other hand, two samples exceed this threshold (321 and 360 mg L^{-1}, respectively). Both wells are located in the extreme southwest of the region, where the context does not favour infiltration; as a result much of the rainwater evaporates and what remains has poor access to the nappe with a higher concentration of chlorides. These conditions concentrate the groundwater in this area and will be discussed later.

According to Wood (2014), the CMB is only effective in estimating the regional recharge rate if the aquifer is both homogeneous and isotropic. However, the addition of a GIS makes it possible to apply the method to a heterogeneous aquifer based on the spatialization of the measured parameters.

Interpolation of the data transforms the point variables into spatial components in raster format as each pixel in the raster has its own value and represents a point recharge rate. It is therefore possible to use Eq. (2). The data from the pixels can be used to create a regional recharge map.

The recharge map created in this way revealed a transition running from east to west. It illustrates very different behaviour in the three basins where the Issil marks a transitional region between the Ourika and Rheraya, and the recharge rates are between 40 and 50 mm/year (10–15 %). The Rheraya basin, located in the east, has very low recharge rates of <20 mm/year (<5 %) in the south, increasing to 30–40 mm/year (10–15 %) towards the north. The Ourika basin sees intense infiltration of 40–100 mm/year (10–25 %); most of its watershed has a recharge rate >80 mm/year (>20 %).

The spatial distribution of the recharge rate is the result of the effect of several different parameters, including the lithology, slope gradient, rock fracturing, land use and soil type. The lithology directly affects the permeability of the rocks and therefore the rate of infiltration. The geology varies in the three basins (Dauteloup 1958; Piqué et al. 2007). The Rheraya basin has a low permeability compared with the rest of the region and consists of continental Permo–Triassic sandstone formations, marly limestone continental facies of Senonian age, and multiple ferro-magnesian rocky outcrops that are almost impermeable. The part of the Ourika basin in the study area is composed of middle Eocene and continental Miocene formations dominated by early-to-middle Quaternary and modern formations, all highly permeable. To the north, up to 80 m of breccia conglomerate promotes infiltration. In addition, the very friable brown soils, which cover almost the entire basin, promote recharge. Apart from its lithological and soil characteristics, lineaments correspond to fault zones and fractures with secondary porosity and high permeability. These lineaments are very important as they promote the passage of water towards the aquifer. Finally, the existence of NE–SW-oriented fractures downstream of the Ourika basin favour recharge as they are perpendicular to the flow of rainwater running from the High Atlas to the adjacent sub-basins.

Conclusion

The development of a recharge rate map for the hydrological year 2011–2012 for the three basins of the Central Haouz Plain revealed three different zones: a zone with very low potential (<20 mm/year); a low recharge zone (30–50 mm/year); and a medium recharge zone (50–100 mm/year). This map of recharge rates over an area of 200 km^2 is a very important tool for the validation of models based on empirical methods. The models can be applied to large areas such as the Haouz Plain (6200 km^2), which has an average width of about 40 km and is oriented N–S with an average length of 150 km in an E–W direction. It remains difficult to use the CMB method over such a large area as a result of the influence of human activity on chloride levels and, in this situation, empirical methods must be used. However, these empirical models must be validated as they provide methods for decision support. This study shows that validation is possible using a direct method that combines prospecting and field surveys supplemented by chemical analyses.

Acknowledgments The authors would like to thank the Analysis and Characterization Centre (CAC, FSSM, UCA, Marrakesh, Morocco), which kindly offered their facilities. We would also like to thank Mrs. Prof. Nouzha Boukhari Hariss and Dr. Layla Elgaini, for their valuable help in the chemical analyses.

References

Abourida A, Vincent S, Errouane S et al (2008) Estimation des volumes d'eau pompés dans la nappe pour l'irrigation (plaine du Haouz, Marrakech, Maroc). Comparaison d'une méthode statistique et d'une méthode basée sur l'utilisation de la télédétection. J Water Sci 22:1–13. doi:10.7202/019820ar

Adar EM, Neuman SP (1988) Estimation of spatial recharge distribution using environmental isotopes and hydrochemical data, II. Application to Aravaipa valley in Southern Arizona, USA. J Hydrol 97:279–302. doi:10.1016/0022-1694(88)90120-5

Adar EM, Neuman SP, Woolhiser DA (1988) Estimation of spatial recharge distribution using environmental isotopes and hydrochemical data, I. Mathematical model and application to synthetic data. J Hydrol 97:251–277. doi:10.1016/0022-1694(88)90119-9

Adelana S, MacDonald A (2008) Applied groundwater studies in Africa: IAH selected papers on hydrogeology. CRC Press, Boca Raton

Ait El Mekki O (2010) Etude de la relation eaux de surface-eaux souterraines au contact de Oued Rheraya avec la plaine du Haouz; Bilan et quantification des apports. Mémoire de 3ème cycle, Université Cadi Ayyad Faculté des Sciences Semlalia

Al-Bassam AM, Al-Rumikhani YA (2003) Integrated hydrochemical method of water quality assessment for irrigation in arid areas: application to the Jilh aquifer, Saudi Arabia. J Afr Earth Sci 36:345–356. doi:10.1016/S0899-5362(03)00046-0

Alcalá FJ, Custodio E (2014) Spatial average aquifer recharge through atmospheric chloride mass balance and its uncertainty in continental Spain. Hydrol Process 28:218–236. doi:10.1002/hyp.9556

Allison GB, Hughes M (1978) The use of environmental chloride and tritium to estimate total recharge to an unconfined aquifer. Soil Res 16:181–195

Allison GB, Stone WJ, Hughes MW (1985) Recharge in karst and dune elements of a semi-arid landscape as indicated by natural isotopes and chloride. J Hydrol 76:1–25. doi:10.1016/0022-1694(85)90088-5

Bazuhair AS, Wood WW (1996) Chloride mass-balance method for estimating ground water recharge in arid areas: examples from western Saudi Arabia. J Hydrol 186:153–159. doi:10.1016/S0022-1694(96)03028-4

Dauteloup J (1958) Carte des roches mères du Haouz de Marrakech. Génie Rural du Sud, Maroc

Dettinger MD (1989) Reconnaissance estimates of natural recharge to desert basins in Nevada, USA, by using chloride-balance calculations. J Hydrol 106:55–78. doi:10.1016/0022-1694(89)90166-2

Diouf OC, Faye SC, Diedhiou M et al (2013) Combined uses of water-table fluctuation (WTF), chloride mass balance (CMB) and environmental isotopes methods to investigate groundwater recharge in the Thiaroye sandy aquifer (Dakar, Senegal). Afr J Environ Sci Technol 6:425–437. doi:10.4314/ajest.v6i11

Dissanayake CB (1991) The fluoride problem in the ground water of Sri Lanka—environmental management and health. Int J Environ Stud 38:137–155. doi:10.1080/00207239108710658

Edmunds WM, Darling WG, Kinniburgh DG (1988) Solute profile techniques for recharge estimation in semi-arid and arid terrain. In: Simmers PDI (ed) Estimation of natural groundwater recharge. Springer, The Netherlands, pp 139–157

Edmunds W, Fellman E, Goni I, Prudhomme C (2002) Spatial and temporal distribution of groundwater recharge in northern Nigeria. Hydrogeol J 10:205–215. doi:10.1007/s10040-001-0179-z

Eriksson E (1985) Principles and Applications of Hydrochemistry, 2012th edn. Springer, Netherlands

Eriksson E, Khunakasem V (1969) Chloride concentration in groundwater, recharge rate and rate of deposition of chloride in the Israel coastal plain. J Hydrol 7:178–197. doi:10.1016/0022-1694(69)90055-9

Fernandes PG, Carreira PM, Bahir M (2009) Mass balance simulation and principal components analysis applied to groundwater resources: essaouira basin (Morocco). Environ Earth Sci 59:1475–1484. doi:10.1007/s12665-009-0133-2

Fouty SC (1989) Chloride mass balance as a method for determining long-term groundwater recharge rates and geomorphic-surface stability in arid and semi-arid regions. Whisky Flat and Beatty, Nevada

Gee GW, Hillel D (1988) Groundwater recharge in arid regions: review and critique of estimation methods. Hydrol Process 2:255–266. doi:10.1002/hyp.3360020306

Gee GW, Zhang ZF, Tyler SW et al (2005) Chloride mass balance cautions in predicting increased recharge rates. Vadose Zone J 4:72–78. doi:10.2113/4.1.72

Glynn PD, Plummer LN (2005) Geochemistry and the understanding of ground-water systems. Hydrogeol J 13:263–287. doi:10.1007/s10040-004-0429-y

Guan H, Love A, Simmons CT et al (2009) Catchment conceptualisation for examining applicability of chloride mass balance method in an area of historical forest clearance. Hydrol Earth Syst Sci Discuss 6:7025–7053

Hagedorn B (2015) Hydrochemical and 14C constraints on groundwater recharge and interbasin flow in an arid watershed: Tule desert, Nevada. J Hydrol 523:297–308. doi:10.1016/j.jhydrol.2015.01.037

Holland HD (1972) The geologic history of sea water—an attempt to solve the problem. Geochim Cosmochim Acta 36:637–651. doi:10.1016/0016-7037(72)90108-1

Huang T, Pang Z (2010) Estimating groundwater recharge following land-use change using chloride mass balance of soil profiles: a case study at Guyuan and Xifeng in the Loess Plateau of China. Hydrogeol J 19:177–186. doi:10.1007/s10040-010-0643-8

Liu G, Wang Y, Zhang Y, Song T (2009) Application of chloride profile and water balance methods in estimating groundwater recharge in Luanjing Irrigation Area, Inner Mongolia/Application des méthodes du profil de chlorure et du bilan hydrique pour l'estimation de la recharge hydrogéologique dans le Périmètre Irrigué de Luanjing, Mongolie Intérieure. Hydrol Sci J 54:961–973. doi:10.1623/hysj.54.5.961

Lo Russo S, Zavattaro L, Acutis M, Zuppi GM (2003) Chloride profile technique to estimate water movement through unsaturated zone in a cropped area in subhumid climate (Po Valley—NW Italy). J Hydrol 270:65–74. doi:10.1016/S0022-1694(02)00278-0

Lu GY, Wong DW (2008) An adaptive inverse-distance weighting spatial interpolation technique. Comput Geosci 34:1044–1055. doi:10.1016/j.cageo.2007.07.010

Marei A, Khayat S, Weise S et al (2010) Estimating groundwater recharge using the chloride mass-balance method in the West Bank, Palestine. Hydrol Sci J 55:780–791. doi:10.1080/02626667.2010.491987

Murphy EM, Ginn TR, Phillips JL (1996) Geochemical estimates of paleorecharge in the Pasco Basin: evaluation of the chloride mass balance technique. Water Resour Res 32:2853–2868. doi:10.1029/96WR01529

Nolan BT, Healy RW, Taber PE et al (2007) Factors influencing ground-water recharge in the eastern United States. J Hydrol 332:187–205. doi:10.1016/j.jhydrol.2006.06.029

Piqué A, Soulaimani A, Hoepffner C et al (2007) Géologie du Maroc, Geode. Imprimerie El Watanya, Marrakesh, Morocco

Rödiger T, Geyer S, Mallast U et al (2014) Multi-response calibration of a conceptual hydrological model in the semiarid catchment of Wadi al Arab, Jordan. J Hydrol 509:193–206. doi:10.1016/j.jhydrol.2013.11.026

Saghravani SR, Yusoff I, Tahir WZWM, Othman Z (2014) Comparison of water table fluctuation and chloride mass balance methods for recharge estimation in a tropical rainforest climate: a case study from Kelantan River catchment, Malaysia. Environ Earth Sci 73:4419–4428. doi:10.1007/s12665-014-3727-2

Scanlon BR (1991) Evaluation of moisture flux from chloride data in desert soils. J Hydrol 128:137–156. doi:10.1016/0022-1694(91)90135-5

Somaratne N, Smettem KRJ (2014) Theory of the generalized chloride mass balance method for recharge estimation in groundwater basins characterised by point and diffuse recharge. Hydrol Earth Syst Sci Discuss 11:307–332. doi:10.5194/hessd-11-307-2014

Sophocleous MA (1991) Combining the soilwater balance and water-level fluctuation methods to estimate natural groundwater recharge: practical aspects. J Hydrol 124:229–241. doi:10.1016/0022-1694(91)90016-B

Stigter TY, Carvalho Dill AMM, Ribeiro L, Reis E (2006) Impact of the shift from groundwater to surface water irrigation on aquifer dynamics and hydrochemistry in a semi-arid region in the south of Portugal. Agric Water Manag 85:121–132. doi:10.1016/j.agwat.2006.04.004

Subyani AM (2004) Use of chloride-mass balance and environmental isotopes for evaluation of groundwater recharge in the alluvial aquifer, Wadi Tharad, western Saudi Arabia. Environ Geol 46:741–749. doi:10.1007/s00254-004-1096-y

Subyani A, Şen Z (2006) Refined chloride mass-balance method and its application in Saudi Arabia. Hydrol Process 20:4373–4380. doi:10.1002/hyp.6172

Takounjou AF, Ngoupayou JRN, Riotte J et al (2010) Estimation of groundwater recharge of shallow aquifer on humid environment in Yaounde, Cameroon using hybrid water-fluctuation and hydrochemistry methods. Environ Earth Sci 64:107–118. doi:10.1007/s12665-010-0822-x

Ting C-S, Kerh T, Liao C-J (1998) Estimation of groundwater recharge using the chloride mass-balance method, Pingtung Plain, Taiwan. Hydrogeol J 6:282–292. doi:10.1007/s100400050151

Wood WW (1999) Use and misuse of the chloride-mass balance method in estimating ground water recharge. Ground Water 37:2–3. doi:10.1111/j.1745-6584.1999.tb00949.x

Wood WW (2014) Interactive comment on "Theory of the generalized chloride mass balance method for recharge estimation in groundwater basins characterised by point and diffuse recharge" by N. Somaratne and K. R. J. Smettem. Hydrol Earth Syst Sci Discuss 11:C19–C21

Wood WW, Sanford WE (1995) Chemical and isotopic methods for quantifying ground-water recharge in a regional, semiarid environment. Ground Water 33:458–468. doi:10.1111/j.1745-6584.1995.tb00302.x

Wood WW, Rainwater KA, Thompson DB (1997) Quantifying macropore recharge: examples from a semi-arid area. Ground Water 35:1097–1106. doi:10.1111/j.1745-6584.1997.tb00182.x

Xiao J, Jin Z-D, Ding H et al (2012) Geochemistry and solute sources of surface waters of the Tarim River Basin in the extreme arid region, NW Tibetan Plateau. J Asian Earth Sci 54–55:162–173. doi:10.1016/j.jseaes.2012.04.009

9

Fluoride occurrence in the groundwater in a coastal region of Andhra Pradesh, India

N. Subba Rao[1] · P. Surya Rao[1] · A. Dinakar[1] · P. V. Nageswara Rao[2] · Deepali Marghade[3]

Abstract Fluoride (F^-) content varies from 0.60 to 1.80 mg/L in the coastal region between Chirala and Ongole of Andhra Pradesh, India. It exceeds the threshold limit of 1.20 mg/L in 20 % of the total groundwater samples. The aim of the present study is to assess the controlling factors of F^- content. The study area experiences a dry climate and is underlain by Charnockite Group of rocks over which the river and coastal alluvium occur. The results of the study identify the four factors that control the high F^- content. First one is related to alkalinity, leading to active dissolution and leaching of F^--bearing minerals, which supports the positive correlation of F^- with pH and HCO_3^-. A longer water residence time in the clays is the second factor, which activates not only solubility and dissolution of F^--bearing minerals, but also anion exchange between F^- and OH^-. Third factor is a result of higher Na^+ due to impact of saline water, ion exchange between Na^+ and Ca^{2+}, and precipitation of $CaCO_3$. This reduces the Ca^{2+} content, causing dissolution of CaF_2 to maintain the chemical equilibria, which is supported by positive correlation between Na^{2+} and F^-. The influence of anthropogenic activities is the last factor, which acts as an additional source of F^-. Thus, the shallow groundwater shows higher content of F^- and the hydrogeochemical facies also support this hypothesis. The study suggests the remedial measures to reduce the F^- content.

Keywords Fluoride · Groundwater · Coastal region · Prakasam district · Andhra Pradesh · India

Introduction

Groundwater is a major source for drinking as well as for agricultural and industrial use. In India, the people suffer mainly from two health disorders due to occurrences of fluoride (F^-) and arsenic (As) in drinking water. Relatively, the health implications caused by F^--contamination are far more wide spread than those of As contamination in the country (Subba Rao 2011). Approximately, 62 million people, including 6 million children suffer from fluorosis in the country due to consumption of F^--contaminated water (Susheela 1999). The endemic fluorosis is prevalent in many states, including Andhra Pradesh, Bihar, Delhi, Gujarat, Haryana, Karnataka, Madhya Pradesh, Maharastra, Odisha, Punjab, Rajasthan, Tamil Nadu, Telangana (a part of erstwhile Andhra Pradesh) and Uttar Pradesh in the country.

Many investigations on geochemistry of F^--bearing groundwater have been carried out in different parts of the world. A detailed study on geochemistry and genesis of F^--containing groundwater in different geological formations of India has been done by Handa (1975), Sahu and Karim (1989), Jacks et al. (1993, 2005), Datta et al. (1996), Agrawal et al. (1997), Gupta et al. (1999, 2005), Das et al. (2003), Chakraborti et al. (2000), Bhosle et al. (2001), Ray et al. (2000), Madhavan and Subramanian (2002), Muraliddharan et al. (2002), Ayoob and Gupta (2006), Misra and Mishra (2006), Kundu and Mandal (2009a, b), Raju

✉ N. Subba Rao
srnandipati@gmail.com

1 Department of Geology, Andhra University, Visakhapatnam 530 003, India

2 Department of Geology, Acharya Nagarjuna University, Nagarjunanagar, Guntur 522 510, India

3 Priyadarshini Institute of Engineering and Technology, Nagpur 440 019, India

et al. (2009) and Vikas et al. (2009). Jha and Jha (1982) explain the natural sources as responsible for F^- contamination in India that causes serious health problems. Ramesam and Rajagopalan (1985) report that the degree of weathering and the leachable F^- in a hard rock terrain is more important in deciding F^- content in the water rather than the mere presence of F^--bearing minerals in the bulk rocks/soils. Fluoride-bearing minerals (fluorite, apatite, biotite, hornblende, etc.) present in hard rock terrains (basalts, granites, gneisses, schists, charnockites, khondalites, etc.) and anthropogenic sources (chemical fertilisers, sewage, sludge, etc.) in India are mainly responsible for higher content of F^- in groundwater with a combination of higher rate of evaporation and longer interaction of water with aquifer materials under alkaline environment (Ramamohana Rao et al. 1993; Wodeyar and Sreenivasan 1996; Chatterjee and Mohabey 1998; Subba Rao et al. 1998a, b, 2012; Sumalatha et al. 1999; Kundu et al. 2001; Saxena and Ahmed 2003; Subba Rao 2003, 2009, 2011; Subba Rao and John Devadas 2003; Subba Rao and Rao 2003; Sujatha 2003; Bhagavan and Raghu 2005; Sreedevi et al. 2006; Madhnure et al. 2007; Shaji et al. 2007; Srinivasamoorthy et al. 2008, 2012; Ramachandran et al. 2012; Reddy et al. 2010; Deepu and Shaji 2011; Chidambaram et al. 2013; Manikandan et al. 2014). All these studies state that high F^- and very low Ca^{2+} and Mg^{2+} or inverse relation between F^- and Ca^{2+} or Mg^{2+} in water are a result of higher Na^+ (due to impact of saline water), prior precipitation of $CaCO_3$ from water and a limited incorporation of F^- in $CaCO_3$ structure so that there is always a net balance of F^- in solution. They also discuss the role of clays that facilitate active solubility and dissolution of F^--containing minerals, and also activate anion exchange between F^- and OH^-. Further, they also explain that high F^- groundwater is associated with high HCO_3^-, Na^+, K^+, SO_4^{2-} and NO_3^- ions. Gao et al. (2007, 2013) point out the processes of F^- activity reduction and the change of ratio Na^+:Cl^- in the groundwater induced by intrusion of saline water, which favours dissolution of F^--bearing minerals at the salt lake area of Yuncheng basin, Northern China. Singaraja et al. (2013), Subba Rao et al. (2014a, b) and Surya Rao et al. (2015) state that the higher Na^+ and pH causing factor for release of F^- into the groundwater is due to the influence of saline water in coastal aquifers of South India.

In Andhra Pradesh, the highest F^--affected villages are approximately 195 in Prakasam district (Eenadu 2013). The first case of endemic fluorosis in the country was reported as long as 1937 in the district (a part of erstwhile Guntur district; Shortt et al. 1937). There are so many unnoticed F^--contaminated groundwater areas in the district, in which the coastal region of Chirala-Ongole is no exception (Fig. 1). The controlling factors of F^- content in the groundwater are not uniform in any area. Thus, the coastal region between Chirala and Ongole is chosen for the present study to assess the controlling factors of F^--bearing groundwater and also to suggest the suitable remedies to reduce the F^--content in the drinking water before one can act intelligently towards the aim of improvement in water quality as well as in health conditions.

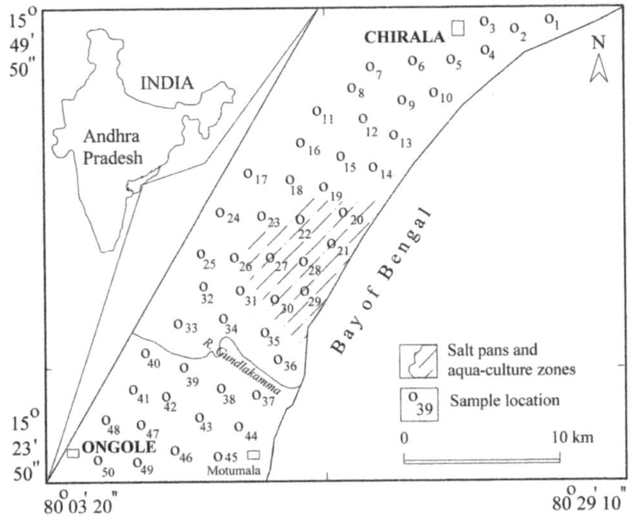

Fig. 1 Location of the coastal region between Chirala and Ongole, Prakasam District, Andhra Pradesh, India

Study area

Location

The study area lies in between 80°29′10″–80°03′20″ East and 15°23′50″–15°49′50″ North with a linear patch extending for about 10 km from the sea coast and covering nearly 40 km in the north–south direction along the coast (Fig. 1). It covers an area of 400 km^2. The area experiences a dry climate, with an annual average temperature of 19.7° to 41.7 °C and receives an annual rainfall of about 900 mm. The southwest monsoon (June to September) contributes 44 % (396 mm), the northeast monsoon (October to December) 50 % (450 mm) and the non-monsoon (January to May) the remaining 6 % (54 mm) of the total rainfall. Cyclones/depressions are not uncommon. The river *Gundalakamma* drains the area (Fig. 2).

Geology

The slope of the study area is gentle towards the southeast. The soils are of silty clay, sand and black cotton types. The sand is dominant in the northern part. The proportion of clay content (silty clay and black cotton soils) increases towards the southern part from the northern part. The occurrence of *kankar* (concretion of $CaCO_3$) in the soil zone is a characteristic feature of dry climate. The southern part of the study area is underlain by Charnockite Group of rocks of the Archaean and the rest of the part by the river and coastal alluvium of the Recent (Fig. 3). Apatite, hornblende and biotite are the main F^--bearing minerals in the host rocks, which occur in and around Ongole.

The sub-surface geology reveals fine, fine to medium and medium sand at depths of less than 0.50, 0.50–16 and 16–22 m from the ground surface, respectively, at Chirala; and fine sand with clay from a depth of less than 3 m, followed by fine to medium sand with clay, quartzite pieces with clay and *kankar*, gravel with sand and clay, and clay with fine sand from depths of 3–10, 10–34, 34–64 and 64–174 m from the surface, respectively, at Motumala (CGWB 2002). As per the well owners, the depth of weathering varies from 1 to 5 m and fracturing extends up to depth of 30 m, which are not uniform in their horizontal and vertical variation.

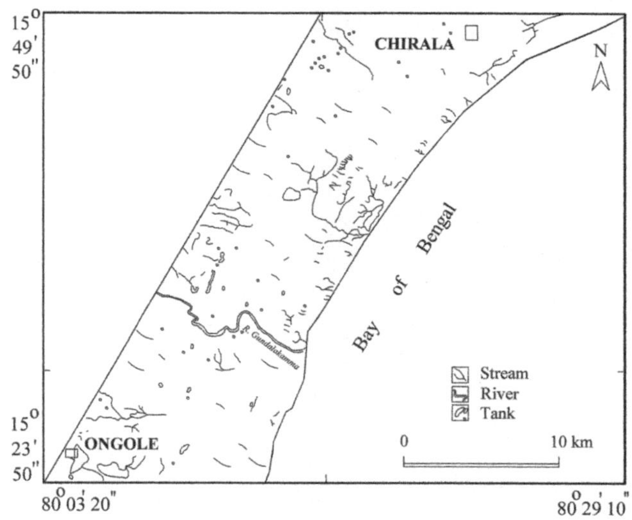

Fig. 2 Drainage of the study area

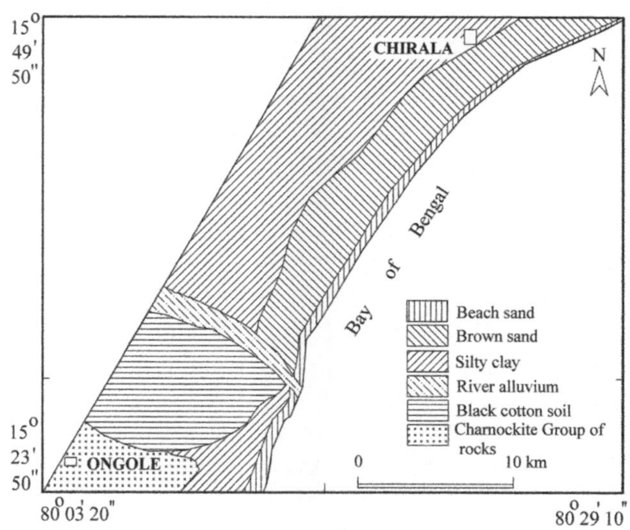

Fig. 3 Geology map of the study area

Hydrogeology

Groundwater is used for drinking and irrigation purposes in the study area. It occurs under unconfined and semi-confined to confined conditions. The water level varies from 7.50 to 13.60 m below ground level (bgl) and the direction of water flow being towards the eastern side (Fig. 4). The occurrence of sand in the northern part develops a good drainage conditions by virtue of its high hydraulic conductivity, while in the southern part the content of clay increases gradually from the northern part, which forms a sluggish drainage condition due to its low hydraulic conductivity. In addition to this, the presence of rocks in the southern part has low hydraulic conductivity, because of limited weathered and fractured portions, which form a poor drainage conditions. The yield is from 215 to 300 m^3/day in filter points/tube wells observed from the area of river alluvium, while it is from 40 to 100 m^3/day in bore wells in the area of rock terrain (CGWB 2013). The transmissivity is in the range of 30–1400 m^2/day in the former area, while it is 1–12 m^2/day in the latter area. It indicates that the movement of groundwater flow is slow in the rock terrain.

Sources of pollution

Dry climate and the occurrence of clay in the soil zone as well as in the sub-surface at different depths cause salinity (which occurs naturally in the geological formations) in the

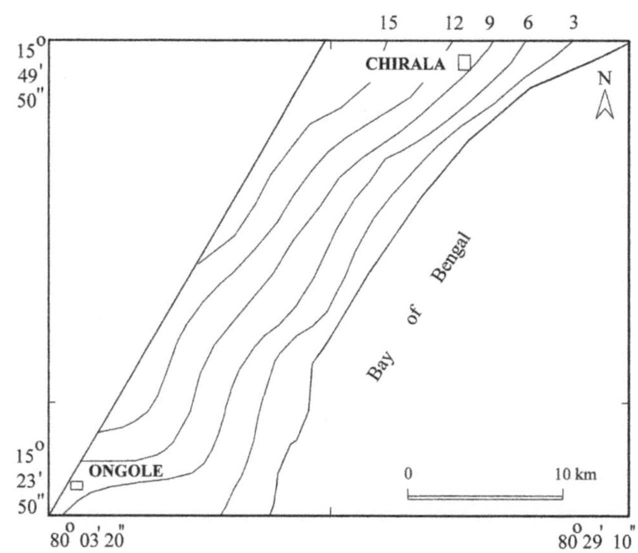

Fig. 4 Depth to groundwater level (m bgl) in the study area

study area (Subba Rao 2014; Subba Rao et al. 2014a, b). Irrigation is intensive and long-term practice, leading to groundwater contamination due to irrigation return flows. Applications of agricultural fertilisers like nitrogen, phosphate and potash varieties, and soil amendments (gypsum) are unlimited for higher crop growth of tobacco, paddy, cotton, groundnut, pulses, etc., causing a deterioration of groundwater quality. Salt pans and aqua-cultural activities are dominant in the central part of the study area (Fig. 1), which are the sources of salinity in the groundwater. Sanitary facilities, such as dumping of domestic wastes and leakage of septic tanks, are poor. They also cause groundwater pollution.

Materials and methods

Water samples from 50 open dug wells were collected in half-litre polythene bottles, after the bottles were soaked in 1:1 HCl for 24 h and were rinsed with distilled water followed by deionised water, during summer 2011 (Fig. 1). The bottles were tightly capped without air gap. Immediately, after collection of the water, the samples are shifted to laboratory for the chemical analysis, following the standard water quality procedures (APHA 1999).

The values of pH and electrical conductivity (EC) were measured in the field, using their portable meters (Systronics make). The EC was used to calculate the total dissolved solids (TDS), following the procedure suggested by Hem (1991). Total hardness (TH) as $CaCO_3$ and calcium (Ca^{2+}) were estimated by volumetric titrations, using standard EDTA. Bicarbonate (HCO_3^-) and chloride (Cl^-) were analysed volumetrically, using standard HCl and $AgNO_3$, respectively. Magnesium (Mg^{2+}) was computed, taking the difference between TH and Ca^{2+}. Sodium (Na^+) and potassium (K^+) were analysed using flame photometer. Sulphate (SO_4^{2-}), nitrate (NO_3^-) and fluoride (F^-), using turbidimetric, colorimetric and ion-selective electrode methods, respectively. All chemical variables (except pH) are expressed in milligrams per litre (mg/L) and also in milliequivalents per litre (me/L).

The analytical precision for the chemical variables was determined by the ionic balance computed between cations (Ca^{2+}, Mg^{2+}, Na^+ and K^+) and anions (HCO_3^-, Cl^-, SO_4^{2-}, NO_3^- and F^-) as $100 \times (\text{cations} - \text{anions})/(\text{cations} + \text{anions})$. This is within the acceptable limit of $\pm$ 5 % (Deutsch 1997).

Results and discussion

The pH varies from 7.2 to 8.2, being an average of 7.81 (Table 1), indicating an alkaline nature of groundwater. The TDS ranges from 850 to 3380 mg/L and its average is 1745.80 mg/L. The groundwater belongs to fresh (TDS <1000 mg/L) in 22 % and brackish (TDS 1000–10,000 mg/L; Fetter 1990) in 78 % of the total samples. The concentrations (mg/L) of Ca^{2+}, Mg^{2+}, Na^+, K^+, HCO_3^-, Cl^-, SO_4^{2-}, NO_3^- and F^- vary from 30 to 75, 40–180, 165–746, 8–110, 310–550, 180–1325, 12–232, 39–123 and 0.60–1.80 with an average of 48.60, 80.90, 399.12, 29.34, 429.00, 623.30, 88.40, 51.84 and 0.98, respectively.

As per the TDS classification (Table 2; Fig. 4), two hydrogeochemical facies, viz., $Na^+ > Mg^{2+} > Ca^{2+}$: $HCO_3^- > Cl^- > SO_4^{2-}$ (where the TDS is less than 1000 mg/L) and $Na^+ > Mg^{2+} > Ca^{2+}$: $Cl^- > HCO_3^- > SO_4^{2-}$ (where the TDS is more than 1000 mg/L) are observed from the study area. The facies is useful for determining the flow pattern and origin of chemical histories of groundwater masses. Thus, the pH, Ca^{2+}, Mg^{2+}, Na^+, K^+, HCO_3^-, Cl^-, SO_4^{2-}, NO_3^- and F^- increase with the increase of TDS (Fig. 5).

A higher Na^+ indicates dissolution of minerals and/or soil salts as well as the influence of anthropogenic and marine sources (Todd 1980; Stallard and Edmond 1983; Subba Rao 2002, 2014; Subba Rao et al. 2014a, b). A dominance of HCO_3^- favours mineral dissolution (Stumm and Morgan 1996). An abundance of Cl^- is a result of leaching from soils and the influence of domestic wastes, salt pans, aqua-cultural activities, etc. (Todd 1980; Subba Rao 2014). Further, the solubility of Na^+ and Cl^- is high (Hem 1991; Drever 1997). Sulphate and NO_3^- are the sources of soils, evaporites, fertilisers, domestic wastes, leakage of septic tanks, etc. (Todd 1980; Drever 1997; Subba Rao 2014). Ferromagnesian minerals and manmade activities (salt pans and aqua-cultural activities) are responsible for higher Mg^{2+} than Ca^{2+} in the groundwater (Hem 1991; Drever 1997; Subba Rao 2014; Subba Rao et al. 2014a, b). A lower K^+ indicates its lower geochemical mobility and its absorption on clays (Hem 1991; Drever 1997). Fluoride minerals, agro-chemicals, etc., are the sources of F^- in the groundwater (Ayoob and Gupta 2006; Ahmed 2014; Rao et al. 2014).

In the present study area (Table 1), the concentration of F^- exceeds 1.20 mg/L in 20 % of the total groundwater samples (18, 20–24, 28, 29, 49 and 50). According to ISI (2012) and WHO (2008), the desirable and permissible limits of F^- in drinking water are 1.00 and 1.50 mg/L, respectively. However, the recommended limit of F^- in the water depends on the climatic conditions of an area (USPHS 1987), because the amount of water consumed and the amount of F^- ingested is influenced primarily by air temperature. Thus, the higher the air temperature, the lower is the allowable concentration of F^-. On the other hand, the intake of F^- below 0.50 mg/L causes dental caries (WHO 2008), while it is above 1.00 mg/L resulting

Table 1 Chemical composition of groundwater

S. no.	pH	TDS (mg/L)	Ca^{2+} (mg/L)	Mg^{2+} (mg/L)	Na^+ (mg/L)	K^+ (mg/L)	HCO_3^- (mg/L)	Cl^- (mg/L)	SO_4^{2-} (mg/L)	NO_3^- (mg/L)	F^- (mg/L)
1	7.4	850	30	40	165	12	450	180	12	21	0.6
2	7.5	950	35	55	173	12	430	220	56	29	0.7
3	7.3	930	40	60	165	14	410	210	45	33	0.6
4	8.2	990	45	55	185	14	460	195	56	46	0.8
5	7.6	960	40	50	190	17	440	215	58	48	0.8
6	7.9	980	50	40	205	13	450	200	55	51	0.9
7	8.2	980	40	45	210	9	450	195	61	50	0.9
8	7.9	960	35	40	220	8	440	225	43	49	0.8
9	8.0	970	40	45	200	12	460	190	71	54	0.9
10	7.6	1000	45	50	190	12	430	240	62	44	0.8
11	7.8	1000	45	50	202	11	430	230	46	52	0.7
12	7.9	1230	45	70	230	19	420	360	66	44	0.7
13	7.3	1430	40	70	290	23	390	460	83	56	0.9
14	7.5	1610	40	60	360	24	380	565	86	44	0.8
15	7.2	1490	50	55	342	27	310	535	92	59	0.6
16	7.4	1430	40	60	321	33	360	505	67	29	0.7
17	7.9	1830	40	80	430	29	450	675	84	36	1.0
18	8.2	1870	40	105	631	61	440	1020	136	82	1.3
19	7.6	2150	70	85	482	34	350	810	110	63	0.9
20	8.1	2930	40	115	720	59	480	1140	153	110	1.6
21	8.2	3055	70	200	600	24	440	1325	84	63	1.4
22	8.2	3000	60	165	676	83	520	1065	210	116	1.6
23	8.2	2930	50	100	735	61	520	1165	185	86	1.5
24	8.3	2600	50	120	625	39	540	990	190	62	1.6
25	8.1	2080	40	85	480	58	450	750	90	66	1.0
26	7.6	2665	40	115	610	92	340	1025	210	75	0.9
27	8.2	3120	70	155	690	36	490	1300	120	54	1.2
28	8.2	3190	75	195	733	62	550	1210	189	96	1.8
29	8.2	3380	75	190	746	110	530	1315	232	123	1.7
30	7.6	2670	40	110	670	39	350	1080	230	66	0.9
31	7.4	2535	50	80	710	26	310	1110	94	43	0.8
32	7.5	1950	50	70	475	18	360	780	53	35	0.8
33	7.3	1560	50	60	360	17	350	590	45	47	0.7
34	8.2	1760	50	80	380	23	480	600	49	33	1.0
35	7.9	1820	50	70	429	39	400	640	56	26	1.0
36	8.1	2015	40	85	480	36	410	765	109	49	1.1
37	8.2	1755	60	55	417	19	430	670	59	49	1.0
38	7.4	1690	50	60	400	10	320	640	53	43	0.8
39	7.6	1500	55	60	350	14	370	565	12	46	0.7
40	7.4	1235	35	75	260	28	310	425	68	36	0.6
41	7.4	1230	55	50	265	26	310	380	96	43	0.6
42	7.7	1430	65	80	324	19	390	525	54	46	0.9
43	7.6	1560	75	70	310	11	410	520	110	21	0.8
44	8.0	1690	75	80	394	23	440	630	61	36	1.0
45	7.9	1625	30	55	390	16	380	610	56	51	0.9
46	7.8	1365	60	80	296	18	420	460	49	33	0.8
47	7.8	1300	40	75	295	32	480	370	110	38	0.8

Table 1 continued

S. no.	pH	TDS (mg/L)	Ca^{2+} (mg/L)	Mg^{2+} (mg/L)	Na^{+} (mg/L)	K^{+} (mg/L)	HCO_3^- (mg/L)	Cl^- (mg/L)	SO_4^{2-} (mg/L)	NO_3^- (mg/L)	F^- (mg/L)
48	7.4	1370	30	70	310	16	390	500	33	40	0.8
49	8.2	1240	60	75	285	10	430	365	49	51	1.4
50	8.2	1430	30	50	260	19	550	425	24	19	1.8
Mean	7.81	1745.80	48.60	80.90	399.12	29.34	429.00	623.30	88.40	51.84	0.98
SD	0.34	716.67	12.90	39.52	182.17	22.37	64.17	346.30	56.30	22.89	0.33
CV	4.35	41.05	26.54	48.85	45.64	76.24	14.96	55.56	63.69	44.16	33.67

SD standard deviation, *CV* coefficient of variation

in fluorosis (USPHS 1987). Therefore, the lowest and highest desirable limits of F^- are considered as 0.60 and 1.20 mg/L, respectively, in the present study area.

From Table 1, it is noted that the concentration of F^- (0.60–1.80 mg/L) is not uniform in the study area. For example, the F^- content varies from 0.60 to 0.90 mg/L in 64 % of the groundwater samples (1 to 16, 19, 26, 30–33, 38–43 and 45–48); from 0.90 to 1.20 mg/L in 16 % of the groundwater samples (17, 25, 27, 34–37 and 44); from 1.20 to 1.50 mg/L in 8 % of the groundwater samples (18, 21, 23 and 49); and from 1.50 to 1.80 mg/L in the remaining 12 % of the groundwater samples (20, 22, 24, 28, 29 and 50). It is also observed that the pH, Mg^{2+}, Na^+, K^+, HCO_3^-, Cl^-, SO_4^{2-} and NO_3^- follow a similar trend of F^- (Table 2).

For assessing the factors that control the F^- content in the groundwater, understanding the influences of climate, slope, soil cover, lithology, anthropogenic activity and marine source on spatial distribution of F^- is very essential. Figure 6 shows the spatial distribution of F^- content in the study area. It indicates a progressive increase of F^- from 0.60 to 0.90, 0.90–1.20, 1.20–1.50 and 1.50–1.80 mg/L towards the southern part from the northern part. Significantly, the higher concentration of F^- (1.20–1.80 mg/L) is observed from the central and southern parts.

Dry climate is known to increase both the solubility and the rate of dissolution of soil salts and/or rock minerals. In addition to this, the climate also increases the rate of evaporation. The slope of the study area is gentle, which leads to slow movement of water. It causes a longer contact time of water with the aquifer material, thereby increasing the solubility of ions. The sand aquifer occurs mainly in the northern part, which reflects the higher rate of water circulation due to its high hydraulic conductivity. Thus, it supports a lesser solubility of ions. The increase of clay content from the northern to the southern part tends to slow water movement, because of its low hydraulic conductivity. This not only causes the increase of ions solubility, but also supports anion exchange. The occurrence of the Charnockite Group of rocks in the southern part (especially in and around Ongole) shows low weathered and fractured portions, thereby causing a low hydraulic conductivity. In addition to this, the rocks contain the F^--bearing minerals (apatite, hornblende and biotite). The influences of salt pans and aqua-cultural activity on the groundwater system are dominant in the central part, while the impact of domestic wastes, leakage of septic tanks and agro-chemicals on the groundwater body is predominant in the remaining part. Thus, the soil–rock–water interaction appears as a prime geogenic factor for a source of F^- and the influences of anthropogenic activities as a supplementary factor for further increase of F^- in the groundwater system. This hypothesis is explained below, using binary diagrams, for understanding the factors that control the F^- content in the groundwater in the study area.

The F^- is plotted against pH, HCO_3^- and Ca^{2+} (Figs. 7, 8, 9). The F^- shows a significant positive correlation (r) with pH ($r = 0.75$) and HCO_3^- ($r = 0.64$), while it has a poor positive relation with Ca^{2+} ($r = 0.32$). Further, the pH has a good positive correlation with HCO_3^- ($r = 0.78$; Fig. 10), as the infiltrated water reacts with soil CO_2, which forms H_2O_3, resulting HCO_3^- and thereby causing a pH (Berner and Berner 1987). On the other hand, the incongruent dissolution of rock weathering causes a release of HCO_3^- (Drever 1997). This indicates a prevailing condition of alkalinity (caused by HCO_3^- only here) in the groundwater, which promotes a mineral dissolution (Stumm and Morgan 1996). Further, an easy accessibility of circulating water caused by rainwater and/or intensive and long-term irrigation practices in the soil and/or weathered zone stimulates the alkalinity more effectively (Wodeyar and Sreenivasan 1996; Subba Rao 2003, 2011; Madhnure et al. 2007). Thus, the alkaline water activates the leaching of F^--bearing minerals (apatite, hornblende and biotite) present in the host rocks. As a result, the chemical relations obviously indicate a greater affinity of F^- with HCO_3^- rather than with Ca^{2+}. Further, it appears that the residence time of water in the aquifer zone could be high, because of the gentle slope and the occurrence of clays in the soil zone as well as in the sub-surface at

Table 2 Mean chemical composition of groundwater in different zones

Zone (Figure 1)	TDS range (mg/L)	pH	TDS (mg/L)	Ca^{2+} (mg/L)	Mg^{2+} (mg/L)	Na^+ (mg/L)	K^+ (mg/L)	HCO_3^- (mg/L)	Cl^- (mg/L)	SO_4^{2-} (mg/L)	NO_3^- (mg/L)	F^- (mg/L)	Na^+: Cl^- (mg/L)	HCO_3^- : Ca^{2+} (mg/L)	Water quality
I	<1000	7.76	960.91	40.45	48.18	199.36	12.18	416.67	209.09	51.36	43.36	0.77	1.47	3.38	Fresh
II	1000–1500	7.64	1360.00	46.54	66.92	301.38	21.85	449.00	451.92	61.77	41.54	0.87	1.03	3.17	Very low brackish
III	1500–2000	7.80	1726.67	50.83	70.42	414.67	24.17	419.33	661.67	70.67	41.92	0.90	0.97	2.70	Low brackish
IV	2000–2500	7.93	2081.67	50.00	85.00	480.67	42.67	423.33	775.00	103.00	59.33	1.00	0.96	2.78	Moderate brackish
V	2500–3000	7.91	2761.43	47.14	115.00	678.00	57.00	437.14	1082.14	181.71	79.71	1.27	0.97	3.05	High brackish
VI	>3000	8.20	3186.25	72.50	185.00	692.25	58.00	502.50	1287.50	156.25	84.00	1.53	0.83	2.28	Very high brackish

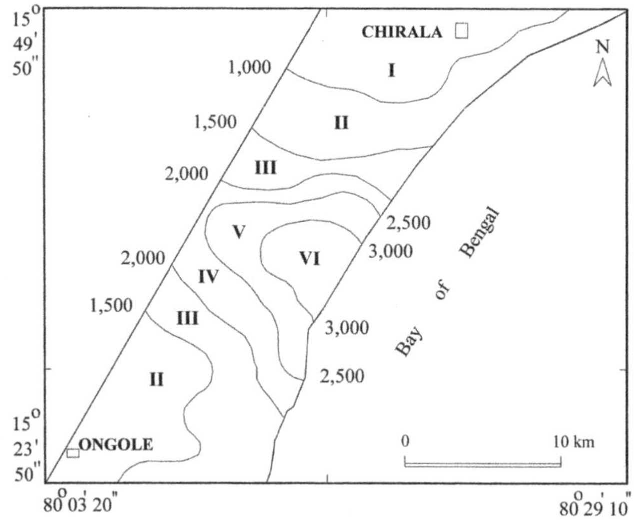

Fig. 5 Spatial distribution of TDS (mg/L) in the study area

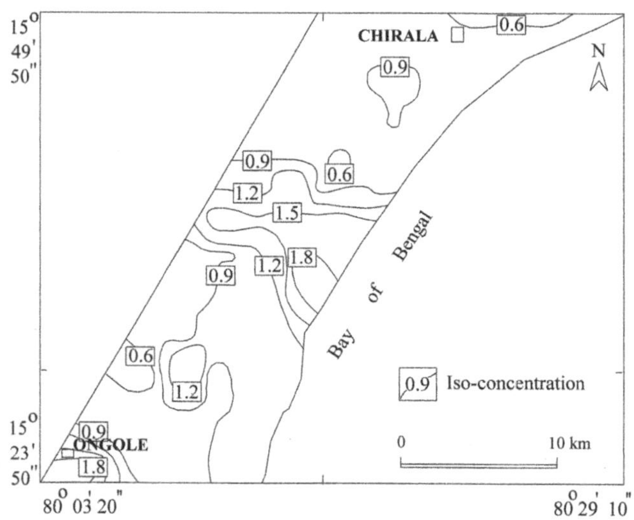

Fig. 6 Spatial distribution of F^- content (mg/L) in the study area

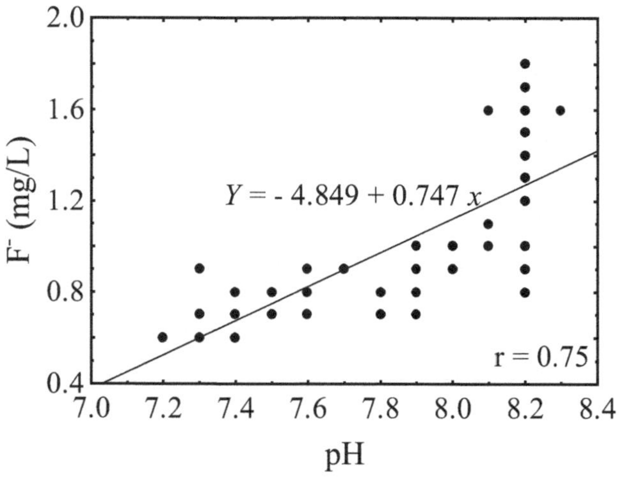

Fig. 7 Correlation between pH and F^-

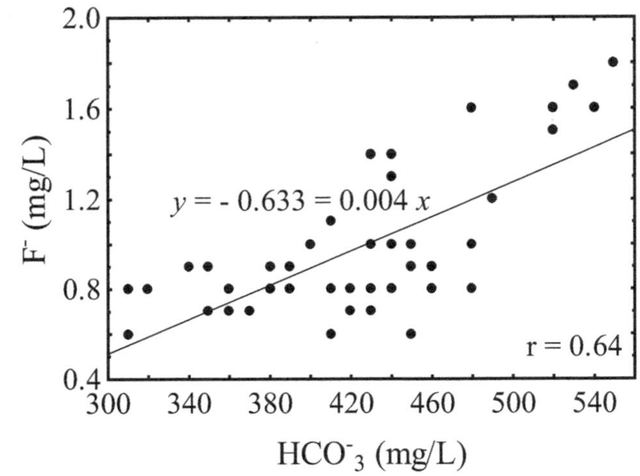

Fig. 8 Correlation between HCO_3^- and F^-

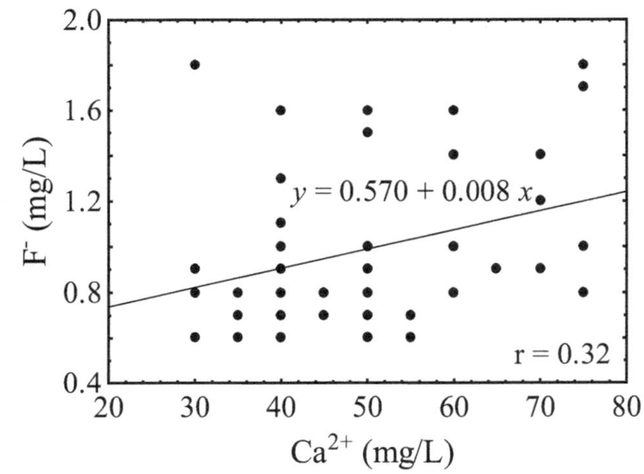

Fig. 9 Correlation between Ca^{2+} and F^-

different depths in the study area. This facilities not only active solubility and dissolution of F^--bearing minerals, but also activate anion exchange between F^- and OH^- (Robinson and Edington 1946; Hem 1991), which favours the increase of F^- content in the groundwater. Therefore, the F^- shows a good correlation with pH and HCO_3^- (Figs. 7, 8) as well as a poor relation with Ca^{2+} (Fig. 9). Similar correlation trends are also observed from other regions, which are discussed in the review literature.

Influences of saline water, chemical activity (ion exchange) and climate (evaporation) also play an important role for increasing F^- content in the groundwater. An insignificant correlation between Na^+ and Ca^{2+} ($r = 0.38$; Fig. 11) is observed from the groundwater due to increase of Na^+ caused by ion exchange between Na^+ and Ca^{2+}, and decrease of Ca^{2+} (Gao et al. 2007, 2013; Singaraja et al. 2013; Subba Rao et al. 2014a, b; Surya Rao et al. 2015). Impacts of saline water (used for salt pans and aquacultural activities), domestic wastes, leakage of septic tanks and clays are responsible for additional source of Na^+ (Todd 1980; Hem 1991; Subba Rao 2014; Subba Rao et al. 2014a, b). On the other hand, the solubility of F^- is low and its dissolution rate is slow (Nordstrom and Jenne 1977). But the solubility of F^- increases with the increase of Na^+ in the groundwater (Apambire et al. 1997; Gao et al. 2007, 2013; Singaraja et al. 2013; Surya Rao et al. 2015). Further, the dry climate increases the rate of solubility and dissolution of minerals present in the host rocks (Hem 1991). It also increases the rate of evaporation, which rises the pH levels due to loss of CO_2 along with water (Handa 1975). This causes a precipitation of $CaCO_3$. The occurrence of *kankar* in the soil zone further supports this hypothesis so that the groundwater in the study area shows an oversaturation with respect to solid phase of $CaCO_3$ (Table 3). This also causes in reduction in Ca^{2+}

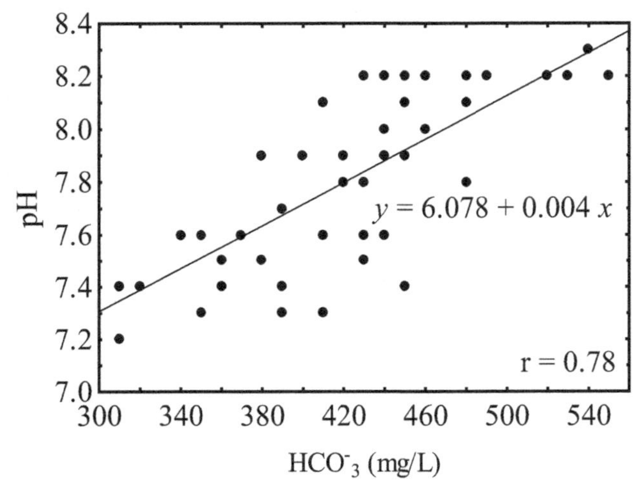

Fig. 10 Correlation between HCO_3^- and pH

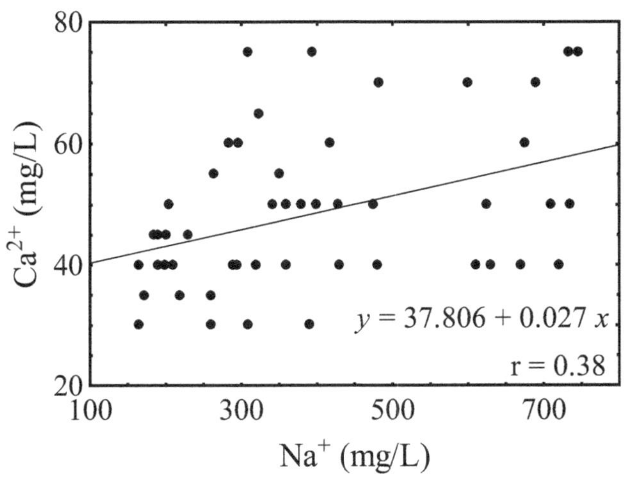

Fig. 11 Correlation between Na^+ and Ca^{2+}

Table 3 Saturation index

Particulars	Saturation index	Remarks
$CaCO_3$	3.12	Precipitation
CaF_2	−2.13	Dissolution

content. As a result, it favours a higher dissolution of CaF_2 from the source material to compensate for requirement of Ca^{2+} ions in the chemical equlibria (Freeze and Cherry 1979). Thus, it shows an under saturation with respect to CaF_2, causing a higher F^- content in the groundwater and thereby the Na^+ shows a good positive correlation with F^- ($r = 0.69$; Fig. 12) and a poor positive relation with Ca^{2+} ($r = 0.38$; Fig. 11), as also discussed earlier in the review literature from other regions. Further, the poor relation between Na^+ and Ca^{2+} is a result of the concurrent increase or decrease in the cations, depending upon the dissolution, precipitation, concentration effects and ion exchange (Admas et al. 2001).

The increase of Na^+:Cl^- ratio from 0.83 to 1.47 (Table 2) indicates modification of $NaCO_3$ water (Na^+:Cl^- >1.0) by dissolution or mixing of NaCl (Na^+:Cl^- <1.0; Dhiman and Keshari 2006; Gao et al. 2007, 2013), while the ratio of HCO_3^-:Ca^{2+} increases from 2.28 to 3.38, which signifies an increase of HCO_3^- (Saxena and Ahmed 2003). The concentration of F^- (0.77–1.53 mg/L) also follows a similar trend of Na^+:Cl^- and HCO_3^-:Ca^{2+} ratios. These infer that the Na^+ and HCO_3^- favour the chemical condition for dissolution of F^- and, therefore, a good positive correlation of F^- with HCO_3^- ($r = 0.64$) and Na^+ ($r = 0.69$) is observed from the study area (Figs. 8, 12), as also reported similar observations (discussed in review literature) from other regions. This is also supported by the hydrogeochemical facies.

A positive correlation of F^- with K^+ ($r = 0.59$), SO_4^{2-} ($r = 0.56$) and NO_3^- ($r = 0.64$) is observed from the study area (Figs. 13, 14, 15), as also discussed similar results from other regions (in review literature). This indicates the influences of uncontrolled usage of agro-chemicals (nitrogen, phosphate and potash fertilisers) and soil amendments (gypsum) for higher crop yields, because the nitrogen variety promotes over-grass shoot growth, phosphate type supports strong root growth, potash variety helps gross withstand stress, drought or disease and gypsum aids to alter the physical or chemical properties of the soils (Todd 1980; Fekri and Kasmaei 2013; Subba Rao 2014). Therefore, these chemicals are responsible for the concentrations of F^-, K^+, SO_4^{2-} and NO_3^- in the groundwater. In addition to this, domestic wastes and leakage of septic tanks contribute K^+ and NO_3^- to the groundwater system (Bhosle et al. 2001; Ayoob and Gupta 2006; Subba Rao 2014; Subba Rao et al. 2014a, b). Thus, the anthropogenic activities appear as a supplementary source of F^- to its further elevation in the groundwater.

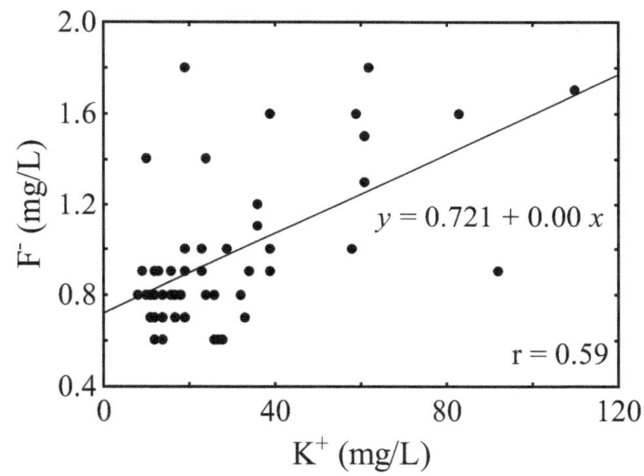

Fig. 13 Correlation between K^+ and F^-

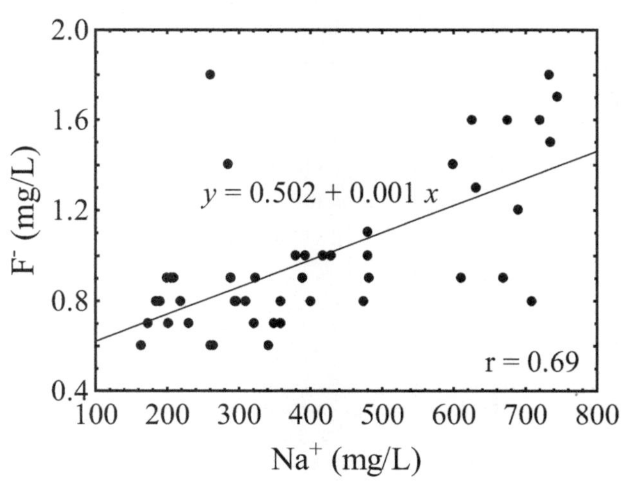

Fig. 12 Correlation between Na^- and F^-

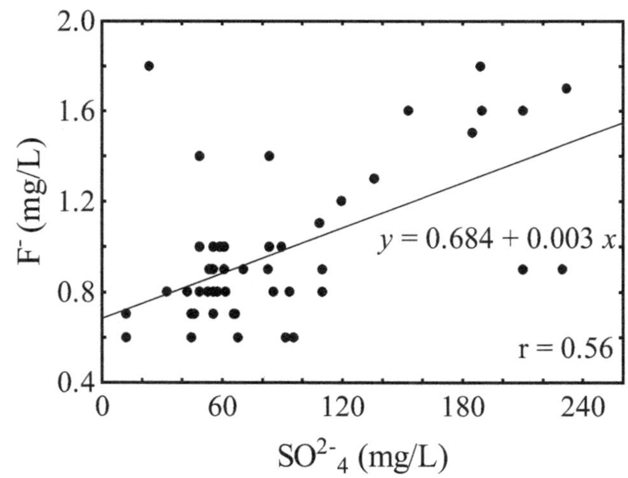

Fig. 14 Correlation between SO_4^{2-} and F^-

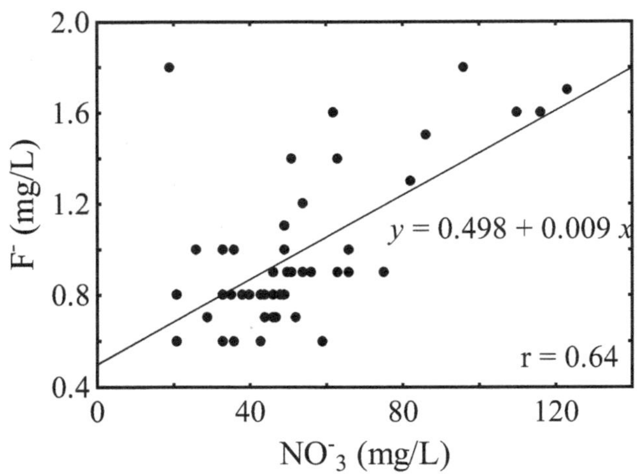

Fig. 15 Correlation between NO_3^- and F^-

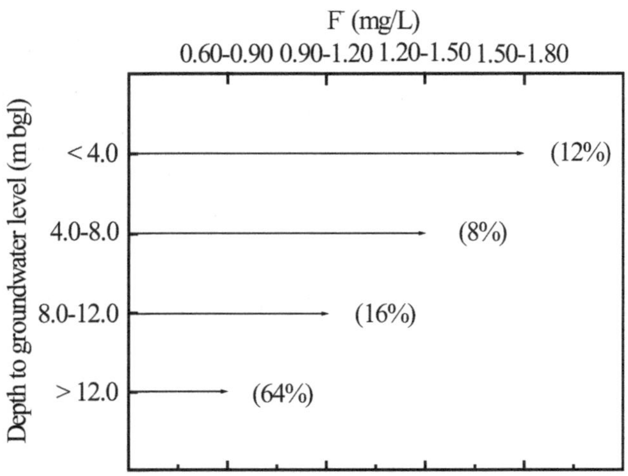

Fig. 16 Fluoride content with depth to groundwater level. Value in bracket denotes percentage of wells

Figure 16 shows that the concentration of F^- increases with the decrease of the depth. It suggests that the factors that control the F^- content in the groundwater are prevalent in the shallow depth rather than those in the deeper level. It is a result of higher rate of solubility, dissolution and leaching of salts present in soil and/or weathered zone, and ion exchange and evaporation in the shallow depth. This is also supported by the above discussion explained for understanding the causative factors that control the F^- content in the groundwater.

Remedial measures

(a) The study area receives 94 % (846 mm) of the total rainfall during monsoon. It can be possible to reduce the F^- content in the groundwater, if a number of rainwater harvesting structures are constructed, as reported in other regions (Narasimha and Pradeep 1997; Bhagavan and Raghu 2005). This is a simple low-cost alternative to the existing high-cost approaches (Nalgonda technique, activated alumina, red mud, montmorillonite, magnesia, etc.).

(b) Calcium food is essential to control the F^- content in the human body.

(c) Application of $CaSO_4$ and CaCl can also reduce the F^- content in the groundwater (Fekri and Kasmaei 2013).

(d) Public awareness on causes of high concentration F^- in drinking water and its health implications is necessary for success of groundwater quality management measures.

(e) The study helps the planning authorities to mitigate the F^- content in the drinking water.

Conclusions

Fluoride varies from 0.60 to 1.80 mg/L in the coastal region, which exceeds the threshold limit of 1.20 mg/L in 20 % of the total groundwater samples. The occurrence of F^--bearing minerals in the country rocks are the main geogenic source of F^- in the groundwater. Alkalinity (pH and HCO_3^-) plays a significant role in dissolution and leaching of F^-. As a result, a positive correlation of F^- with pH and HCO_3^- is observed. The clays lead to a longer contact of water with aquifer material for active solubility and dissolution of F^--bearing minerals, and are also responsible for anion exchange between F^- and OH^-. The influences of saline water, ion exchange between Na^+ and Ca^{2+}, and the precipitation of $CaCO_3$ due to evaporation reduce the Ca^{2+} content, which favour to dissolve CaF_2 to maintain the chemical equilibria. Thus, there is a positive correlation between Na^+ and F^-. Anthropogenic activities act as supplementary factor for further elevation of F^- in the groundwater. This is supported by positive correlation of F^- with K^+, SO_4^{2-} and NO_3^-. These factors are dominant in shallow depth so that the groundwater shows higher content of F^- in the shallow depth than that in the deeper depth. Hydrogeochemical facies also support this hypothesis. Appropriate management measures are suggested to reduce the F^- in the groundwater.

Acknowledgments The author (NSR) acknowledges the financial support (DST/WAR-W/WSI/08/2010) received from the Department of Science and Technology, New Delhi to carry out this work. Prof. P. Bhanumurthy, Co-Principal Investigator and Mr. G. Vidyasagar, Research Scholar is thanked for their help during the work. The author (NSR) thanks anonymous reviewers and editor of the journal for their critical and constructive suggestions in bringing the manuscript in the present form.

References

Admas S, Tredoux G, Harris C, Titus R, Pietersen K (2001) Hydrochemical characteristics of aquifers near Sutherland in the Western Karoo, South Africa. J Hydrol 241:91–103

Agrawal V, Vaish AK, Vaish P (1997) Groundwater quality-focus on fluoride and fluorosis in Rajasthan. Curr Sci 73:743–746

Ahmed AA (2014) Fluoride in Quaternary groundwater aquifer, Nile Valley, Luxor, Egypt. Arab J Geosci 7:3069–3083

Apambire WB, Boyle DR, Michel FA (1997) Geochemistry, genesis and health implications of fluoriferous groundwaters in the upper regions of Ghanna. Environ Geol 33:13–24

APHA (1999) Standard methods for the examination of water and wastewater. American Public Health Association, Washington, DC

Ayoob S, Gupta AK (2006) Fluoride in drinking water: a review on the status and stress effects. Crit Rev Environ Sci Tech 36:433–487

Berner EK, Berner RA (1987) The global water cycle, geochemistry and environment. Prentice-Hall, New Jersey

Bhagavan SVBK, Raghu V (2005) Utility of check dams in dilution of fluoride concentration in groundwater and the resultant analysis of blood serum and urine of villagers, Anantapur District, Andhra Pradesh, India. Environ Geochem Health 27:97–108

Bhosle AB, Narkhede RK, Rao Balaji, Patil PM (2001) Studies on the fluoride content of Godavari river water at Nanded. Econ Environ Conserv 7:341–344

CGWB (2002) Basic data reports of exploratory wells drilled in Prakasam District. Central Ground Water Board, Southern Region, Hyderabad

CGWB (2013) Groundwater brochure, Prakasam district, Andhra Pradesh (AAP-2012-13). Central Ground Water Board, Southern Region, Hyderabad

Chakraborti D, Chanda CR, Samanta G, Chowdhury UK, Mukherjel SC, Pal AB, Sharma B, Mahanta KJ, Ahmed HA, Sing B (2000) Fluorosis in Assam, India. Curr Sci 78:1421–1423

Chatterjee MK, Mohabey NK (1998) Potential fluorosis problems around Chandidongri, Madhya Pradesh, India. Environ Geochem Health 20:1–4

Chidambaram S, Bala Krishna Prasad M, Manivannan R, Karmegam U, Singaraja C, Anandhan P, Prasanna MV, Manikandan S (2013) Environmental hydrogeochemistry and genesis of fluoride in groundwaters of Dindigul District, Tamilnadu (India). Environ Earth Sci 68:333–342

Das B, Talukdar J, Sarma S, Gohain B, Dutta RK, Das HB, Das SC (2003) Fluoride and other inorganic constituents in groundwater of Guwahati, Assam, India. Curr Sci 85:657–661

Datta PS, Deb DL, Tyagi SK (1996) Stable isotope (^{18}O) investigations on the processes controlling fluoride contamination of groundwater. J Contam Hydrol 24:85–96

Deepu TR, Shaji E (2011) Fluoride contamination in groundwater resources of Chittur block, Palghat district, Kerala, India—a health risk. In: Disaster, Risk and vulnerability conference 2011. School of Environmental Sciences, Mahatma Gandhi University, India in association with the Applied Geoinformatics for Society and Environment, Germany, 12–14 March 2011

Deutsch WJ (1997) Groundwater geochemistry fundamentals and application to contamination. Lewis Publishers, Boca Raton

Dhiman SD, Keshari AK (2006) GIS assisted inverse geochemical modeling for plausible phase transfers in aquifers. Environ Geol 50:1211–1219

Drever JI (1997) The geochemistry of natural waters, 3rd edn. Prentice Hall, Englewood **436 p**

Eenadu (2013) Relief to fluoride-affected areas in budget 2013 (daily news paper in Telugu), p 6

Fekri M, Kasmaei LS (2013) Fluoride pollution in soils and waters of Koohbanan region, southeastern Iran. Arab J Geosci 6:157–161

Fetter CW (1990) Applied hydrogeology. CBS Publishers & Distributors, New Delhi **592 p**

Freeze RA, Cherry JA (1979) Groundwater. Prentice-Hall, New Jersey, p 603

Gao X, Wang Y, Li Y, Guo Q (2007) Enrichment of fluoride in groundwater under the impact of saline water intrusion t the salt lake area of Yuncheng basin, Northern China. Environ Geol 53:795–803

Gao XB, Zhang FC, Wang C, Wang YX (2013) Coexistence of high fluoride fresh and saline groundwaters in the Yuncheng Basin, northern China. Proc Earth Planet Sci 7:280–283

Gupta MK, Singh V, Rajwanshi P, Agarwal M, Rai K, Srivastava S, Shrivastav R, Dass S (1999) Groundwater quality assessment of Tehsil Kheragarh Agra (India) with special reference to fluoride. Environ Monit Assess 59:272–285

Gupta SK, Deshpande RD, Agarwal M, Raval BR (2005) Origin of high fluoride in groundwater in the North Gujarat-Cambay region India. Hydrogeol J 13:596–605

Handa BK (1975) Geochemistry and genesis of fluoride-containing groundwater in India. Ground Water 13:275–281

Hem JD (1991) Study and interpretation of the chemical characteristics of natural water. In: United States Geological Survey professional paper 2254, Scientific Publisher, Jodhpur

ISI (2012) Indian standard drinking water—specification. Indian Standard Institute, IS: 10500

Jacks G, Rajagopalan K, Alveteg T, Johnsson M (1993) Genesis of high-F groundwaters, Southern India. Appl Geochem (supplementary) 2:241–244

Jacks G, Bhattacharya P, Chaudhary V, Singh KP (2005) Controls on the genesis of some high-fluoride groundwaters in India. Appl Geochem 20:221–228

Jha LB, Jha M (1982) Fluoride pollution in India. Int J Environ Stud 19:225–230

Kundu MC, Mandal B (2009a) Agricultural activities influence nitrate and fluoride contamination in drinking groundwater of an intensively cultivated district in India. Water Air Soil Pollut 198:243–252

Kundu MC, Mandal B (2009b) Assessment of potential hazards of fluoride contamination in drinking groundwater of an intensively cultivated district in West Bengal, India. Environ Monit Assess 152:97–103

Kundu N, Panigrahi MK, Tripathy S, Munshi S, Powell MA, Hart BR (2001) Geochemical appraisal fluoride contamination of groundwater in Nayagarh District of Orissa, India. Ind J Environm Health 41:451–460

Madhavan N, Subramanian V (2002) Fluoride in fractioned soil samples of Ajmer District, Rajasthan. J Environ Monit 4:821–822

Madhnure P, Sirsikar DY, Tiwari AN, Ranjan B, Malpe DB (2007) Occurrence of fluoride in the groundwaters of Pandharkawada area, Yavatmal District, Maharashtra, India. Curr Sci 92:675–679

Manikandan S, Chidambaram S, Ramanathan AL, Prasanna MV, Karmegam U, Singaraja C, Paramaguru P, Jainab I (2014) A study on the high fluoride concentration in the magnesium-rich waters of hard rock aquifer in Krishnagiri district, Tamilnadu, India. Arab J Geosci 7:273–285

Misra AK, Mishra A (2006) Study of quaternary aquifers in Ganga Plain, India: focus on groundwater salinity, fluoride and fluorosis. J Hazard Mater 144:438–448

Muralidharan D, Nair A, Satyanarayana U (2002) Fluoride in shallow aquifers in Rajgarh Teshil of Churu District, Rajashtan—an arid environment. Curr Sci 83:699–702

Narasimha RT, Pradeep R (1997) Hydrogeological conditions and optimum well discharges in granitic terrain in parts of Nalgonda District, Andhra Pradesh, India. J Geol Soci India 49:61–74

Nordstrom DK, Jenne EA (1977) Fluoride solubility in selected geothermal waters. Geochim Cosmochim Acta 41:175–188

Raju JN, Dey S, Das K (2009) Fluoride contamination in groundwater of Sonbhadra District, Uttar Pradesh, India. Curr Sci 96:979–985

Ramachandran M, Chidambaram S, Ulaganthan K, Paluchamy A, Sivaji M, Hameed S (2012) Mapping of fluoride ions in groundwater of Dindigul district, Tamilnadu, India—using GIS technique. Arab J Geosci 5:433–439

Ramamohana Rao NV, Suryaprakasa Rao K, Schuiling RD (1993) Fluorine distribution in waters of Nalgonda District, Andhra Pradesh, India. Environ Geol 21:84–89

Ramesam V, Rajagopalan K (1985) Fluoride ingestion into the natural water of hard rock areas, Peninsular India. J Geol Soci India 26:125–132

Rao PN, Rao AD, Bhargav JS, Siva Sankar K, Sudharshan G (2014) Regional appraisal of the fluoride occurrence in groundwaters of Andhra Pradesh. J Geol Soc India 84:483–493

Ray D, Rao RP, Bhoi AV, Biswas AK, Ganguly AK, Sanyal PI (2000) Physico-chemical quality of drinking water in Rohtas District of Bihar. Environ Monit Assess 61:387–398

Reddy DV, Nagabhushanam P, Sukhija BS, Reddy AGS, Smedley PL (2010) Fluoride dynamics in the granitic aquifer of the Wailapally watershed, Nalgonda District, India. Chem Geol 269:278–289

Robinson WD, Edington G (1946) Fluorine in soils. Soil Sci 61:341–353

Sahu NK, Karim MA (1989) Fluoride incidence in groundwaters in Amreli District, Gujarat. J Geol Soci India 33:450–456

Saxena VK, Ahmed S (2003) Inferring the chemical parameters for the dissolution of fluoride in groundwater. Environ Geol 43:731–736

Shaji E, Bindu Viju J, Thambi DS (2007) High fluoride in groundwater of Palghat District, Kerala. Curr Sci 92:240–245

Shortt HE, McRobert GR, Barnard TW, Mannadinayer AS (1937) Endemic fluorosis in Madras Presidency, India. Ind J Med Res 25:553–561

Singaraja C, Chidambaram S, Anandhan P, Prasanna MV, Thivya C, Thilagavathi R (2013) A study on the status of fluoride ion in groundwater of coastal hard rock aquifers of south India. Arab J Geosci 6:4167–4177

Sreedevi PD, Ahmed S, Made B, Ledoux E, Gandolfi JM (2006) Association of hydrological factors in temporal variations of fluoride concentration in a crystalline aquifer in India. Environ Geol 50:1–11

Srinivasamoorthy K, Chidambaram S, Vasanthavigar M (2008) Geochemistry of fluorides in Salem district, Tamilnadu, India. J Envron Geol 16:25

Srinivasamoorthy K, Vijayaraghavan K, Vasanthavigar M, Sarma Subramanya, Chidambaram S, Paluchamy A, Manivannan R (2012) Assessment of groundwater quality with special emphasis on fluoride contamination in crystalline bed rock aquifers of Mettur region, Tamilnadu, India. Arab J Geosci 5:83–94

Stallard RF, Edmond JM (1983) Geochemistry of the Amazon River—the influence of the geology and weathering environment on the dissolved load. J Geophy Res 88:9671–9688

Stumm W, Morgan JJ (1996) Aquatic chemistry. Wiley-Interscience, New York

Subba Rao N (2002) Geochemistry of groundwater in parts of Guntur District, Andhra Pradesh, India. Environ Geol 41:552–562

Subba Rao N (2003) Groundwater quality-focus on fluoride concentration in rural parts of Guntur District, Andhra Pradesh, India. Hydrol Sci J 48:835–847

Subba Rao N (2009) Fluoride in groundwater, Varaha River Basin, Visakhapatnam District, Andhra Pradesh, India. Environ Monit Assess 152:47–60

Subba Rao N (2011) High-fluoride groundwater. Environ Monit Assess 176:637–645

Subba Rao N (2014) Development and management of groundwater resources in a coastal region: a study from Prakasam District, Andhra Pradesh. In: Presented at 7th DST PAC-WTI on 10th January 2014. IIT, Bhubaneswar

Subba Rao N, John Devadas D (2003) Fluoride incidence in groundwater in an area of Peninsular India. Environ Geol 45:243–253

Subba Rao N, Rao AT (2003) Fluoride in groundwater in a developing area of Guntur District, Andhra Pradesh, India. J Appl Geochem 5:99–100

Subba Rao N, Krishna Rao G, John Devadas D (1998a) Variation of fluoride in groundwater of crystalline terrain. J Environ Hydrol 6:1–5

Subba Rao N, Prakasa Rao J, Nagamalleswara Rao B, Niranjan Babu P, Madhusudhana Reddy P, John Devadas D (1998b) A preliminary report on fluoride content in groundwater of Guntur Area, Andhra Pradesh, India. Curr Sci 75:887–888

Subba Rao N, Subrahmanyam A, Babu Rao G (2012) Fluoride-bearing groundwater in Gummanampadu sub-basin, Guntur District, Andhra Pradesh, India. Environ Earth Sci 70:575–586

Subba Rao N, Vidyasagar G, Surya Rao P, Bhanumurthy P (2014a) Chemistry and quality of groundwater in a coastal region of Andhra Pradesh. Appl Water Sci (accepted), India

Subba Rao N, Vidyasagar G, Surya Rao P, Bhanumurthy P (2014b) Assessment of hydrogeochemical processes in a coastal region: application of multivariate statistical model. J Geol Soc India 84:494–500

Sujatha D (2003) Fluoride levels in the groundwater of the south eastern part of Ranga Reddy District, Andhra Pradesh, India. Environ Geol 44:587–591

Sumalatha S, Ambika SR, Prasad SJ (1999) Fluoride contamination status of groundwater in Karnataka. Curr Sci 76:730–734

Surya Rao P, Subba Rao N, Dinakar A (2015) Fluoride in the groundwater in Etcherla Mandal, Srikakulam District, Andhra Pradesh. J Geosci Res 1:53–59

Susheela AK (1999) Fluorosis management programme in India. Curr Sci 77:1250–1256

Todd DK (1980) Groundwater hydrology. Wiley, New York

USPHS (1987) Drinking water standards. United States Pubic Health Services Publication, Washington, DC

Vikas C, Kushwaha RK, Pandit MK (2009) Hydrochemical status of groundwater in district Ajmer (NW India) with reference to fluoride distribution. J Geol Soc India 73:773–784

WHO (2008) Guidelines for drinking water quality. World Health Organization, Geneva

Wodeyar BK, Sreenivasan G (1996) Occurrence of fluoride in the groundwater and its impact in Peddavankahalla Basin, Bellary District, Karnataka, India—a preliminary study. Curr Sci 70:71–74

10

Modeling of the adsorptive removal of arsenic(III) using plant biomass: a bioremedial approach

Palas Roy[1,2] · Uttiya Dey[2] · Soumya Chattoraj[2] · Debasis Mukhopadhyay[3] · Naba Kumar Mondal[2]

Abstract In the present work, the possibility of using a non-conventional finely ground (250 μm) *Azadirachta indica* (neem) bark powder [AiBP] has been tested as a low-cost biosorbent for the removal of arsenic(III) from water. The removal of As(III) was studied by performing a series of biosorption experiments (batch and column). The biosorption behavior of As(III) for batch and column operations were examined in the concentration ranges of 50–500 μg L^{-1} and 500.0–2000.0 μg L^{-1}, respectively. Under optimized batch conditions, the AiBP could remove up to 89.96 % of As(III) in water system. The artificial neural network (ANN) model was developed from batch experimental data sets which provided reasonable predictive performance ($R^2 = 0.961$; 0.954) of As(III) biosorption. In batch operation, the initial As(III) concentration had the most significant impact on the biosorption process. For column operation, central composite design (CCD) was applied to investigate the influence on the breakthrough time for optimization of As(III) biosorption process and evaluation of interacting effects of different operating variables. The optimized result of CCD revealed that the AiBP was an effective and economically feasible biosorbent with maximum breakthrough time of 653.9 min, when the independent variables were retained at 2.0 g AiBP dose, 2000.0 μg L^{-1} initial As(III) concentrations, and 3.0 mL min^{-1} flow rate, at maximum desirability value of 0.969.

Keywords Arsenic(III) removal · Low-cost adsorbent · Biosorption experiments · Artificial neural network model · Central composite design

✉ Naba Kumar Mondal
nkmenvbu@gmail.com

Palas Roy
palaschem@gmail.com

Uttiya Dey
uttiya.1988@gmail.com

Soumya Chattoraj
soumya_chattoraj@yahoo.co.in

Debasis Mukhopadhyay
mukhopadhyay.debasis@yahoo.co.in

[1] Department of Chemistry, B.N. Mahavidyalaya (Itachuna), The University of Burdwan, Burdwan 713104, West Bengal, India

[2] Environmental Chemistry Laboratory, Department of Environmental Science, The University of Burdwan, Burdwan, West Bengal 713104, India

[3] Department of Economics, B.N. Mahavidyalaya (Itachuna), The University of Burdwan, Burdwan, West Bengal 713104, India

Abbreviations

AiBP	*Azadirachta indica* bark powder
ANN	Artificial neural network
ANOVA	Analysis of variance
CCD	Central composite design
D–R	Dubinin–Radushkevich
RSM	Response surface methodology
Ag(–DDC)	Silver diethyl dithiocarbamate
SEM	Scanning electron microscopy
2FI	Two factor interactions

Introduction

Arsenic contamination in aqueous environment has become a global concern because of its serious environmental and health hazards (Mamun et al. 2009; Ranjan et al. 2009) in several developing regions. Arsenic can enter into streams, lakes, rivers, and groundwater (Mamun et al. 2009) through natural and human activities (Maiti et al. 2010). The existence of arsenic in groundwater, and eventually in drinking water can be dangerous because of the large quantity of water consumed by the human beings in long term (Mohan and Pittman 2007; Mamun et al. 2009).

High level of arsenic in drinking water sourced from groundwater has been reported in 21 countries (Mohan and Pittman 2007; Roy et al. 2014), especially Argentina, Bangladesh, India, Mexico, Mongolia, Thailand, and Taiwan, where a large proportion of groundwater is contaminated with arsenic at levels from 100 to over 2000 $\mu g\ L^{-1}$ (Guo and Chen 2005).

In groundwater, arsenic is typically present in one of the two oxidation states: arsenite [As(III)] and arsenate [As(V)] of which As(III) is predominant and is generally more difficult to remove than As(V) (Kamala et al. 2005). Furthermore, As(III) is more mobile in groundwater and 25–60 times more toxic than As(V) (Ratna Kumar et al. 2004).

Due to the high toxic and delirious effects on human health (Guo and Chen 2005; Giri et al. 2011), the World Health Organization (WHO) has revised the safe limit for arsenic in drinking water from 50 to 10 $\mu g\ L^{-1}$ (Maiti et al. 2010; Kamsonlian et al. 2012). This new regulation has posed a challenge for researchers to develop new technologies capable of selectively removing arsenic present at very low level, especially arsenic(III) [As(III)].

Development of effective technologies to remove arsenic from water drew great attention over the last two decades. Several physiochemical techniques, such as adsorption, ion exchange, lime softening, reverse osmosis, coagulation, and precipitation, have been applied to remove arsenic from aqueous system (Bhakat et al. 2007; Mohan and Pittman 2007; Ranjan et al. 2009; Giri et al. 2011; Kamsonlian et al. 2012). Among these removal technologies, adsorption technology is mainly used because it is very simple and cost effective (Bhattacharya et al. 2006; Ranjan et al. 2009; Kamsonlian et al. 2012). Besides, it is also popular due to availability of a wide range of adsorbents.

Among all types of conventional and non-conventional adsorbents, there has been an increasing tendency to use numerous inactive biological materials, or biosorbents, for the removal of arsenic from contaminated environment. However, only a limited number of biosorbents (without chemical modification) have been examined for their efficacy to remove arsenic from contaminated solutions (Baig et al. 2010; Sari and Tuzen 2010; Giri et al. 2011; Prasad et al. 2011; Sari et al. 2011; Kamsonlian et al. 2012; Pennesi et al. 2012; Wu et al. 2012; Giri et al. 2013; Nigam et al. 2013; Raj et al. 2013; Saqib et al. 2013).

The powdered form of brown-colored mature stem bark of the *Azadirachta indica* (neem) is an excellent example of such type of a non-conventional biosorbent. In recent years, this low-cost biosorbent has drawn much attention to the investigators due to its wide availability all over the world. The removal behavior of Zn(II) (Bhattacharya et al. 2006; Arshad et al. 2008; King et al. 2008; Naiya et al. 2009), Cd(II) (Tiwari et al. 1999; Naiya et al. 2009), Cr(VI) (Bhattacharya et al. 2008; Kumar and Phanikumar 2013) and dyes (Srivastava and Rupainwar 2010, 2011; Sadhukhan et al. 2014) from aqueous solutions on the *A. indica* bark powder (AiBP) had previously been investigated but not a single research has been reported on the removal of arsenic by AiBP.

In view of these attributes, it is of great interest to study the possibility for application of AiBP in the removal of arsenic from aqueous environment. With this goal in mind, the present study deals with a series of biosorption experiments (batch and column) to assess the potentiality of AiBP for removal of arsenic(III) from contaminated water. For sensitivity analysis and optimization abilities, the biosorption experiments have been statistically modeled using artificial neural network (ANN) and central composite design (CCD). The models were also applied to study the individuals as well as the combined effect of different variables influencing the biosorption process. As a whole a complete study for the economic, easy and eco-friendly solution of catastrophic arsenic pollution has been done, that can be implemented in the developing regions. Results obtained from this study are presented and discussed.

Materials and methods

Reagents and apparatus

Chemicals of analytical grade were procured from M/S, Merck India Ltd., and used in the study without further purification. All reagents and standards were prepared using double distilled water. Sodium arsenite ($NaAsO_2$) was used for the preparation of standard As(III) solution, and the required As(III) concentrations for experiments were prepared by serial dilution of standard solution on daily basis. All borosil glassware were cleaned by being soaked in 15 % HNO_3 and rinsed with double distilled water. Different pH of As(III) solution were generated by the addition of 0.1 N H_2SO_4 or NaOH solutions whenever necessary.

Preparation of biosorbent

The brown-colored neem (*A. indica*) bark used in the present study was collected from the university campus,

Burdwan University, Burdwan. The collected *A. indica* bark was thoroughly washed with double distilled water to remove muddy materials and soaked in 0.1 N NaOH to remove lignin-based color materials followed by 0.1 N H_2SO_4 (King et al. 2008; Naiya et al. 2009). The washed *A. indica* bark was dried in sun for 15 days. The resulting product was directly used as biosorbent. The dried bark was then cut into small pieces and ground to powder with kitchen grinder. For this experimental study, the powdered materials were sieved to obtain particle size of 250 μm prior to use as biosorbent without any further treatment. The physiochemical characterization of the AiBP biosorbent has been reported elsewhere (Srivastava and Rupainwar 2011; Sadhukhan et al. 2014).

Analytical determination of arsenic

Estimation of As(III) was done spectrophotometrically by silver diethyl dithiocarbamate method (Gupta and Sankararamakrishnan 2010; Kamsonlian et al. 2012; Roy et al. 2013a) with minimum detectable amount of 1 μg (Bhakat et al. 2007). This method is based on collection of arsine, AsH_3 generated, through hydride formation of As(III), in silver diethyl dithiocarbamate [Ag(–DDC)] in pyridine solution and on subsequent reaction of AsH_3 with Ag(–DDC) to form a colored $As(-DDC)_3$ complex (Hautfenne 1980; Arbab-Zavar and Hashemi 2000).

$$AsH_3 + 6Ag(-DDC) \rightarrow 6Ag + As(-DDC)_3 + 3H(-DDC)$$

Each sample was analyzed thrice and the results were found reproducible within ±3 % error limit. Calibration was carried out daily with a freshly prepared arsenic standard, before analysis. Experiment done with control biosorbent indicated no release of arsenic by the biosorbent. Blank experiments were also conducted to ensure that no adsorption was taking place on the walls of the apparatus used.

Experimental setup

In the batch operation, the effect of different parameters (i.e., pH, contact time, initial concentration of As(III), dose of biosorbent, stirring rate, and temperature) on biosorption of As(III) was studied. The 100 mL solution of As(III) was taken in each Erlenmeyer flask separately. After pH adjustment, a known quantity of dried AiBP was added and the As(III) bearing suspensions was kept under magnetic stirring until the equilibrium was reached. After shaking, the suspension was allowed to settle down and filtered through a (Whatman No. 42 grade) filter paper. The filtrate was collected and subjected for arsenic estimation using Ag(–DDC) method. The arsenic concentrations before and after biosorption were recorded, and then the percent arsenic biosorption (removal) by the biosorbent was computed by using the following equation:

$$\text{Percent biosorption (Removal)} = [(C_i - C_e)/C_i] \times 100 \quad (1)$$

where C_i and C_e are the initial and final concentration of As(III) in the solution.

The arsenic uptake loading capacity ($q_e = \mu g\ g^{-1}$) of AiBP for different concentrations of As(III) at equilibrium was also determined as

$$qe = \left(\frac{Ci - Ce}{M}\right) \times V \quad (2)$$

where V is the volume of solution (L) and M is the mass of the biosorbent (g) used.

The experimental datasets, which were obtained from the batch operation, were used as inputs to the ANN model to provide the reasonable predictive performance of the biosorbent.

Fixed-bed column operation was conducted using borosil glass columns of 3-cm internal diameter and 50-cm length. The column was packed with different amounts of AiBP, in order to achieve different bed heights, between two supporting layers of 1-cm of glass wool to prevent the floating of biosorbent.

The column studies were conducted to evaluate the effects of different variables [initial As(III) concentration, flow rate, and biosorbent dose] on the breakthrough time of As(III) biosorption by the biosorbent using CCD in response surface methodology (RSM). As(III) solution having the desired initial concentration was adjusted to the optimum batch pH for the maximum removal of arsenic, and pumped through the column bed at the desired flow rate in an up-flow mode. The flow rates were chosen to have sufficient amount of effluent per 30 min for the practical purpose of analyzing arsenic concentration. Operation of the column was stopped when the breakthrough time is achieved, the time at which the arsenic concentration in the effluent reached below 50 μg L^{-1}. All of the experiments were carried out in triplicate at room temperature (30 °C) and the mean values were taken for analysis.

Results and discussion

Batch operation

Effect of initial As(III) concentration

The biosorption behavior of arsenic(III) was studied in the concentration range of 50–500 μg L^{-1} initially at pH 6.0.

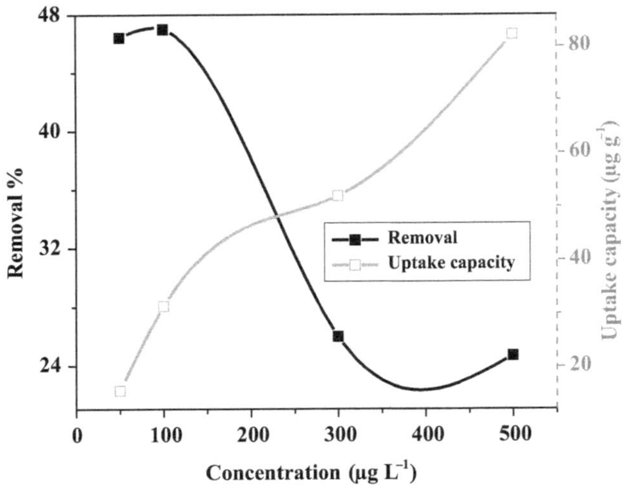

Fig. 1 Effect of initial arsenic(III) concentration on the biosorption of As(III) by AiBP. Dose = 0.15 g, contact time = 20 min, pH = 6.0, temperature = 30 °C, stirring rate = 500 rpm

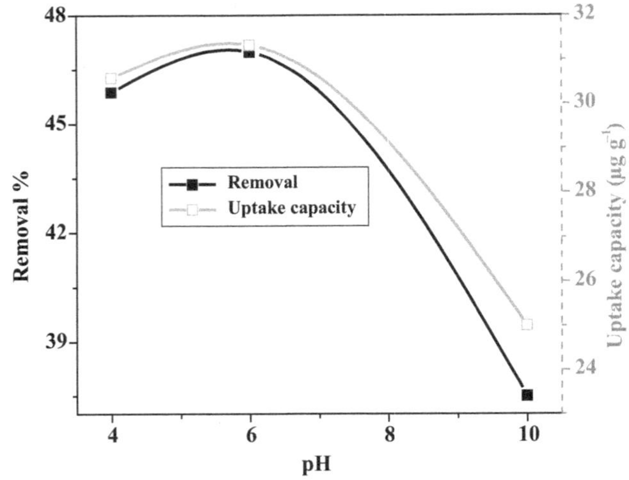

Fig. 2 Effect of pH on the biosorption of As(III) by AiBP. Initial arsenic(III) concentration = 100 µg L^{-1}, dose = 0.15 g, contact time = 20 min, temperature = 30 °C, stirring rate = 500 rpm

In general, the removal percentage of As(III) on AiBP biosorbent was initially increased with the increasing initial concentration of arsenic reaching the optimum level of 46.99 % at 100 µg L^{-1} arsenic concentration. Thereafter, the percentage of removal showed sharp decrease (Fig. 1). But the actual amount of arsenic adsorbed per unit mass of biosorbent was increased with increasing initial concentration in the test solution. This increase was probably due to decrease in resistance to the uptake of solute from solution with increasing initial arsenic concentration (Bhaumik et al. 2011; Suresh et al. 2012; Das et al. 2013).

Effect of pH

Figure 2 represents the percentage removal (or uptake capacity) as a function of the pH at optimum concentration (100 µg L^{-1}) of As(III). The optimum removal of 46.99 % (31.32 µg g^{-1}) was obtained at the pH of 6.0. In the experimental pH range of 4.0–10.0, the predominant arsenite species are H_3AsO_3 and $H_2AsO_3^-$ (Ratna Kumar et al. 2004; Guo and Chen 2005; Kamala et al. 2005; Giri et al. 2011). At optimum pH of 6.0, arsenite is probably absorbed through specific adsorption between the neutral species (H_3AsO_3) and positively charged surface sites as the zeta potential: pH_{ZPC} of AiBP = 6.80 (Srivastava and Rupainwar 2011). And the reduced trend of removal at higher pH might be attributed to an increase of negatively charged arsenite species ($H_2AsO_3^-$) and negatively charge surface sites (Giri et al. 2011).

Effect of biosorbent dose

In studying the influence of the dry weight of AiBP biomass on the removal efficiency at pH 6.0 with initial As(III) concentration of 100 µg L^{-1}, it was found that the removal efficiency of As(III) increased steadily (Fig. 3) with increasing biosorbent doses (0.05–0.30 g). No remarkable increment in the removal tendency was noted thereafter on further increasing the biosorbent dose. Increasing removal with biosorbent dose might be attributed to increase in biosorbent surface area with subsequent increment of more active binding sites (King et al. 2008; Kamsonlian et al. 2012; Das et al. 2014).

Effect of contact time

The effect of contact time on the biosorption of As(III) on AiBP biosorbent (0.25 g dose) was studied in the duration of 10–30 min at pH 6.0 with initial As(III) concentration of 100 µg L^{-1}. The percentage removal of arsenic showed a rapid and steady increase up to 20 min, and thereafter, no significant increase was recorded. It seems possible that an equilibrium had set up in 20 min (Fig. 4). The fast biosorption rate at the initial stage may be explained by the abundant availability of active binding sites on the biosorbent surface (Srivastava and Rupainwar 2011; Das et al. 2013; Saqib et al. 2013). However, with a lapse of time, the remaining unsaturated sites cannot be occupied because of repulsive force between the solute (arsenic) on

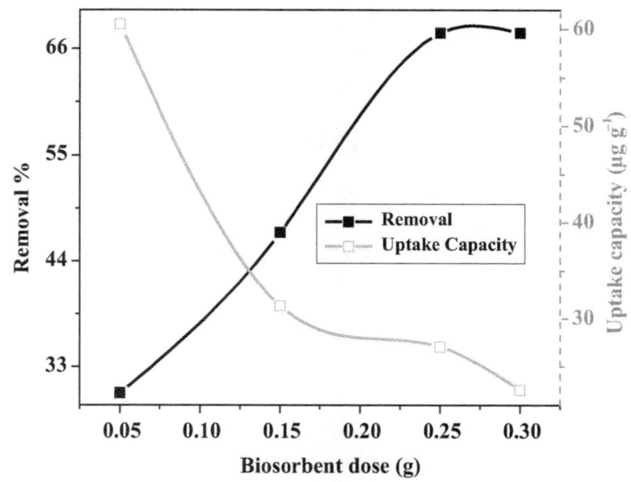

Fig. 3 Effect of biosorbent dose on the biosorption of As(III) by AiBP. Initial arsenic(III) concentration = 100 μg L^{-1}, contact time = 20 min, pH = 6.0, temperature = 30 °C, stirring rate = 500 rpm

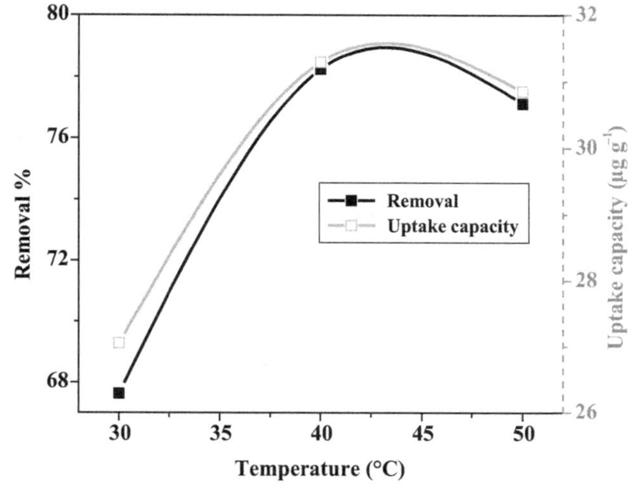

Fig. 5 Effect of temperature on the biosorption of As(III) by AiBP. Initial arsenic(III) concentration = 100 μg L^{-1}, dose = 0.25 g, contact time = 30 min, pH = 6.0, stirring rate = 500 rpm

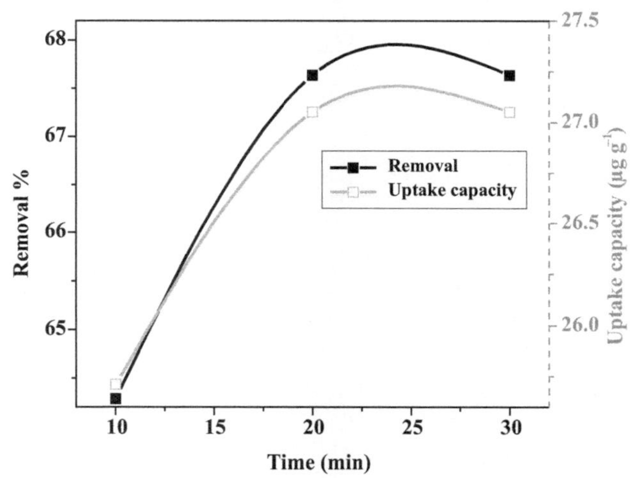

Fig. 4 Effect of contact time on the biosorption of As(III) by AiBP. Initial arsenic(III) concentration = 100 μg L^{-1}, dose = 0.25 g, pH = 6.0, temperature = 30 °C, stirring rate = 500 rpm

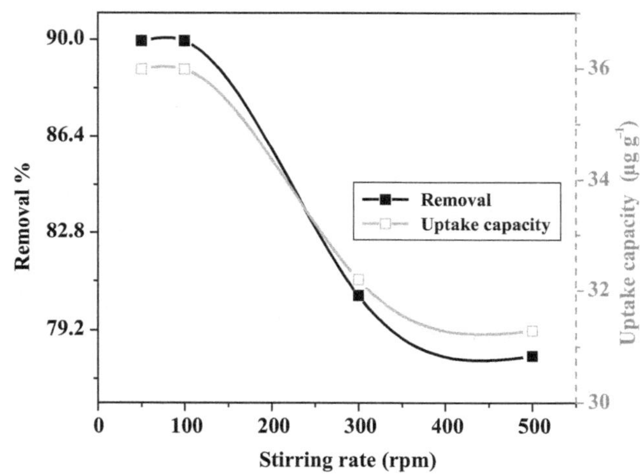

Fig. 6 Effect of stirring rate on the biosorption of As(III) by AiBP. Initial arsenic(III) concentration = 100 μg L^{-1}, dose = 0.25 g, contact time = 30 min, pH = 6.0, temperature = 40 °C

the solid and bulk phases (Rajesh Kannan et al. 2010; Roy et al. 2014).

Effect of temperature

The influence of temperature on the removal of As(III)by AiBP was measured at different temperatures ranging from 30 to 50 °C. Figure 5 displays that the percent removal of As(III) was increased with temperature, from 67.63 to 78.24 % when temperature was increased from 30 to 40 °C. The removal probably reached equilibrium at 40 °C and thereafter slight decrease was noted. This explains that at very high temperature AiBP loses its biosorption power through denaturation (Rajesh Kannan et al. 2010; Kamsonlian et al. 2012).

Effect of stirring rate

The effect of different stirring rate was studied with varying speeds from 50 to 500 rpm, and it appears that the percentage of As(III) removal was increased with decreasing stirring rate (Fig. 6). An increasing stirring rate did not give enough time for adsorbate and biosorbent surfaces to interact with each other and also resulted in detachment of loosely bounded adsorbates. However, maximum removal of 89.96 % occurred as below as 100 rpm, while no increment was noted thereafter. This is because all the binding sites have been utilized and no binding sites were left for further biosorption (Kanwal et al. 2012; Roy et al. 2013a).

Artificial neural network (ANN) modeling

In the last decade, ANN modeling has been successfully applied for estimating and predicting biosorption properties that are function of many variables and parameters. ANN is a statistical or computational model for processing of information based on the connectionist approach. This model has the ability to learn from existing data and adopt to map a set of input parameters into a set of output parameters, without knowing the intricate relationship among them. ANN can be trained to identify patterns and extract trends in imprecise and complicated non-linear data (Giri et al. 2011; Roy et al. 2013a, 2014). As biosorption is a complex non-linear process, neural network is found suitable for prediction of arsenic biosorption properties. Neural network toolbox of SPSS–17 mathematical software was used to predict the biosorption properties of AiBP under batch operation. Figure 7 shows the ANN model used in this study.

A total of 21 experimental datasets, which were obtained from batch operation, were used to develop a three-layer feed-forward neural network model by applying hyperbolic tangent function under the standardized method. Out of these 21 datasets, 66.7 % were used to train the network, and remaining 33.3 % were used for testing and validation of the ANN model. There were six neurons [viz., pH, initial As(III) concentration, biosorbent dose, contact time, stirring rate and temperature] in the input layer whereas two neurons in the output layer (removal efficiency and uptake capacity). The 7-4-2 ANN (including bias neuron) model is found to be working satisfactorily with an average relative error of 0.352 and sum square error of 10.065 during testing phase, indicating that the model is able to predict the biosorption process with reasonable accuracy.

The performances of optimized ANN are shown in Figs. 8 and 9. In these Figures, experimental and ANN-predicted values are compared for both uptake capacity and removal of As(III), respectively. The values of R^2 (0.961; 0.954) very close to 1 for each case show an excellent agreement between the experimental and the ANN-predicted values.

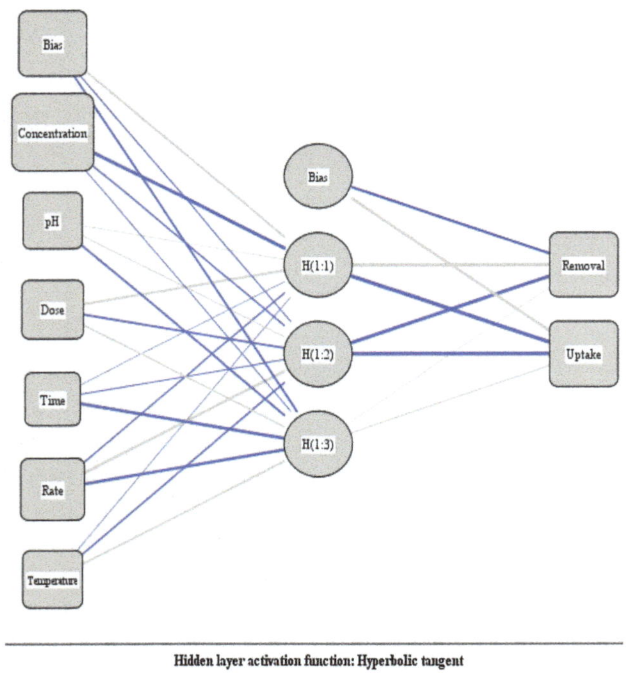

Fig. 7 Neural network architecture of As(III)

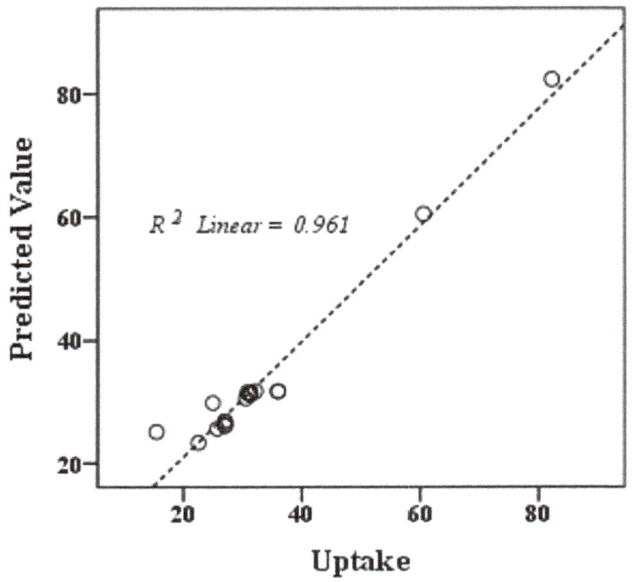

Fig. 8 Comparison of As(III) uptake for the experimental data and the simulation results in the training step of batch operation

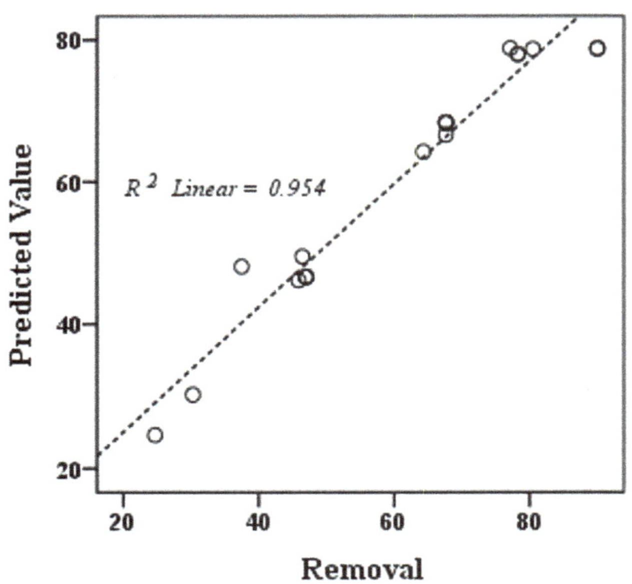

Fig. 9 Comparison of As(III) removal for the experimental data and the simulation results in the training step of batch operation

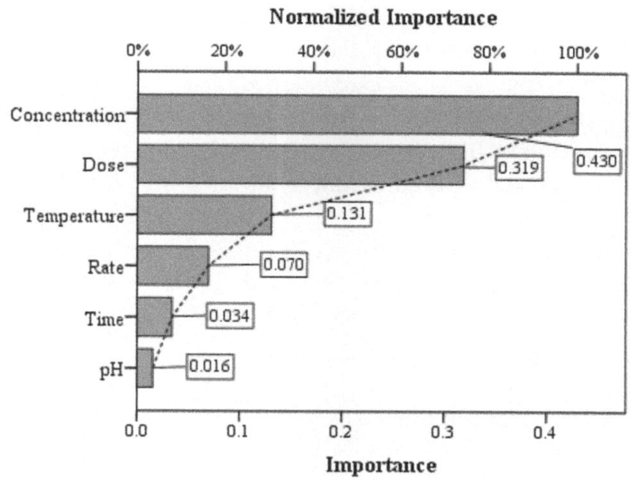

Fig. 10 Importance analysis of the various operating (input) parameters

The importance analysis for the developed network was also performed in order to determine the most significant parameter (Raj et al. 2013; Roy et al. 2014). It is reveled from the importance analysis (Fig. 10) that the initial concentration was the most significant parameter followed by biosorbent dose, temperature, stirring rate, contact time, and pH. The influence percentages of these parameters on the output were 100.0, 74.2, 30.5, 16.3, 8.0, and 3.7 %, respectively.

Equilibrium isotherms and kinetics study

An adsorption (biosorption) isotherm represents the equilibrium relationship between the adsorbate concentration in the liquid phase and that on the adsorbents surface at a given condition. A number of isotherms have been developed to describe equilibrium relationships. In the present study, the Langmuir, Freundlich, Temkin, and Dubinin–Radushkevich (D–R) models were used to describe the equilibrium isotherms. The illustration of the equilibrium adsorption plots and the summarized isotherm models are shown in Fig. 11 and Table 1.

Fig. 11 Four adsorption equilibrium plots: Langmuir, Freundlich, Temkin, and D–R isotherm. Dose = 0.15 g, contact time = 20 min, pH = 6.0, temperature = 30 °C, stirring rate = 500 rpm

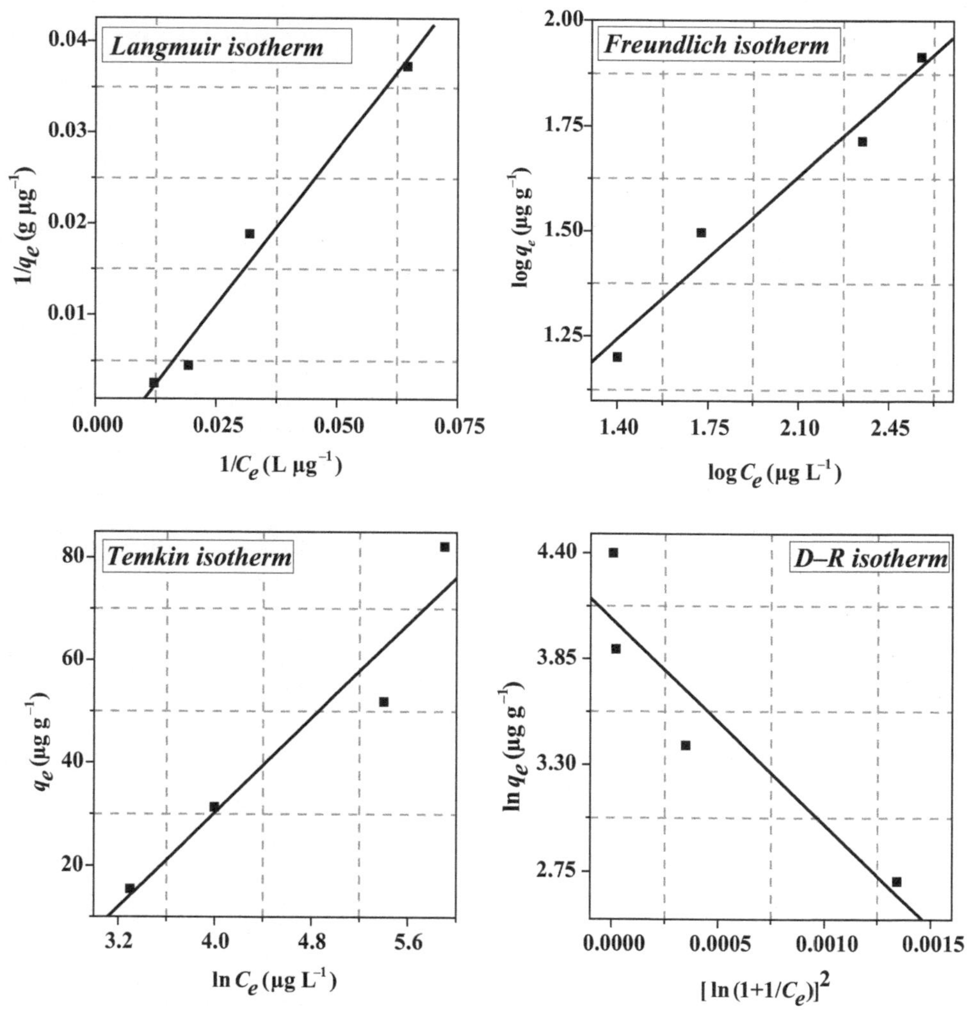

From Table 1, it is observed that the Langmuir isotherm showed good fit to the experimental equilibrium data than the Freundlich, Dubinin–Radishkevich, and Temkin isotherm equation for arsenic(III) biosorption according to the values of R^2. It is also seen from Table 1 that the Langmuir maximum biosorption capacity q_{max} (μg g^{-1}) is 166.7 and the equilibrium constant K_L (L μg^{-1}) is 0.009. The separation factor (R_L) values are 0.695, 0.532, 0.275, and 0.185 while initial As(III) concentrations are 50, 100, 300, and 500 μg L^{-1}, respectively. All the R_L values were found to be less than one and greater than zero indicating the favorable biosorption of As(III) by AiBP. The Freundlich constant K_F indicates the adsorption capacity of the biosorbent and the value of K_F is 2.679 μg g^{-1}. Furthermore, the value of 'n' at equilibrium is 1.757. The value of n lies in between 1 to 10 also represents a favorable biosorption. From Temkin constant, b_T related to adsorption binding energy for As(III) is found 111.4 J mol^{-1}, clearly denying to follow the ion-exchange mechanism. In addition, D–R isotherm shows the value of the adsorption energy (E) of 54.72 J mol^{-1}. The estimated value of E (<8 kJ mol^{-1}) has been indicated toward the physisorption process (Baig et al. 2010; Das et al. 2014; Roy et al. 2014).

The pseudo-first-order and pseudo-second-order kinetic models were tested to investigate the rate of biosorption of As(III) by AiBP (Baig et al. 2010; Chattoraj et al. 2014). The linearized form of adsorption kinetics and their constants are presented in Table 2.

From the table, it is confirmed that the arsenic(III) biosorption followed the pseudo-second-order reaction. It is also clear from Table 2 that the pseudo-second-order kinetic model showed excellent linearity with high correlation coefficient ($R^2 = 0.999$) at 100 μg L^{-1} As(III) concentration in comparison to the first-order kinetic model. Furthermore, the calculated q_e value also agrees with the experimental value (31.3 μg g^{-1}) in the case of pseudo-second-order kinetic model.

Thermodynamic parameters

In order to describe thermodynamic properties of the biosorption of As(III) by AiBP; enthalpy change (ΔH°), Gibbs free energy change (ΔG°), and entropy change (ΔS°) were calculated by using equations as shown in Table 3.

From Table 3, it is clear that the biosorption process is thermodynamically feasible and spontaneous in nature as ΔG° values are negative at all the temperature studied. The more negative value of ΔG° with increasing in temperature suggests that higher temperature makes the biosorption easier. Again, the positive ΔH° value implies the endothermic nature of the biosorption. The type of biosorption can be explained in terms of the magnitude of ΔH°. The enthalpy or the heat of adsorption ranging from

Table 1 Isotherm data for biosorption of As(III) by AiBP

Adsorption isotherms	Equations	Parameters (unit)	Values	R^2
Langmuir isotherm	$\frac{1}{q_e} = \frac{1}{q_{max} K_L C_e} + \frac{1}{q_{max}}$	q_{max} (μg g^{-1})	166.7	0.977
		K_L (L μg^{-1})	0.009	
Freundlich isotherm	$\log q_e = \log K_F + \frac{1}{n} \log C_e$	K_F (μg g^{-1})	2.679	0.955
		n	1.757	
Temkin isotherm	$q_e = \frac{RT}{b_T}(\ln A_T + \ln C_e)$	A_T (L μg^{-1})	14.15	0.930
		b_T (J mol^{-1})	111.4	
D–R isotherm	$\ln q_e = \ln q_{max} - \frac{1}{2E^2} \times \left[RT \ln\left(1 + \frac{1}{C_e}\right)\right]^2$	q_{max} (μg g^{-1})	169.8	0.908
		E (J mol^{-1})	54.72	

q_{max} is the maximum biosorption capacity; K_L, K_F, A_T and b_T are different biosorption constants; n is the heterogeneity factor; E is the mean free energy of biosorption per mole of the adsorbate; T is the temperature (K), and R is the ideal gas constant (8.3145 J mol^{-1} K^{-1})

Table 2 Kinetic parameters for biosorption of As(III) by AiBP

Kinetic Models	Equations	Parameters (unit)	Values	R^2
Pseudo-first-order	$\ln q_e - q_t = \ln q_e - k_1 t$	q_e (μg g^{-1})	4.93	0.875
		k_1 (min^{-1})	0.546	
Pseudo-second-order	$\frac{t}{q_t} = \frac{1}{k_2 q_e^2} + \frac{t}{q_e}$	q_e (μg g^{-1})	33.2	0.999
		k_2 (g μg^{-1} min^{-1})	0.969	

q_t is the biosorption capacity at time t; k_1 and k_2 are the first-order and second-order rate constant, respectively

Table 3 Thermodynamic parameters for biosorption of As(III) by AiBP

Thermodynamic parameters	Equations	Values (kJ mol^{-1})
Standard free energy	$\Delta G^0 = -RT \ln K_C$	
303 K		−1.857
308 K		−2.538
313 K		−3.330
318 K		−3.297
323 K		−3.263
Standard enthalpy change	$\ln K_C = \frac{\Delta S^0}{R} - \frac{\Delta H^0}{RT}$	19.95
Standard entropy change		72.9×10^{-3}

K_C is the equilibrium constant, T and R are as defined in previous equation

2.1 to 20.9 kJ mol^{-1} corresponds to physical adsorption whereas ranging from 20.9 to 418 kJ mol^{-1} is regarded as chemical adsorption. Therefore, the ΔH° value (19.95 kJ mol^{-1}) confirms that the biosorption process of As(III) by AiBP occurred due to physisorptions. Furthermore, the positive ΔS° value also corresponds to an increase in the degree of freedom of the adsorbed species during the biosorption process (Baig et al. 2010; Prasad et al. 2011; Das et al. 2014).

Column operation

Central composite design (CCD)

The RSM was applied to determine the optimized set of operational variables in the arsenic(III) biosorption process for the column study. The CCD in RSM was designed for a set of three independent variables, viz., initial As(III) concentration, flow rate, and biosorbent dose, to investigate their influence on the breakthrough time for continuous fixed-bed biosorption of As(III) by AiBP. The experimental range of the variables with their unit and notation used in CCD is given in Table 4.

The central composite design was developed by "Design–Expert" software. According to this design, a total of 20 experiments were performed in duplicate to the CCD matrix as per Table 5.

Table 4 Independent variables and their levels employed in the CCD

Variables	Units	Notations	Level of variables	
			Low	High
Initial concentration	µg L^{-1}	A	500.0	2000.0
Flow rate	mL min^{-1}	B	3.0	7.0
Biosorbent dose	g	C	2.0	6.0

Evaluation of the fitting process model and ANOVA

To detect and suggest a valid model, the actual responses were fitted with existing linear, two factor interactions (2FI), cubic and quadratic model by CCD. Based on suggestion, the quadratic model was selected to continue the progress (Table 6).

The selected quadratic model was validated by a few numbers of statistical evidences in the analysis of variance (ANOVA). The evidences were Fisher variation ratio (F value), probability value (p value), lack of fit, coefficient of determination R-squared (R^2_{d}), adjusted R-squared(R^2_{Adj}), predicted R-squared (R^2_{Pred}), and adequate precision. Adequate precision is a signal to-noise ratio, which compares the range of the predicted values at the design points to the average prediction error. The ratios greater than 4 indicate adequate model discrimination. R^2_{Adj} and the R^2_{Pred} are measurements of the amount of variation around the mean and new explained data, respectively. The p value represents the degree of significance of each variable while F value is a statistically valid measure of how well the factors describe the variation in the data about its mean (Abdollahi et al. 2012; Kumar and Phanikumar 2013; Chattoraj et al. 2014; Roy et al. 2014; Sadhukhan et al. 2014).

The selected model was validated by ANOVA which provided evidences such as high F value (20202.14), very low p value (<0.0001), non-significant lack of fit, the high coefficient of R-squared (0.9999), adjusted R-squared (0.9999), predicted R-squared (0.9998), and the adequate precision (491.63). Moreover, Fig. 12 shows the actual values versus predicted values of the As(III) biosorption for column study, which indicates an excellent agreement between actual and predicted values. As observed, the validity (significance and adequacy) of the model was confirmed by the reasonable evidence.

The empirical relationships relating the breakthrough time to the tested independent variables have been expressed in terms of unit less regression coefficient by the quadratic model is given by the following equation:

$$\begin{aligned} \text{Breakthrough time (min)} &= +629.87 - 161.24\mathrm{A} \\ &- 201.96\mathrm{B} + 230.59\mathrm{C} - 90.64\mathrm{AB} - 106.44\mathrm{AC} \\ &- 53.89\mathrm{BC} + 3.14\mathrm{A}^2 + 38.89\mathrm{B}^2 + 27.97\mathrm{C}^2 \end{aligned} \tag{3}$$

where A: initial concentration, B: flow rate, and C: biosorbent dose are in coded factors.

A regression analysis of the model equation (Table 7) shows that the main as well as the interaction effects of initial As(III) concentration, flow rate, and biosorbent dose were highly significant ($p < 0.0001$). Also, the square effect of flow rate and biosorbent dose was highly significant ($p < 0.0001$).

Table 5 CCD for three used independent variables and the observed response (breakthrough time: min) for As(III)

Standard order	Run order	Factor 1 A: initial concentration (μg L^{-1})	Factor 2 B: flow rate (mL min^{-1})	Factor 3 C: biosorbent dose (g)	Response Breakthrough time (min)
4	1	500.0	3.0	2.0	581.0
18	2	500.0	3.0	6.0	1366.3
15	3	500.0	5.0	2.0	483.8
2	4	500.0	5.0	6.0	1159.4
13	5	500.0	7.0	2.0	466.2
14	6	500.0	7.0	6.0	1032.2
3	7	1250.0	3.0	2.0	615.8
7	8	1250.0	3.0	6.0	1174.8
11	9	1250.0	5.0	2.0	428.2
10	10	1250.0	5.0	6.0	890.6
5	11	1250.0	7.0	2.0	320.2
20	12	1250.0	7.0	6.0	673.0
1	13	2000.0	3.0	2.0	652.2
8	14	2000.0	3.0	6.0	1013.0
9	15	2000.0	5.0	2.0	374.2
17	16	2000.0	5.0	6.0	623.4
16	17	2000.0	7.0	2.0	175.8
6	18	2000.0	7.0	6.0	315.4
19	19	500.0	3.0	4.0	943.0
12	20	500.0	3.0	4.0	945.4

Table 6 Sequential model sum of squares

Source	Sum of squares	*df*	Mean square	*F* value	*p* value, Prob > *F*	
Mean vs total	10130195	1	10130195			
Linear vs mean	1862252	3	620750.6	38.51	<0.0001	
2FI vs linear	251448.7	3	83816.23	169.3	<0.0001	
Quadratic vs 2FI	6319.651	3	2106.55	180.7	<0.0001	*Suggested*
Cubic vs quadratic	58.73597	5	11.74719	1.015	0.4937	Aliased
Residual	57.86444	5	11.57289			
Total	12250332	20	612516.6			

Interaction effect of biosorption process variables

Contour plots: In order to study the interaction among the different independent variables and their corresponding effect on the response, contour plots were drawn (Figs. 13, 14, 15). Contour plot is the projection of the response surface as a two-dimensional plane. This analysis gives a better understanding of the influence of variables and their interaction on the response.

The combined effect of initial As(III) concentration and flow rate on breakthrough time is shown in the contour plot of Fig. 13. The breakthrough time decreases with increase in both the initial As(III) concentration as well as the flow rate within the experimental range. Such behavior can be explained by the fact that all biosorbents have a limited number of active sites which become saturated at a certain concentration. With increasing inlet arsenic(III) concentration, the active sites on the biosorbent surface become more quickly saturated and hence the breakthrough time is decreased (Bhakat et al. 2007; Ranjan et al. 2009; Das et al. 2014). Also at higher flow rates, the residence time of the arsenic(III) solution in the column decreases. The insufficient residence time decreases the biosorption efficiency of As(III) on the biosorbent surface or diffusion into the pores of the biosorbent, leaving the column before equilibrium has reached with result in earlier breakthrough time (Gupta and Sankararamakrishnan 2010; Chowdhury et al. 2013; Saha et al. 2012; Roy et al. 2013a).

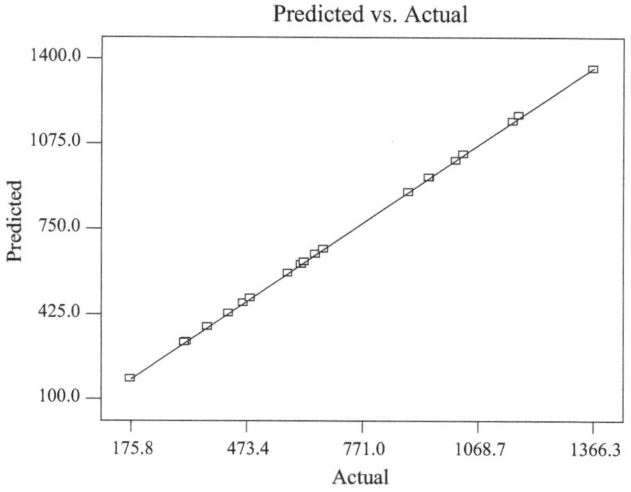

Fig. 12 Plot of actual response versus predicted response for column operation of As(III) biosorption by AiBP

Table 7 Regression Analysis using CCD

Model terms	Coefficient estimate	Standard error	*F* value	*p* value, Prob > *F*
A	−161.24	0.986	26757	<0.0001
B	−201.69	0.986	41866	<0.0001
C	230.59	0.805	82086	<0.0001
AB	−90.64	1.207	5636	<0.0001
AC	−106.44	0.986	11660	<0.0001
BC	−53.89	0.986	2989	<0.0001
A^2	3.14	1.707	3.386	0.0956
B^2	38.89	1.707	518.9	<0.0001
C^2	27.97	3.244	74.31	<0.0001

Figure 14 depicts the interaction effect of initial As(III) concentration and biosorbent dose on the breakthrough time. Within the experimental range, the response function, i.e., the breakthrough time decreases with increasing initial As(III) concentration. On the contrary, the breakthrough time increases with increasing biosorbent dose. The observed trend can be attributed to the fact that increase in biosorbent doses in larger beds provides a greater number of active sites for As(III) with increase in the surface area of biosorbent (Bhakat et al. 2007; Ranjan et al. 2009; Roy et al. 2013a). The arsenic(III) has more time to contact with the biosorbent resulting in delayed breakthrough (Chowdhury et al. 2013; Saha et al. 2012; Roy et al. 2014).

The contour plot in Fig. 15 shows the combined effect of flow rate and biosorbent dose on the breakthrough time for continuous biosorption of As(III) by AiBP. The interactive effect of flow rate and biosorbent dose showed a significant impact on the breakthrough time. The breakthrough time decreases with increasing flow rate while it increases with increasing biosorbent dose. Such breakthrough pattern can be explained in terms of residence time of the arsenic ion in the column as well as the availability of active sites, as already mentioned.

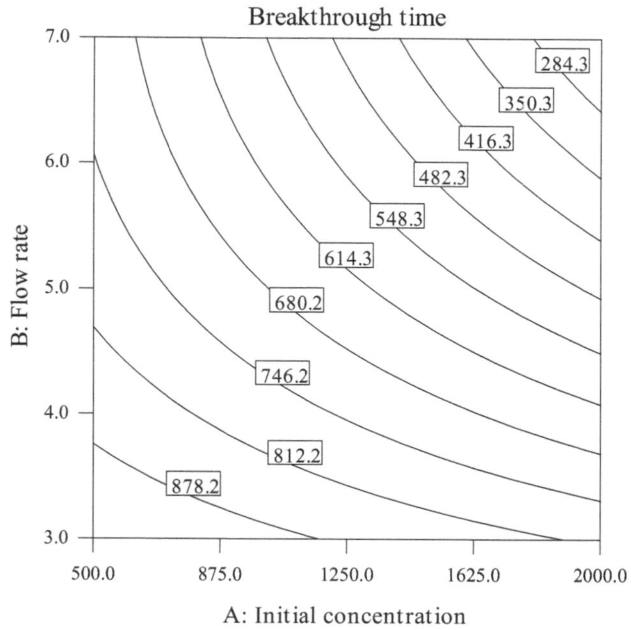

Fig. 13 Contour plot showing the combined effect of initial arsenic(III) concentration and flow rate on the breakthrough time for column operation of As(III) biosorption by AiBP. Biosorbent dose of 4.0 g

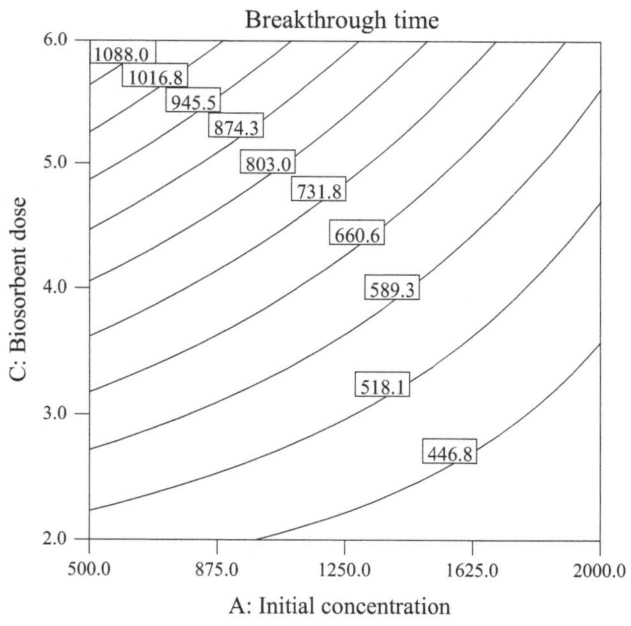

Fig. 14 Contour plot showing the combined effect of initial arsenic(III) concentration and biosorbent dose on the breakthrough time for column operation of As(III) biosorption by AiBP. Flow rate of 5.0 mL min^{-1}

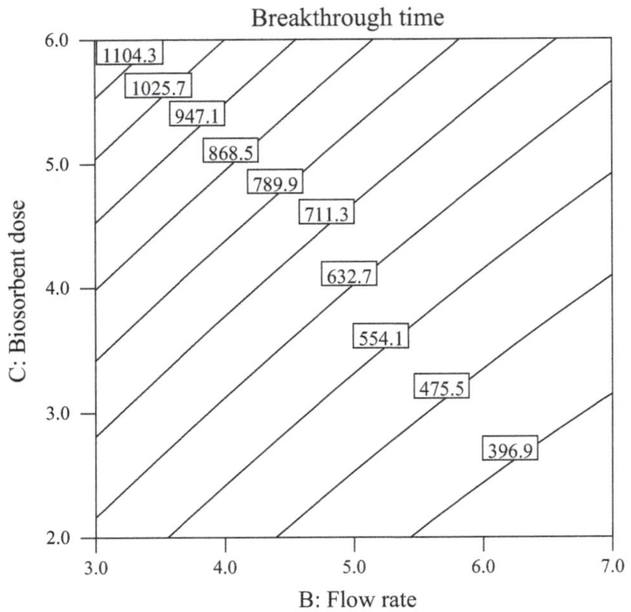

Fig. 15 Contour plot showing the combined effect of flow rate and biosorbent dose on the breakthrough time for column operation of As(III) biosorption by AiBP. Initial arsenic(III) concentration of 1250.0 μg L^{-1}

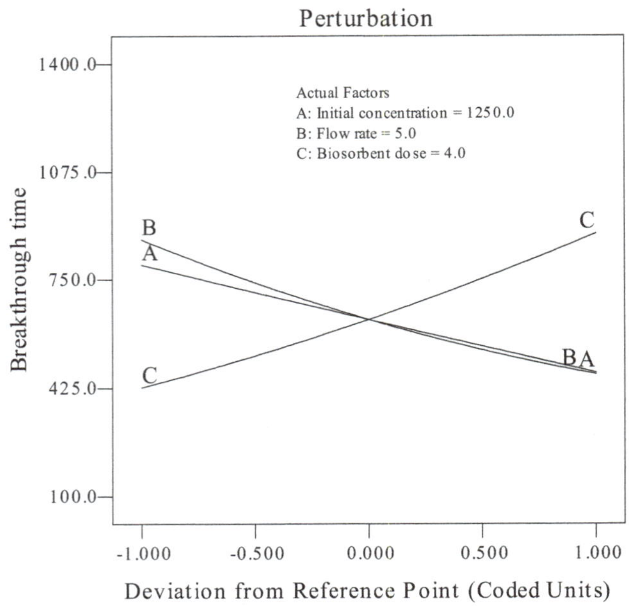

Fig. 16 Perturbation plot showing the effect of the tested variables on breakthrough time

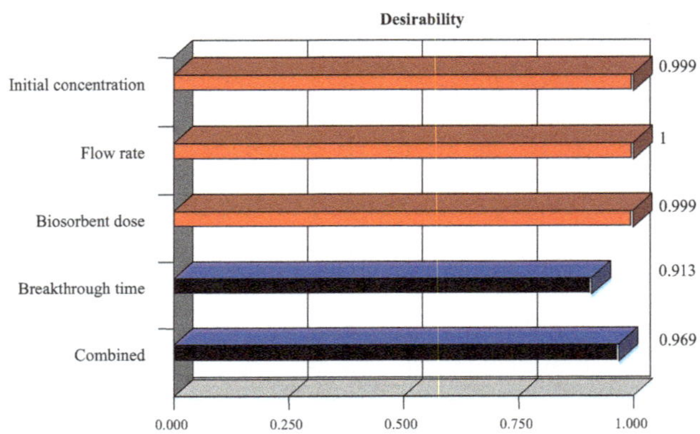

Fig. 17 Bar graph for optimization procedure

Perturbation plot: The effects of initial As(III) concentration, flow rate, and biosorbent dose on the breakthrough time were evaluated. The individual effect of independent variables including initial As(III) concentration (A), flow rate (B), and biosorbent dose (C) was found by perturbation plot. A perturbation plot does not show the effect of interactions and it is like one factor-at-a-time experimentation. The perturbation plot as in Fig. 16 helps comparing the effect of all independent variables at a particular point in the design space. The response is plotted by changing only one factor over its range while holding the other factors constant. A steep slope or curvature in a factor shows that the response is sensitive to that factor (Kumar and Phanikumar 2013; Roy et al. 2014; Sadhukhan et al. 2014). A relatively flat line shows insensitivity to change in that particular factor. The present result revealed that breakthrough time is more sensitive to biosorbent dose followed by flow rate and initial concentration.

Optimization using the desirability function

The possible goals in the Design–Expert software's are to maximize, minimize, target, in range and set to an exact value (factors only). In numerical optimization, the desired goal was preferred for each variable and response from menu.

Figure 17 demonstrates the desirability values of optimization procedure in which the criterion was set as "minimum" for AiBP dose, "maximum" for initial As(III) concentration, "in range" for flow rate and the goal was set as "maximum" to analyze economically viable optimal condition. The objective of this process was to find the maximum breakthrough time by utilizing less amount of biosorbent dose. In this, the desirability value ranges from 0.913 to 1 for individual variables and 0.969 for combination of all the variables.

By seeking from 100 starting points in the response surface changes, the best local maximum breakthrough time out of 34 solutions was found to be 653.9 min when the independent variables retained at 2.0 g AiBP dose, 2000.0 μg L^{-1} initial As(III) concentrations, and 3.0 mL min^{-1} flow rate at maximum desirability value of 0.969. The optimized result obtained from CCD coincides well with experimental value, suggesting that the AiBP may be an effective and economically feasible biosorbent for the removal of As(III) from water.

Scanning electron microscopy (SEM) analysis

SEM analysis is a useful tool for the study of the surface morphology of the biosorbent. The SEM micrographs (20 kV; 10 μm) at 1,000 magnifications for AiBP surface before and after As(III) biosorption at the optimization conditions of column operation are shown in Figs. 18 and 19, respectively. Figure 18 clearly shows the presence of porous, rough, and irregular surface morphology of the unloaded AiBP. The surface roughness of an unloaded biosorbent indicated the availability of a tremendous surface area of AiBP meant for high adsorption capacity in the biosorption process (Kamsonlian et al. 2012; Das et al. 2014; Roy et al. 2014). Contrary to this, after biosorption, the biosorbent surface seemed to have decreased porosity due to the heavy impregnation of As(III) onto the surface of the AiBP biomass (Fig. 19).

Fig. 18 SEM micrograph of AiBP before treatment of As(III)

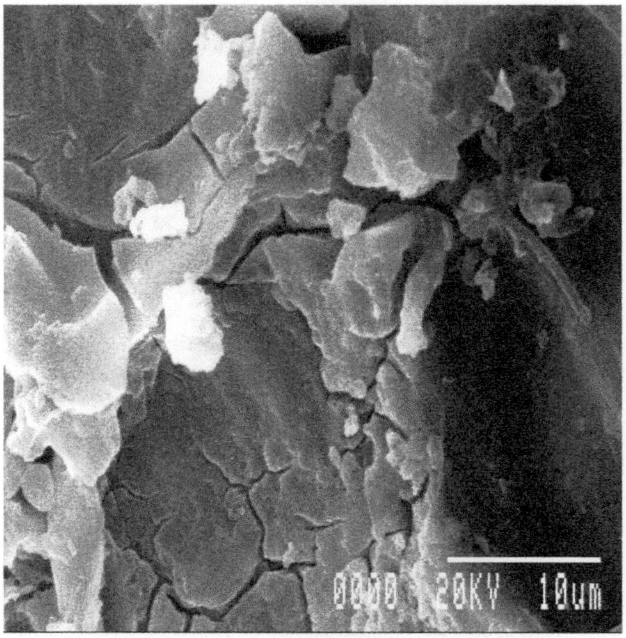

Fig. 19 SEM micrograph of AiBP after treatment of As(III) in column operation

Fixed-bed column performance using real contaminated groundwater (CGW)

The proposed indigenous biosorbent AiBP was satisfactorily tested for the removal of arsenic from real CGW sample. Hundred ten natural CGW samples were collected from an arsenic polluted region of Purbasthali Block-II in the district of Burdwan, West Bengal, India (Roy et al. 2013b). Out of 110 tubewell water samples, the maximum concentration of inorganic arsenic (total) was 261 μg L^{-1} as recorded in our earlier study. Removal of arsenic from the CGW sample (total arsenic = 261 μg L^{-1}, As(III) = 112 μg L^{-1}, and As(V) = 149 μg L^{-1}) by fixed-bed column experiment was performed at optimal condition of 3.0 mL min^{-1} flow rate and 2.0 g AiBP dose. After treating 15 L of CGW through the proposed column operation, the results on water quality before and after biosorption under study are shown in Table 8. The results

Table 8 Physico-chemical parameters of water and removal of As(III) by AiBP

Parameters (Unit)	Before biosorption	After biosorption
pH	7.2	6.9
Electrical conductivity (μS cm^{-1})	3.8	3.4
Total dissolve solid (mg L^{-1})	536.8	455.7
Carbonate (mg L^{-1})	40.5	29.6
Bicarbonate (mg L^{-1})	220.3	176.3
Calcium (mg L^{-1})	126.2	87.6
Magnesium (mg L^{-1})	43.4	25.9
Chloride (mg L^{-1})	30.2	21.8
Sodium (mg L^{-1})	38.4	36.6
Potassium (mg L^{-1})	2.7	2.3
Iron (mg L^{-1})	14.6	5.1
Sulfate (mg L^{-1})	3.8	2.4
Phosphate (mg L^{-1})	1.4	0.3
Fluoride (mg L^{-1})	0.5	BDL
Antimony (mg L^{-1})	0.3	BDL
Total arsenic (μg L^{-1})	261	33.2
Arsenic(III) (μg L^{-1})	112	12.3
Arsenic(V) (μg L^{-1})	149	20.9

BDL below detection limit

of column operation using real CGW sample, containing both As(III) and As(V) ions, showed that biosorbent is effective to remove both arsenic species from natural CGW. However, the concentration of arsenic (33.2, 12.3, and 20.9 μg L^{-1}) in treated sample was higher than the safe level recommended by WHO. An additional biosorption step can be employed when single-stage biosorption is not enough to reduce the concentration of a pollutant to the recommended safe limit. Therefore, the present study recommends a two-stage biosorption by use of AiBP to reduce the arsenic concentration below 10 μg L^{-1}.

Conclusion

AiBP was found to be efficacious in the removal of arsenic(III) from water. The effect of different process parameters (i.e., solution pH, contact time, initial concentration of As(III), AiBP dose, stirring rate, and temperature) in batch operation were studied for the biosorption of As(III) on AiBP. ANN model was applied upon batch experimental values to archive the reasonable predictive performance of the biosorbent; additionally the importance analysis of ANN provided that the initial concentration was the most significant parameter for the batch operation. The model was also found to be working satisfactorily as evidenced with average relative error of 0.352 and sum square error of 10.065 during testing phase to predict the biosorption efficiency with reasonable accuracy. The study on equilibrium biosorption of batch operation revealed that Langmuir isotherm model gave the best fit to experimental data. The nature of biosorption of As(III) by AiBP was physisorption as inferred from the D–R isotherm model. The biosorption is pseudo-second-order, endothermic, and spontaneous. For column operation, CCD was applied to investigate the influence on the breakthrough time for optimization of As(III) biosorption process and evaluation of interacting effects of three different operating variables. A correlation coefficient (R^2) value of 0.999, model F value of 20202.14, and its low p value (<0.0001) along with higher adequate precision (491.63) indicated the fitness of the response surface quadratic model developed in the present study. The optimized result of CCD revealed that the AiBP was supposed to be an effective and economically feasible biosorbent for the removal of As(III) from the aqueous system. The effectiveness of the biosorbent is also critically examined by treating a contaminated groundwater sample collected form an arsenic affected area of Purbasthali Block-II, Burdwan, West Bengal, India. Overall, it is concluded that the data presented in this investigation can be further extrapolated for designing and establishing an efficient arsenic removal scheme while treating the groundwater that might have contaminated.

Acknowledgments The authors are thankful to the Department of Environmental Science, University of Burdwan, West Bengal, India for providing laboratory facilities for the work.

References

Abdollahi Y, Zakaria A, Matori KA, Shameli K, Jahangirian H, Rezayi M, Abdollahi T (2012) Interactions between photodegradation components. Chem Cent J 6:100. doi:10.1186/1752-153X-6-100

Arbab-Zavar MH, Hashemi M (2000) Evaluation of electrochemical hydride generation for spectrophotometric determination of As(III) by silver diethyldithiocarbamate. Talanta 52:1007–1014

Arshad M, Zafar MN, Younis S, Nadeem R (2008) The use of neem biomass for the biosorption of zinc from aqueous solutions. J Hazard Mater 157:534–540

Baig JA, Kazi TG, Shah AQ, Kandhro GA, Afridi HI, Khan S, Kolachi NF (2010) Biosorption studies on powder of stem of *Acacia nilotica*: removal of arsenic from surface water. J Hazard Mater 178:941–948

Bhakat PB, Gupta AK, Ayoob S (2007) Feasibility analysis of As(III) removal in a continuous flow fixed bed system by modified calcined bauxite (MCB). J Hazard Mater B139:286–292

Bhattacharya AK, Mandal SN, Das SK (2006) Adsorption of Zn(II) from aqueous solution by using different adsorbents. Chem Eng J 123:43–51

Bhattacharya AK, Naiya TK, Mandal SN, Das SK (2008) Adsorption, kinetics and equilibrium studies on removal of Cr(VI) from aqueous solutions using different low-cost adsorbents. Chem Eng J 138:529–541

Bhaumik R, Mondal NK, Das B, Roy P, Pal KC (2011) Predicting iron adsorption capacity and thermodynamics onto calcareous soil from aqueous solution by linear regression and neural network modeling. Univers J Environ Res Technol 1:486–499

Chattoraj S, Mondal NK, Das B, Roy P, Sadhukhan B (2014) Biosorption of carbaryl from aqueous solution onto *Pistia stratiotes* biomass. Appl Water Sci 4:79–88

Chowdhury S, Chakraborty S, Saha PD (2013) Response surface optimization of a dynamic dye adsorption process: a case study of crystal violet adsorption onto NaOH-modified rice husk. Environ Sci Pollut Res 20:1698–1705

Das B, Mondal NK, Roy P, Chattaraj S (2013) Equilibrium, kinetic and thermodynamic study on chromium(VI) removal from aqueous solution using *Pistia stratiotes* biomass. Chem Sci Trans 2:85–104

Das B, Mondal NK, Bhaumik R, Roy P (2014) Insight into adsorption equilibrium, kinetics and thermodynamics of lead onto alluvial soil. Int J Environ Sci Technol 11:1101–1114

Giri AK, Patel RK, Mahapatra SS (2011) Artificial neural network (ANN) approach for modelling of arsenic(III) biosorption from aqueous solution by living cells of *Bacillus cereus* biomass. Chem Eng J 178:15–25

Giri AK, Patel RK, Mahapatra SS, Mishra PC (2013) Biosorption of arsenic(III) from aqueous solution by living cells of *Bacillus cereus*. Environ Sci Pollut Res 20:1281–1291

Guo X, Chen F (2005) Removal of arsenic by bead cellulose loaded with iron oxyhydroxide from groundwater. Environ Sci Technol 39:6808–6818

Gupta A, Sankararamakrishnan N (2010) Column studies on the evaluation of novel spacer granules for the removal of arsenite and arsenate from contaminated water. Bioresour Technol 101:2173–2179

Hautfenne A (1980) Standard methods for the analysis of oils, fats and derivatives, 6th Edn. 1st Supplement: Part 2, IUPAC, Great Britain

Kamala CT, Chu KH, Chary NS, Pandey PK, Ramesh SL, Sastry AR, Sekhar KC (2005) Removal of arsenic(III) from aqueous solutions using fresh and immobilized plant biomass. Water Res 39:2815–2826

Kamsonlian S, Suresh S, Majumder CB, Chand S (2012) Biosorption of As(III) from contaminated water onto low cost palm bark biomass. Int J Curr Eng Technol 2:153–158

Kanwal F, Rehman R, Mahmud T, Anwar J, Ilyas R (2012) Isothermal and thermodynamical modeling of chromium(III) adsorption by composites of polyaniline with rice husk and saw dust. J Chil Chem Soc 57:1058–1063

King P, Anuradha K, Lahari SB, Prasanna Kumar Y, Prasad VS (2008) Biosorption of zinc from aqueous solution using *Azadirachta indica* bark: equilibrium and kinetic studies. J Hazard Mater 152:324–329

Kumar MPS, Phanikumar BR (2013) Response surface modelling of Cr^{6+} adsorption from aqueous solution by neem bark powder: box–behnken experimental approach. Environ Sci Pollut Res 20:1327–1343

Maiti A, Basu JK, De S (2010) Development of a treated laterite for arsenic adsorption: effects of treatment parameters. Ind Eng Chem Res 49:4873–4886

Mamun AA, Maan FR, Zahiarh AK, Yehya MA, Mohammed ARS, Alam MZ, Muyibi SA, Faris IA, Azin I (2009) Optimisation of arsenic adsorption from water by carbon nanofibers grown on powdered activated carbon impregnated with nickel. J Appl Sci 9:3180–3183

Mohan D, Pittman CU Jr (2007) Arsenic removal from water/wastewater using adsorbents–a critical review. J Hazard Mater 142:1–53

Naiya TK, Chowdhury P, Bhattacharya AK, Das SK (2009) Saw dust and neem bark as low-cost natural biosorbent for adsorptive removal of Zn(II) and Cd(II) ions from aqueous solutions. Chem Eng J 148:68–79

Nigam S, Gopal K, Vankar PS (2013) Biosorption of arsenic in drinking water by submerged plant: *Hydrilla verticilata*. Environ Sci Pollut Res 20:4000–4008

Pennesi C, Veglio F, Totti C, Romagnoli T, Beolchini F (2012) Nonliving biomass of marine macrophytes as arsenic(V) biosorbents. J Appl Phycol 24:1495–1502

Prasad KS, Srivastava P, Subramanian V, Paul J (2011) Biosorption of As(III) ion on *Rhodococcus* sp. WB-12: biomass characterization and kinetic studies. Sep Sci Technol 46:2517–2525

Raj KR, Kardam A, Arora JK, Srivastava S (2013) An application of ANN modeling on the biosorption of arsenic. Waste Biomass Valorization 4:401–407

Rajesh Kannan R, Rajasimman M, Rajamohan N, Sivaprakash B (2010) Equilibrium and kinetic studies on sorption of malachite green using *Hydrilla verticillata* biomass. Int J Environ Res 4:817–824

Ranjan D, Talat M, Hasan SH (2009) Rice polish: an alternative to conventional adsorbents for treating arsenic bearing water by up-flow column method. Ind Eng Chem Res 48:10180–10185

Ratna Kumar P, Chaudhari S, Khilar KC, Mahajan SP (2004) Removal of arsenic from water by electrocoagulation. Chemosphere 55:1245–1252

Roy P, Mondal NK, Bhattacharya S, Das B, Das K (2013a) Removal of arsenic(III) and arsenic(V) on chemically modified low-cost adsorbent: batch and column operation. Appl Water Sci 3:293–309

Roy P, Mondal NK, Das B, Das K (2013b) Arsenic contamination in groundwater: a statistical modeling. J Urban Environ Eng 7:24–29

Roy P, Mondal NK, Das K (2014) Modeling of the adsorptive removal of arsenic: a statistical approach. J Environ Chem Eng 2:585–597

Sadhukhan B, Mondal NK, Chattoraj S (2014) Biosorptive removal of cationic dye from aqueous system: a response surface methodological approach. Clean Techn Environ Policy 16:1015–1025

Saha PD, Chakraborty S, Chowdhury S (2012) Batch and continuous (fixed-bed column) biosorption of crystal violet by *Artocarpus heterophyllus* (jackfruit) leaf powder. Colloids Surf B Biointerfaces 92:262–270

Saqib ANS, Waseem A, Khan AF, Mahmood Q, Khan A, Habib A, Khan AR (2013) Arsenic bioremediation by low cost materials derived from blue pine (*Pinus wallichiana*) and walnut (*Juglans regia*). Ecol Eng 51:88–94

Sari A, Tuzen M (2010) Biosorption of As(III) and As(V) from aqueous solution by lichen (*Xanthoria parietina*) biomass. Sep Sci Technol 45:463–471

Sari A, Uluozlu OD, Tuzen M (2011) Equilibrium, thermodynamic and kinetic investigations on biosorption of arsenic from aqueous solution by algae (*Maugeotia genuflexa*) biomass. Chem Eng J 167:155–161

Srivastava R, Rupainwar DC (2010) Liquid phase adsorption of indigo carmine and methylene blue on neem bark. Desalin Water Treat 24:74–84

Srivastava R, Rupainwar DC (2011) A comparative evaluation for adsorption of dye on neem bark and mango bark powder. Indian J Chem Technol 18:67–75

Suresh S, Srivastava VC, Mishra IM (2012) Adsorptive removal of aniline by granular activated carbon from aqueous solutions with catechol and resorcinol. Environ Technol 33:773–781

Tiwari D, Mishra SP, Mishra M, Dubey RS (1999) Biosorptive behaviour of mango (*Mangifera indica*) and neem (*Azadirachta indica*) bark for Hg^{2+}, Cr^{3+} and Cd^{2+} toxic ions from aqueous solutions: a radiotracer study. Appl Radiat Isot 50:631–642

Wu Y, Wen Y, Zhou J, Dai Q, Wu Y (2012) The characteristics of waste *Saccharomyces cerevisiae* biosorption of arsenic(III). Environ Sci Pollut Res 19:3371–3379

11

A new approach using coagulation rate constant for evaluation of turbidity removal

Mukheled Al-Sameraiy[1]

Abstract Coagulation–flocculation–sedimentation processes for treating three levels of bentonite synthetic turbid water using date seeds (DS) and alum (A) coagulants were investigated in the previous research work. In the current research, the same experimental results were used to adopt a new approach on a basis of using coagulation rate constant as an investigating parameter to identify optimum doses of these coagulants. Moreover, the performance of these coagulants to meet (WHO) turbidity standard was assessed by introducing a new evaluating criterion in terms of critical coagulation rate constant (k_c). Coagulation rate constants (k_2) were mathematically calculated in second order form of coagulation process for each coagulant. The maximum (k_2) values corresponded to doses, which were obviously to be considered as optimum doses. The proposed criterion to assess the performance of coagulation process of these coagulants was based on the mathematical representation of (WHO) turbidity guidelines in second order form of coagulation process stated that (k_2) for each coagulant should be $\geq$ (k_c) for each level of synthetic turbid water. For all tested turbid water, DS coagulant could not satisfy it. While, A coagulant could satisfy it. The results obtained in the present research are exactly in agreement with the previous published results in terms of finding optimum doses for each coagulant and assessing their performances. On the whole, it is recommended considering coagulation rate constant to be a new approach as an indicator for investigating optimum doses and critical coagulation rate constant to be a new evaluating criterion to assess coagulants' performance.

Keywords Coagulation rate constant · Kinetics modeling · Coagulants' performance · Turbidity removal

Introduction

Coagulation is the process of destabilizing colloids, aggregating them, and binding them together for ease of sedimentation. This is achieved by adding simple salts or by charge neutralization, resulting in a tendency in the aggregates (coagula) to be small and dense. Flocculation is limited to the cases where polymer bridging is the dominant mechanism and aggregates (flocs) tend to be larger and more open in structure. In the area of water and wastewater treatment, coagulation refers to destabilization by the dosing of appropriate additives and flocculation refers to the formation of aggregates usually by some form of fluid motion (Gregory 2006).

The coagulation mechanisms depending on the physical and chemical properties of the solution, pollutant and coagulant include charge neutralization, double layer compression, bridging and sweep (Holt et al. 2002).

Coagulation and flocculation constitute the backbone processes in most water and advanced wastewater treatments plants (Wang et al. 2005) in which they are of great importance in solid–liquid separation practice (Yukselen and Gregory 2004) and widely used to remove turbidity in the form of suspended as well as colloidal materials from synthetic turbid water (Jadhav and Mahajan 2013; Kılıç and Hoşten 2010) and raw surface water (Esther et al. 2013; Zainal-Abideen et al. 2012).

The expression of kinetic was early introduced for the time evolution of particle aggregates on the basis of a mean filed of theory of identical particles in a matrix. The rate of aggregation of particles depends on physical properties of

✉ Mukheled Al-Sameraiy
mukheled73@gmail.com; mahas20@bath.ac.uk

1 Environmental Research Center, University of Technology, Baghdad, Iraq

the system such as shape and size of particles, composition, interactive forces, frequency of collision and probability of sticking once collision has occurred. These factors can vary based on many conditions such as temperature, solution pH or other chemical properties that affect particle correlations, and presence of an external field (Smoluchowski 1917).

Modeling is a valuable engineering tool in both design and operation of treatment plants. It also helps to develop a better understanding of the treatment processes and provide a significant potential for solving operational problems as well as reducing operational cost in a specific treatment processes. Coagulation kinetics modeling of water treatment plants can be used for process optimization and testing of control strategies to meet effluent quality requirements at a reasonable cost. The kinetics of coagulation has been investigated for several kinetics functional parameters such as order of coagulation process, coagulation rate constant, coagulation time, coagulation period, collision factor for Brownian transport and collision efficiency. In addition, various forms of the rate law equation have been analytically solved (Okolo et al. 2014; Nnaji et al. 2014; Chukwudi and Onukwuli 2011; Ani et al. 2011).

The objective of the current research was to develop a new approach using coagulation rate constant determined on the basis of experimental results reported by Al-Sameraiy (2012) as an indicator to investigate the optimum doses of date seeds and alum coagulants, and introducing a new evaluating criterion in terms of critical coagulation rate constant to assess the performance of coagulation process of these coagulants in achieving the minimum level of residual turbidity to satisfy WHO turbidity standard guidelines.

Materials and methods

The three selected levels of bentonite synthetic turbid water namely: low (75 NTU), medium (150 NTU) and high (300 NTU) as well as the doses of natural coagulant (date seeds, DS) and inorganic coagulant (alum, A) in the range (10–100 mg/L) under various settling times from 30 to 120 min using a standard jar test apparatus to simulate coagulation, flocculation and sedimentation processes were carried out by Al-Sameraiy (2012). In the current research, his experimental data findings were used to investigate coagulation kinetics modeling in terms of second order of coagulation process form for coagulation rate constant and critical coagulation rate constant.

Coagulation kinetics modeling

It is well known that the fundamentals of the rate of aggregation started from the classic work of Smoluchowski (1917). After a period of the aggregation of particles, size distribution and shape of these particles can be changed basically. To describe the aggregation rate of particles count based on the Brownian controlled and rapid coagulation, the general differential equation can be shown below:

$$-\frac{\mathrm{dc}}{dt} = \mathrm{k}\mathrm{c}^{\alpha} \quad (1)$$

where α is the order of coagulation process, k is the coagulation rate constant in (L/mg·min), c is the total concentration of constituent particles in (mg/L) at time t (min).

To simplify and solve the Eq. (1), the theoretical values of the order of coagulation process (α) are in the range of ($1 \leq \alpha \leq 2$) (Chukwudi et al. 2009; WST 2003). Basically, Elimelech et al. (1995) proposed that the aggregation process is predominant in the form of second order by which the collision is proportional to the product of concentrations of two colliding species. Moreover, in real and empirical practice, extensive studies such as (Ani et al. 2012; Ifeanyi et al. 2012; Ugonabo et al. 2013) used ($\alpha = 2$) and found that it was more logical in representing primarily the aggregation rate of particles count on the basis of Brownian controlled and rapid coagulation process. Based on the above aforementioned, substituting ($\alpha = 2$) in the Eq. (1) and integrating it with the following boundary conditions, at initial condition ($t = 0$, $c = c_0$) and at final condition ($t = t$, $c = c$) to obtain Eq. (2):

$$\frac{1}{c} = \mathrm{k}_2 t + \frac{1}{c_0} \quad (2)$$

Rearranging the Eq. (2) to get Eq. (3), which is used to calculate the values of (k_2) mathematically in the current research:

$$\mathrm{k}_2 = \frac{1/c - 1/c_0}{t} \quad (3)$$

where k_2 is the second order coagulation rate constant.

The total concentration of bentonite constituent particles of three levels of synthetic turbid water as blank before treatment (c_0) and after treatment (c) used in Eq. (3) could be expressed in either (mg/L) and the unit of (k_2) would be (L/mg·min) or (NTU) as turbidity reading (WST 2003; Ni'am et al. 2007) and the unit of (k_2) would be ($\mathrm{NTU}^{-1}\cdot\mathrm{min}^{-1}$).

Results and discussion

Coagulation rate constant as an investigating parameter for optimum coagulant dose

Dosage is one of the most important parameters, which has been considered to determine the optimum conditions for the coagulation and flocculation. Basically, insufficient dosage or overloading would result in the poor performance in flocculation. Therefore, it was crucial to determine the optimum dosage in order to minimize the dosing cost and obtain the optimum performance in treatment (American Water Works Association (AWWA) 2000). In conventional water treatment practice, finding the optimum dose for each coagulant is a problem that must be solved and determined by empirical experiments. The experimental results obtained and reported by Al-Sameraiy (2012) are briefly summarized in Tables 1, 2, 3 based on the best settling time of (120 min). It is clear to see from these tables that the values of residual turbidity of DS coagulant are higher than that of A coagulant at all the investigated doses from 10 to 100 mg/L for three tested levels of low, medium and high synthetic turbid water. This variation in performance of these coagulants is expected due to every coagulant has ability in terms of coagulation activity for turbidity removal. In consequence, this ability plays a vitally important role in investigating the optimum dose for coagulation–flocculation process (Al-Sameraiy 2012). The objective of this section is only to use coagulation rate constant (k_2) as an investigating parameter to identify optimum doses of (DS and A) coagulants in spite of their performances in terms of satisfying the (WHO) turbidity standard of drinking water.

In the current research, the experimental results recorded in Tables 1, 2, 3 were used to find the values of second order coagulation rate constant (k_2) in which these values

Table 1 Investigated optimum dose for (DS and A) coagulants as a function of residual turbidity according to WHO turbidity standard at low synthetic turbid water (Al-Sameraiy 2012)

Investigated doses (mg/L)	DS coagulant	Optimum dose (mg/L)	WHO turbidity standard ≤5 NTU	A coagulant	Optimum dose (mg/L)	WHO turbidity standard ≤ 5NTU
	Residual turbidity (*c*) in (NTU) after settling time (120 min)					
10	11.40	30	Not satisfied	2.10	10	Satisfied
20	7.90			5.87		
30	5.20			8.13		
40	7.40			7.76		
50	6.45			8.68		
60	8.20			7.23		
70	8.98			9.36		
80	9.70			7.56		
90	9.16			9.81		
100	10.30			9.41		

Table 2 Investigated optimum dose for (DS and A) coagulants as a function of residual turbidity according to WHO turbidity standard at medium synthetic turbid water (Al-Sameraiy 2012)

Investigated doses (mg/L)	DS coagulant	Optimum dose (mg/L)	WHO turbidity standard ≤5 NTU	A coagulant	Optimum dose (mg/L)	WHO turbidity standard ≤5 NTU
	Residual turbidity (*c*) in (NTU) after settling time (120 min)					
10	25.90	60	Not satisfied	6.00	20	Satisfied
20	23.50			4.18		
30	21.80			5.80		
40	20.60			8.59		
50	20.80			11.58		
60	14.90			11.00		
70	17.00			11.38		
80	17.60			10.90		
90	26.40			10.40		
100	27.00			10.61		

Table 3 Investigated optimum dose for (DS and A) coagulants as a function of residual turbidity according to WHO turbidity standard at high synthetic turbid water (Al-Sameraiy 2012)

Investigated doses (mg/L)	DS coagulant	Optimum dose (mg/L)	WHO turbidity standard ≤5 NTU	A coagulant	Optimum dose (mg/L)	WHO turbidity standard ≤5 NTU
	Residual turbidity (c) in (NTU) after settling time (120 min)					
10	72.70	80	Not satisfied	7.70	60	Satisfied
20	70.70			7.00		
30	65.60			6.50		
40	60.50			6.00		
50	63.00			5.50		
60	57.20			4.41		
70	52.70			5.10		
80	50.20			7.10		
90	57.20			7.20		
100	60.50			7.70		

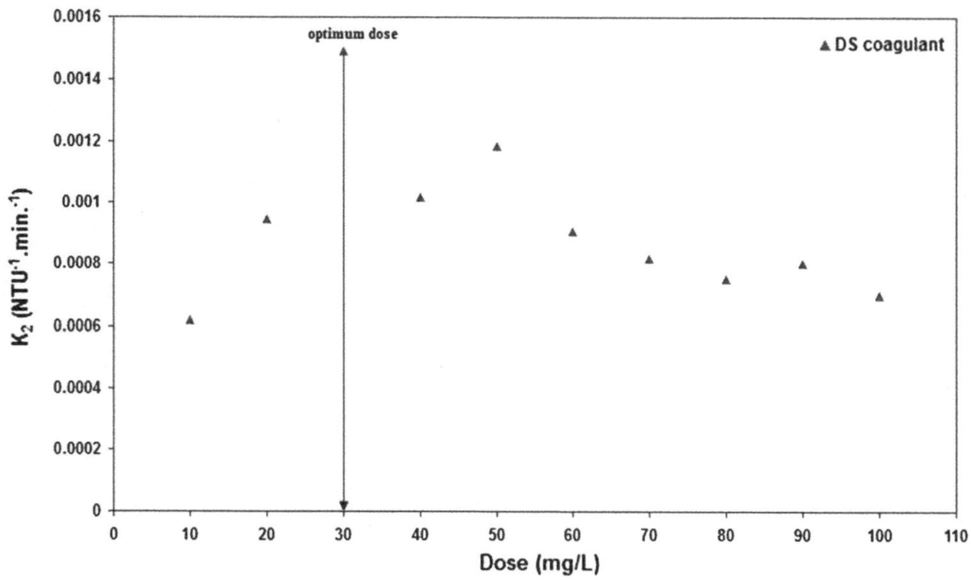

Fig. 1 The relationship between (k_2) and DS coagulant doses at low synthetic turbid water

of (k_2) for DS and A coagulants were mathematically determined by substituting the residual turbidities of these coagulants for every level of low, medium and high synthetic turbid water at the end of settling time (120 min) directly in Eq. (3). The relationship between calculated (k_2) versus the doses of DS and A coagulants were shown in Figs. 1, 2, 3, 4, 5, 6.

In low synthetic turbid water, Fig. 1 shows that the maximum value of (k_2) for DS coagulant was 0.00149 (NTU^{-1}·min^{-1}) corresponded to the dose of 30 mg/L. This dose is to be considered an optimum dose compared with other doses. Figure 2 indicates that the maximum value of (k_2) for A coagulant was reported as 0.00386 (NTU^{-1}·min^{-1}) occurred at the dose of 10 mg/L, which is an optimum dose for alum coagulation process.

In medium synthetic turbid water, Figs. 3, 4 demonstrate that the highest values of (k_2) for DS and A coagulants were found to be 0.00050 and 0.00194 (NTU^{-1}·min^{-1}) that corresponded to the doses of 60 and 20 mg/L respectively. These doses are the optimum doses for DS (60 mg/L) and A (20 mg/L) coagulation process.

Similar results were obtained with high synthetic turbid water as shown in Figs. 5, 6. The maximum value of (k_2) for DS coagulant was 0.00014 (NTU^{-1}·min^{-1}) in which it occurred at dose of 80 mg/L, which represents an optimum dose. Whilst Fig. 6 states that the highest value of (k_2) for A coagulant was recorded as 0.00186 (NTU^{-1}·min^{-1}) and corresponded to the optimum dose of 60 mg/L.

In general, it can be seen from the results reported in Figs. 1, 2, 3, 4, 5, 6 that the values of (k_2) for each of DS

Fig. 2 The relationship between (k_2) and A coagulant doses at low synthetic turbid water

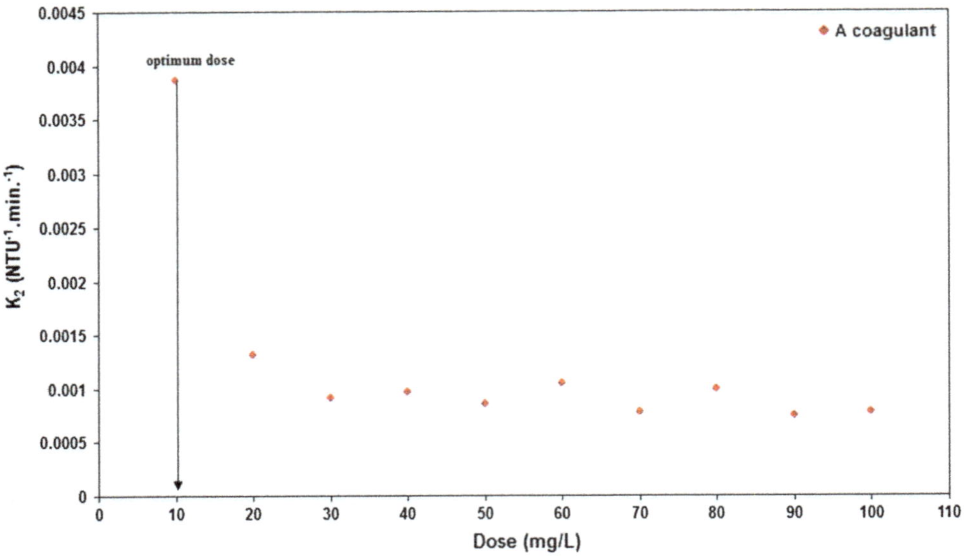

Fig. 3 The relationship between (k_2) and DS coagulant doses at medium synthetic turbid water

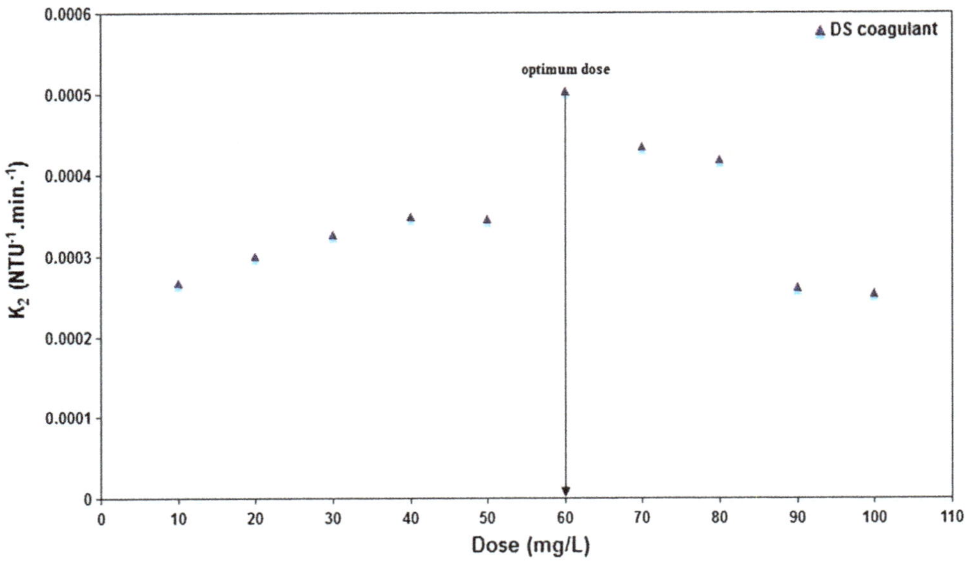

Fig. 4 The relationship between (k_2) and A coagulant doses at medium synthetic turbid water

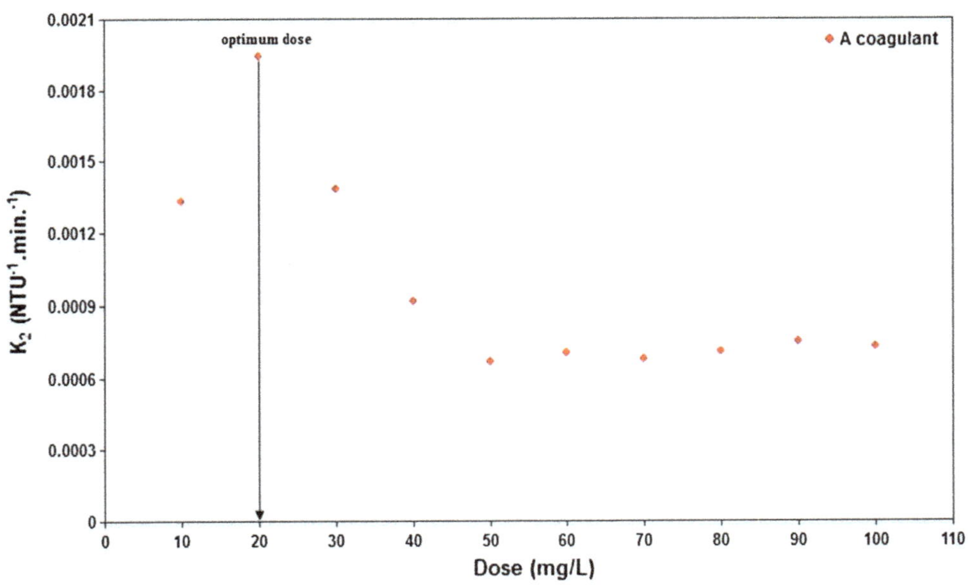

Fig. 5 The relationship between (k_2) and DS coagulant doses at high synthetic turbid water

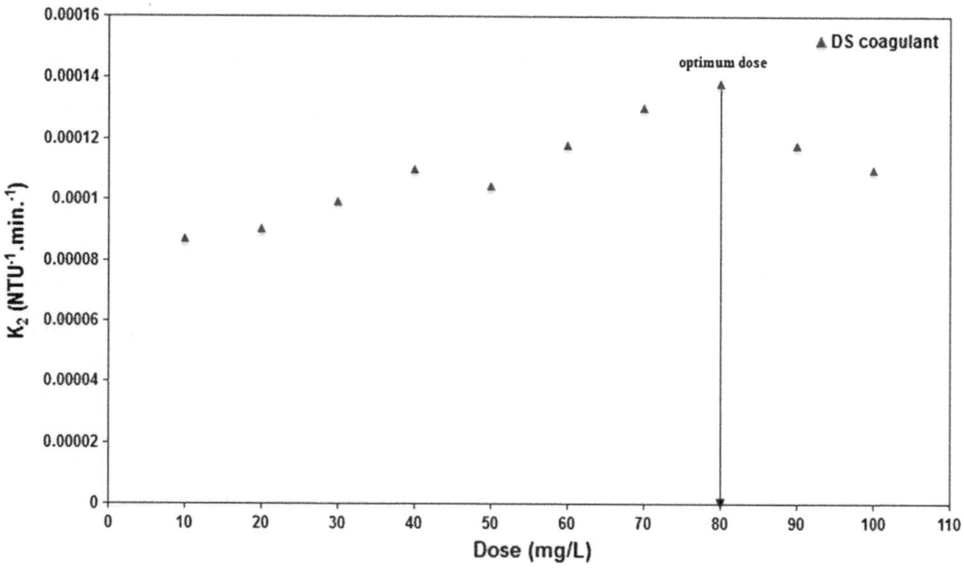

Fig. 6 The relationship between (k_2) and A coagulant doses at high synthetic turbid water

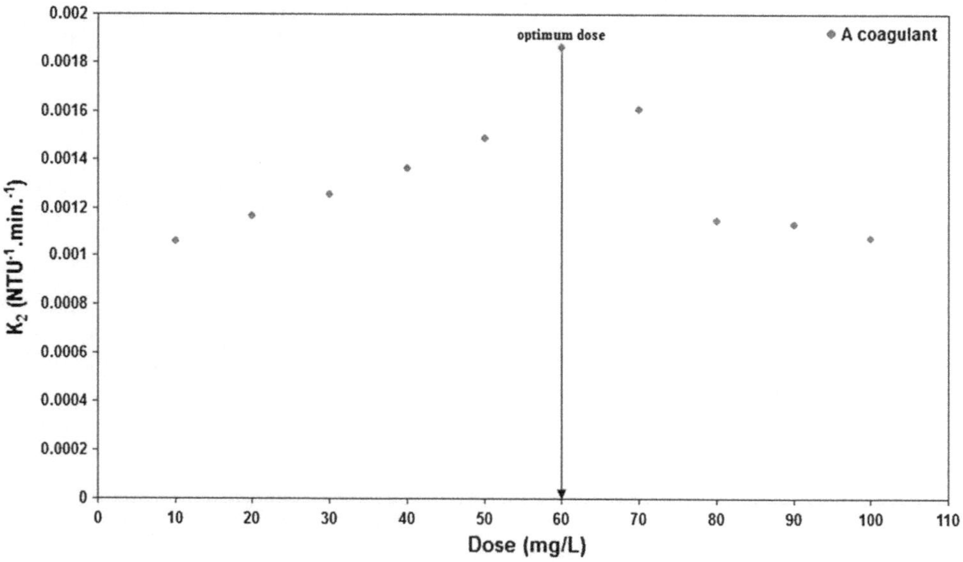

and A coagulants increased gradually with an increase in the initial coagulant doses to reach a maximum value at a certain dose, which represents an optimum dose and then decreased steadily to reach a minimum values. This behavior is expected in which it could be explained by the fact that the optimal dose of coagulant in suspension causes larger amount of solid to aggregate and settle. However an overoptimal amount coagulant would cause the aggregated particles to re-disperse in the suspension and would also disturb particle settling (Al-Sameraiy 2012; Chaudhuri and Khairuldin 2009; Divakaran and Pillai 2002). On the other hand, in terms of residual turbidity, it decreases with increasing the coagulant doses, but after a certain dose, the suspension showed a tendency to restabilize and further increase in coagulant doses adversely affected turbidity removal as residual turbidity (Al-Sameraiy 2012; Diaz et al. 1999). The optimum doses of each coagulant in these figures were selected on a basis of the maximum values of (k_2), which corresponded to these doses.

It is obvious from the results reported that the numeric values of coagulation rate constant (k_2) depends basically on the several affecting parameters including: the initial turbidity level of synthetic turbid water, type of coagulant and coagulant doses. This is expected mathematically to get different values of (k_2) when used Eq. (3) on the basis of second order form of coagulation process. Consequently, the values of (k_2) would vary with the type of coagulant used, initial turbidity level of synthetic turbid water and coagulant dose. Table 4 shows the relationship between the second order coagulation rate constant and the affecting parameters.

Table 4 Second order coagulation rate constant (k_2) as a function of several affecting parameters

Synthetic turbid water (NTU)	k_2 ($NTU^{-1}{\cdot}min^{-1}$)	
	DS coagulant	A coagulant
Low (75)	0.00149	0.00386
Medium (150)	0.00050	0.00194
High (300)	0.00014	0.00186

In terms of real applications and process scale-up, it is of importance to mention that the values of (k_2) for each coagulant in the current research should be specifically selected based on the initial turbidity level of synthetic water involving low (75 NTU), medium (150 NTU) and high (300 NTU). Accordingly, it is not required to modify or correct the order of coagulation process.

The results obtained in Figs. 1, 2, 3, 4, 5, 6 are exactly in accordance with the experimental data reported in Tables 1, 2, 3 relating to find the optimum doses for each coagulant by which the using of both coagulation rate constant (k_2) in the current research and residual turbidity obtained by Al-Sameraiy (2012) leading to identify the optimum doses for each coagulant accordingly. These findings using the highest numeric values of coagulation rate constant (k_2) gave a successful indicator to investigate the optimum doses for each coagulant of coagulation process. The order of coagulation process, $\alpha = 2$ showed that an excellent representation in terms of satisfaction of the experimental results in the current research.

Critical coagulation rate constant as an evaluating criterion for coagulants' performance

World Health Organization (WHO) guidelines recommend the upper limit of turbidity for finished water is 5 NTU (WHO 1996, 2006, 2008). In terms of mathematical representation of this recommendation, critical coagulation rate constant (k_c) could be introduced according to the following assumption:

c: is equal to (5 NTU) referring to the upper limit of (WHO) water turbidity (WHO 1996, 2006, 2008).

c_0: is the concentration of synthetic turbid water levels including: low (75 NTU), medium (150 NTU) and high (300 NTU) (Al-Sameraiy 2012).

t: is the end of predetermined settling time and equal to (120 min) (Al-Sameraiy 2012).

Substituting the values of (c, c_0, t) in Eq. (3) to obtain the values of (k_c) as shown in Table 5:

To assess the performance of coagulation process for each coagulant of date seeds and alum at their optimum doses in terms of satisfying (WHO) turbidity drinking water guidelines (WHO 1996, 2006, 2008), the following criterion was proposed:

Table 5 Critical coagulation rate constant (k_c) as a function of synthetic turbid water levels

Synthetic turbid water	c_0 (NTU)	k_c ($NTU^{-1}{\cdot}min^{-1}$)
Low	75	0.00156
Medium	150	0.00161
High	300	0.00164

$$(k_2) \text{ for each coagulant} \geq (k_c) \quad \text{for each level of synthetic turbid water} \tag{4}$$

The validity of the proposed criterion was tested with the results obtained by Al-Sameraiy (2012) in terms of residual turbidity of coagulation process as shown in Tables 1, 2, 3 in order to meet (WHO) turbidity standard ($\leq$5 NTU) of drinking water.

Based on the results shown in Figs. 1, 2, 3, 4, 5, 6, it can be clearly observed that the maximum values of (k_2) corresponded to the optimum doses for each coagulant. These values of (k_2) at their predetermined optimum doses are assessed according to the proposed criterion (Eq. 4) in Figs. 7, 8, 9.

In low synthetic turbid water, Fig. 7 shows the value of (k_2) for DS coagulant at the optimum dose (30 mg/L) is 0.00149 ($NTU^{-1}{\cdot}min^{-1}$) lower than (k_c) value of 0.00156 ($NTU^{-1}{\cdot}min^{-1}$) as shown in Table 5 and thus it could not satisfy the proposed criterion (Eq. 4). This means that its performance could not meet (WHO) turbidity standard. Similarly, according to Table 1, it produced residual turbidity of 5.2 NTU higher than 5 NTU as (WHO) turbidity standard. While for alum coagulant, the value of (k_2) at the optimum dose (10 mg/L) is 0.00386 ($NTU^{-1}{\cdot}min^{-1}$) higher than (k_c) value of 0.00156 ($NTU^{-1}{\cdot}min^{-1}$) and it could successfully satisfy the proposed criterion. As a result, the performance of A coagulant could meet (WHO) turbidity standard. Also, as shown in Table 1, it produced residual turbidity of 2.1 NTU and satisfied (WHO) turbidity standard accordingly.

In medium synthetic turbid water, Fig. 8 states that the (k_2) values for DS and A coagulants at their optimum doses of 60 and 20 mg/L are 0.00050 and 0.00194 ($NTU^{-1}{\cdot}min^{-1}$) lower and higher than (k_c) value of 0.00161 ($NTU^{-1}{\cdot}min^{-1}$) as shown in Table 5 respectively. DS coagulant could not satisfy the proposed criterion. Likewise, according to Table 2, it produced residual turbidity of 14.9 NTU higher than the upper limit of (WHO) turbidity standard. On the other hand, alum coagulant could effectively satisfy the proposed criterion. In the same way as shown in Table 2, it satisfied (WHO) turbidity standard by producing residual turbidity of 4.18 NTU.

Fig. 7 (k_c) as an evaluating parameter compared with (k_2) for DS and A coagulants at predetermined optimum doses and low synthetic turbid water

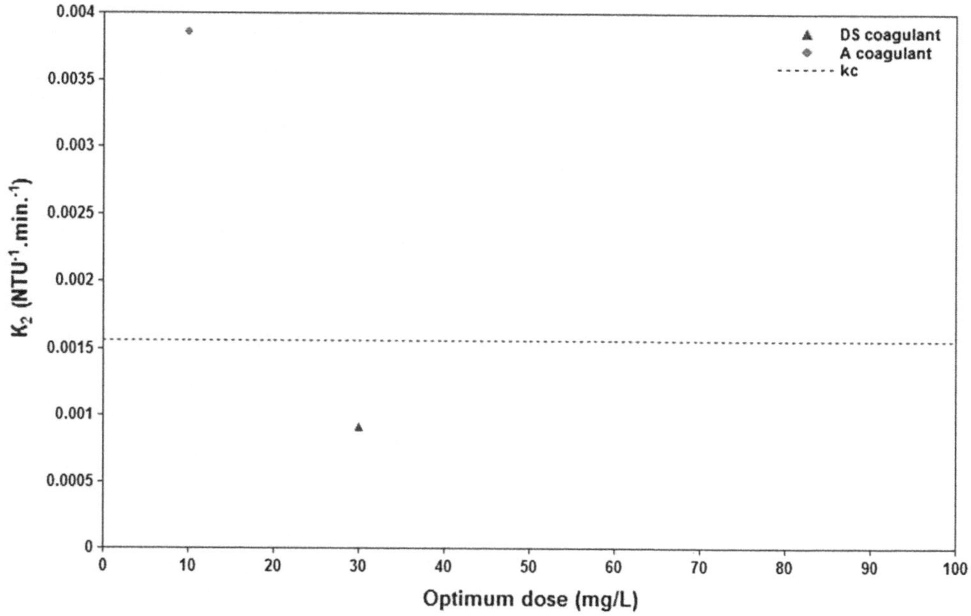

Fig. 8 (k_c) as an evaluating parameter compared with (k_2) for DS and A coagulants at predetermined optimum doses and medium synthetic turbid water

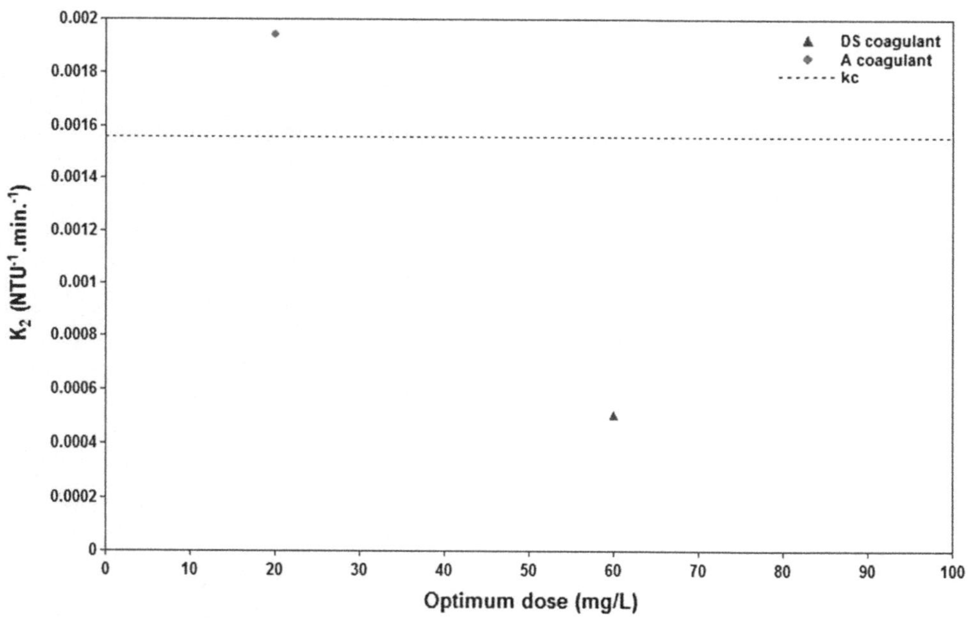

Similar observations were reported with high synthetic water as shown in Fig. 9. At the optimum dose (80 mg/L) of DS coagulant, the value of (k_2) is 0.00014 ($NTU^{-1} \cdot min^{-1}$) smaller than (k_c) value of 0.00164 ($NTU^{-1} \cdot min^{-1}$) as shown in Table 5. Consequently, its performance could not meet (WHO) turbidity standard since it could not satisfy the proposed criterion. Also, according to Table 3, it produced residual turbidity of 50.2 NTU higher than the standard value of turbidity according to (WHO) guidelines. Whilst at the optimum dose (60 mg/L) of A coagulant, the value of (k_2) is 0.00186 ($NTU^{-1} \cdot min^{-1}$) greater than (k_c). Hence, it could efficiently satisfy the proposed criterion. In the same way as shown in Table 3, it produced residual turbidity of 4.41 NTU and satisfied (WHO) turbidity standard accordingly.

Conclusion

The results obtained in the current research in terms of the second order coagulation rate constant as an investigating parameter to find the optimum doses of coagulation process

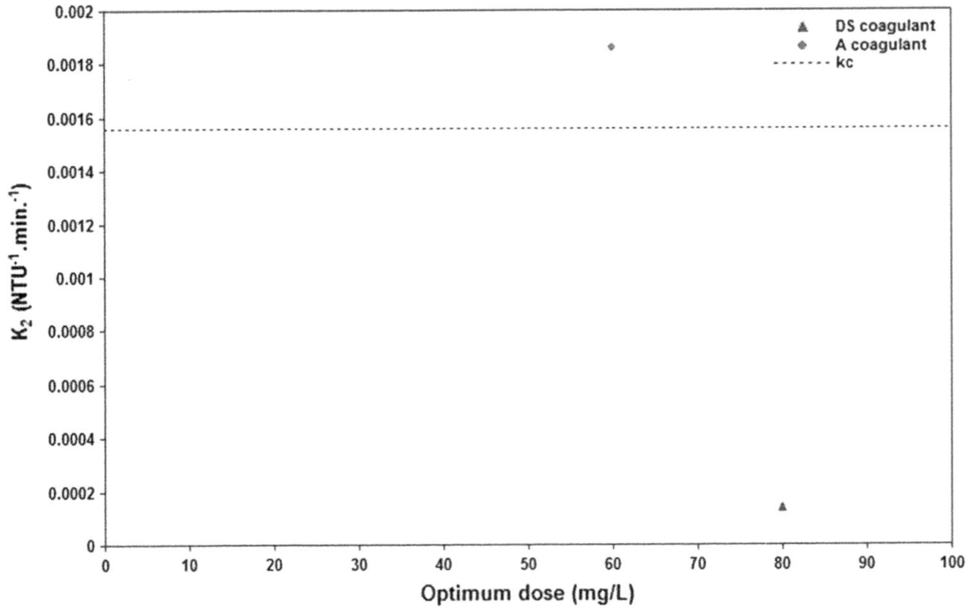

Fig. 9 (k_c) as an evaluating parameter compared with (k_2) for DS and A coagulants at predetermined optimum doses and high synthetic turbid water

for date seeds and alum coagulants as well as the critical coagulation rate constant as an evaluating criterion to assess the performance these coagulants were tested and validated with the published results by Al-Sameraiy (Al-Sameraiy 2012). They are exactly in accordance with his findings. Therefore, it is reasonable to consider this approach in the current search to be a new approach in coagulation kinetics modeling for monitoring the performance of water treatment plants to meet effluent quality requirements.

References

Al-Sameraiy M (2012) A novel water pretreatment approach for turbidity removal using date seeds and pollen sheath. J Water Resour Prot 4:79–92

American Water Works Association (AWWA) (2000) Operational control of coagulation and filtration processes. Manual M37, 2nd edn. Denver

Ani JU, Chukwudi MM, Onukwuli OD (2011) Coagulation-flocculation performance of snail shell biomass for waste water purification. N Y Sci J 4(2):81–90

Ani JU, Nnaji NJ, Okoye COB, Onukwuli OD (2012) The coagulation performance of Okra Mucilage in an industrial effluent by Turbidimetry. Int J Chem Sci 10(3):1293–1308

Chaudhuri M, Khairuldin PSAB (2009) Coagulation-clarification of coloured water by natural coagulant (*Moringa oleifera*) seed extract. Nat Environ Pollut Technol 8(1):137–139

Chukwudi MM, Onukwuli OD (2011) Coag-flocculation studies of *Afzelia Bella* Coagulant (ABC) in Coal Effluent using single and simulated multi angle nephelometry. J Miner Mater Character Eng 10(3):279–298

Chukwudi MM, Nnaji PC, Onukwuli OD (2009) Coag-flocculation kinetics and functional parameters response of Periwinkle shell coagulant (PSC) to pH variation in organic rich coal effluent medium. Nat Sci 7(6):1–18

Diaz A, Rincon N, Escorihuela A, Fernandez N, Forster CF (1999) A preliminary evaluation of turbidity removal by natural coagulants indigenous to Venezuela. Process Biochem 35(3):391–395

Divakaran R, Pillai SVN (2002) Flocculation of river silt using chitosan. Water Res 36(9):2414–2418

Elimelech MJ, Gregory J, Jia X, Williams RA (1995) Particle deposition and aggregation: measurement, modeling, and simulation. Butterworth-Heinemann Ltd, Oxford

Esther YA, Cokou IN, Agbangnan P, Koudoro AY, Dèdjiho CA, Aïna MP, Mama D, Sohounhloué DC-K (2013) Evaluation of the *Opuntia dillenii* as natural coagulant in water clarification: case of treatment of highly turbid surface water. J Water Resour Protect 5:1242–1246

Gregory J (2006) Particles in water: properties and processes. IWA Pub., CRC Press Taylor & Francis, Boca Raton, London

Holt PK, Barton GW, Wark M, Mitchell CA (2002) a quantitative comparison between chemical dosing and electrocoagulation. Colloids Surf A: Physicochem Eng Aspect 211:233–248

Ifeanyi UV, Chukwudi MM, Okechukwu OD (2012) Effect of Coag-Flocculation kinetics on telfairia occidentalis seed coagulant (TOC) in Pharmaceutical Wastewater. Int J Multidisciplin Sci Eng 3(9):22–33

Jadhav MV, Mahajan YS (2013) Investigation of the performance of chitosan as a coagulant for flocculation of local clay suspensions of different turbidities. KSCE J Civil Eng 17(2):328–334

Kılıç MG, Hoşten C (2010) A comparative study of electrocoagulation and coagulation of aqueous suspensions of kaolinite powders. J Hazard Mater 176:735–740

Ni'am MF, Othman F, Sohaili J, Fauzia Z (2007) Removal of Cod and turbidity to improve wastewater quality using electrocoagulation technique. Malays J Anal Sci 11(1):198–205

Nnaji PC, Okolo BI, Chukwudi MM (2014) Nephelometric performance evaluation of oxidized starch in the treatment of coal Washery Effluent. Nat Resour 5:79–89

Okolo BI, Nnaji PC, Menkiti MC, Ugonabo VI, Onukwuli OD (2014) Parametric response evaluation for *Xanthosoma* spp. Induced Coag-Flocculation of Brewery Effluent. Green Sustain Chem 4:7–14

Smoluchowski MV (1917) Verucheiner Mathemtischen Theorie der Koagulations KineticKolloider Lousungen. Int J Res Phys Chem Chem Phys 92:29–168

Ugonabo VI, Chukwudi MM, Atuanya CU, Onukwuli DO (2013) Comparative studies on Coag-flocculation Kinetics of pharmaceutical industry effluent by Achatina maginata shell biomass and aluminum sulphate. Int J Eng Technol 13(2):134–147

Wang LK, Hung Y-T, Shammas NK (2005) Physicochemical treatment processes. Handbook of environ mental engineering, 3rd edn. The Humana Press Inc., Totowa

Water Specialist Technology (WST) (2003) About coagulation and flocculation. Information Bulletins, Sanford, pp 1–10

World Health Organization (WHO) (1996) Water quality monitoring—a practical guide to the design and implementation of freshwater quality studies and monitoring programs. In: Bartram J, Balance R (eds). Published on behalf of United Nations Environment Programme and the World Health Organization© UNEP/WHO. ISBN 419 22320

World Health Organization (WHO) (2006) Guidelines for drinking water quality, vol 1, 1–3rd edn. Recommendations, WHO Library Cataloguing-in-Publication Data, Electronic version for the Web. ISBN 9241546964 **(NLM classification: WA 675)**

World Health Organization (WHO) (2008) Guidelines for drinking water quality, vol 1, 3rd edn. Recommendations, WHO Library Cataloguing-in-Publication Data [electronic resource]. ISBN 9789241547611 (WEB version) **(NLM classification: WA 675)**

Yukselen MA, Gregory J (2004) The effect of rapid mixing on the break-up and re-formation of flocs. J Chem Technol Biol 79:782–788

Zainal-Abideen M, Aris A, Yusof F, Abdul-Majid Z, Selamat A, Omar SI (2012) Optimizing the coagulation process in a drinking water treatment plant–comparison between traditional and statistical experimental design jar tests. Water Sci Technol 65(3):496–503

The application of Dempster–Shafer theory of evidence for assessing groundwater vulnerability at Galal Badra basin, Wasit governorate, east of Iraq

Alaa M. Al-Abadi[1]

Abstract The process of delineating areas that are more susceptible to pollution from anthropogenic sources has become an important issue for groundwater resources management and land-use planning. In this study, an attempt was made to delineate aquifer vulnerability zones for nitrate contamination at Galal Badra basin, east of Iraq using Dempster–Shafer method of evidence in GIS platform. First, an inventory map of the wells with elevated nitrate concentration (>3 mg/L) was prepared. The map showed that there are 63 wells with elevated nitrate concentrations in the study area. These data were partitioned randomly into two sets, for training and testing. The partition criterion was 70/30, 44 wells for training and 19 wells for testing. Then, the most influencing evidential thematic factors in determining aquifer vulnerability were selected depending on the availability of data. These factors were groundwater depth, hydraulic conductivity, slope, soil, and land use land cover (LULC). The spatial association between well locations and evidential thematic layers was investigated by means of mass functions (belief, disbelief, uncertainty, and plausibility) of Dempster–Shafer method. The integrated belief function was used to produce groundwater aquifer vulnerability index (GVI) for the study area. The pixel values of GVI were reclassified into five categories: very low, low, moderate, high, and very high using Jenks classification scheme. The very low–low zones cover 32 % (209 km^2). These classes mainly concentrate on the eastern parts of the study area and occupy small zone in the central part. The moderate zone extends over an area of 42 % (279 km^2) and mainly encompasses the western part of the study area. The high–very high zones cover 26 % (170 km^2) and these zones concentrate on the central part of the study area. The results indicate that the aquifer system in the study area is moderately vulnerable to contamination by nitrate. The model was validated by using relative operating characteristic technique. The success and prediction rates for area under the curve (AUC) were 0.86 and 0.77, respectively, indicating that the model has good capability to delineate aquifer vulnerability zones for nitrate contamination in the study area.

Keywords Aquifer vulnerability · Dempster–Shafer method of evidence · GIS · Galal Badra area · Iraq

Introduction

Groundwater is a primary life-giving resource. Its availability is an essential component in socio-economic development, human evolution, poverty reduction, and ecological diversity. Groundwater often provides a reliable source of water where surface water is unavailable or inadequate. Thus, it is essential to manage groundwater resources in sustainable mode to ensure its quality and quantity for a long period. To properly manage and protect groundwater reservoirs, especially shallow water-bearing layers, it is necessary to delineate areas where groundwater may be more vulnerable to pollution. Analysis of aquifer vulnerability is an important tool for groundwater management and provides basic information for facilitating proper planning and protection of groundwater resources (Majandang and Sarapirome 2013). The term vulnerability is defined as the degree to which human or environmental

✉ Alaa M. Al-Abadi
alaaatiaa@gmail.com

1 Department of Geology, College of Sciences, University of Basra, Basra, Iraq

systems are likely to experience harm due to perturbation of stress, and can be identified for a specified system, hazard, or group of hazards (Popescu et al. 2008). In groundwater hydrology, vulnerability assessments typically describe the susceptibility of the water table, a particular aquifer, or water well to contaminants that can reduce the groundwater quality (Liggett and Talwar 2009). Two terms are used to describe groundwater vulnerability: intrinsic and specific. Intrinsic vulnerability is the natural susceptibility to contamination based on the physical characteristics of the environment. On the other hand, specific vulnerability is defined as an accounting for the transport properties of a particular contaminant or a group of contaminants through the subsurface. In general, three different methods can be used to assess groundwater vulnerability namely, index and overly, statistical, and process-based methods. In overly and index methods, factors which are believed to have an influence on the movement of pollutants such as geology, soil, slope, hydraulic conductivity are mapped. These factors are assigned weights and rates depending on their importance on controlling pollutants movement. The resultant maps are linearly summed to produce a map of vulnerability index of an area. The groundwater vulnerability produced by such methods is generally qualitative and relative. Several overly and index methods have been developed. The most common are the DRASTIC (Aller et al. 1987), the GOD (Foster 1987), the AVI (Van Stempvoort et al. 1993), the SINTACS (Civita 1994), and the EPIK (Doerfliger and Zwahlen 1997). The process-based models use simulation models to estimate time of travel, concentration of contaminant, and duration of contamination to quantify areas of high and low susceptibility to pollution. Some of these models are designed to simulate migration of contaminants through unsaturated zone, saturated zone, and unsaturated–saturated zones. Process-based models are not commonly used to assess vulnerability because they are constrained by data shortage, computational difficulty, and the expertise required to implement them. Statistical methods are used to quantify the risk of groundwater pollution by determining the statistical dependence between observed contamination and observed land uses that are potential source of contamination (Harter and Walker 2001). Once the statistical relationship is attained, the model can be used to predict the probability of contamination risk. The main advantage of this method is that the statistical significance can be explicitly calculated. There are only few studies that have used these methods to quantify groundwater vulnerability around the world. For example, Arthur et al. (2007) implemented a Bayesian-probabilistic weights-of-evidence (WOE) technique to generate a series of maps reflecting the relative aquifer vulnerability of Florida's principle aquifer system in United States of America (USA). They used WOE to explore the relationship between several evidential hydrogeological themes (such as soil hydraulic conductivity, density of karst features, thickness of aquifer confinement, and hydraulic head difference) and ambient groundwater parameters in wells that reflect relative degree of vulnerability. The same technique was used by Masetti et al. (2007) to assess aquifer vulnerability to occurrence of elevated nitrate concentration in the Province of Milan (northern Italy). Uhan et al. (2010) used outputs of three models (GROWA, SWAT, and FEFLOW) as evidential themes for assessing aquifer vulnerability for nitrate concentrations using WOE technique in Lower Savinja Valley (Slovenia). They concluded that WOE model was capable to indicate regional groundwater nitrate distribution and enable spatial prediction of the probability for nitrate groundwater concentrations. Mair and El-Kadi (2013) successfully applied bivariate logistic regression technique (LR) to assess the groundwater vulnerability to contamination in Hawaii, USA. Sorichetta et al. (2013) used multivariate WOE and LR methods for assessing groundwater vulnerability in the Milan District, Italy. They concluded that these methods were suitable for evaluating aquifer vulnerability for nitrate contamination.

The Dempster–Shafer theory (DST) of evidence (also known as evidential belief functions EBF's) is a generalization of the Bayesian theory of subjective probability. It has a relative flexibility to accept uncertainty and the ability to combining beliefs from multiple source of evidence (Thiam 2005). In Earth sciences, the application of this method is still limited. This method has been used for mineral potential mapping (Moon 1990; An et al. 1992; Carranza and Hale 2003; Carranza et al. 2005), landslide susceptibility (Park 2011; Mohammady et al. 2012; Bui et al. 2012; Lee et al. 2012; Pourghasemi et al. 2013), and groundwater potential mapping (Nampak et al. 2014; Mojaji et al. 2014).

To our knowledge, application of this method for assessing aquifer vulnerability for nitrate contamination has never been investigated. Due to the high mobility and solubility, nitrate NO_3^- always exists in groundwater under oxidizing conditions (Almasri and Ghabayen 2008). In general, source of nitrate in groundwater can be classified into point and non-point sources (Alagha et al. 2013). The non-point source of nitrate includes fertilizer, manures, and return flows from irrigation, while the point sources include septic system and cesspits. Groundwater contamination by nitrate causes many diseases such as methemoglobinemia, which at severe cases may result in brain damage and death (Cissé and Mao 2008). Thus, the main objective of this study is to evaluate the applicability of the DST for GIS-based aquifer vulnerability analysis. A case study of the Galal Badra area in central Iraq was conducted to explore the application of this method for assessing specific aquifer vulnerability.

The basic principles of DST

The DST is a generalization of the Bayesian theory of subjective probability. Whereas the Bayesian theory requires probabilities for each question of interest, DST allow us to base degrees of belief for one question on probabilities for a related question (Dempster 1968). The detailed mathematical description of the DST is outside of this study, only a brief description of the theory synthesized from works of An et al. (1992), Carranza and Hale (2003), and Park (2011) was reviewed here.

A frame of discernment, for an event propositions exist such that $\Theta = \{A_1, A_2, \ldots, A_n\}$ which is a set of mutually exclusive and exhaustive proposition, is first established. Then, a mass function $[m(A)]$ assigns belief committed to each proposition as shown below.

$$m = 2^{\Theta} \rightarrow [0, 1] \tag{1}$$

with

$$m(\varphi) = 0 \tag{2}$$

and

$$\sum_{A \subset \Theta} m(A) = 1, \tag{3}$$

where φ is the empty set and A is a subset of Θ. The function m is a measure of belief committed to each possibility (Wally 1987). The belief (Bel) and plausibility (Pls) functions are defined based on the mass functions as follows:

$$\text{Bel}(H) = \sum_{A \subset H} m(A) \tag{4}$$

$$\text{Pls}(H) = \sum_{A \cap H \neq \varphi} m(A) \tag{5}$$

where for every $H \subset \Theta$, Bel(H) is a measure of the total amount of beliefs committed exactly to every subset of H by m. Pls(H) represents the degree to which the evidence remains plausible (Park 2011).

According to An et al. (1992), Eqs. (4) and (5) represent the lower and upper probabilities. These probabilities have the following properties:

$$\text{Bel}(H) \leq \text{Pls}(H) \tag{6}$$

$$\text{Pls}(H) = 1 - \text{Bel}(\bar{H}) \tag{7}$$

where $\bar{H}$ is the negation of H. $Bel(\bar{H})$ is called the disbelief function.

The difference between Pls(H)and Bel(H) indicates the degree of uncertainty. When the degree of uncertainty equals 0, then $\text{Bel}(H) + \text{Bel}(\bar{H}) = 1$, which is a Bayesian probability (An et al. 1992).

The core part of the application of the DST to assess specific aquifer vulnerability is to define mass functions using quantitative relationship between the well locations having elevated nitrate concentrations (>3 mg/L) and factors which control aquifer vulnerability such as depth of groundwater table, hydraulic conductivity, slope, soil, land use land cover. The mass functions in this study were derived from likelihood ratio functions. Suppose that there are ℓ multiple spatial thematic layers in an area where wells with elevated nitrate concentration existed, then each thematic layer is regarded as evidence $E_i (i = 1, 2, \ldots, \ell)$ for the target proposition T_p. If E_{ij} is the jth class attribute of the evidence E_{ij} and frequency distribution function of positive and opposite target prepositions, the likelihood ratio is (Park 2011)

$$\lambda(T_p)_{E_{ij}} = \frac{\frac{N(L \cap E_{ij})}{N(L)}}{\frac{N(E_{ij}) - N(L \cap E_{ij})}{N(A) - N(L)}} \tag{8}$$

where $N(L \cap E_{ij})$: number of wells that occurred in E_{ij}

$N(L)$ total number of existed wells with elevated nitrate concentrations in the study area

$N(E_{ij})$ number of pixels in E_{ij}

$N(A)$ total number of pixels in the study area.

The Bel function can be calculated as

$$\text{Bel} = \frac{\lambda(T_p)_{E_{ij}}}{\sum \lambda(T_p)_{E_{ij}}}. \tag{9}$$

The likelihood ratio for supporting the opposite target proposition is calculated as

$$\lambda(\bar{T}_p)_{E_{ij}} = \frac{\frac{N(L) - N(L \cap E_{ij})}{N(L)}}{\frac{N(A) - N(L) - N(E_{ij}) + N(L \cap E_{ij})}{N(A) - N(L)}}. \tag{10}$$

The Dis function is calculated as

$$\text{Dis} = \frac{\lambda(\bar{T}_p)_{E_{ij}}}{\sum \lambda(\bar{T}_p)_{E_{ij}}}. \tag{11}$$

The uncertainty (Unc) and plausible (Pls) values are obtained using Eqs. (12) and (13)

$$\text{Unc} = 1 - \text{Dis} - \text{Bel} \tag{12}$$

$$Pls = 1 - Dis. \tag{13}$$

The values of Bel and Pls range between 0 and 1.

Once the mass functions are calculated for all the used factors, the Dempster's rules of the combination is used to obtain the four integrated mass functions (Dempster 1968). These rules have both commutative and associative attributes such that different groupings or orderings of evidence combinations do not affect the final results (George and Pal 1996 in Mogaji et al. 2014). The Dempster's rules for combining the two used factor maps A and B are as follows (Carranza et al. 2005; Mogaji et al. 2014):

$$Bel_X = \frac{Bel_A Bel_B + Bel_A Unc_B + Bel_B Unc_A}{\beta} \quad (14)$$

$$Dis_X = \frac{Dis_A Dis_B + Dis_A Unc_B + Dis_B Unc_A}{\beta} \quad (15)$$

$$Unc_X = \frac{Unc_A Unc_B}{\beta} \quad (16)$$

$$Pls_X = Unc_X + Bel_X, \quad (17)$$

where

Bel_X Lower degree of belief for each layer of parameters type of range

Dis_X Degree of disbelief for each layer of parameters type or range

Unc_X Degree of uncertainty for each layer of parameters type or range

X The A, B, ..., E denoting each parameters types.

The denominator β is called the normalization factor. It is also called the degree of conflict, which measures the conflict between the pieces of evidence (George and Pal 1996). The β is written mathematically as

$$\beta = 1 - Bel_A Dis_B - Dis_A Bel_B. \quad (18)$$

The study area

The study area extends over an area of 657 km^2 and lies between 32°59′31.43″ and 33°12′30.00″ latitude and 45°50′52.22″ and 46°12′42.68″ longitude in the north-eastern Wasit governorate at central east part of Iraq, Fig. 1. It is bounded by Iraqi–Iranian borders (Hamrin hills) from the east, wadi Galas from north, and hor Al-Shiwach from east and south. The main cities within the study area are Badra, Zurbatiyah, and Jassan. Badra city is located in the central part of the area while Zurbatiyah and Jassan are located 12 and 20 km in the north and south of the Badra city, respectively. Relief is low with only a few isolated hills rising above the general level of the plain in the east (Parsons 1956). Elevation in the study area ranges from 6 to 691 m with an average of ~45 m above mean sea level. The climate of the study area is characterized by hot, dry summer, cold winter, and a pleasant spring and fall. Approximately 90 % of the annual rainfall occurs between November and April, most of it in the winter months from December to March. The remaining 6 months are dry and hot. According to the recorded meteorological data in Badra station (north of the study area) for the period (1994–2013), the monthly maximum, minimum, and average temperatures are 37.8, 10.4, and 24.56 °C, respectively. The area receives an average mean annual rainfall of approximately 212 mm/year with uneven rainfall distribution between plain and mountain parts. The major stream in the study area is Galal Badra River. The mean monthly discharge of this river is 2.5 and 1000 m^3/s in drought and flood period, respectively (Al-Shammary 2006). Due to the prolonged drought conditions and intermittent nature of the streams in the study area, most of the farmers depend on the groundwater for their irrigation needs.

Rocks in the study area range in age from Middle Miocene to Recent. In the western portion, the younger rocks are exposed and increasingly become old to the east. Most of the area is covered by rocks of alluvial and lacustrine origin, Pliocene or younger in age. The stratigraphic succession is composed of Fatha, Injana, Muqdadiyah formations in addition to the Quaternary deposits. The Quaternary deposits mainly consist of mixture of gravel, sand, silt, and conglomerates of post Pliocene deposits. A brief description of these units is provided in Table 1. Approximately 80 % of the study area is covered with Quaternary deposits. Tectonically, the platform of the Iraqi territory is divided into two basic units, the stable and unstable shelf (Jassim and Goff 2006). The stable shelf is characterized by reduced thickness of the sedimentary cover and by the lack of folding, while the unstable shelf has a thick and folded sedimentary cover. The folds are arranged in narrow long anticlines and broad flat synclines (Al-Sayab et al. 1982). The greater parts of the study are located in the stable shelf (Mesopotamian plain) and only a small part extends over the unstable shelf close to the Iraqi–Iranian border (folded zone). There are many folds and faults in the study area. The bigger one is Shbichia–Najaf fault.

Two major aquifer systems exist within the study area. The first one represents the shallow unconfined aquifer consisting mainly of layers of sand, gravel with overlapping clay, and silt (Al-Abadi 2015b). This hydrogeological unit is located within the Quaternary lithological layers. The second hydrogeological unit is Muqdadiyah water-bearing layer. The aquifer condition of this unit is confined/semi-confined. The regional groundwater flow is from northeast to southwest. The hydraulic characteristic of the two units was estimated by Al-Shammary (2006) by means of pumping test. For the unconfined aquifer, the hydraulic conductivity, transmissivity, and specific yield were 6.3 m/d, 228.43 m^2/d, and 0.042, respectively. For the confined aquifer, the values were 3.5 m/d, 81.07 m^2/d, and 0.0017 for hydraulic conductivity, transmissivity, and storage coefficient, respectively.

Fig. 1 Location map of the study area

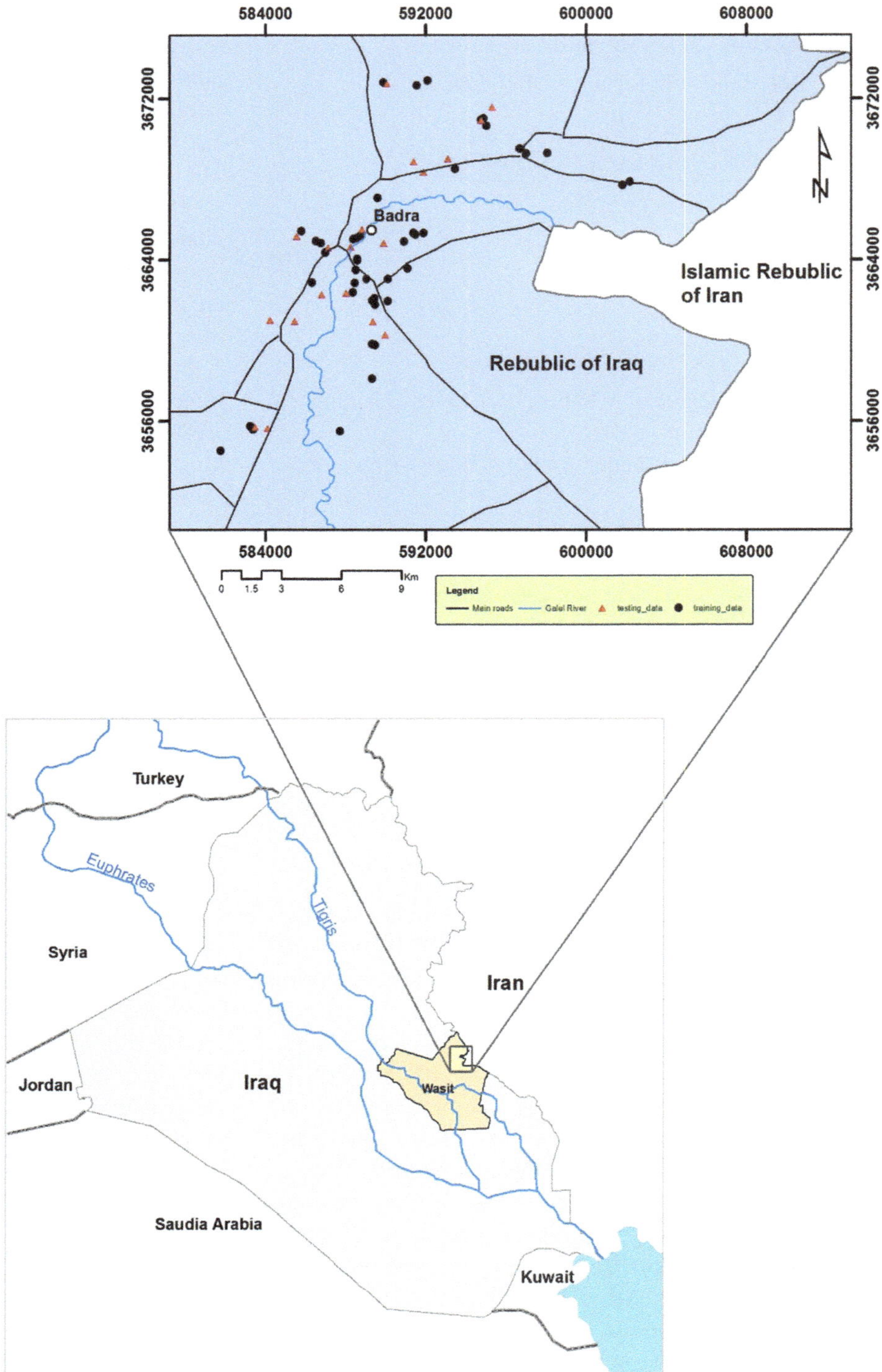

Table 1 Description of the lithological units in the study area

Formation	Age	Environment	Description
Jeribe	Middle Miocene	Lagoonal (back reef)	Recrystallized, dolomitized, massive limestone
Fatha	Middle Miocene	Shallow marine	Anhydrite, mudstone, and thin limestone
Injana	Upper Miocene	Sub-marine	Red or gray colored silty marl or clay stones and purple silt stones
Muqdadiyah	Pliocene	Continental	Gravely sandstone, sandstone, and red mudstone
Quaternary	Pleistocene—Holocene	Continental	Mixture of gravel, sand, silt, and conglomerate

Table 2 Information and sources of data used in this study

Name of evidential factor	Type of data	GIS data type
Groundwater depth	Borehole records[a]	Points
Hydraulic conductivity	Borehole records[a]	Points
Slope	ASTER-GDEM[b]	Raster
Soil	Hard copy[c]	Polygon
LULC	Landsat 8 imagery[d]	Raster

[a] Borehole records (including lithological logs and pumping tests results) were obtained from the General Commission of Groundwater/Iraq and Al-Shammary (2006) work

[b] ASTER-DEM was downloaded from web location (http://gdem.ersdac.jspacesystems.or.jp/search.jsp)

[c] Soil map of Iraq 1958 and Soil map of the world obtained from FAO-UNESCO 2005

[d] Landsat 8 multi-bands imagery (path 167-37 in 6/2/2015) was download from USGS web site (http://earthexplorer.usgs.gov/)

Table 3 Data used for constructing maps of groundwater depth and aquifer hydraulic conductivity

Well location (UTM)		Well name	Well depth (m)	Groundwater depth (m)	Hydraulic conductivity (m/s)
Easting	Northing				
609,382.36	3,670,037.71	w1	60	162.00	2.17
599,476.93	3,668,422.55	w2	95	104.00	2.55
587,399.22	3,663,101.97	w4	70	61.10	2.88
589,606.93	3,659,795.62	w5	54	62.00	11.46
587,933.17	3,655,591.80	w6	84	39.00	1.91
585,244.25	3,654,551.73	w7	60	26.00	2.99
594,676.69	3,655,746.94	w8	74	42.40	4.80
605,111.45	3,645,504.24	w9	70	16.00	1.08
615,758.63	3,645,592.84	w10	60	29.00	1.38
586,243.56	3,685,298.17	w12	70	81.00	0.62
589,102.96	3,683,722.36	w13	90	75.10	0.41
588,287.56	3,673,797.35	w14	60	60.00	1.40
597,930.54	3,668,342.43	w15	66	108.60	4.41
597,514.18	3,667,786.75	w16	43	101.10	11.34
599,278.44	3,667,558.11	w17	25	114.90	11.24
597,858.68	3,665,115.59	w18	30	112.30	13.43
588,012.26	3,661,228.72	w20	60	57.50	7.03
586,661.44	3,667,376.49	w21	60	50.00	0.99
586,038.11	3,664,599.07	w22	54	49.00	45.02
587,583.14	3,662,826.43	w23	54	60.60	3.24
589,953.98	3,664,357.09	w24	54	70.00	2.13
588,168.77	3,664,002.05	w25	66	71.50	1.26
590,201.71	3,662,788.61	w26	50	65.00	2.49
589,430.27	3,662,103.94	w27	54	58.00	2.67
584,925.55	3,658,398.79	w28	56	40.50	0.61
581,443.09	3,646,758.36	w29	92	32.80	0.20
612,992.58	3,646,731.16	w32	63	32.00	5.36
Average			62.00	65.98	5.37
Minimum			25	16	0.20
Maximum			95	162	45.02

Borehole records were obtained from the General Commission of Groundwater/Iraq and Al-Shammary (2006) work

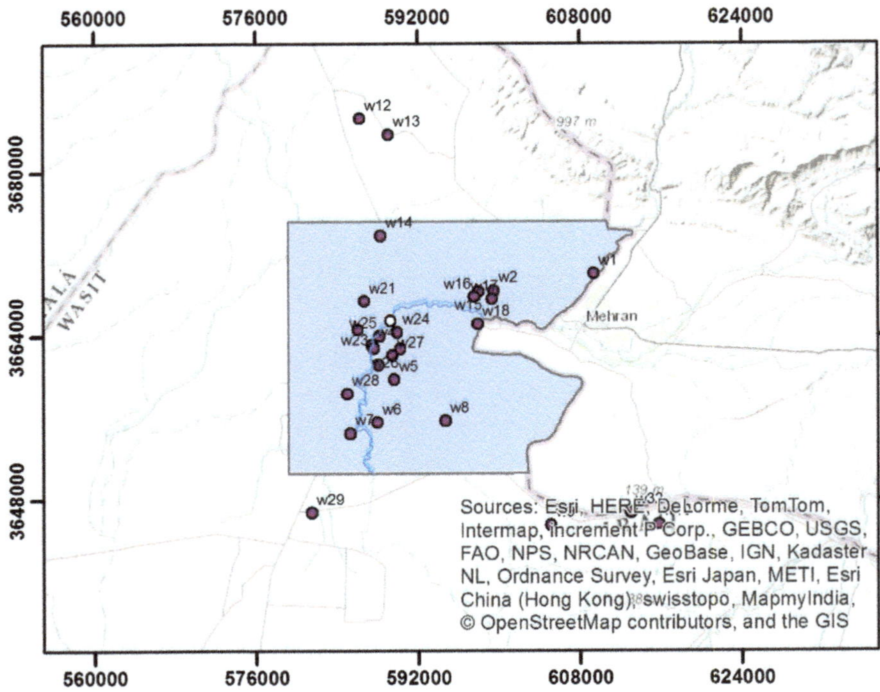

Fig. 2 Locations of wells used to produce maps of groundwater depths and hydraulic conductivity

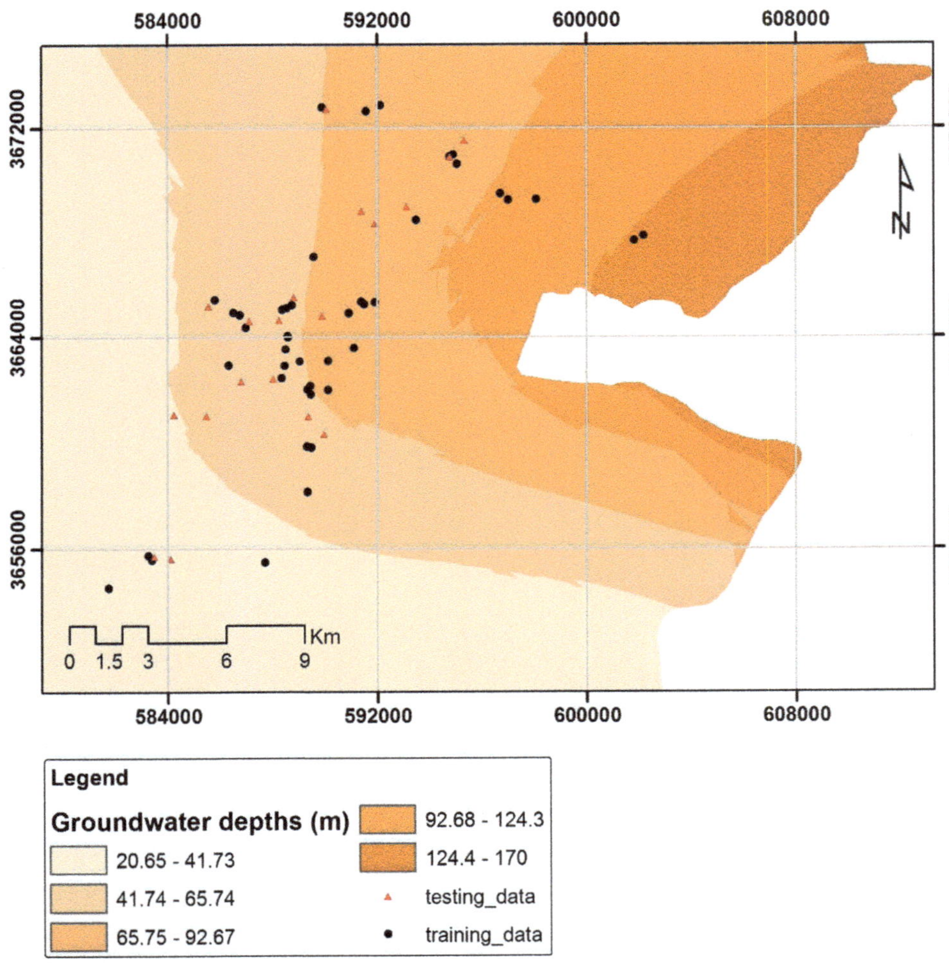

Fig. 3 Spatial distribution of groundwater depths in the study area

Fig. 4 Spatial distribution of hydraulic conductivity over the study area

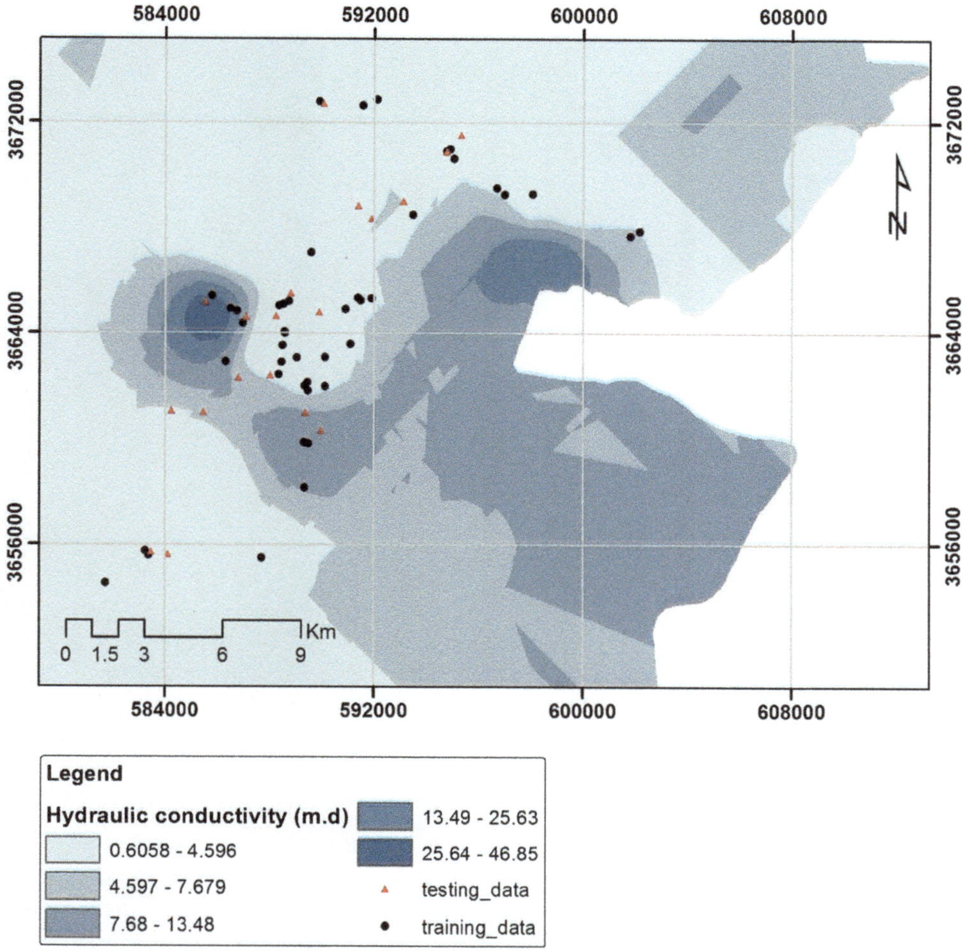

Generating of evidential thematic layers

To assess aquifer vulnerability for nitrate contamination in the study area, two main steps were implemented. In the first step, an inventory of wells with elevated nitrate concentrations was prepared. The nitrate levels in most ambient groundwater in Iraqi aquifers are very low, generally much less than 1 mg/L (Jabar Al-Saydi, Expert, Head of Groundwater Commission of Groundwater/Basra Branch, personal communication). Therefore, the presence of nitrate in groundwater greater than 3 mg/L usually reflects the impact of human activities on well water quality. The total number of wells in the interested area is 102. From these, 63 wells with elevated nitrate concentration (3 mg/L) were selected. The 63 wells were randomly partitioned into two sets, 44 wells (70 %) for training and 19 wells (30 %) for testing. In the second step, the evidential thematic layers of five factors influencing groundwater vulnerability were selected and mapped depending on the availability of data and literature review. These layers were depth of groundwater level, hydraulic conductivity, slope percentage, soil, and land use land cover (LULC). The sources of these layers are explained in Table 2. All evidential thematic layers were prepared as raster with 30 × 30 m cell size using different types of tools in ArcGIS 10.2 commercial software such as Geostatistical extension, Spatial Analyst extension, Image Classification tool, and ArcTool box.

Depth of groundwater represents the depth of groundwater levels, both for confined or unconfined aquifer with comparison to the ground surface. Its value is important because it determines the thickness of material through which a contaminant must travel before reaching water-bearing layers. In addition, attenuation capacity increases as the depth to groundwater increases. The presence of confining layers (low permeability layers) limits the travel of pollutants into an aquifer (Aller et al. 1987). Deeper water levels imply lesser chance for contaminants to enter (Piscopo 2001). The data used for drawing groundwater depths for the shallow aquifer in the study area were taken from the General Commission of Groundwater/Wasit Branch, Iraq, and Al-Shammary (2006) work, Table 3. The data include locations of the wells, well depths, groundwater depths (m), and hydraulic conductivities (m/s). The

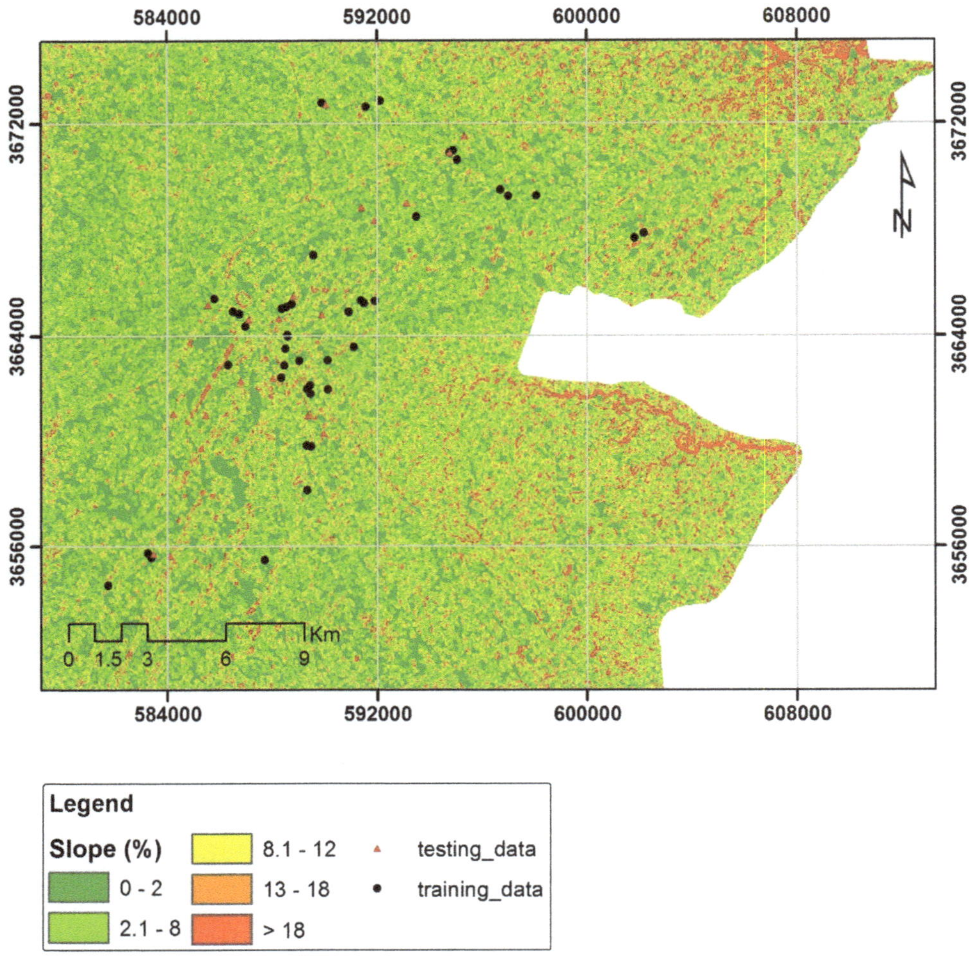

Fig. 5 Slope (%) in the study area

well locations covered the interested region and beyond, Fig. 2, and were different from the data used for the rest of the analysis. The minimum, maximum, and average depth of groundwater are 16, 162, and 65.98 m, respectively. The spatial distribution of the groundwater depth is shown in Fig. 3. Ordinary kriging technique in geostatistical extension of ArcGIS 10.2 was used to interpolate groundwater depths data after a comprehensive exploratory data analysis, i.e., investigate the data normality and trend detection. The greater part of the study area (about 69 %) has a groundwater depth greater than 30 m which implies that the aquifer systems are relatively protected from contamination at the ground surface. The groundwater depth map, Fig. 3, shows that the central part of the study area has a relatively shallow depth while the eastern parts have a greater depth. The groundwater depths increases from the west toward the east corresponding with the elevation increase in the same direction. To use the map of groundwater depth in further analysis, the raster map of this factor is classified into five categories based on the Jenks (natural break) classification system. Selection of this classification scheme is based on literature reviews and author's experience of study area and its condition (Al-Abadi 2015a).

Hydraulic conductivity is defined as the rate at which the aquifer materials transmit water, which in turn, controls the rate at which groundwater will flow under a given hydraulic gradient. The rate at which the groundwater flows also controls the rate at which a contaminant moves away from the point at which it enters the aquifer system. Higher rates represent higher susceptibility to contamination. Evidential thematic layer of the hydraulic conductivity was prepared based on data provided by Al-Shammary (2006), Table 3 and Fig. 4. The hydraulic conductivity values were also interpolated using ordinary kriging interpolation technique and then reclassified into five categories using Jenks scheme too. Figure 4 shows that the 77 % of the total study area have low hydraulic conductivities with an average of 4.3 m/day.

Slope is a rise or fall of land surface. It is an important factor for groundwater vulnerability assessment because it controls the likelihood that pollutants will runoff or remain on the surface to allow contaminants' percolation to the saturated zone. Slopes that provide a greater opportunity

Fig. 6 HSG in the study area

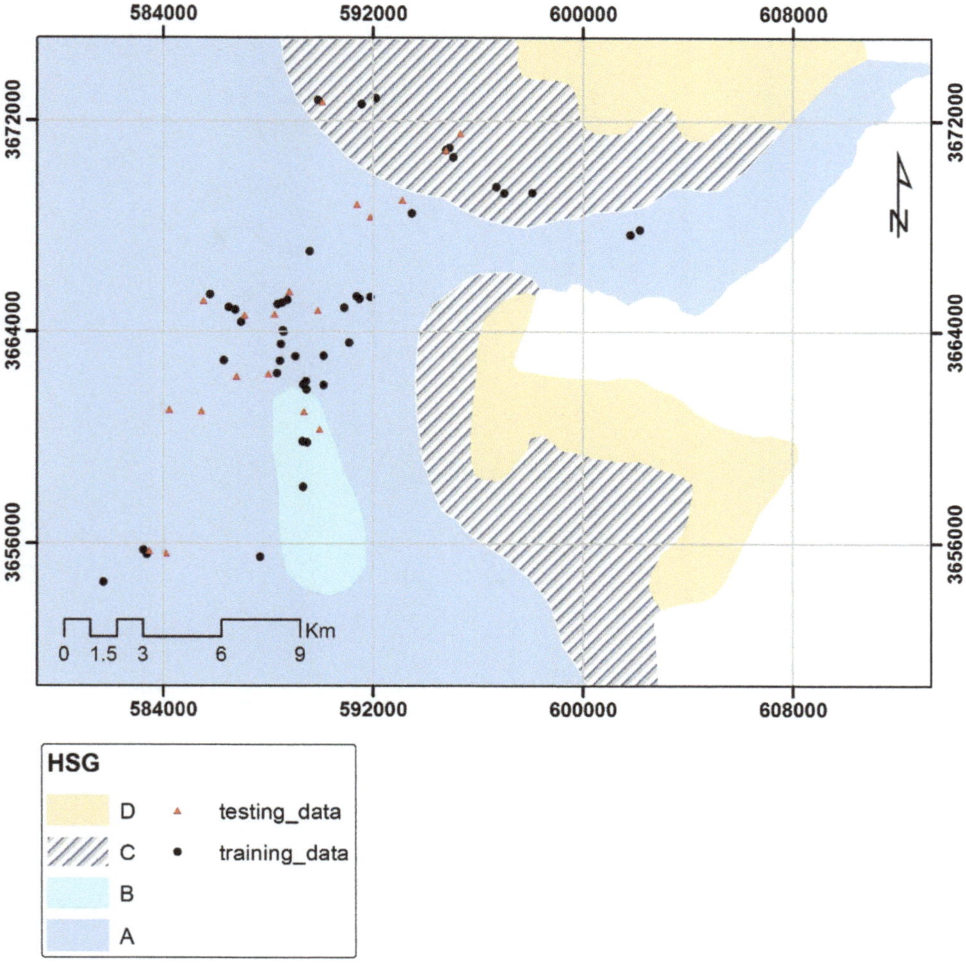

for contaminants to infiltrate will be associated with higher groundwater pollution potential (Aller et al. 1987). Slope also influences soil development, and therefore, has an impact on contaminants attenuation (Babiker et al. 2005). To prepare thematic layer of slope, the Advanced Spaceborne Thermal Emission and Reflection Radiometer (ASTER) Global Digital Elevation Model (GDEM) (http://gdem.ersdac.jspacesystems.or.jp/search.jsp) was used. The ASTER-GDEM was developed by the Ministry of Economy of Japan and the United States National Aeronautics and Space Administration (NASA). The spatial resolution of the ASTER-GDEM is approximately 30 m. Four tiles were downloaded from the previous web location, merged to new mosaic raster, clipped for the study area, reprojected using UTM WGS 1984 38 N projected coordinate system, and fill sinks. The treated raw DEM was then used to derive slope raster using Spatial Analyst extension of ArcGIS 10.2. The resulting raster values were then classified into five categories: 0–2 % (23 %), 2–8 % (51 %), 8–12 % (15 %), 12–18 % (8 %), and >18 % (3 %), Fig. 5. The greater part of the study area (about 74 %) has low slope values (0–8 %), indicating that the area is sensitive to contamination on the ground surface.

Soil refers to the uppermost portion of the unsaturated zone characterized by significant biological activities. Soil has an impact on the amount of recharge, which can infiltrate to the groundwater, and hence contaminants' movement (Piscopo 2001). The presence of fine-textured material such as silt and clay can decrease relative soil permeability and restrict contaminants' migration (Aller et al. 1987). The attenuation processes such as biodegradation, filtration, sorption, and volatilization may be significant if the soil zone is thick enough. Soils are classified into four hydrologic soil groups (HSG's) to indicate the minimum rate of infiltration for bare soil after prolonged wetting. The four hydrologic soils groups are A, B, C, and D, where A generally has the greatest infiltration rate (smallest runoff potential) and D has the smallest infiltration rate (greatest runoff potential). The HSG map of the study area was prepared by digitizing the hard copy of the soil map of Iraq and few soil textures available in the work of Al-Shammary (2006), Fig. 6. From this map, it is

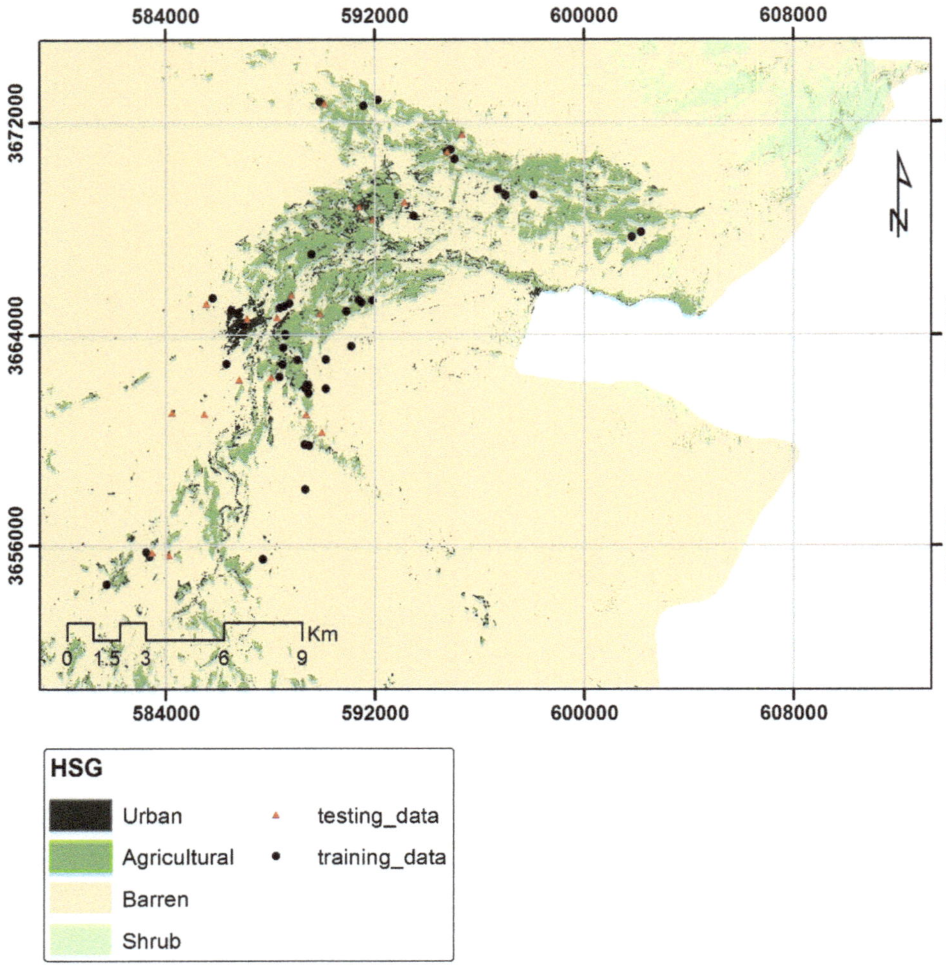

Fig. 7 LULC categories in the study area

obvious that the major portion of the study area (about 65 %) has high infiltration rate (A and B groups).

Land cover defines the biophysical state of the earth's surface and immediate subsurface, thus embracing the soil, material, vegetation, and water status. Land use on the other hand is a description of how people utilize land and socio-economic activity. There are two primary methods for capturing information on LULC: analysis of remotely sensed imagery and filed survey. The LULC map in this study was prepared using remote sensing data of Landsat 8. The raw image acquired in 6/2/2015 was first download from the official web site of USGS (United State of Geological Survey) (http://earthexplorer.usgs.gov/). Seven bands of the raw image (bands 1–7) were merged to create new raster, enhance radiometry, clipped for the study area, and then classified using supervised maximum likelihood approach by Image Classification tool in ArcGIS 10.2. Four LULC classes were found in the study area after being compared with ground truth: urban, agriculture, barren, and shrub, Fig. 7. The Barren and Shrub classes encompassed an area of 595 km^2 (90 %). Only 62 km^2 of the study area (10 %) was covered by urban and agricultural classes.

Results and discussion

As previously mentioned, the five evidential thematic layers were prepared as raster comprising of 30 × 30 m cell size. The number of wells per each class of a specific thematic layer was determined through multi-stage procedure. In the first stage, the evidential theme was reclassified. After that, it was converted to polygon. The resultant polygon was interested with training wells layer using tabulate intersection command to produce a table containing the number of wells for each class in the specific evidential thematic layer. The total number of pixels of the study area and the number of pixels of each class of a factor were determined directly from the attribute tables of a reclassified raster layer. The attribute table for each reclassified raster layer has a column from which the number of pixels of each class is directly determined. Summation of the pixels for all classes gives the total number of pixels of the study area.

The Bel, Dis, Unc, and Pls functions of the DST are summarized in Table 4. The detail procedure to calculate these functions is given in the previous section; an example

Table 4 Values of DS mass functions for classes of groundwater vulnerability factors

Factor	Class	Number of class pixel	Pixels (%)	Number of wells	Well (%)	Frequency ratio	DS mass functions			
							Bel	Dis	Unc	Pls
Slope (%)	0–2	170,762	0.23	11	0.25	1.07	0.31	0.199	0.487	0.801
	2–8	374255	0.51	26	0.59	1.15	0.34	0.171	0.491	0.829
	8–12	112,920	0.15	6	0.14	0.88	0.26	0.208	0.534	0.792
	12–18	54,586	0.07	1	0.02	0.30	0.09	0.215	0.696	0.785
	>18	17,657	0.02	0	0	0	0.00	0.208	0.792	0.792
Groundwater depth (m)	20.65–41.73	243,837	0.33	4	0.09	0.27	0.06	0.27	0.674	0.729
	41.74–65.74	172,013	0.24	21	0.48	2.03	0.41	0.14	0.452	0.864
	65.75–92.67	142,911	0.20	14	0.32	1.63	0.33	0.17	0.501	0.832
	92.68–124.3	110,638	0.15	3	0.07	0.45	0.09	0.22	0.690	0.782
	124.4–170	60,782	0.08	2	0.05	0.55	0.11	0.21	0.682	0.793
Hydraulic conductivity (m/d)	0.60–4.59	377,458	0.52	35	0.80	1.54	0.27	0.09	0.644	0.912
	4.59–7.67	185,621	0.25	1	0.02	0.09	0.02	0.27	0.712	0.727
	7.68–13.48	150,568	0.21	5	0.11	0.55	0.10	0.23	0.672	0.768
	13.49–25.63	13,986	0.02	3	0.07	3.56	0.62	0.20	0.182	0.802
	25.64–46.85	2548	0	0	0.00	0.00	0.00	0.21	0.791	0.791
HSG	A	452,221	0.62	32	0.73	1.17	0.27	0.19	0.548	0.814
	B	21,471	0.03	3	0.07	2.32	0.52	0.25	0.226	0.750
	C	161,317	0.22	9	0.20	0.93	0.21	0.27	0.525	0.735
	D	95,171	0.13	0	0.00	0.00	0.00	0.30	0.701	0.701
LULC	Urban	9582.5	0.01	9	0.20	15.59	0.79	0.12	0.093	0.880
	Agricultural	59,129.5	0.08	13	0.30	3.65	0.18	0.11	0.702	0.886
	Barren	641,348.5	0.88	22	0.50	0.57	0.03	0.61	0.359	0.388
	Shrub	20,119.5	0.03	0	0.00	0.00	0.00	0.15	0.847	0.847

of calculation is presented here for groundwater depth class 2; the number of wells in the class (=26), total number of training boreholes in the study area (=44), number of pixels in the considered class (=374,255), total number of pixels in the study area (730,180). Therefore,

$$\lambda(T_p)_{\text{class}\,2} = (26/44)/[374255 - 26/(730180 - 44)] = 1.15$$

$$\lambda(\bar{T}_p)_{\text{class}\,2} = ((44 - 26)/44)/(((730180 - 44 - 374255) + 26)/(730180 - 44)) = 0.84.$$

The other $\lambda(T_p)$ for 1, 3, 4, and 5 classes were 1.07, 0.88, 0.30, and 0, respectively. Thus, $\sum\lambda(T_p)_{\text{slope}} = 3.41$. The Bel function was then calculated using Eq. 9 as

$$\text{Bel} = \frac{1.15}{3.41} = 0.34.$$

The values of $\lambda(\bar{T}_p)$ for classes 1, 3, 4, and 5 were 0.98, 1.02, 1.06, and 1.02, respectively. Therefore, $\sum\lambda(\bar{T}_p)_{\text{slope}} = 4.92$. The Dis function were calculated from Eq. 11

$$\text{Dis} = \frac{0.84}{4.92} = 0.171.$$

The Unc and Pal functions was calculated using Eqs. 12 and 13

$$\text{Unc} = 1 - 0.34 - 0.171 = 0.49$$

$$\text{Pls} = 1 - 0.171 = 0.83.$$

For the groundwater depth factor, high Bel (0.41) and low Dis (0.33) values were found in the ranges of 41.7–65.7 m and 65.7–92.67 m which indicate that these classes have positive associations with aquifer vulnerability. The remaining classes have minor effect on vulnerability due to low values of Bel and high values of Dis. In the case of hydraulic conductivity, the range of 13.49–25.63 has the highest Bel value (0.62) and the lowest Dis value (0.20) indicating the highest probability of contamination by nitrate. The other classes have relatively low Bel values indicating that these classes play a minor role in the control of contamination processes in the study area. For the slope factor, slope angle in the range of 20–30 % has the highest Bel and the lowest Dis values indicating the highest probability of contamination, followed by slope range of 0–2 % and then of 8–12 %. For the remaining slope ranges, Bel values are low referring to the low probability of contamination by

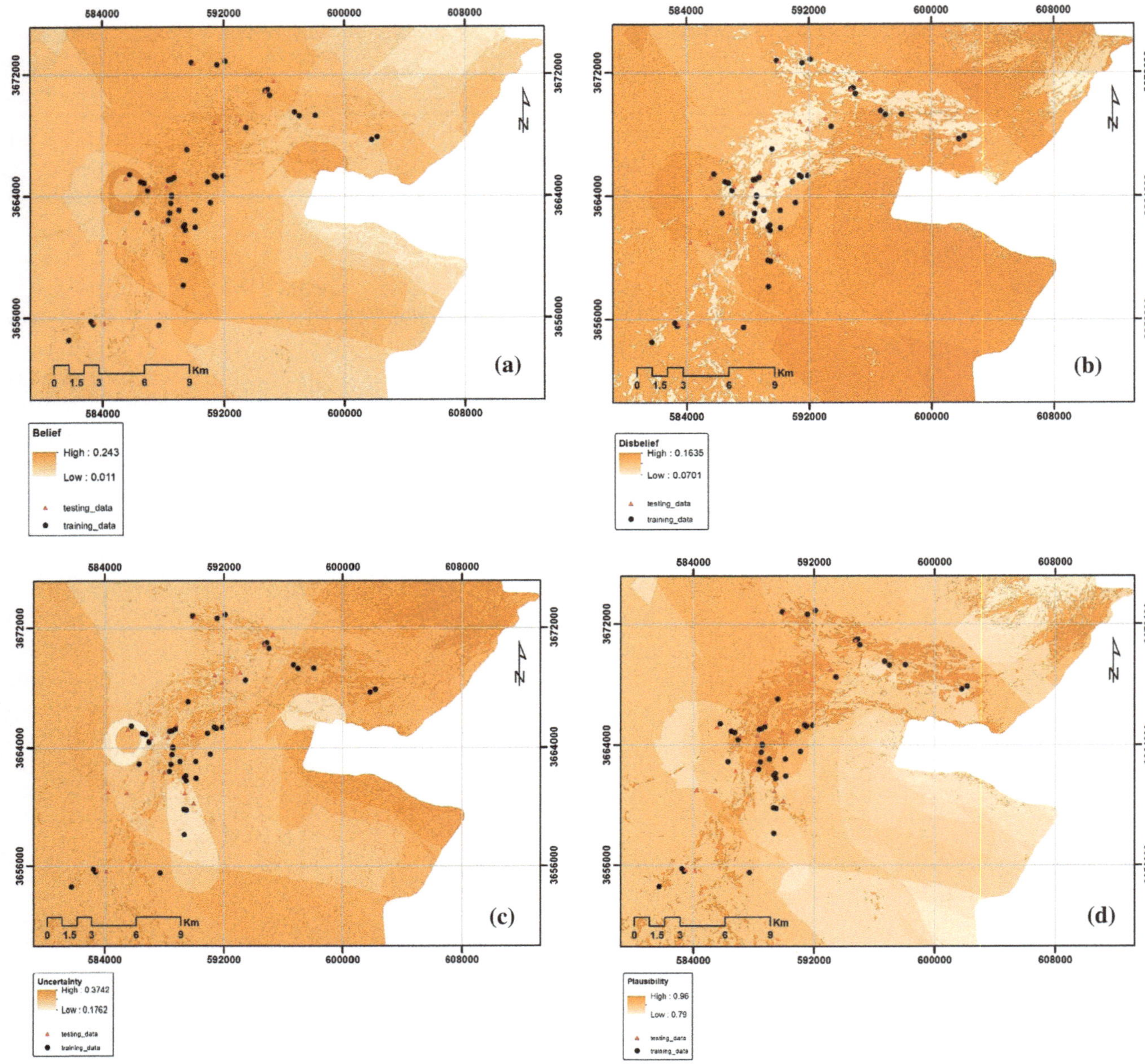

Fig. 8 Integrated BEF map **a** Bel, **b** Dis, **c** Unc, and **d** Pls

Table 5 Areas covered by GVI classes

GVI class	Area (%)	Area (km^2)
Very low	0.09	61
Low	0.22	148
Moderate	0.42	279
High	0.24	157
Very high	0.02	12

nitrate. For the soil factor, the highest values of Bel and the lowest values of Dis are associated with A and B groups. These groups have higher infiltration rates and thus they are more vulnerable to contamination. The low Bel and high Dis values for other groups indicate that the probability of contamination is low. In the case of LULC, there are high Bel and low Dis values for urban and agricultural categories, reflecting the high probability of contamination by nitrate for these categories. High probability of contamination in these LULC is due to increase in human activity and population growth. As the high value of Bel is correlated with urban and agricultural cause, the major sources of nitrate in the groundwater may be latrines and manures.

The integrated results are shown in Fig. 8. Comparison between the belief map, Fig. 8a, and the disbelief map,

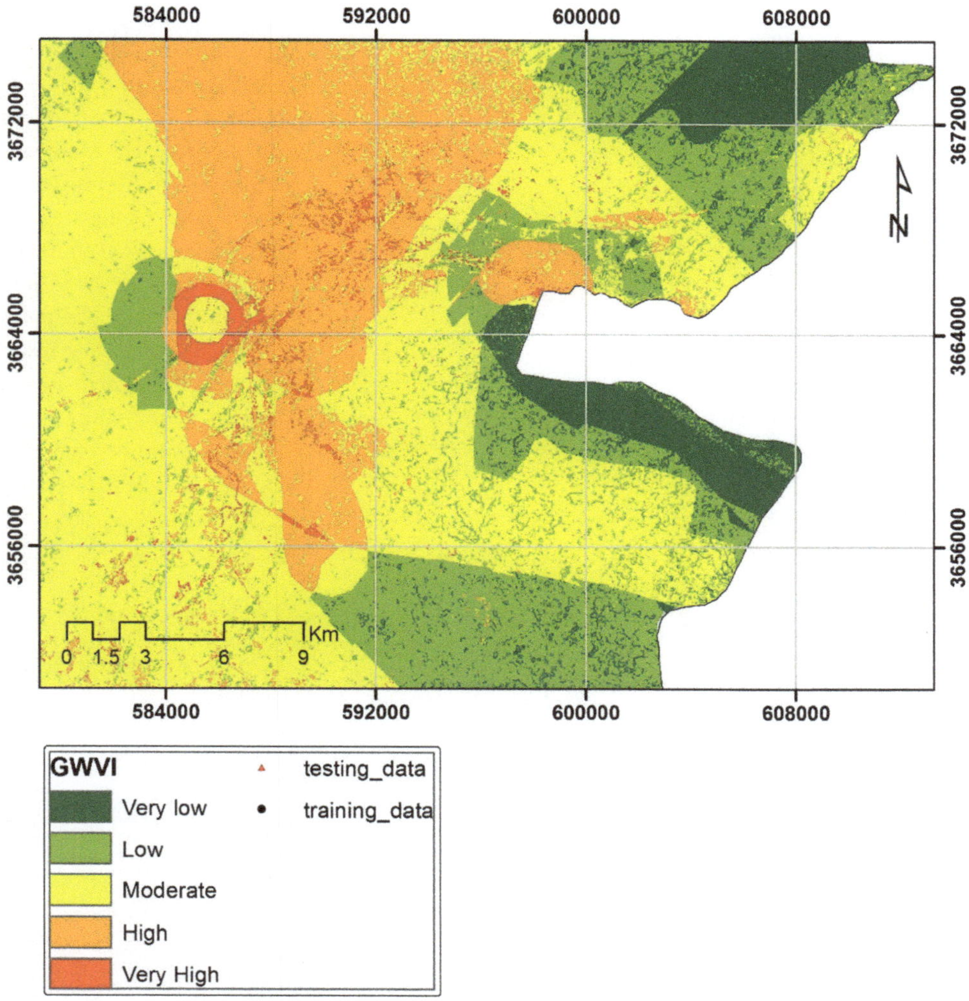

Fig. 9 GVI classes in the study area

Table 6 The relation between AUC and model prediction accuracy (after Yesilnacar (2005))

AUC	Prediction accuracy
0.5–0.6	Poor
0.6–0.7	Average
0.7–0.8	Good
0.8–0.9	Very good
0.9–1	Excellent

Fig. 8b, exhibits that belief values are high for the area where disbelief values are low and vice versa. The areas with high belief and low disbelief indicate high vulnerability for contamination by nitrate. The uncertainty map shows lack of information support uncertainty for vulnerability. As indicated from comparison between Fig. 8c and Fig. 8a, the uncertainty values are high in areas with low values of Belief. On the other hand, the plausibility map, Fig. 8d, shows high values for areas where both belief and uncertainty values are high. The integrated belief function map was used in this study as groundwater vulnerability index (GVI). The pixel values of GVI were reclassified into five categories: very low, low, moderate, high, and very high using Jenks classification scheme, Table 5 and Fig. 9. The very low–low zones cover 32 % (209 km^2). These classes mainly concentrate in the eastern parts of the study area and occupy small zone in the central part. The moderate zone extends over an area of 42 % (279 km^2) and mainly encompasses the western part of the study area. The high–very high zones cover 26 % (170 km^2) and these zones concentrate in the central part of the study area. The results indicate that the aquifer systems in the study area are moderately vulnerable to contamination by nitrate; thus it needs a wise plan to protect groundwater quality.

The next step in the analysis is to validate the results. Any predictive model (deterministic or stochastic) requires validation before it is used in prediction purposes. Without validation process the model will have no scientific significance (Chung and Fabbri 2003). In this context, the receive operating characteristic (ROC) curve is usually used for examining the quality of deterministic and

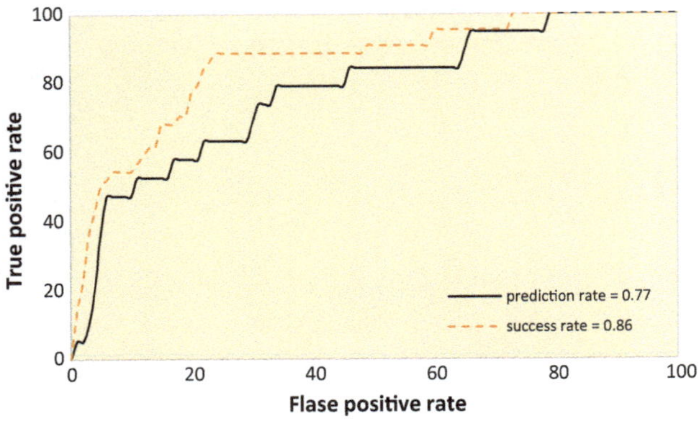

Fig. 10 Validation results using ROC technique

probabilistic detection and forecast system (Swets 1988). The area under the ROC curve (AUC) characterizes the quality of a forecast system by describing the system's ability to anticipate correctly the occurrence or non-occurrence of pre-defined "event" (Negnevitsky 2002). The quantitative–qualitative relationship between AUC and prediction accuracy is given in Table 6. The AUC was obtained for both the training (success rate) and testing (prediction rate) using ROC module in IDRISI software, Fig. 10. The success rate is important to explain how well the resulting GVI map classified the area of existing borehole locations. The success rate results were obtained by comparing the training well locations (32) with the GVI map. The AUC was 0.86. On the other hand, the prediction rate used a measure of performance as a predictive rule (Yesilnacar and Topal 2005; Pradhan et al. 2010a). It has only used the testing data set to explore the predictive capability of the model. The AUC for prediction rate was 0.77. These results indicate that DS has good capability (Table 6) to delineate groundwater vulnerability zones in the study area.

Conclusions

Groundwater is a very important renewable resource for drinking, agricultural, industrial, and other purposes. Therefore, it is vital that the use of groundwater should be carefully managed in terms of both quantity and quality. In recent years, delineation of areas that are more vulnerable to contamination is an essential step for managing aquifer system. In this study, the vulnerability of shallow aquifer for nitrate contamination in Galal Badra basin, east of Iraq was evaluated using DST of evidence in GIS framework. In the first stage of this study, an inventory map of the wells locations with elevated nitrate concentrations was prepared. After that, these wells were split into two sets: training and testing. In the second stage, the evidential thematic layers were prepared. Five factors namely groundwater depth, hydraulic conductivity, slope, soil, and LULC were selected for modeling the relationship between training well locations and factor classes using mass functions of DS method. The Bel function was combined according to Dempster rules to produce aquifer vulnerability index of the study area. The results of application of the method were validated using ROC. The prediction of the model was 87 % for success rate and 77 % for prediction rate. So, the performance of the map made using DST was good. The results of this study could be used by planners and decision makers to protect groundwater aquifer in the study area. The prediction accuracy of the method could be increased by adding other thematic layers if they are available or by combining multi-methods to produce more accurate picture of the vulnerability status in the study area.

Compliance with ethical standards

Conflict of interest The author declares that he has no conflict of interest.

References

Al-Abadi AM (2015a) Groundwater potential mapping at northeastern Wasit and Missan governorates, Iraq using a data-driven weights of evidence technique in framework of GIS. Environ Earth Sci. doi:10.1007/s12665-015-4097-0

Al-Abadi AM (2015b) Modeling of groundwater productivity in northeastern Wasit Governorate Iraq by using frequency ratio and Shannon's entropy models. Appl Water Sci. doi:10.1007/s13201-015-0283-1

Alagha JS, Said MD, Mogheir Y (2013) Modeling of nitrate concentration in groundwater using artificial intelligence approach—a case study of Gaza coastal aquifer. Environ Monit Assess 186:35–45. doi:10.1007/s10661-013-3353-6

Aller L, Bennett T, Lehr J, Petty R, and Hackett G (1987) DRASTIC: a standardized system for evaluation ground water pollution potential using hydrogeological settings. National Water Well Association, Dublin, Ohio and Environmental Protection Agency, Ada, Ok.EPA-600/2-87-035

Almasri M, Ghabayen S (2008) Analysis of nitrate contamination of Gaza Coastal Aquifer, Palestine. J Hydrol Eng 13:132–140

Al-Sayab A, Al-Ansari N, Al-Rawi D, Al-Jassim J, Al-Omari F, Al-Shaikh Z (1982) Geology of Iraq (in Arabic). Mosul University, Mosul

Al-Shammary SH (2006) Hydrogeology of Galal Basin-Wasit east, Iraq. unpublished Ph.D thesis, Baghdad, Iraq

An P, Moon WM, Bonham-Carter GF (1992) On knowledge-based approach on integrating remote sensing, geophysical and geological information. Proceedings of the international geoscience and remote sensing symposium (IGARSS '92). Houston, Texas, pp 34–38

Arthur J, Wood HA, Baker AE, Cichon JR, Raines GL (2007) Development and implementation of a Bayesian aquifer vulnerability assessment in Florida. Nat Resour Res 16:93–107. doi:10.1007/s11053-007-9038-5

Babiker IS, Mohammed AA, Hiyama T, Kato K (2005) A GIS-base DRASTIC model for assessing aquifer vulnerability in Kakamigahara heights, Gifu prefecture, central Japan. Sci Total Environ 345:127–140. doi:10.1016/j.scitotenv.2004.11.005

Bui DT, Pradhan B, Lofman O, Revhaug Dick OD (2012) Spatial prediction of landslide hazards in Hoa Binh province (Vietnam): a comparative assessment of the efficacy of evidential belief functions and fuzzy logic models. Catena 96:28–40. doi:10.1016/j.catena.2012.04.001

Carranza EJM, Hale M (2003) Evidential belief functions for data-driven geologically constrained mapping of gold potential, Baguio district, Philippines. Ore Geol Rev 22:117–132

Carranza EJM, Woldai T, Chikambwe EM (2005) Application of data-driven evidential belief functions to prospectivity mapping for aquamarine-bearing pegmatites, Lundazi district, Zambia. Nat Resour Res 14:47–63

Chung CF, Fabbri AG (2003) Validation of spatial prediction models for landslide hazard mapping. Nat Hazards 30:451–472

Cissé IA, Mao X (2008) Nitrate: health effect in drinking water and management for water quality. Environ Res 2:311–316

Civita M (1994) La carte della vulnerbilità deli aquifer alí inquinamento: teoria & pratica (in Italian). Pitagora Editrica, Bologan

Dempster AP (1968) A generalization of Bayesian inference. J R Stat Soc Series 30:205–247

Doerfliger N, and Zwahlen F (1997) EPIK: a new method for outlining of protection areas in karstic environment. In: Günay G, Johnson AL (eds) International symposium and field seminar on "karst waters and environmental impacts". Antalya, Turkey. Balkema, Rotterdam. pp 117–123

Foster SD (1987) Fundamental concepts in aquifer vulnerability pollution risk and protection strategy. In: Van Duijvedbooden W, van Waegeningh HG (eds) Vulnerability of soil and groundwater to pollutions: proceedings and information. TNO Committee on Hydrological Research, The Hague, pp 69–86

George T, Pal NR (1996) Quantification of conflict in Dempster–Shafer framework: a new approach. Int J Gen Syst 24(4):407–423

Harter T, Walker LG (2001) Assessing vulnerability of groundwater. US Natural Resources Conservation Service, Washington

Jassim SZ, Goff JC (2006) Geology of Iraq. Prague and Moravian Museum, Brno, Czech Republic, Dolin **431p**

Jože Uhan, Vižintin G, Pezdic J (2010) Groundwater nitrate vulnerability assessment in alluvial aquifer using process-based models and weights-of-evidence method: lower Savinja Valley case study (Slovenia). Environ Earth Sci 64:97–105. doi:10.1007/s12665-010-0821-y

Lee S, Hwang J, Park I (2012) Application of data-driven evidential belief functions to landslide susceptibility mapping in Jinbu, Korea. Catena 100:15–30. doi:10.1016/j.catena.2012.07.014

Liggett JE, Talwar S (2009) Groundwater vulnerability assessments and integrated water resource management. Watershed Manag Bull 13(1):18–29

Mair A, El-Kadi A (2013) Logistic regression modeling to assess groundwater vulnerability to contamination in Hawaii, USA. J Contam Hydrol 153:1–23. doi:10.1016/j.jconhyd.2013.07.004

Majandang J, Sarapirome S (2013) Groundwater vulnerability assessment and sensitivity analysis in Nong Rua, Khon Kaen, Thailand, using a GIS-based SINTACS model. Environ Earth Sci 68:2025–2039. doi:10.1007/s12665-012-1890-x

Masetti M, Poli S, Sterlacchini S (2007) The use of the weights-of-evidence modeling technique to estimate the vulnerability of groundwater to nitrate contamination. Nat Resour Res 16:109–119. doi:10.1007/s11053-007-9045-6

Mogaji KA, Lim HS, Abdullah K (2014) Regional prediction of groundwater potential mapping in a multifaceted geology terrain using GIS-based Demspter–Shafer model. Arab J Geosci. doi:10.1007/s12517-014-1391-1

Mohammady M, Pourghasemi HR, Pradhan B (2012) Landslide susceptibility mapping at Golestan Province, Iraq: a comparison between frequency ratio, Dempster–Shafer, and weights-of-evidence models. J Asian Earth Sci 61:221–236

Moon WM (1990) Integration of geophysical and geological data using evidential belief function. IEEE Trans Geosci Remote Sens 28:711–720

Nampak H, Pradhan B, Manap MA (2014) Application of GIS based data driven evidential belief function model to predict groundwater potential zonation. J Hydrol 513:283–300. doi:10.1016/j.jhydrol.2014.02.053

Negnevitsky M (2002) Artificial intelligence—a guide to intelligent systems. Addison-Wesley Co, Great Britain

Park NW (2011) Application of Dempster–Shafer theory of evidence to GIS-based landslide susceptibility analysis. Environ Earth Sci 62:367–376. doi:10.1007/s12665-010-0531-5

Parsons RM (1956) Ground-water resources of Iraq. Khanaqin-Jassan area, vol. 1. Development Board, Ministry of Development Government of Iraq. Baghdad

Piscopo G (2001) Groundwater vulnerability map, explanatory notes, Castlereagh Catchment. NSW Department of Land and Water Conservation, Australia

Popescu IC, Gardin N, Brouyére S and Dassargues A (2008) Groundwater vulnerability assessment using physically-based modeling: From challenges to pragmatic solutions. In: Refsgaard JC, Kovar K, Harder E, Nygard E (eds) ModelCARE 2007 proceedings, calibration and reliability in groundwater modeling, Denmark (**IAHS Publication no. 320)**

Pourghasemi H, Pradhan B, Gokceoglu C, Moezzi KD (2013) A comparative assessment of prediction capabilities of Dempster–Shafer and weights-of-evidence models in landslide susceptibility mapping using GIS. Geomat Nat Hazards Risk 4:93–118

Pradhan B, Lee S, Buchroithner MF (2010) Remote sensing and GIS-based landslide susceptibility analysis and its cross-validation in three test areas using a frequency ratio model. Photogramm Fernerkun 1:17–32. doi:10.1127/1432-8364/2010/0037

Sorichetta A, Ballabio C, Masetti M, Robinson GR Jr, Sterlacchini S (2013) A comparison of data-driven groundwater assessment methods. Ground Water 51:866–879. doi:10.1111/gwat.12012

Swets JA (1988) Measuring the accuracy of diagnostic systems. Science 240:1285–1293

Thiam AK (2005) An evidential reasoning approach to land degradation evaluation: Dempster–Shafer theory of evidence. Trans GIS 9:507–520

Van Stempvoort D, Ewert D, Wassenaar L (1993) Aquifer vulnerability index: a GIS-compatible method for groundwater vulnerability mapping. Can Water Resour J 18(1):25–37

Wally P (1987) Belief function representation of statistical evidence. Ann Stat 15:1439–1485

Yesilnacar EK (2005) The application of computational intelligence to landslide susceptibility mapping in Turkey. Ph.D Thesis, Department of Geomatics, the University of Melbourne, pp 423

Yesilnacar E, Topal T (2005) Landslide susceptibility mapping: a comparison of logistic regression and neural networks methods in a medium scale study, Hendek region (Turkey). Eng Geol 79:251–266

13

Nanofiltration membranes of poly(styrene-*co*-chloromethylstyrene)-*grafted*-DGEBA reinforced with gold and polystyrene nanoparticles for water purification

Ayesha Kausar[1] · Muhammad Siddiq[2]

Abstract The matrix material for nanofiltration membranes was prepared through chemical grafting of poly (styrene-*co*-chloromethylstyrene) (PSCMS) to DGEBA using hexamethylenediamine as linker. The phase inversion technique was used to form PSCMS-*g*-DGEBA membranes. This effort also involves the designing of gold nanoparticles and its composite nanoparticles with polystyrene microspheres as matrix reinforcement. The nanoporous morphology was observed at lower filler content and there was formation of nanopattern at increased nanofiller content. The tensile strength was improved from 32.5 to 35.2 MPa with the increase in AuNPs-PSNPs loading from 0.1 to 1 wt%. The glass transition temperature was also enhanced from 132 to 159 °C. The membrane properties were measured via nanofiltration set-up. Higher pure water permeation flux, recovery, and salt rejection were measured for novel membranes. PSCMS-*g*-DGEBA/AuNPs-PSNPs membrane with 1 wt% loading showed flux of 2.01 mL cm^{-2} min^{-1} and salt rejection ratio of 70.4 %. Efficiency of the gold/polystyrene nanoparticles reinforced membranes for the removal of Hg^{2+} and Pb^2 was found to be 99 %. Novel hybrid membranes possess fine characteristics to be utilized in industrial water treatment units.

Keywords Nanofiltration · Poly(styrene-*co*-chloromethylstyrene) · Epoxy · Phase inversion · Pure water flux

Introduction

The membrane segregation processes are extensively employed for the water treatment in various industries such as chemical, food, nuclear, and pharmaceutical productions. The membrane processes offer an attractive alternate to the traditional systems for the ultra pure water production, desalination, pathogen exclusion from water, and solid–liquid isolation in water treatment. The membrane technology is an attractive method in water purification industries (Arockiasamy et al. 2013). The membrane separation process has high functional stability and is easy to control and integrate in larger industrial practices (Kumar et al. 2013). The important types of membrane processes presently employed are ultrafiltration (UF), nanofiltration (NF), and reverse osmosis (RO). Generally, the RO technique is based on salt ion repellent capacity of thick membrane, so, allowing the flow of water molecules (Greenlee et al. 2009). The process is usually controlled through solution-diffusion with a sufficiently large external pressure to overcome the salt water osmotic pressure (Evuti and Lawal 2011). On the other hand, UF membranes permit the separation of intermediate molecular weight pollutants and certain heavy metals. The effectiveness of both the membrane processes depends on their design and properties (Yuan and Zydney 2000). Among them, NF technique is known for smaller pore size of membrane and optimal purification efficacy (Dalwani et al. 2010). At present, there are two extensively used categories of membranes for water detoxification, i.e., polymeric

✉ Ayesha Kausar
asheesgreat@yahoo.com

1 Nanosciences and Catalysis Division, National Centre For Physics, Quaid-i-Azam University Campus, Islamabad 44000, Pakistan

2 Department of Chemistry, Quaid-i-Azam University, Islamabad, Pakistan

membranes (polysulfone, polyamide, polyfurane, etc.), and ceramic/inorganic composite (Bowen et al. 1997; Chen et al. 2001). The polymeric membranes are of primary significance due to their flexibility and low costs. However, these membranes experience the lack of thermal and chemical stability. In this regard, the hybrid polymeric membranes are well established in desalination due to straightforward module design, improved selectivity, and permeability (Bowen and Mohammad 1998; Bowen and Welfoot 2002). The desalination operates through various transference mechanisms depending mostly on the membrane pore sizes. The hybrid membranes are quite robust to process conditions such as long-term stability at elevated temperature under hydrothermal conditions, continuous operation over prolonged time periods without any decline of selectivity, and acid stability of the material. Therefore, the beneficial use of hybrid polymer/inorganic materials has been exploited in membrane technology (Gevers et al. 2005). In this area, various inorganic fillers such as titania (Ebert et al. 2004), silica (Sadeghi et al. 2008), and carbon molecular sieves (Vu et al. 2003) were employed to fabricate hybrid thin films. For the removal of pollutants from ground water, hybrid nanofiltration seems to be a suitable method capable of removing hardness, sulfates, chlorides, and bacteria.

Removal of heavy metals is an essential area of research in membrane filtration technology. An important property presented by modified surfaces of noble metal nanoparticles is the diagnosis of heavy metals. Initial studies regarding the interaction between metal ions and noble metal nanoparticles were carried out in 1990s (Henglein 1998). A number of adsorbents have been industrialized for the elimination of mercury from water such as activated carbon (Namasivayam and Periasamy 1993) and functionalized clay (Asyhar 2014). Likewise, there is literature available regarding the nanomaterial-based mercury elimination from contaminated water (Lu and Liu 2007). One of these methods involves the use of biomolecule-modified gold nanoparticles (Ono and Togashi 2004). In a comparable attempt, the peptide-modified gold nanoparticles were used for colorimetric recognition of Pb^{2+} and Hg^{2+} (Slocik et al. 2008). However, the hybrid nanofiltration membrane has been less explored in literature compared with other membranes types. The important factors affecting the morphology and performance of membranes are the type and the concentration of nanofiller (Ebert et al. 2004). In this exploration, the matrix poly(styrene-*co*-chloromethylstyrene) was chemically grafted to epoxy (bisphenol A diglycidyl ether) using hexamethylenediamine (PSCMS-*g*-DGEBA). The phase inversion technique was used to form nanofiltration membranes loaded with gold/polystyrene composite nanoparticle (AuNPs-PSNPs). The noble metal nanoparticles were introduced as reinforcement to remove several toxic metal ions from water. The influence of the inclusion of various AuNPs-PSNP contents on membrane performance and structure was studied using pertinent methods and techniques. To the best of our knowledge, epoxy blend membranes of poly(styrene-*co*-chloromethylstyrene)-*grafted*-DGEBA have been prepared for the first time. Novelty of the research also lies in the reinforcement of the membranes with the gold composite nanoparticles. Epoxy has been employed in water filtration membranes (Du et al. 2011). In this research, the cured DGEBA was first blended with poly(styrene-*co*-chloromethylstyrene). Afterwards, the composite membranes of poly(styrene-*co*-chloro-methylstyrene)-*grafted*-DGEBA with the nanoparticles were formed. After blending of the composite nanoparticles in cured epoxy blend structure, there are least chances of the gold nanoparticles to move into the media relative to reported gold nanoparticle-polymer composites (Coulston et al. 2011). The fabrication strategy opted for the AuNPs-PSNPs resulted in fine morphology compared with the other reported methods (Li et al. 2011). The spherical shape and high surface area of AuNPs-PSNPs increased their compatibility with the nanofiller. The membranes simply formed by poly(styrene-*co*-chloro-methylstyrene)-*grafted*-DGEBA may have lower mechanical properties. Therefore, the nanoparticles were used as matrix component to increase the tensile and other physical properties of the hybrids (Susrutha et al. 2012). The mechanical properties of the hybrid membranes were superior to that of the gold-reinforced chitosan bio-nanocomposites (Patel et al. 2014).

Experimental

Materials

Poly(styrene-*co*-chloromethylstyrene) ($M_w \sim 100{,}000$), was purchased from Aldrich. Styrene ($\geq$99 %), obtained from Aldrich was purified by distillation under reduced pressure. Poly(vinylpyrrolidone) (PVP K-30) was purchased from Acros Organics. Bisphenol A diglycidyl ether (DGEBA, $\geq$95.0 %), hexamethylenediamine (HDA, 98 %), 2-2′-azoisobutyronitrile (AIBN, $\geq$98 %), $HAuCl_4$ (10 nm particle size), 4-(4-aminophenylsulfonyl)benzenamine (APSBA) (97 %), ammonium thiocyanate (98 %), and sodium carbonate were obtained from Aldrich and used as received.

Characterization techniques

The infrared (IR) spectra were recorded using a Fourier transform infrared (FTIR) Spectrometer, Model No. FTSW 300 MX, manufactured by BIO-RAD, California, USA

(4 cm^{-1} resolution). PerkinElmer's LAMBDA 45 UV/Vis system was used to detect gold nanoparticles in the filtrate. The morphology, size, and distribution of AuNPs-PSNPs were deliberated through transmission electron microscopy. Scanning electron microscopy was employed to study the cross-section and internal fractured surface structure of PSCMS-*g*-DGEBA/AuNPs-PSNPs membranes. The SEM images were obtained by Scanning Electron Microscope S-4700 (Japan Hitachi Co. Ltd.) and TEM images were taken by Transmission Electron Microscope H-800 (Japan Hatachi Co. Ltd). The particle size and the particle size distribution were measured by Zetasizer Nano ZS particle analyzer (UK Malvern Instruments Ltd.) by means of dynamic light scattering (DLS). Brunauer Emmett Teller (BET) surface area measurement of membranes was carried out using a Micromeritics ASAP (Accelerated Surface Area and Porosimetry) 2000, (Micromeritics Instruments, Co., Georgia, USA). The measurement of surface area with this instrument was based on physisorption and desorption of a gas (nitrogen) at the surface and in the pores of the sample. The amount of gas adsorbed was calculated from measurement of differential gas pressure. Before analysis, the membrane samples were prepared by degassing under vacuum at 30 °C for 2 h. The tensile tests were conducted with dog-bone-shape samples on a Testomeric materials testing machine M500–30CT at a speed of 500 mm/min according to ASTM D412. To analyze the water content, membranes were soaked in distilled water for 24 h followed by mopping with blotting paper and weighing. The wet membranes were first kept at 100 °C for 24 h, and subsequently weighed to obtain the dry weight using an electronic balance. The percent solvent content was evaluated using the following equation:

$$\% \text{ Solvent content} = \frac{W_1 - W_2}{W_2} \times 100,$$

where W_1 = Wet membrane weight (g) and W_2 = Dry membrane weight (g).

To analyze the membrane porosity, samples were soaked in distilled water for 24 h followed by mopping with blotting paper and weighing. The membranes were dried at 100 °C for 24 h and weighed. Porosity was calculated using the following equation:

$$\text{Porosity} = \frac{W_1 - W_2}{A_h},$$

where W_1 = Wet membrane weight (g), W_2 = Dry membrane weight (g), and A_h = Membrane area (cm^2).

Pure water flux of membrane was measured using membrane cell with cross-flow mode. The permeate was collected every 10 min for 1 h. Membranes after compaction were subjected to pure water flux estimation at trans-membrane pressure of 345 kPa. The permeability was measured under steady-state flow. Pure water flux was calculated as follows:

$$J_w = \frac{Q}{At},$$

where Q = Quantity of permeate collected (L), J_w = Pure water flux (mLcm^{-2} min^{-1}), A = Membrane area (cm^{-2}), t = Sampling time (min).

The length and width of a piece of wet membrane, immersed in a water bath, were measured using a vernier caliper. Then the membrane was heated at 100 °C for 24 h. The dry membrane's length and width were measured. The shrinkage ratio was calculated using the following equation:

$$\text{Shrinkage ratio} = \left[1 - \left(\frac{ab}{a_o b_o}\right)\right] \times 100,$$

where A = length of dry membrane (m), b = breadth of dry membrane (m), a_o = length of wet membrane (m), b_o = breadth of wet membrane (m).

Rejection indicates the amount of components rejected by membranes. It shows the separation efficiency of the components by the membrane, which is calculated as

$$R_e = 1 - \frac{C_p}{C_f},$$

where R_e = Rejection, C_p = Concentration of components in permeate (g/m^3), C_f = Concentration of components in feed water (g/m^3).

Recovery is also known as productivity. Here also the membrane cell with cross-flow mode was used. According to mass balance, the feed flow is equal to the sum of concentrate flow and permeates flow. Recovery was calculated as follows:

$$\gamma = \frac{Q_p}{Q_f} \times 100,$$

where Γ = Recovery (%), Q_p = Permeate flow (m^3/h), Q_f = Feed flow (m^3/h).

The solid–liquid extraction tests of nitrates or chloride salts of Fe^{3+}, Co^{2+}, Ni^{2+}, Cu^{2+}, Cd^{2+}, Pb^{2+}, and Hg^{2+} were carried out using neat PSCMS-*g*-DGEBA membrane and PSCMS-*g*-DGEBA/AuNPs-PSNPs 1 membranes. After filtration, the concentration of each cation in the liquid phase was determined by inductively coupled plasma mass spectrometry (ICP-MS, Agilent 7500). Consecutive dilutions of sample aliquots with ultra-pure water/nitric acid (5 % v/v) were performed to reach concentrations in the range of the calibration curve (0–40 ppb), and thus direct information regarding the extraction percentages of metal ions by membranes was obtained. The effectiveness of membranes for the extraction of environmentally toxic heavy metal ions was studied in terms of distribution coefficient (Kd′) and solid–liquid extraction selectivity ($\alpha_{S,L}$). Kd′ is the measure of the capacity of material to extract cations under competitive conditions, where *%E is* extraction percentage, *V* is volume of the solution (liters), and *n* the molar amount of polymer structural units.

$$\mathrm{Kd}' = \left(\frac{\%E}{100 - \%E}\right)\left(\frac{V}{n}\right),$$

$\alpha_{S,L}$ is the ratio of two distribution coefficients. Hg^{2+} cation was taken as the reference, as it has the highest distribution coefficient (Kd'_{M2}).

$$\alpha_{S,L} = \frac{\mathrm{Kd}_{M1}}{\mathrm{Kd}_{M2}} = \frac{\mathrm{Kd}'_{M1}}{\mathrm{Kd}'_{M2}}$$

Synthesis of polystyrene microspheres (PSNPs)

Initially, PVP stabilizer (2.6 g) was dissolved in 2-propanol (160 mL) at 70 °C. AIBN initiator (0.2 g) was then dissolved in styrene (20 g) and added to the above mixture with stirring. The polymerization was allowed to proceed at 70 °C for 24 h. Subsequently, the PS microspheres were centrifuged and washed with ethanol and deionized water (Wu et al. 2010).

Synthesis of gold nanoparticle-coated polystyrene nanocomposite particles (AuNPs-PSNPs)

0.025 g Na_2CO_3 was dissolved in 100 mL deionized water and added to 16.4 mL $HAuCl_4$ aqueous solution (3×10^{-3} M) and stirred for 0.5 h. Then 0.1 g PS (10 wt%) emulsion and 0.25 mL formaldehyde (35 wt%) were added to the above mixture and stirred for 1 h. The resulting AuNPs-PSNPs nanoparticles were separated out by centrifugation and re-dispersion in ethanol and deionized water to remove the residual free nanoparticles.

Preparation of the hybrid membranes

For the preparation of membranes, the homogenous solution of poly(styrene-*co*-chloromethylstyrene) matrix and DGEBA were mixed in the ratio of 90:10 wt% in 10 mL DMAc. Afterwards, 1 wt% of HDA was added as cross-linking agent to the above mixture (Fig. 1). The mixture was refluxed at 80 °C for 6 h. The desired amount of AuNPs-PSNPs was added and reaction flask was transferred to ultrasonic bath cleaner for 6 h. The homogeneous solutions obtained were cast onto glass plates using casting knife. The prepared films were dipped into deionized water as non-solvent. In this step, the exchange of solvent and non-solvent led to the membrane formation. After finishing phase separation step and membrane formation, the membranes were kept in deionized water for 24 h to remove remaining solvent. Finally, the membranes were dried at 80 °C for 24 h before testing (Buonomenna et al. 2007). The membrane properties were measured using the experimental nanofiltration set-up (Fig. 2). FTIR (KBr, cm^{-1}) (Fig. 3): 3401 (O–H stretch), 3254, 1519 (N–H stretch and bend), 3005 (aromatic C–H stretch), 2911 (aliphatic C–H stretch), 1420 cm^{-1} (C–N stretch), 1299 (C–O stretch).

Results and discussion

Mechanical features

The PSCMS-*g*-DGEBA graft copolymer was reinforced with gold/polystyrene composite nanoparticles in the range of 0.1–1 wt%. As shown in Fig. 4 and Table 1, there was improvement in mechanical properties of PSCMS-*g*-DGEBA/AuNPs-PSNPs 0.1–1 membranes. There was an increasing trend in the tensile stress, tangent modulus, and toughness of the membranes, while the elongation at break was decreased in the novel membranes. Consequently, the elongation at break of the PSCMS-*g*-DGEBA/AuNPs-PSNPs 0.1–1 membranes was decreased from 4.6 to 3.1 % with the increasing filler content. The tensile strength of unfilled PSCMS-*g*-DGEBA matrix was found to be lower (31.2 MPa) compared with the composite membranes, whereas the membranes filled with AuNPs-PSNPs nanoparticles demonstrated enhancement in the tensile strength from 32.5 to 35.2 MPa. In the series, the tensile strength was higher for PSCMS-*g*-DGEBA/AuNPs-PSNPs 1 hybrid with 1 wt% filler content (35.2 MPa) compared with the PSCMS-*g*-DGEBA/AuNPs-PSNPs 0.1–0.5 membranes (32.5–34.2 MPa) with lower filler amount. The addition of nanoparticles resulted in modest increase in the tensile strength. This may be related to moderately uniform distribution of nanoparticles with their adding amounts. As revealed by the fractured surfaces in Fig. 7 c and d (PSCMS-*g*-DGEBA/AuNPs-PSNPs 0.1), improvement in the tensile properties was accompanied by a transition from a brittle non-uniform fracture (Fig. 6a) for PSCMS-*g*-DGEBA/AuNPs-PSNPs 1 to uniform patterned structure. Therefore, the mechanism behind the slight improvement in the mechanical properties of the composite was actually reliant on the alteration of the inconsistent fracture surface towards the unvarying structure. Moreover, the higher nanoparticle content actually provided greater interface for the polymer chains, compared with the lower AuNPs-PSNPs level. The enhancement was due to the fine miscibility of the particles with the matrix (Ashraf et al. 2014). Furthermore, there was fine enhancement in the tensile modulus for the PSCMS-*g*-DGEBA/AuNPs-PSNPs 1 membrane (2.9 GPa) relative to PSCMS-*g*-DGEBA/

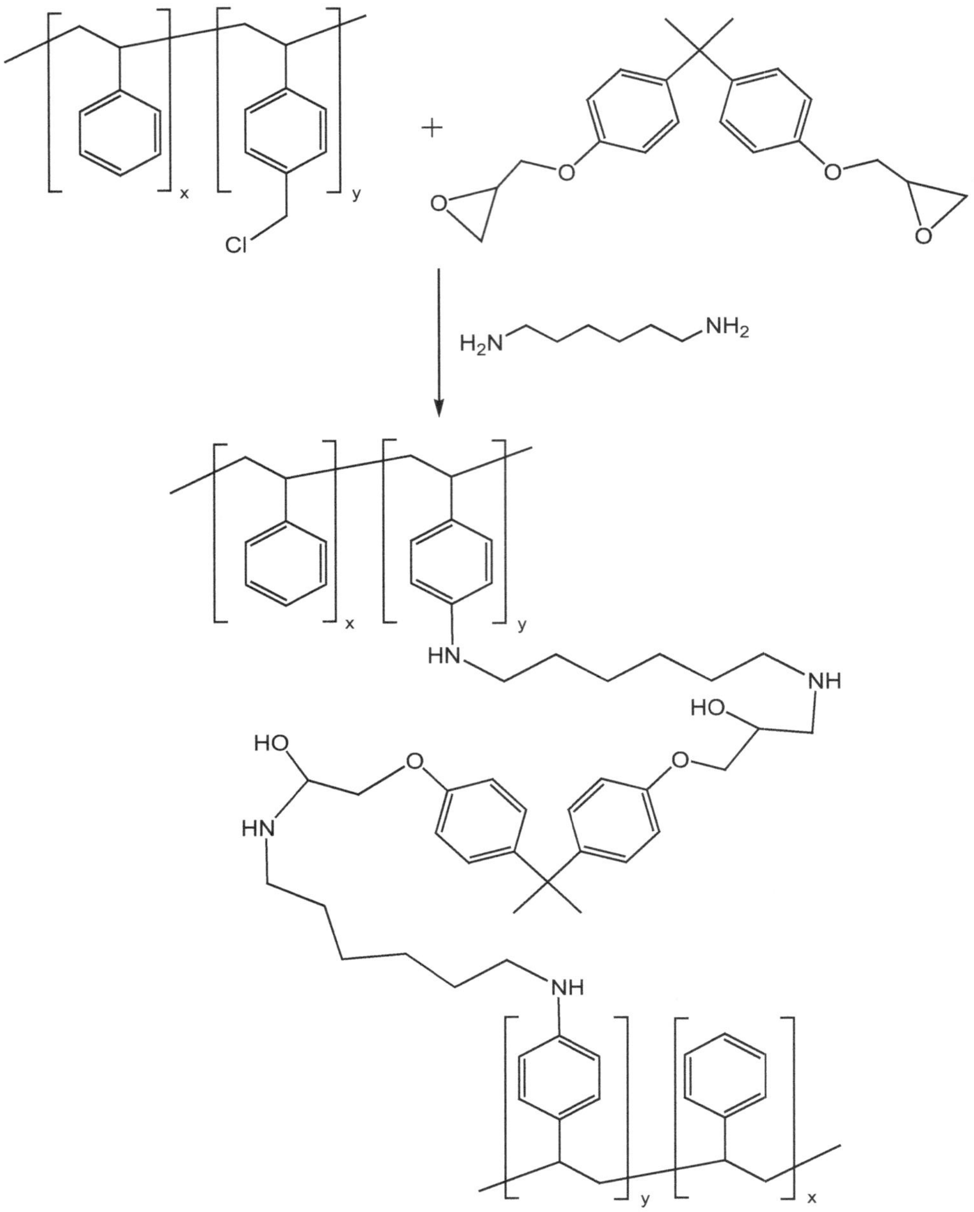

Fig. 1 Schematic illustration for the formation of PSCMS-*g*-DGEBA

AuNPs-PSNPs 0.1 (1.8 GPa), PSCMS-*g*-DGEBA/AuNPs-PSNPs 0.3 (2.2 GPa), and PSCMS-*g*-DGEBA/AuNPs-PSNPs 0.5 (2.6 GPa) and neat matrix (1.5 GPa). Moreover, AuNPs-PSNPs caused an improvement in the toughness of the system from 333 to 466 Jm^{-3}. Here again the toughness of the composites was found to be higher than the base matrix (78 Jm^{-3}). The physical adhesion between the filler particles and the matrix actually operated as a key factor in enhancing the tensile characteristics.

Morphological investigation

The TEM image of gold nanoparticles has revealed round shape and uniform distribution (Fig. 5a). The AuNPs were also analyzed with Zetasizer Nano ZS particle analyzer and found to have size of 10–20 nm. Figure 5b shows the polystyrene microspheres physically embedded with the gold nanoparticles. The AuNPs-PSNPs composite nanoparticles also acquired spherical shape in the size range of 30–50 nm having narrow distribution. Due to high specific surface area, the composite nanoparticles offer ample interphase for the polymer/filler interaction. Thus, the polymer segments aligned with the adjoining filler particles may form well-oriented structure. Accordingly, the inside particulars of the composites turned out to be of higher strength and modulus as explained in the preceding section. The FESEM images of PSCMS-g-DGEBA/AuNPs-PSNPs 0.1, PSCMS-*g*-DGEBA/AuNPs-PSNPs 0.3, and PSCMS-*g*-DGEBA/AuNPs-PSNPs 0.5 are given in Fig. 6a–e. The PSCMS-g-DGEBA 0.1 with 0.1 wt%

Fig. 2 Experimental nanofiltration setup

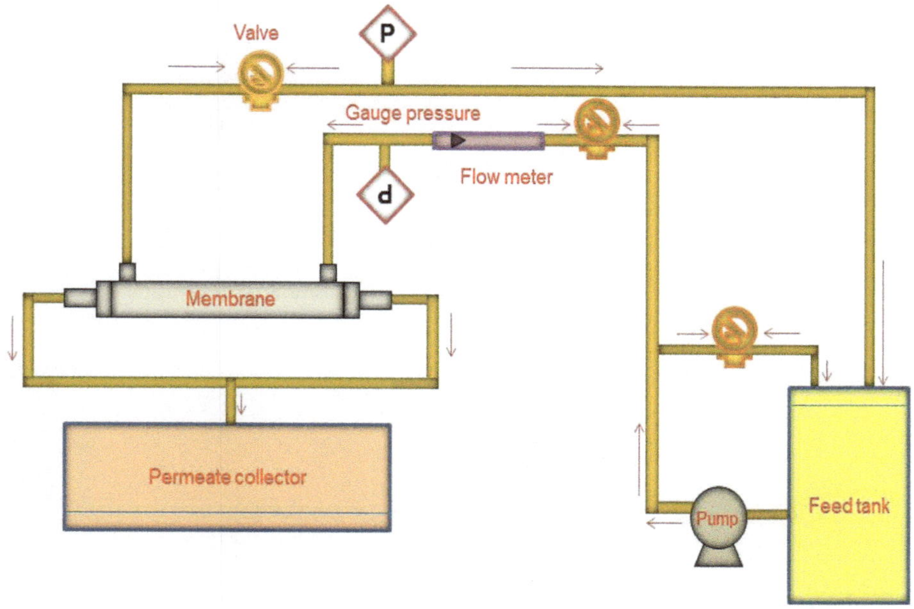

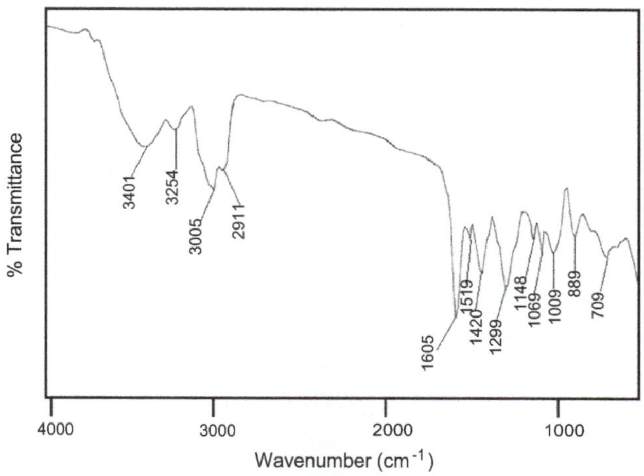

Fig. 3 FTIR spectrum of PSCMS-*g*-DGEBA

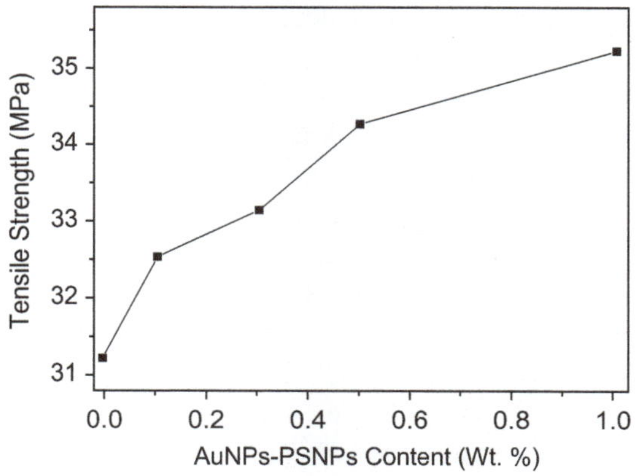

Fig. 4 Drift of tensile strength *vs.* AuNPs-PSNPs in PSCMS-*g*-DGEBA/AuNPs-PSNPs membranes

AuNPs-PSNPs composite nanoparticles revealed porous morphology but the nanopores were not homogeneously dispersed (Fig. 6a–c). The micrograph in Fig. 6d (PSCMS-*g*-DGEBA/AuNPs-PSNPs 0.3) showed somewhat regular porous morphology. It appeared that the inclusion of 0.5 wt% filler formed few circular areas with the uniformly distributed nanopores (Fig. 6e). The cross-section morphology of PSCMS-*g*-DGEBA/AuNPs-PSNPs 1 membrane showed sponge-like micro-pores (Fig. 7a, b). The increase in the nanopores was also obvious in Fig. 7c and d compared with PSCMS-*g*-DGEBA/AuNPs-PSNPs 0.1–0.5. The appropriate amount of membrane porosity is usually important to increase the filtration competence (Jing et al. 2008). On the other hand, the extremely high porosity and larger size pores may unfavorably impact the membrane performance because pores might not be closed effectively and the membrane tends to shrink. The pore size of the membranes was calculated using BET analyzer. There was increase in the pore size of the membranes with the nanoparticle addition. PSCMS-*g*-DGEBA has mean pore size of 50 nm. PSCMS-*g*-DGEBA/AuNPs-PSNPs 0.1 has pore size of 80 nm, PSCMS-*g*-DGEBA/AuNPs-PSNPs 0.3 has pore size of 90 nm, PSCMS-*g*-DGEBA/AuNPs-PSNPs 0.5 has pore size of 100 nm, and PSCMS-*g*-DGEBA/AuNPs-PSNPs 1 has mean pore size of 120 nm.

Membrane properties and performance

At constant operating pressure (410 kPa), the pure water flux of PSCMS-*g*-DGEBA and PSCMS-*g*-DGEBA/AuNPs-PSNPs 0.1–1 membranes upon compaction was measured every 1 h. During compaction, the water flux was

Table 1 Mechanical properties of PSCMS-*g*-DGEBA and PSCMS-*g*-DGEBA/AuNPs-PSNPs membranes

Membrane type	Tensile Strength (MPa) ± 0.02	Elongation at break ± 0.01 (%)	Tensile modulus (GPa) ± 0.03	Toughness (Jm^{-3}) ± 0.02
PSCMS-*g*-DGEBA	31.2	1.7	1.5	78
PSCMS-*g*-DGEBA/AuNPs-PSNPs 0.1	32.5	4.6	1.8	333
PSCMS-*g*-DGEBA/AuNPs-PSNPs 0.3	33.1	4.1	2.2	378
PSCMS-*g*-DGEBA/AuNPs-PSNPs 0.5	34.2	3.6	2.6	421
PSCMS-*g*-DGEBA/AuNPs-PSNPs 1	35.2	3.1	2.9	466

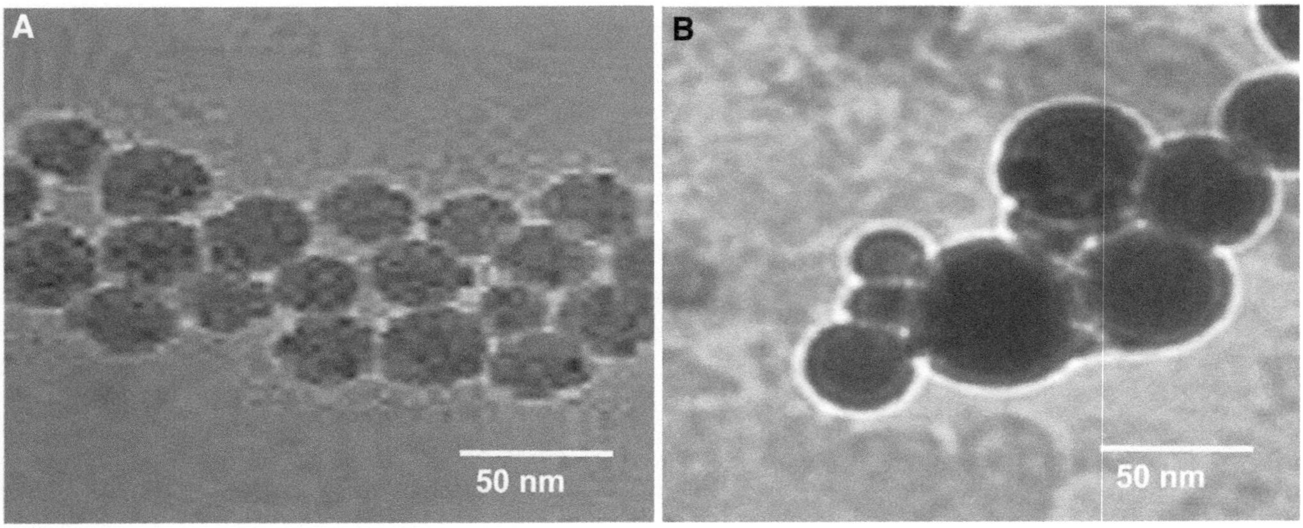

Fig. 5 TEM micrographs of **a** AuNPs at 50 nm; **b** AuNPs-PSNPs at 50 nm

high initially, declined gradually, and reached a steady state after 2–3 h of compaction. This preliminary decrease in flux might be due to the fact that the membrane pores are being compacted leading to uniform pore size and steady-state water flux. The new membranes after compaction were subjected to a pressure of 340 kPa for the measurement of pure water flux every 1 h. The porosity, % solvent content, shrinkage ratio, flux, recovery, and salt rejection tests were performed for the evaluation of membrane properties and performance. The porosity measurement of PSCMS-*g*-DGEBA and PSCMS-*g*-DGEBA/AuNPs-PSNPs 0.1-1 membranes is given in Table 2. The unfilled PSCMS-*g*-DGEBA membrane possesses lower porosity ~0.12 g/cm^2 in water and also low values in other solvents. Moreover, there was steady increase in the membrane porosity in different solvents (water, ethanol, methanol, and propanol). This increase was due to enhancement in the hydrophilic nature of the membranes with nanofiller addition. As the hydrophilicity of solvent was decreased (moving from water to propanol), the membrane porous nature was also found to decrease. To assess the hydrophilic nature of the membranes, % solvent content of membranes was calculated (Table 3). The % solvent content had increasing tendency with composite nanoparticle loading. Both the porosity and % solvent content were maximum for PSCMS-*g*-DGEBA/AuNPs-PSNPs 1 and were found minimum for neat PSCMS-*g*-DGEBA. Similar to the membrane porosity, the % solvent content was also observed to decline when moving from water to less polar solvent depending upon the solvent hydrophilicity. The porosity and % water content of PSCMS-*g*-DGEBA/AuNPs-PSNPs 1 were calculated as 1.42 g/cm^2 and 3.11 %, respectively. Furthermore, the higher porosity of PSCMS-*g*-DGEBA/AuNPs-PSNPs 1 resulted in lower shrinkage ratio of the membrane. Table 4 shows the trend of shrinkage ratio of the membranes. As the AuNPs-PSNPs filler content increased from 0.1 to 1 wt%, the shrinkage ratio was found to decrease rapidly. Moving from neat PSCMS-*g*-DGEBA to PSCMS-*g*-DGEBA/AuNPs-PSNPs 0.1–1 membranes, there was decrease in the shrinkage ratio from 13.2 to 10.1 due to increase in nanoporous nature of the membranes. Generally, the measurement of pure water flux is an important criterion to study the performance of nanofiltration membranes. Owing to the increased membrane porosity, % solvent content, and hydrophilicity, the flux of the membranes was also supposed to increase. As

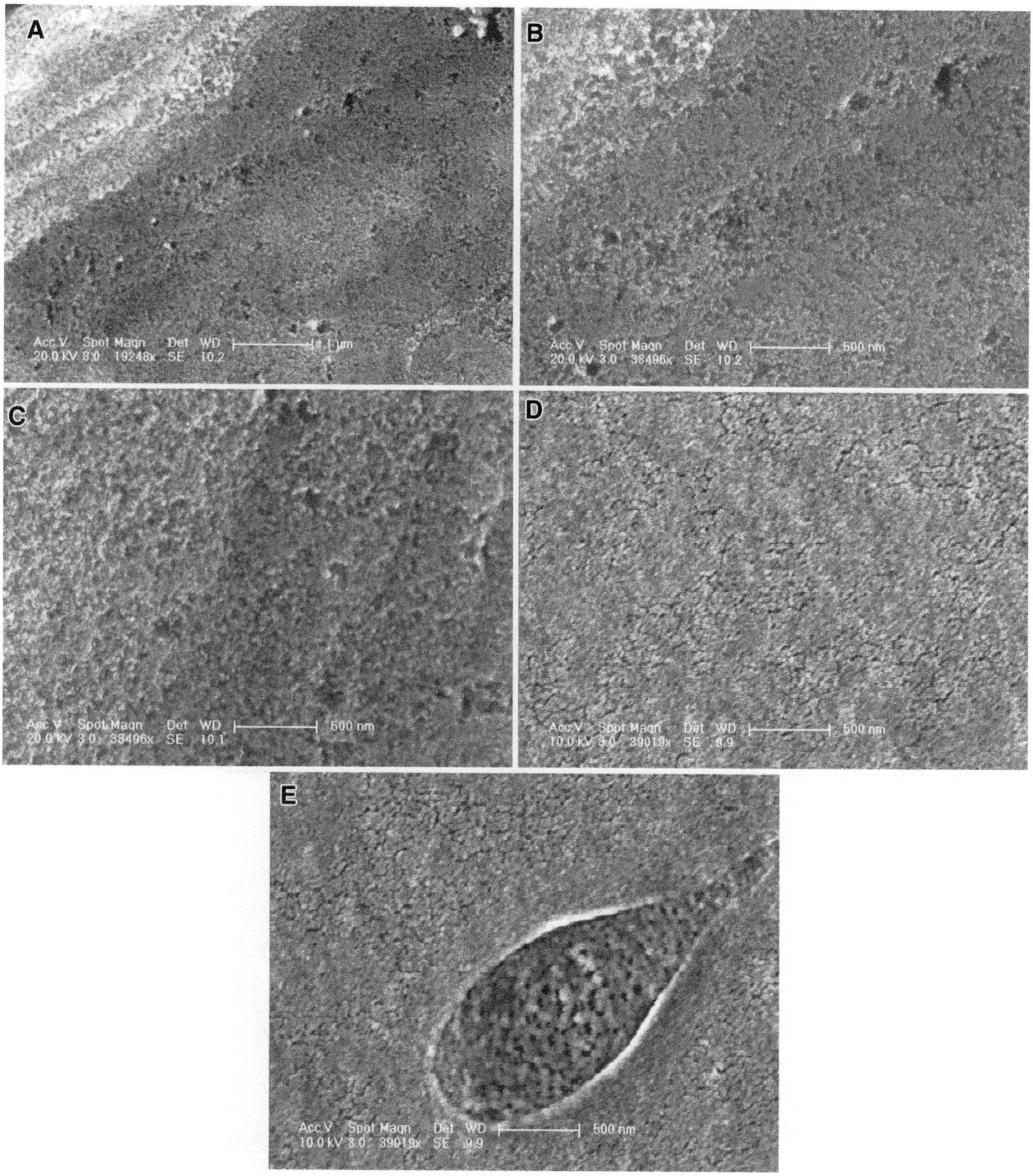

Fig. 6 FESEM micrographs of **a** PSCMS-*g*-DGEBA/AuNPs-PSNPs 0.1 at 1 μm; **b** PSCMS-*g*-DGEBA/AuNPs-PSNPs 0.1 at 500 nm; **c** PSCMS-*g*-DGEBA/AuNPs-PSNPs 0.1 at 500 nm; **d** PSCMS-*g*-DGEBA/AuNPs-PSNPs 0.3 at 500 nm; **e** PSCMS-*g*-DGEBA/AuNPs-PSNPs 0.5 at 1 μm

expected, the PSCMS-*g*-DGEBA/AuNPs-PSNPs 1 membrane showed higher flux of 2.01 $mLcm^{-2}\ min^{-1}$ (Table 5). The flux of PSCMS-*g*-DGEBA, PSCMS-*g*-DGEBA/AuNPs-PSNPs 0.1, PSCMS-*g*-DGEBA/AuNPs-PSNPs 0.3, and PSCMS-*g*-DGEBA/AuNPs-PSNPs 0.5 membranes was found to be relatively lower as 1.51, 1.62, 1.77, and 1.87 $mLcm^{-2}\ min^{-1}$, respectively. The NaCl concentration was 35,000 mg/L and osmotic pressure was

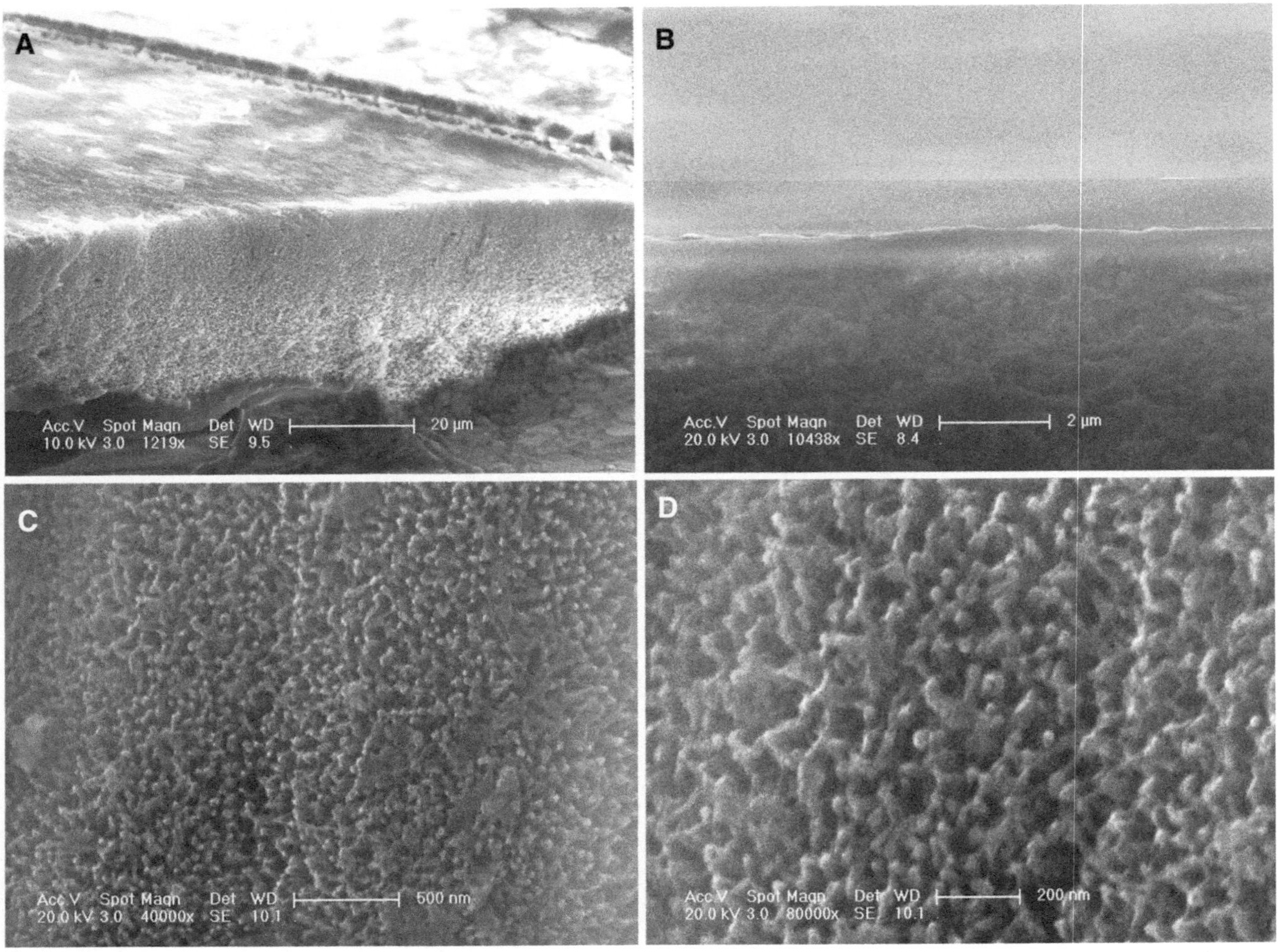

Fig. 7 FESEM micrographs of **a** PSCMS-*g*-DGEBA/AuNPs-PSNPs 1 at 20 μm; **b** PSCMS-*g*-DGEBA/AuNPs-PSNPs 1 at 2 μm; **c** PSCMS-*g*-DGEBA/AuNPs-PSNPs 1 at 500 nm; and **d** PSCMS-*g*-DGEBA/AuNPs-PSNPs 1 at 200 nm

Table 2 Porosity of PSCMS-*g*-DGEBA and PSCMS-*g*-DGEBA/AuNPs-PSNPs membranes in different solvents

Membrane type	Porosity (g/cm^2)			
	Water	Ethanol	Methanol	Propanol
PSCMS-*g*-DGEBA	0.12	0.11	0.07	0.05
PSCMS-*g*-DGEBA/AuNPs-PSNPs 0.1	0.56	0.43	0.13	0.07
PSCMS-*g*-DGEBA/AuNPs-PSNPs 0.3	0.69	0.55	0.32	0.09
PSCMS-*g*-DGEBA/AuNPs-PSNPs 0.5	0.98	0.67	0.39	0.13
PSCMS-*g*-DGEBA/AuNPs-PSNPs 1	1.42	0.91	0.45	0.19

Table 3 % Solvent content of PSCMS-*g*-DGEBA and PSCMS-*g*-DGEBA/AuNPs-PSNPs membranes in different solvents

Membrane type	Water	Ethanol	Methanol	Propanol
PSCMS-*g*-DGEBA	2.11	1.23	0.99	0.78
PSCMS-*g*-DGEBA/AuNPs-PSNPs 0.1	2.43	1.32	1.12	0.87
PSCMS-*g*-DGEBA/AuNPs-PSNPs 0.3	2.65	1.45	1.22	0.98
PSCMS-*g*-DGEBA/AuNPs-PSNPs 0.5	2.76	1.77	1.32	1.17
PSCMS-*g*-DGEBA/AuNPs-PSNPs 1	3.11	1.89	1.43	1.26

Table 4 Shrinkage ratio of PSCMS-*g*-DGEBA and PSCMS-*g*-DGEBA/AuNPs-PSNPs membranes in different solvents

Membrane type	Water	Ethanol	Methanol	Propanol
PSCMS-*g*-DGEBA	13.2	9.15	8.21	7.12
PSCMS-*g*-DGEBA/AuNPs-PSNPs 0.1	12.9	8.31	8.14	6.57
PSCMS-*g*-DGEBA/AuNPs-PSNPs 0.3	12.1	7.98	7.97	6.44
PSCMS-*g*-DGEBA/AuNPs-PSNPs 0.5	10.6	7.52	7.16	5.93
PSCMS-*g*-DGEBA/AuNPs-PSNPs 1	10.1	7.1	6.99	5.33

Table 5 Pure water flux, salt rejection, recovery of PSCMS-*g*-DGEBA and PSCMS-*g*-DGEBA/AuNPs-PSNPs membranes

Membrane type	Pure water flux ($mL\ cm^{-2}\ min^{-1}$)[a]	Salt rejection (%)	Recovery (%)
PSCMS-*g*-DGEBA	1.51	52.6	71.2
PSCMS-*g*-DGEBA/AuNPs-PSNPs 0.1	1.62	58.1	75.1
PSCMS-*g*-DGEBA/AuNPs-PSNPs 0.3	1.77	62.4	78.5
PSCMS-*g*-DGEBA/AuNPs-PSNPs 0.5	1.87	65.3	80.6
PSCMS-*g*-DGEBA/AuNPs-PSNPs 1	2.01	70.4	83.8

[a] Applied pressure = 50 bar

397 psi. The membrane performance was also evaluated by its salt (NaCl) rejection efficiency. The rejection ratio was increased up to 70.4 % with higher loading of 1 wt% compared with 0.1 wt% filler (58.1 %). The membrane recovery of PSCMS-*g*-DGEBA/AuNPs-PSNPs 1 was also maximum ~83.8 %. The overall performance of PSCMS-*g*-DGEBA/AuNPs-PSNPs membranes was complimentary to be used in the industrial water purification systems (Jing et al. 2008). The membranes prepared in this study showed higher flux than the reported high-flux nanofiltration membranes of poly(styrenesulfonate)/protonated poly(allylamine). Pure water flux for poly(styrenesulfonate) membranes at 4.8 bar was 1.6 $m^3/(m^2day)$ (Malaisamy and Bruening 2005). However, PSCMS-*g*-DGEBA/AuNPs-PSNPs membranes showed lower flux compared with some commercial nanofiltration membranes such as MPF 50. The commercial membranes have a flux as high as 1600 l/m^2 h at a 20 bar transmembrane pressure (Van der Bruggen et al. 2002). In this research, we have employed phase inversion technique to obtain skinned porous asymmetric membranes. The nanoparticles had a notable effect on the membrane structure, as shown by SEM micrographs. The inclusion of AuNPs-PSNPs improved the membrane performance via affecting several membrane properties. The phase separation was observed in these membranes. Pores are usually formed from the gap between matrix and filler spherical particles, as appeared in these membranes. Generally, when salty water is in contact with the surface of a membrane, an interfacial pure water layer is formed at the solution/membrane interface (Mehwish et al. 2014). The interfacial water layer is few nanometers thick. In the presence of a pore, the diameter of the layer is smaller. The pure water layer will flow through the pore under the pressure applied on the salt solution and appear on the other side of the membrane as permeate (He et al. 2011). If the pore size is larger then the salt solution will flow in the central region of the pore contaminating the permeate. So the mechanism for the desalination from such membranes is based on the presence of the nanopores. Similar mechanism was operative for the removal of other contaminants from the polluted water. Second, membranes containing higher amount of gold nanoparticles become more porous and macrovoids are diminishing at higher concentrations. Therefore, the denser membranes revealed intrinsically higher flux (Vanherck et al. 2011; Cobley and Xia 2009). The membrane stability was also investigated in terms of the persistence of nanoparticles inside the membrane structure during filtration process. The composite nanoparticles of gold were covered with the polystyrene shell. Moreover, these nanoparticles were reinforced inside the epoxy blend matrix and cannot be easily released into the filtrate. For the confirmation, permeate was tested for the presence of gold nanoparticles. The occurrence of gold nanoparticles in permeate was determined using UV–VIS technique. The solution was characterized by a Perkin Elmer UV–VIS spectrophotometer for the typical large peak at 530 nm, corresponding to the plasmon absorbance band of the AuNPs. A strong peak at 530 nm was not observed in the UV–VIS spectrum indicating the absence of the nanoparticles in the filtrate. This means that the AuNPs did not turn into the secondary pollution during water purification.

Extraction of toxic metal ions

In PSCMS-*g*-DGEBA, the thiourea moieties were introduced in the matrix backbone. Thiourea-based polymers have been reported for the extraction of toxic metal ions

Table 6 Solid–liquid extraction of various metal cations from aqueous solution by PSCMS-*g*-DGEBA and PSCMS-*g*-DGEBA/AuNPs-PSNPs 1

Salt	%*E*		Kd′/L mol^{-1}		$\alpha \times 10^3$	
	PSCMS-*g*-DGEBA	PSCMS-*g*-DGEBA/AuNPs-PSNPs 1	PSCMS-*g*-DGEBA	PSCMS-*g*-DGEBA/AuNPs-PSNPs 1	PSCMS-*g*-DGEBA	PSCMS-*g*-DGEBA/AuNPs-PSNPs 1
$Fe(NO_3)_3$	35	38	18	29	8	9
$Cu(NO_3)_2$	36	68	20	95	9	1×10^1
$Co(NO_3)_2$	36	68	20	95	9	1×10^1
$Cd(NO_3)_2$	36	68	20	95	9	1×10^1
$Ni(NO_3)_2$	36	68	20	95	9	1×10^1
$Pb(NO_3)_2$	95	99	30×10^2	30×10^2	1×10^3	1×10^3
$Hg(NO_3)_2$	95	99	30×10^2	30×10^2	1×10^3	1×10^3

from water (Kausar and Hussain 2013a, b). The PSCMS-*g*-DGEBA/AuNPs-PSNPs membranes were tested in terms of distribution coefficient (Kd′) and solid–liquid extraction selectivity ($\alpha_{S,L}$) (Duhart et al. 2001). Table 6 shows the results obtained for the extraction of toxic metal ions using neat PSCMS-*g*-DGEBA and PSCMS-*g*-DGEBA/AuNPs-PSNPs 1 membranes. The extraction percentage for Fe^{3+}, Co^{2+}, Ni^{2+}, Cu^{2+}, and Cd^{2+} was 68 % in the case of PSCMS-*g*-DGEBA/AuNPs-PSNPs 1. However, the efficiency of the membranes was exceptional for Pb^{2+} and Hg^{2+}, i.e., the screening extraction level was close to 100 % from aqueous solution. The higher extraction percentage, towards the elimination of heavy metal ions, was due to the combined affect of thiourea moieties and gold nanoparticles (Calderón et al. 2007). Particularly, high extraction level of lead and mercury shows the potential of novel selective ion transport membranes for decontamination of water.

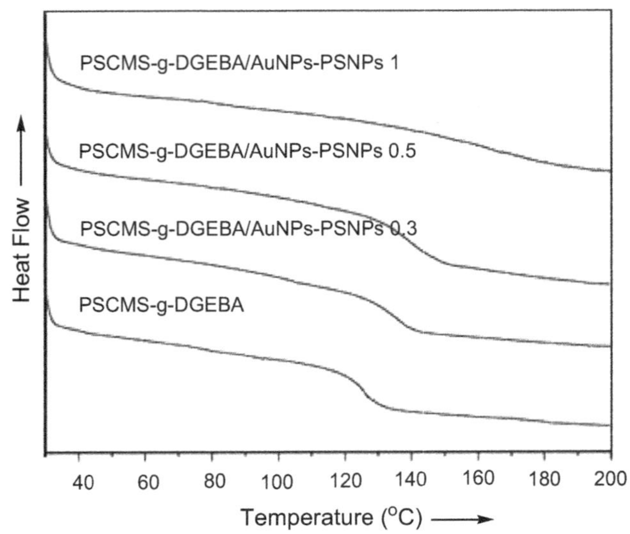

Fig. 8 DSC curves of PSCMS-*g*-DGEBA with varied AuNPs-PSNPs content

Analysis of glass transition temperature

Differential scanning calorimetry was used to study the glass transition temperature of the membranes. The neat PSCMS-*g*-DGEBA matrix revealed glass transition temperature of 123 °C which is sufficiently higher due to to the cross-linked structure. As shown in Fig. 8, the glass-transition temperature of pure matrix was further increased with the addition of AuNPs-PSNPs. The results depicted that the polymer-based composite nanoparticles developed interaction with the matrix to further enhance its crosslinking structure. The segmental T_g of PSCMS-*g*-DGEBA/AuNPs-PSNPs 0.1-1 was recorded in the range of 132–159 °C. The PSCMS-*g*-DGEBA/AuNPs-PSNPs 0.1, PSCMS-*g*-DGEBA/AuNPs-PSNPs 0.3, PSCMS-*g*-DGEBA/AuNPs-PSNPs 0.5, and PSCMS-*g*-DGEBA/AuNPs-PSNPs 1 membranes possess glass transition temperature of 132, 137, 140, and 159 °C, respectively. Consequently, the DSC curves showed clear shift towards higher temperature with the inclusion of nanofiller. The increase in the glass transition for the system depicted the increase in the structural rigidity of PSCMS-*g*-DGEBA due to better interaction with the filler.

Conclusions

The epoxy grafted poly(styrene-*co*-chloromethylstyrene) was prepared using diamine linker and solution polymerization technique. The spherical gold/polystyrene composite nanoparticles having size of 30–50 nm were designed for the reinforcement of the membranes. The composite nanoparticles were loaded in varying concentration of 0.1–1 wt%. The mechanical behavior of the AuNPs-PSNPs filled membranes was found to improve with nanoparticle loading. Predominantly in 1 wt% loaded PSCMS-*g*-DGEBA/AuNPs-PSNPs 1, the cross-linked polystyrene/epoxy showed fine compatibility and strong interactions with the composite nanoparticles, thus resulting in adequately enhanced mechanical properties. Another unique advantage of the composite nanofiller was the increase in glass transition temperature of the membranes compared with the pure matrix. In surface morphology analysis, the entire embedding character of nanoparticles in the nanoporous membrane was observed. Moreover, the PSCMS-*g*-DGEBA/AuNPs-PSNPs 1 membrane exhibited fine behavior in the removal of Pb^{2+} and Hg^{2+} from water. Study of membrane characteristics such as porosity, flux, shrinkage ratio, recovery, and rejection revealed that the gold/polystyrene composite nanoparticles reinforced novel hybrids are well suited for water purification set-up.

References

Arockiasamy DL, Alam J, Alhoshan M (2013) Carbon nanotubes-blended poly(phenylene sulfone) membranes for ultrafiltration applications. Appl Water Sci 3:93–103

Ashraf R, Kausar A, Siddiq M (2014) High-performance polymer/nanodiamond composites: synthesis and properties. Iran Polym J 23:531–545

Asyhar R (2014) Simple method for the removal of mercury(II) from its aqueous solution using aluminum as a reducing agent. Chem Mater Res 6:2224–3224

Bowen WR, Mohammad AW (1998) Diafiltration by nanofiltration: prediction and optimization. AIChE J 44:1799–1812

Bowen WR, Welfoot JS (2002) Modelling the performance of membrane nanofiltration–critical assessment and model development. Chem Engineer Sci 57:1121–1137

Bowen WR, Mohammad AW, Hilal N (1997) Characterisation of nanofiltration membranes for predictive purposes—use of salts, uncharged solutes and atomic force microscopy. J Membr Sci 126:91–105

Buonomenna MG, Macchib P, Davolic M, Driolia E (2007) Poly(vinylidene fluoride) membranes by phase inversion: the role the casting and coagulation conditions play in their morphology, crystalline structure and properties. Eur Polym J 43:1557–1572

Calderón V, Serna F, García F, de la Peña JL, García JM (2007) Selective solid–liquid extraction of cations using solid-phase polyamides with crown ether moieties as cation host units. J Appl Polym Sci 106:2875–2884

Chen X, Yang H, Gu Z, Shao Z (2001) Preparation and characterization of HY zeolite-filled chitosan membranes for pervaporation separation. J Appl Polym Sci 79:1144–1149

Cobley CM, Xia YN (2009) Gold and nanotechnology. Elements 5:309–313

Coulston Roger J, Jones Samuel T, Lee Tung-Chun, Appel Eric A, Oren A (2011) Scherman. Supramolecular gold nanoparticle—polymer composites formed in water with cucurbit[8]uril. Chem Commun 47:164–166

Dalwani M, Benes NE, Bargeman G, Stamatialis D, Wessling M (2010) A method for characterizing membranes during nanofiltration at extreme pH. J Membr Sci 363:188–194

Du NUL, Abu AB, Azahari B, Ariff ZM, Chujo Y (2011) Porous epoxy microparticles prepared by an advanced aqueous method. Mater Lett 65:1655–1658

Duhart A, Dozol JF, Rouquette H, Deratani A (2001) Selective removal of cesium from model nuclear waste solutions using a solid membrane composed of an unsymmetrical calix[4]arenebiscrown-6 bonded to an immobilized polysiloxane backbone. J Membr Sci 185:145–155

Ebert K, Fritsch D, Koll J, Tjahjawiguna C (2004) Influence of inorganic fillers on the compaction behaviour of porous polymer based membranes. J Membr Sci 233:71–78

Evuti AM, Lawal M (2011) Recovery of coagulants from water works sludge: a review. Adv Appl Sci Res 2:410–417

Gevers LEM, Vankelecom IFJ, Jacobs PA (2005) Zeolite filled polydimethylsiloxane (PDMS) as an improved membrane for solventresistant nanofiltration (SRNF). Chem Commun 19:2500–2502

Greenlee LF, Lawler DF, Freeman BD, Marrot B, Moulin P (2009) Reverse osmosis desalination: water sources, technology, and today's challenges. Water Res 43:2317–2348

He J, Lin X-M, Chan H, Vukovi L, Kral P, Jaeger HM (2011) Diffusion and filtration properties of self-assembled gold. nanocrystal membranes. Nano Lett 11:2430–2435

Henglein A (1998) Colloidal silver nanoparticles: photochemical preparation and interaction with O_2, CCl_4, and some metal ions. Chem Mater 10:444–450

Jing M, Lingling L, Guohua C, Congjie G, Shengxiong D (2008) Preparation of *N, O*-carboxymethyl chitosan (NOCC) composite nanofiltration membranes and its rejection performance for the fermentation effluent from a wine factory. J Chem Eng 16:209–213

Kausar A, Hussain ST (2013a) New generation of thermally stable and conducting poly(azomethine-ester)s: nano-blend formation with polyaniline. Polym Int 62:1442–1450

Kausar A, Hussain ST (2013b) Synthesis and properties of poly(thiourea-azo-naphthyl)/multi-walled carbon nanotube composites. J Plast Film Sheet 30:6–27

Kumar R, Sharma K, Tiwary KP, Sen G (2013) Polymethacrylic acid grafted psyllium (Psy-*g*-PMA): a novel material for waste water treatment. Appl Water Sci 3:285–291

Li Y, Pan Y, Zhu L, Wang Z, Su D, Xue G (2011) Facile and controlled fabrication of functional gold nanoparticle-coated polystyrene composite particle. Macromol Rapid Commun 32:1741–1747

Lu Y, Liu J (2007) Smart nanomaterials inspired by biology: dynamic assembly of error-free nanomaterials in response to multiple chemical and biological stimuli. Acc Chem Res 40:315–323

Malaisamy R, Bruening ML (2005) High-flux nanofiltration membranes prepared by adsorption of multilayer polyelectrolyte membranes on polymeric supports. Langmuir 21:10587–10592

Mehwish N, Kausar A, Siddiq M (2014) Advances in polymer-based nano-structured membranes for water treatment. Polym Plast Technol Engineer 53:1290–1316

Namasivayam C, Periasamy K (1993) Carbon for mercury(I1) removal from aqueous solution. Water Res 27:1663–1668

Ono A, Togashi H (2004) Highly selective oligonucleotide-based sensor for mercury(II) in aqueous solutions. Angew Chem Int Ed 43:4300–4302

Patel NG, Kumar A, Jayawardana VN, Woodworth CD, Yuya PA (2014) Fabrication, nanomechanical characterization, and cytocompatibility of gold-reinforced chitosan bio-nanocomposites. Mater Sci Eng C Mater Biol Appl 44:336–344

Sadeghi M, Khanbabaei G, Dehaghani AHS, Sadeghi M, Aravand MA, Akbarzade M, Khatti S (2008) Gas permeation properties of ethylene vinyl acetate-silica nanocomposite membranes. J Membr Sci 322:423–428

Slocik JM, Zabinski JS Jr, Phillips DM, Naik RR (2008) Colorimetric response of peptide-functionalized gold nanoparticles to metal ions. Small 4:548–551

Susrutha B, Ram S, Tyagi AK (2012) Effects of gold nanoparticles on rheology of nanofluids containing poly(vinylidene fluoride) molecules. J Nanofluids 1:120–127

Van der Bruggen B, Geens J, Vandecasteele C (2002) Fluxes and rejections for nanofiltration with solvent stable polymeric membranes in water, ethanol and n-hexane. Chem Engineer Sci 57:2511–2518

Vanherck K, Vankelecoma I, Verbiest T (2011) Improving fluxes of polyimide membranes containing gold nanoparticles. J Membr Sci 373:5–13

Vu DQ, Koros WJ, Miller SJ (2003) Mixed matrix membranes using carbon molecular sieves:I. Preparation and experimental results. J Membr Sci 211:311–334

Wu Y, Zhang Y, Xu J, Chen M, Wu L (2010) One-step preparation of PS/TiO_2 nanocomposite particles via miniemulsion polymerization. J Colloid Interface Sci 343:18–24

Yuan W, Zydney AL (2000) Humic acid fouling during ultrafiltration. Environ Sci Technol 34:5043–5050

14

Water quality monitoring of an international wetland at Harike, Punjab and its impact on biological systems

Jasmit Kaur[1] · Harpreet Walia[1] · Samson Okongo Mabwoga[1] · Saroj Arora[1]

Abstract The present study entails the investigation of mutagenic and genotoxic effect of surface water samples collected from 13 different sites of the Harike wetland using the histidine reversion point mutation assay in *Salmonella typhimurium* (TA98) strain and plasmid nicking assay using *pBR322*, respectively. The physicochemical characterization of water samples using different parameters was conducted for water quality monitoring. Heavy metal analysis was performed to quantify the toxic components present in water samples. It was observed that although the water samples of all the sites demonstrated mutagenic as well as genotoxic activity, the effect was quite significant with the water samples from sites containing water from river Satluj, i.e., site 1 (upstream Satluj river), site 2 (Satluj river) and site 3 (reservoir Satluj). The high level of pollution due to industrial effluents and agricultural run-off at these sites may engender the genotoxicity and mutagenicity of water samples.

Keywords Harike · Ramsar · Ames · Plasmid nicking · Monitoring

Introduction

Rivers, lakes and wetlands are major sources of potable water in the country. Contamination of these water resources with harmful chemicals is occurring at an alarming rate in India during the past few decades (Alam et al. 2009, 2010). Massive industrialization, urbanization, excessive utilization of pesticides in agriculture is the reason that threatens aquatic ecosystem and human beings (White and Rasmussen 1998). Punjab is mainly dependent upon surface water systems (i.e., rivers and wetlands) for water resources. Surface water system is used as source of drinking as well as agricultural, recreational and religious purposes. Wetlands are subjected to multiple environmental pressures such as water abstraction, industrial and domestic effluents, agricultural run-off, altered hydrology, habitat degradation and overexploitation of the resources. Surface water quality of the state is largely influenced by both natural as well as anthropogenic inputs resulting from industrial and domestic wastewater, as well as agricultural sources. The surface water quality is one of the important factors that directly concern the health of humans and aquatic organisms. The higher incidence of cancer and other health effects in the state can be related to the toxic components present in the water (Mathur et al. 2005). These toxic components can cause DNA damage, thus they are mutagenic in nature. Genotoxic testing of complex industrial effluents has demonstrated the presence of many unidentified and unregulated toxicants that are potential carcinogens. This has raised concern associated with their release into the water bodies (Houk 1992; Claxton et al. 1998; Tabrez and Ahmad 2011). The standard chemical analysis of the water samples can rarely provide an adequate assessment of the genotoxicity potential of complex mixture present in surface water (Ohe et al. 2004). However, bioassays can effectively assess the genotoxicity of complex mixtures without the requirement of prior information about their chemical composition. There are number of reports suggesting the correlation between pollution level and mutagenicity (Hubert et al. 2012). A number of animal, bacterial and plant-based assays have been developed for the estimation of mutagenic and genotoxic potential of water samples (Vargas et al. 1995;

✉ Saroj Arora
sarojarora.gndu@gmail.com

[1] Department of Botanical and Environmental Sciences, Guru Nanak Dev University, Amritsar, Punjab 143005, India

Tabrez and Ahmad 2011). The in vivo toxicity analysis using small mammals is a very long and expensive procedure, hence some in vitro analysis for routine evaluation of genotoxicity of water has been recommended. Among them, Ames test takes a prominent position, which detects the induced mutations in bacteria culture (Maron and Ames 1983; Ames 1984). On the other hand, amongst in vitro tests that evaluate the DNA damage, the plasmid nicking assay has been widely used and provides a valid indicator of genotoxicity (Mortelmans and Zeiger 2000). Therefore, this study was conducted to gain insight into the mutagenicity and genotoxicity of water samples by Ames and plasmid nicking assay, respectively, along with the physicochemical and heavy metal analysis for water quality assessment.

Materials and methods

Chemical and reagents

Salmonella typhimurium TA 98 strain was obtained from IMTC (Institute of Microbial Technology, Chandigarh). *pBR322* plasmid DNA was obtained from GeNei^{TM}. All other chemicals used were of analytical grade.

Study area

Harike wetland is the largest freshwater wetland of the country, covering an area of about 8435 ha (Fig. 1). It is an internationally recognized site falling into four districts viz. Amritsar, Firozpur, Kapurthala and Jalandhar in Punjab, India, between the latitudes 31°06′N and 31°12′N and longitudes 74°55′E and 75°05′E. It is about 12 km long and 11 km in width (Mabwoga et al. 2010). The wetland came into existence in 1952 due to the formation of barrage at the confluence of river Sutlej and Beas at Harike Pattan, with the aim of storing and providing irrigation and drinking water to parts of southern Punjab, and to the adjoining state Rajasthan. The wetland is having high ecological value as it supports rare, vulnerable and endangered floral and faunal species and attracts large populations of avifauna during the winters. The wetland was declared as wildlife sanctuary in 1982 and in 1990; it was designated as Ramsar site and included in the list of Wetlands of International Importance under the Ramsar Convention (Ramsar Convention Secretariat 2006; Anonymous 2008). In 2000, the Sanctuary came under "Wild Life Act" and fishing was totally banned.

Over the years, wetland has been facing a threat from a developing economy with increasing pressure on the resources (Chopra et al. 2001; Jain et al. 2008). The wetland is getting degraded due to intense siltation, weed invasion, agricultural run-off and pollution brought in by river Sutlej and Beas (via Buddha Nullah, Chitti Bein, Kala Sanghian drain and Kali Bein (Dhillon et al. 1996; Ladhar 2002; Kaur et al. 1999). The impact of the polluted water is spread to a wide area of southern Punjab and Rajasthan, through the Harike waterworks.

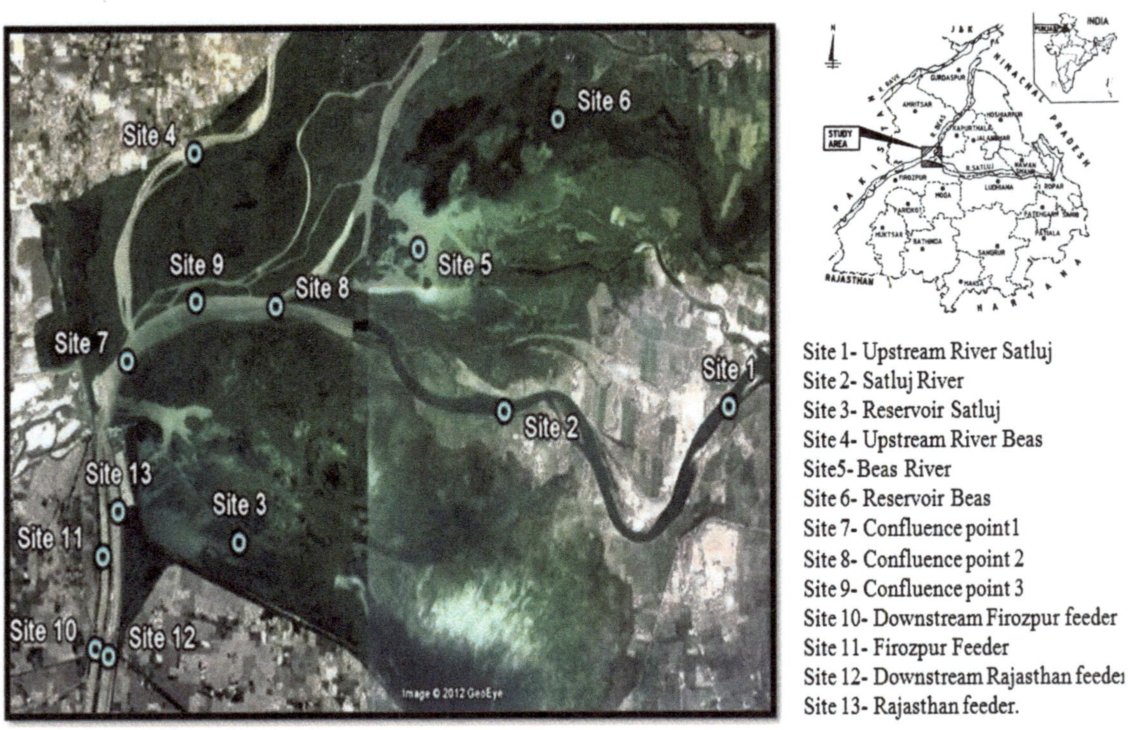

Fig. 1 Map of Harike wetland showing sampling area

Water sampling

Harike wetland is divided into two major parts, i.e., wetland area and feeder area. Wetland area comprises two rivers, i.e., Satluj and Beas, reservoirs for both the rivers, i.e., Reservoir Satluj (contains maximum water from river Satluj) and Reservoir Beas (contains maximum water from river Beas), and confluence point of both the rivers. Feeder area comprises two feeders called Rajasthan feeder and Firozpur feeder which supply water to Rajasthan and southern part of Punjab. To cover all the wetland and feeder areas, 13 sites of Harike wetland were fixed as shown in Fig. 1. Sampling sites in terms of latitude/longitude were determined using Global Positioning System (GPS). Garb sampling was carried out before Monsoon to eliminate the potential effect of rainfall. Samples were collected in sterile glass bottles, 20 cm below the surface of water as prescribed in the standard methods (APHA 1998). Samples were tested for the mutagenicity and DNA damaging assay as soon as possible. Prior to the mutagenicity assay, test samples were filter sterilized by passing through 0.22 μm filters to remove turbidity. For chemical analysis, samples were processed according to the methods described in APHA. Samples were stored in deep freezer at 4 °C until used.

Ames assay

The *S. typhimurium* mutagenicity assay given by Maron and Ames (1983) was performed on filter-sterilized water samples using TA98 strain that is meant for detecting the presence of frame-shift mutagens. Each sample was tested in triplicate. The twofold increase in the revertant colonies as compared to negative control is considered as mutagenic (Mortelmans and Zeiger 2000). The ratio and percentage mutagenicity of samples was calculated as compared with negative control.

Plasmid nicking assay

The effect of water sample on the integrity of supercoiled DNA was determined using the plasmid nicking assay described by Tabrez and Ahmad (2011). In this assay, *pBR322* DNA is used as a model to study the impact of pollutants on DNA.

Physicochemical analysis

Physicochemical, biological and microbiological analysis of water samples collected from different points of wetland as presented in Fig. 1 was conducted according to methods provided by standard methods (APHA 1998). The temperature, pH and electrical conductivity (EC) were measured in situ, and fixation of DO was carried out immediately after the collection of samples.

Heavy metal analysis

Heavy metal analysis was conducted in water samples collected from 13 different sites Harike wetland as described in standard methods (APHA 1998). Digestion solution was prepared in 1:3 ratios of HCl and HNO_3, respectively. Samples were digested with digestion solution in fume hood. Control was prepared using distilled water (APHA 1998). The concentration of different heavy metals was measured using AAS (Atomic Absorption Spectrophotometer).

Statistical analysis

The experimental results of mutagenicity, genotoxicity assays and physicochemical parameters were analyzed using SPSS software (SPSS-16). Data were expressed as mean, standard deviation, range, percentage mutagenicity and percentage genotoxicity. All the comparisons were made with controls.

Results

Mutagenicity assay

The results of Ames plate incorporation assay are presented in Table 1. The samples from site 1 and 3 showed mutagenicity level greater than 90 % as compared to negative control. The samples from the sites 7, 8 and 9 showed 85 % percentage mutagenicity, while the remaining 7 samples showed less than 80 % mutagenicity. However, it was found that in general, all the samples showed mutagenicity responses from three- to tenfold greater than negative control, thus representing significant mutagenicity indicating the presence of frame-shift mutagens in the water samples.

Plasmid nicking assay

The *pBR322* supercoiled DNA (form I) upon interaction with free radicals, mutagens or other DNA intercalating agents is converted to either open circular DNA (form II) by damaging of single strand of DNA or to linear DNA (form III) as a result of damage to both the strands of DNA. In some cases, there was formation of smear indicating the fragmentation of DNA molecule (Table 2).

The results showed that water samples collected from sites 1, 2, 3 and 8 of Sutlej river system had greater DNA damaging effect, thus was more polluted in terms of

Table 1 Mutagenic effect of water samples from different sites of Harike wetland in TA 98 tester strain of *Salmonella typhimurium* in Ames assay

S. no.	Locations	Mean/± SE	Range	Ratio	% age mutagenicity
1	Spontaneous	25.33 + 0.88	24–27	–	–
2	Tap water (−ve control)	36.66 + 2.03	33–40	–	–
3	NPD (+ve control)	1420 + 48	1324–1469	38.7	97.42
4	Site 1	386.67 + 5.2	378–396	10.5	90.5
5	Site 2	365 + 4.36	357–372	9.9	89.9
6	Site 3	390 + 7.4	376–401	10.6	90.6
7	Site 4	173 + 3.79	167–180	4.7	78.8
8	Site 5	162 + 8.3	150–178	4.4	77.37
9	Site 6	175.33 + 4.09	169–183	4.8	79.09
10	Site 7	263.67 + 4.6	256–272	7.2	86.09
11	Site 8	256 + 8.18	245–272	6.9	85.67
12	Site 9	259.67 + 6.38	248–270	7.1	85.88
13	Site 10	137.67 + 6.94	126–150	3.7	73.37
14	Site 11	148.67 + 4.81	142–158	4.1	75.34
15	Site 12	147 + 4.73	138–154	4	75.06
16	Site 13	135.67 + 8.06	124–151	3.7	72.97

Table 2 Genotoxic effect of water samples from different sites of Harike wetland in plasmid nicking assay

S. no.	Location	Form I	Form II	Form III	Total	% DNA damage
1	Control	2823.67 (79.66)	479.33 (13.52)	241.67 (6.82)	3544.67	20.34
2	Site 4	2488.67 (55.67)	1224.33 (27.39)	757.67 (16.95)	4470.67	44.33
3	Site 5	3037.33 (65.87)	880 (19.08)	693.67 (15.04)	4611	34.13
4	Site 6	2977.67 (46.63)	2093.33 (32.78)	1314.33 (20.58)	6385.33	53.37
5	Site 1	557.67 (31.38)	328.67 (18.49)	891 (50.13)	1777.34	68.62
6	Site 10	3318 (40.38)	3349 (40.75)	1550.33 (18.87)	8217.33	59.62
7	Site 2	522.3 (17.17)	894.67 (29.41)	1624.67 (53.41)	3041.64	82.83
8	Site 3	728.7 (11.4)	2460 (38.49)	3202 (50.10)	6390.7	88.59
9	Control	3344.67 (81.34)	517 (12.57)	250.33 (6.08)	4112	18.66
10	Site 11	2071.33 (23.74)	3345 (38.33)	3310 (37.93)	8726.33	76.26
11	Site 12	3318.33 (57.27)	771.67 (13.31)	1704.67 (29.42)	5794.67	42.73
12	Site 13	2527 (28.3)	3120.67 (34.95)	3280.33 (36.74)	8928	71.69
13	Site 7	386 (44.09)	100.67 (11.5)	388.67 (44.40)	875.34	55.9
14	Site 8	647.67 (10.24)	2830.33 (44.74)	2847.67 (45.02)	6325.67	89.76
15	Site 9	1998 (24.94)	2835.33 (35.39)	3178.33 (39.67)	8011.66	75.06

Within bracket the amount of DNA in % was shown

genotoxicity, than from sites 4, 5 and 6 of Beas river, as supercoiled DNA got affected to different levels, when incubated with water from both rivers separately. The results are further strengthened by the fact that damage to supercoiled DNA by water of river Sutlej is lesser at confluence point of both rivers and Feeders (at site 7, 10, 12 and 13). This indicates the diluting effect once the river converges at the confluence and the damaging agents in water samples from Sutlej got moderated. The density of DNA of different forms (i.e., Form I, II and III) was calculated using Quantity One 1-D Analysis Software. The percentage DNA damage was estimated by comparing the density of DNA in form II and form III with total amount of DNA loaded (Table 2).

Table 3 The mean values of physicochemical parameters of water samples from 13 different sites of Harike wetland

Parameters	Site 1	Site 2	Site 3	Site 4	Site 5	Site 6	Site 7
BOD (mg/l)	105.3 ± 4.6	100 ± 2	96 ± 4	57.3 ± 0.58	55 ± 1	49.7 ± 1.5	68.7 ± 1.2
COD (mg/l)	265.2 ± 2.7	252.2 ± 1.59	241.9 ± 1.35	97.6 ± 1.37	96.3 ± 0.83	97.16 ± 0.76	214.16 ± 0.76
pH	8.2 ± 0.17	8.57 ± 0.06	7.77 ± 0.25	7.57 ± 0.12	7.57 ± 0.15	7.2 ± 0.25	7.5 ± 0.23
Temp (°C)	28.8 ± 1.76	30.3 ± 0.58	32.1 ± 0.93	26.3 ± 0.58	27.5 ± 1.3	26.3 ± 1.5	33.6 ± 0.53
DO (mg/l)	7.23 ± 0.25	7.53 ± 0.21	7.47 ± 0.31	8.13 ± 0.15	8.3 ± 0.3	7.9 ± 0.36	8.17 ± 0.31
Turbidity (NTU)	20.3 ± 4.51	14.7 ± 2.52	14.3 ± 4.51	224.7 ± 4.04	221.7 ± 3.79	166.7 ± 6.03	44.3 ± 4.04
TS (mg/l)	269 ± 27.62	272.6 ± 11.01	295 ± 17.35	346.3 ± 25.11	457 ± 34.77	339.3 ±33.23	226.7 ± 9.07
TSS (mg/l)	44 ± 5.6	59.3 ± 10.9	83.3 ± 5.5	153.3 ± 6.7	183.67 ± 13	107 ±7	40.3 ± 10.2
TDS (mg/l)	225 ± 18	213.3 ± 11	211.67 ± 10	193 ± 18	273.3 ± 11	232.3 ± 24	186.3 ± 12
F (mg/l)	0.74 ± 0.006	0.64 ± 0.059	0.497 ± 0.0057	0.0663 ± 0.047	0.1066 ± 0.0115	0.094 ± 0.0052	0.504 ± 0.0441
NO3 (mg/l)	1.3236 ± 0.0345	1.279 ± 0.0259	1.256 ± 0.0524	0.315 ± 0.0259	0.6966 ± 0.0057	0.5233 ±0.04	1.05 ± 0.0458
SO_4 (mg/l)	26.1 ± 0.17	26.9 ± 0.79	26.7 ± 0.66	26.1 ± 0.96	26.9 ± 0.66	24.27 ±0.46	29.87 ± 0.81
P (mg/l)	0.04 ± 0.0021	0.031 ± 0.0017	0.05 ± 0.002	0.015 ± 0.0001	0.02 ± 0.0003	0.0149 ±0.0016	0.0862 ± 0.11
Cl (mg/l)	50 ± 2	48.7 ± 3.05	46 ± 5.29	32 ± 5.2	27.67 ± 4.04	26.67 ± 3.2	32.67 ± 2.52
EC (m/S/m)	0.2366 ± 0.0272	0.2466 ± 0.0166	0.26 ± 0.015	0.2566 ± 0.0145	0.2233 ± 0.012	0.23 ± 0.0288	0.2633 ± 0.012
Ca (mg/l)	96.5 ± 3.32	81.1 ± 1.61	76.87 ± 1.63	67.23 ± 2.44	66.9 ± 3	58.63 ± 1.4	62.8 ± 1.06
Mg (mg/l)	6.13 ± 0.42	8.45 ± 0.66	15.03 ± 1.15	20.93 ± 1.55	18.06 ± 1.9	21.53 ± 1.39	13.83 ± 0.97
Hardness (mg/l)	266.22 ± 9.67	237.29 ± 4.84	253.84 ± 5.93	254.09 ± 8.01	241.28 ± 12.79	235.08 ± 8.57	213.78 ± 6.44
T. Coli. (MPN/100 ml)	8.66E + 03 ± 5.77E + 02	7.66E + 03 ± 1.15E + 03	8.00E + 03 ± 1.00E + 03	1.73E + 03 ± 4.50E + 02	2.40E + 03 ± 6.10E + 02	2.30E + 03 ± 2.51E + 02	9.70E + 02 ± 2.10E + 02

Parameters	Site 8	Site 9	Site 10	Site 11	Site 12	Site 13
BOD (mg/l)	70.3 ± 1.5	70 + 2	38.3 ± 0.5	43.7 ± 2.1	40.7 ± 1.1	46.3 ± .6
COD (mg/l)	221.5 ± 3.12	212.13 ± 3.35	100.26 ± 2.05	95.93 ± 0.31	98.67 ± 1.15	95.9 ± 0.98
pH	7.83 ± 0.3	7.83 ± 0.12	7.83 ± 0.2	7.93 ± 0.15	7.63 ± 0.06	8.03 ± 0.12
Temp (°C)	29.5 ± 2.29	32.77 ± 0.68	26.7 ± 2.1	29.3 ± 3.2	30.7 ± 1.15	31 ± 1.7
DO (mg/l)	8.13 ± 0.06	7.73 ± 0.21	7.3 ± 0.31	7.4 ± 0.21	7.5 ± 0.3	7.27 ± 0.38
Turbidity (NTU)	35 ± 3.61	65 ± 3	31.3 ± 3.21	25.7 ± 2.52	36.7 ± 9.87	42.3 ± 3.51
TS (mg/l)	232.7 ± 20.23	248 ± 13.11	378 ± 17.09	377 ± 6.56	498.3 ± 16.07	443.3 ± 11.06
TSS (mg/l)	62 ± 7.5	71.3 ± 5.9	194.7 ± 3.2	212.7 ± 4	273 ± 7.2	238 ± 1.5
TDS (mg/l)	170.67 ± 10	176.67 ± 15	183.3 ± 20	164.3 ± 8	225.3 ± 5.5	205.3 ± 9.4
F (mg/l)	0.4913 ± 0.0023	0.691 ± 0.0831	0.31 ± 0.0173	0.13 ± 0.036	0.242 ± 0.035	0.125 ± 0.025
NO3 (mg/l)	1.158 ± 0.0386	1.0366 ± 0.08	0.2106 ± 0.012	0.2206 ± 0.0179	0.206 ± 0.0122	0.228 ± 0.03
SO_4 (mg/l)	29.57 ± 0.74	25.3 ± 0.52	24.3 ± 0.4	27.1 ± 0.17	26.43 ± 0.4	25.16 ± 0.29
P (mg/l)	0.0256 ± 0.0047	0.0195 ± 0.0005	0.0323 ± 0.002	0.0213 ± 0.0023	0.0333 ± 0.0042	0.0134 ± 0.0042
Cl (mg/l)	27.67 ± 3.06	24.67 ± 3.06	30 ± 1	30.67 ± 2.8	26.67 ± 2.1	29 ± 1.7

Table 3 continued

Parameters	Site 8	Site 9	Site 10	Site 11	Site 12	Site 13
EC (m/S/m)	0.2 ± 0.011	0.2433 ± 0.0088	0.2466 ± 0.0145	0.2433 ± 0.0185	0.2433 ± 0.0185	0.24 ± 0.0115
Ca (mg/l)	64.4 ± 1.51	67.4 ± 0.53	75.87 ± 2.65	71.67 ± 1.07	73.87 ± 0.91	69.43 ± 1.4
Mg (mg/l)	19.96 ± 1.17	21.23 ± 1.63	15.57 ± 1.5	8.07 ± 1.56	17.37 ± 2.15	12.3 ± 1.13
Hardness (mg/l)	243.03 ± 8.49	255.74 ± 5.58	253.54 ± 3.68	212.15 ± 6.35	255.96 ± 10.49	224.03 ± 2.37
T. Coli. (MPN/100 ml)	8.60E + 02 ± 2.70E + 02	7.10E + 02 ± 9.00E + 01	4.00E + 03 ± 1.00E + 03	3.10E + 03 ± 3.61E + 02	1.47E + 03 ± 2.10E + 02	2.30E + 03 ± 2.65E + 02

Physicochemical analysis

The water quality data of Harike wetland system for 19 parameters from 13 different sites are given in Table 3. The values from river Satluj were higher compared to those from river Beas in terms of organic and inorganic pollution (i.e., BOD and COD). The highest value of BOD and COD was from river Satluj at site 1 (105, 265.2 mg/l) and site 2 (100, 252.2 mg/l), respectively, and lowest from river Beas at site 4 (57.3, 97.6 mg/l) and site 5 (55, 96.3 mg/l). The water from river Beas was more turbid than water from river Satluj. The value of turbidity and total suspended solids (TSS) was highest in river Beas at site 4 (224.7 NTU, 153.3 mg/l) and at site 5 (221.7 NTU, 183.6 mg/l), respectively. River Satluj had lowest value of turbidity and total suspended solids at site 1 (20.3 NTU, 44 mg/l) and site 2 (14.7 NTU, 59.3 mg/l). The values of BOD and COD at the confluence point were intermediate, as the water from river Beas dilutes the pollutants in the wetland area carried by river Satluj, i.e., at site 7 (68.7, 214.1 mg/l) and site 9 (70, 212.3 mg/l). The average DO level in the wetland area was between 7.2 and 8.3 mg/l. The pH level of water samples was slightly alkaline and varied in between 7.20 and 8.57. There was not much variation in temperature of the samples from all the sites. The other parameters including F, P, SO_4^{2-} and NO_3^- were found within Indian Standard specification for drinking water IS: 10500.

Heavy metal analysis

The heavy metal analysis data of Harike wetland from different sites are given in Table 4. The highest concentration of heavy metals was found in sites 1, 2, and 3, i.e., samples collected from river Satluj area, while the lowest concentration of heavy metals was found in sites 10, 11, 12 and 13, i.e., samples collected from both the feeders (Rajasthan and Firozpur feeder, respectively). The concentration of heavy metals in both the reservoirs is less than their respective rivers. The concentration of metals was higher than the drinking water permissible limit provided by WHO.

Discussion

Pollution of aquatic systems occurs from multiple sources. Wetlands are very vulnerable ecosystems of the world. Even low-level pollution of these ecosystems can cause adverse effects to the ecology (Hildebrandt et al. 2008). There are number of analytical methods that have been used widely for the evaluation of mutagenicity and genotoxicity (Egito et al. 2007; Gupta and Ahmad 2012). In this study, Harike wetland is taken as an area for the analysis.

Table 4 Heavy metal analysis of water samples collected from different sites of Harike wetland

S. no.	Sites	Lead (ppm)	Chromium (ppm)	Zinc (ppm)	Iron (ppm)	Copper (ppm)	Mercury (ppm)
1	Control	0.0	0.00012 ± 0.001	0.00009 ± 0.006	0.0	0.00013 ± 0.0003	0.0
2	Site 1	1.5702 ± 0.003	0.4706 ± 0.001	5.9288 ± 0.0002	9.7671 ± 0.022	5.2281 ± 0.002	0.3612 ± 0.002
3	Site 2	2.2002 ± 0.008	0.4067 ± 0.001	5.0836 ± 0.0001	7.3388 ± 0.001	4.4911 ± 0.0008	0.0818 ± 0.001
4	Site 3	1.9498 ± 0.003	0.3288 ± 0.0005	5.0614 ± 0.0003	7.0332 ± 0.001	4.2892 ± 0.0006	0.0348 ± 0.006
5	Site 4	0.6194 ± 0.003	0.0642 ± 0.0005	3.6264 ± 0.0002	7.0553 ± 0.002	2.9853 ± 0.001	ND
6	Site 5	0.6215 ± 0.003	0.0525 ± 0.0003	3.5092 ± 0.0003	5.0326 ± 0.001	3.0248 ± 0.0006	ND
7	Site 6	0.7589 ± 0.002	0.0812 ± 0.002	2.3289 ± 0.0006	5.9726 ± 0.0003	2.0009 ± 0.0031	ND
8	Site 7	0.2701 ± 0.0004	0.0165 ± 0.0092	0.9446 ± 0.002	3.3289 ± 0.0006	0.6656 ± 0.001	0.0097 ± 0.0214
9	Site 8	0.3875 ± 0.002	0.0482 ± 0.003	2.0001 ± 0.002	4.1332 ± 0.0004	1.7219 ± 0.002	0.0115 ± 0.012
10	Site 9	0.4367 ± 0.0003	0.0423 ± 0.002	1.7456 ± 0.002	3.9221 ± 0.0003	1.2113 ± 0.002	0.0205 ± 0.003
11	Site 10	0.0943 ± 0.002	0.0098 ± 0.002	0.6398 ± 0.003	1.2791 ± 0.002	0.3914 ± 0.003	ND
12	Site 11	0.06191 ± 0.002	0.0284 ± 0.002	0.6477 ± 0.0004	1.6733 ± 0.008	0.3289 ± 0.0006	ND
13	Site 12	0.0826 ± 0.0004	0.0118 ± 0.004	1.2675 ± 0.002	2.094 ± 0.002	0.2371 ± 0.003	ND
14	Site 13	0.1032 ± 0.002	0.0162 ± 0.0003	0.8845 ± 0.002	2.156 ± 0.002	0.2995 ± 0.0004	ND
Permissible limit in drinking water by WHO in ppm		0.01	0.05	3	0.3	2	0.001

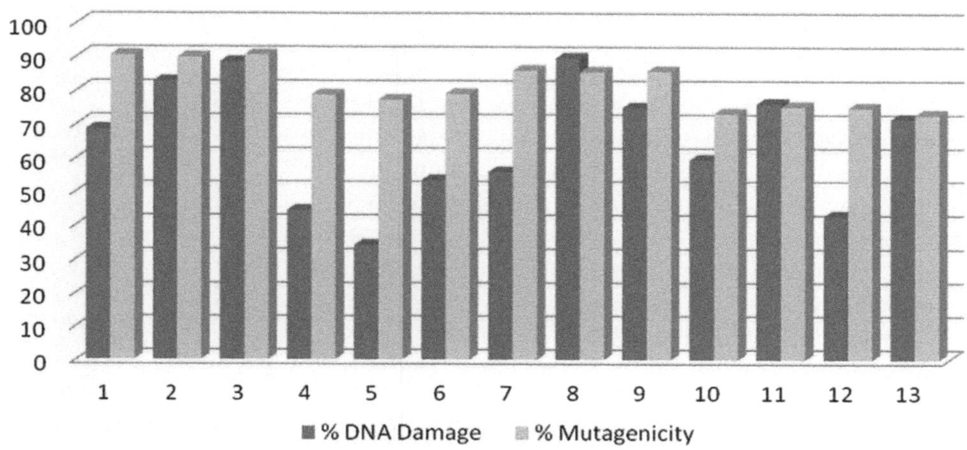

Fig. 2 Graph showing the comparison between % age DNA damaging effect in plasmid nicking assay and % age mutagenicity in Ames assay

Wetland is surrounded by agricultural field used for cultivation of wheat, rice and other seasonal crops. Over the years, the wetland has been facing a threat from a developing economy with increasing pressure on the resources (Chopra et al. 2001; Jain et al. 2008). As a result, many parts of southern Punjab and Rajasthan are facing a number of health problems. The plants and animals residing in the wetland are also under great pressure of pollutants. To assess mutagenic/genotoxic effect as well as to understand the composition of pollution load of Harike wetland, the present research plan was framed using short-term Ames *Salmonella* histidine reversion assay and plasmid nicking assay along with determining physicochemical analysis and heavy metal content.

The results of this study suggest that maximum pollution into Harike wetland was brought by river Satluj (site 1 and 2) which was contaminated with toxic chemicals causing mutations in bacterial strains and as well as single and double stranded breaks in plasmid DNA. These assays provide valuable biomarkers for the assessment of poisonous chemicals mixed in our water bodies and their effect on the various components of our ecosystems. (Tabrez and Ahmad 2011; Siddiqui et al. 2011a, b). The risk of mutations caused by these agents inhabiting in the wetlands will pose threat on their existence. (Dearfield et al. 2002). In addition to their direct ingestions, these mutagenic substances can get biomagnified in the food chains and can pose serious health effect on the coming generations (Wu 2005).

Moreover, the bioassays used for estimating genotoxicity and mutagenicity of water sample used in our study works on different principles as in Ames test (with Strain TA 98 of *Salmonella typhimurium*), the frame-shift mutagens are detected, while plasmid nicking assay (using supercoiled *pBR322* DNA) represents the breakage of supercoiled DNA into single and double strand DNA.

These two systems have different sensitivity level towards contaminants. In this analysis, similar trend has been found in both the assays. The water samples with maximum revertant colonies, i.e., having high percentage level of mutagenicity also cause maximum genotoxicity (Fig. 2). The results of the two assays support each other except for site 4, site 5 and site 12 that showed high mutagenicity level but low genotoxicity. A critical analysis of results showed that the mutagenic effect of water samples from the different sites was higher than their genotoxic effect except for site 8 and 11, which showed slightly higher values of genotoxicity as compared to mutagenicity.

Physicochemical analysis is widely used as a measure of pollution load in wastewater and natural waters. The level of BOD, COD and TDS was higher at the sites mainly containing the water from river Satluj. The level of turbidity and TSS was higher at the sites that contain water from river Beas which mainly includes the soil particles, which may be due to erosion occurring at the upstream Beas and run-off from the adjacent agriculture fields (Fig. 1). The high level of organic pollution (BOD and COD) and nutrients (NO_3, SO_4 and P) is the reason for excessive growth of water hyacinth. Heavy metal analysis also showed that maximum toxic components in the wetland are brought in by river Satluj (site 1 and 2). The results of genotoxicity and mutagenicity compared with physicochemical parameters and heavy metal analysis were found to be correlated with each other as the sites showing higher level of pollution and maximum number and concentrations of heavy metals are also causing high mutagenicity and genotoxicity (site 1 and site 2).

The results of mutagenicity, genotoxicity, physicochemical and heavy metal analysis suggest that the water present in the Harike wetland is contaminated with different types of organic pollutants and toxic contaminants that may be causing the mutations and DNA damage. There is also the possibility of presence of other organic and inorganic pollutants in water samples yet to be identified.

Conclusion

Wetlands rank among the most productive and sensitive ecosystems. They provide habitat to many rare, endangered species and are very important for human population. Thus, adverse effect of pollution on these ecosystems is attracting attention in recent years. In the present study, it has been found that water samples collected from Harike wetland are contaminated with different toxic pollutants. The results of mutagenic/genotoxic analysis showed that pollutants present in wetland are capable of causing mutations and DNA damage in bacterial strain and plasmid DNA. Physicochemical parameters and heavy metal analysis quantify different organic and inorganic pollutants and identify the toxic components, respectively.

Acknowledgments The authors gratefully acknowledge the research facilities provided by Guru Nanak Dev University (GNDU) Amritsar, Punjab, under UPE, CPEPA (UGC) and financial assistance from the Department of Science and Technology (DST), New Delhi under INSPIRE Program.

References

Alam ZM, Ahmad S, Malik A, Ahmad M (2009) Genotoxicity and mutagenicity potential of agriculture soil irrigated with tannery effluents at Jajmau (Kanpur), India. Arch Environ Contam Toxicol 57(3):463–476

Alam ZM, Ahmad S, Malik A, Ahmad M (2010) Mutagenicity and genotoxicity of tannery effluents used for irrigation at Kanpur, India. Ecotoxicol Environ Saf 73(7):1620–1628

Aleem A, Malik A (2003) Genotoxicity of water extracts from the river Yamuna and Mathura India. Environ Toxicol 18:69–77

Ames BN (1984) The detection of environmental mutagens and potential carcinogens. Cancer 53:2034–2040

Anonymous. The list of wetlands of international importance. Retrieved from Convention on Wetlands (Ramsar, Iran, 1971). http://www.ramsar.org/pdf/sitelist.pdf. Accessed 1 Aug 2008

APHA (1998) Standard methods for the examination of water and wastewater, 20th edn. APHA, Washington, DC

Chopra R, Verma VK, Sharma PK (2001) Mapping, monitoring and conservation of Harike wetland ecosystem, Punjab, India through remote sensing. Int J Remote Sens 22:89–98

Claxton LD, Houk VS, Hughes TJ (1998) Genotoxicity of industrial wastes and effluents. Mutat Res 410:237–243

Dearfield KL, Cimino MC, McCarrill NE, Mauer I, Valcovic LR (2002) Genotoxicity risk assessment: a proposed classification strategy. Mutat Res 521:121–135

Dhillon SS, Kaur H, Bath KS, Mander G (1996) Aquatic animal diversity and its conservation at Harike wetland. Environ Ecol 14(3):619–623

Egito LCM, Medeiros MDG, Medeiros SRBD, Agnez-Lima LF (2007) Cytotoxicity and genotoxicity potential of surface water from Pitimbu river, northeastern/RN Brazil. Genet Mol Biol 30(2):435–441

Gupta AK, Ahmad M (2012) Assessment of cytotoxic and genotoxic potential of refinery waste effluent using plant, animal and bacterial systems. J Hazard Mat 201–202:92–99

Hildebrandt A, Guillamon M, Lacorte S, Tauler R, Barceló D (2008) Impact of pesticide used in agriculture and vineyards to surface and groundwater quality (North Spain). Water Res 42:3315–3326

Houk VS (1992) The genotoxicity of industrial wastes and effluents: a review. Mutat Res 277:91–138

Hubert FV, Moisan KH, Munschy C, Tronczynski J (2012) Mutagenicity and genotoxicity of suspended particulate matter in the Seine river estuary. Mutat Res 741:7–12

Jain KS, Sarkar A, Garg V (2008) Impact of declining trend of flow on Harike wetland. Water Res Manag 22:409–421

Kaur H, Dhillon SS, Bath KS, Mander G (1999) Physico-chemical characteristics of water of Buddha Nallah. Indian J Environ Prot 3(1):27–30

Ladhar SS (2002) Status of ecological health of wetlands in Punjab, India. Aquat Ecosyst Health 5:457–465

Mabwoga SO, Chawla A, Thukral AK (2010) Assessment of water quality parameters of the Harike wetland in India, a Ramsar site, using IRS LISS IV satellite data. Environ Monit Assess 170:117–128

Maron DM, Ames BN (1983) Revised method for the *Salmonella* mutagenicity assay. Mutat Res 113:173–215

Mathur N, Bhatnagar P, Nagar P, Bijarnai MK (2005) Mutagenicity assessment of effluents from textile/dye industries of Sanganer, Jaipur (India): a case study. Ecotoxicol Environ Saf 61:106–113

Mortelmans K, Zeiger E (2000) The Ames Salmonella/microsome mutagenicity assay. Mutat Res 455:29–60

Ohe T, Watanabe T, Wakabayashi K (2004) Mutagens in surface water: a review. Mutat Res 567:109–149

Ramsar Convention Secretariat (2006) The Ramsar Convention manual: a guide to the convention on wetlands (Ramsar, Iran, 1971), 4th edn. Switzerland, Gland

Siddiqui AH, Ahmad M (2003) The *Salmonella* mutagenicity of industrial, surface and ground water samples of Aligarh region of India. Mutat Res 541:31–52

Siddiqui AH, Tabrez S, Ahmad M (2011a) Validation of plant based bioassays for toxicity testing of Indian waters. Environ Monit Assess 179(1–4):241–253

Siddiqui AH, Tabrez S, Ahmad M (2011b) Short term in vitro and in vivo genotoxicity testing system for some water bodies of northern India. Environ Monit Assess 180(1–4):87–95

Simeonov V, Stratis JA, Samara C, Zachariadis G, Voutsa D, Anthemidis A, Sofoniou M, Kouimtzis Th (2003) Assessment of surface water quality Northern Greece. Water Res 37:4119–4124

Singh KP, Malik A, Sinha S (2005) Water quality assessment and apportionment of pollution sources of Gomti river (India) using multivariate techniques—a case study. Anal Chim Acta 538:355–374

Tabrez S, Ahmad M (2011) Oxidative stress-mediated genotoxicity of wastewaters collected from two different stations in northern India. Mutat Res 726:15–20

Vargas VM, Guidobono RR, Jordao C, Henriques JA (1995) Use of two short term tests to evaluate the genotoxicity of river water treated with different concentration and extraction procedures. Mutat Res 34:331–352

White PA, Rasmussen JB (1998) The genotoxic hazards of domestic wastes in surface waters. Mutat Res 410:223–236

Wu JY (2005) Assessing surface water quality of the Yangtze Esturay with genotoxicity data. Mar Pollut Bull 50(12):1661–1667

15

Development of efficiency improved polymer-modified TiO_2 for the photocatalytic degradation of an organic dye from wastewater environment

Murugan Sangareswari[1] · Mariappan Meenakshi Sundaram[1]

Abstract In this study, the photocatalytic activity of polypyrrole-TiO_2 nanocomposite was studied experimentally for the degradation of methylene blue (MB) dye under simulating solar light irradiation. To improve the photocatalytic activity of TiO_2 under sunlight irradiation, conducting polymers such as polypyrrole (PPy) and its derivatives are generally used as photosensitizers. The PPy-TiO_2 nanocomposite was prepared by the chemical oxidative polymerization method. The prepared nanocomposite showed better photocatalytic activity than bare TiO_2 under sunlight irradiation for the degradation of MB dye. The prepared nanocomposite was subjected to characterization techniques such as SEM-EDAX, FT-IR, UV-DRS, XRD, TGA and PL spectral analysis. Different influencing operating parameters like initial concentration of dye, irradiation time, pH and amount of PPy-TiO_2 nanocomposite used have also been studied. The optical density of the dye degradation was measured by UV–Visible spectrophotometer. The repeatability of photocatalytic activity was also tested. A plausible mechanism was proposed and discussed on the basis of experimental results.

Keywords PPy-TiO_2 nanocomposite · Methylene blue · Sun light · Scanning Electron Microscopy · Photocatalysis

✉ Mariappan Meenakshi Sundaram
drmmsundaram61@gmail.com

[1] Centre for Research and Post Graduate Studies in Chemistry, Ayya Nadar Janaki Ammal College (Autonomous), Sivakasi 626124, Tamil Nadu, India

Introduction

In current years, pollution from dye wastewater has become a severe ecological problem due to the enormous and increasing uses of a variety of dyes (Kangwansupamonkon et al. 2010), When these compounds are discharged to the major water bodies without any previous treatment, they can cause disaster to the ecological equilibrium in the environment as these molecules have carcinogenic and mutagenic properties in the direction of water organisms and thus produce treat to a human being at the end of these food chains (Muthirulan et al. 2012a, b). Methylene blue is a heterocyclic aromatic substance having IUPAC name 3,7-bis(dimethylamine)–phenothiazine-5-iumchloride with the molecular formula $C_{16}H_{18}N_3SCl$. It has many uses in a selection of different fields, such as environmental science and chemistry (Eskizeybek et al. 2012). The physico-chemical properties of methylene blue (MB) dye are illustrated in Table 1.

Methylene blue dye is a cationic dye. The pzc of TiO_2 is 6.8. For pH values less than $pH_{pzc,}$ the surface becomes positively charged, and for pH values greater than pH_{pzc} the TiO_2 surface will be negatively charged. Hence, it is absorbed the positively charged MB dye species easily in basic pH.

Semiconductor photocatalysis shows a potential approach for solving worldwide environmental pollution issues. Titanium dioxide (TiO_2), as an important semiconductor, has been widely investigated in the photocatalytic field, due to its abnormal chemical and physical behaviours. On the other hand, the high charge recombination rate in TiO_2 considerably limits its photocatalytic application (Ibrahim and Halim 2008; Maeda 2012; Ochiai and Fujishima 2012; Park et al. 2013).

To develop the photocatalytic activity of TiO_2, there has been significant progress in the construction of new

Table 1 Physico-chemical properties of MB dye

Dye	Methylene blue
Characteristics	Cationic, water soluble
Molecular formula	$C_{16}H_{18}N_3SCl$
Molecular weight	319.85 g/mol
Structure	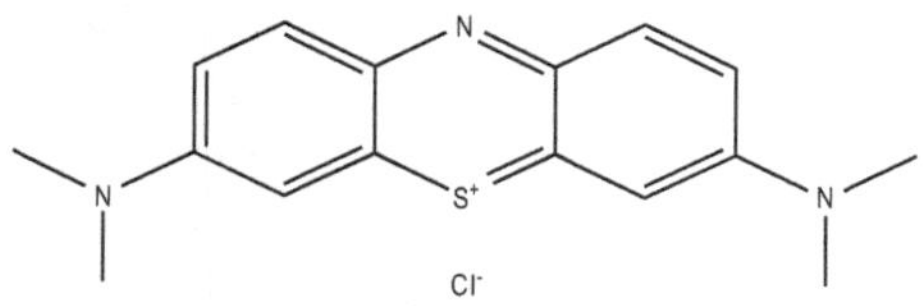 3,7-bis (dimethylamino) - phenothiazine-5-ium chloride
λ_{max}	665 nm
Applications	Redox indicator, peroxide generator, sulphide analysis, biology, photodynamic treatment of cancer

efficient materials by coupling TiO_2 with other organic, inorganic, polymeric materials, such as TiO_2/zeolite, TiO_2/activated carbon, TiO_2/Al_2O_3, TiO_2/SiO_2 and TiO_2/polymer materials. These composites really show photocatalytic development to a certain level. Among them, TiO_2/polymer composite paying more attention than others. The reason is that apart from outstanding mechanical property and large surface area, polypyrrole has electrical and electronic properties.

In recent times, some studies have been published on the combination of conductive polymers and TiO_2 to get better performance of UV light and sunlight activities (Muthirulan et al. 2013; Hou et al. 2011; Xu et al. 2012; Zhang et al. 2008; Denga et al. 2012; Zhang et al. 2006; Li et al. 2008; Wang et al. 2008; Kandiel et al. 2009; Wang et al. 2009). Moreover, many conjugated polymers also are efficient electron donors and good hole transporters upon visible light excitation. Therefore, conjugated polymers with large band gap inorganic semiconductors obtain awareness for optical, electronic, photocatalytic and photoelectric conversion applications.

In spite of many reward of TiO_2, its photocatalytic water-splitting efficiency under solar energy is still somewhat low, mainly owing to the following reasons; First, the photo-generated electrons in the CB of TiO_2 can recombine with the VB holes rapidly to liberate energy in the outward appearance of blocked heat or photons; Second, the decomposition of water into hydrogen and oxygen is a chemical reaction with large positive Gibb's free energy ($\Delta G = 237$kJ/mol), thus the backward reaction (recombination of hydrogen and oxygen into water) easily proceeds; Third, the band gap of TiO_2 is about 3.2 eV. Since UV light only accounts for in the region of 4 % of solar energy, while visible light contributes about 50 %, the lack of ability to make use of visible light confines the effectiveness of TiO_2 in solar photocatalytic hydrogen production.

To solve the above-mentioned trouble and to make solar photocatalytic by TiO_2 sufficient, constant efforts have been made by adding electron donating organic polymer (PPy). Adding electron donors or sacrificial reagents to react with the photo-generated VB holes is an effective quantify to enhance the electron–hole separation, resulting in higher quantum efficiency. However, the disadvantage of this system is required to continuously add electron donors (organic molecules) to maintain the reaction since they will be inspired during the photocatalytic reaction. (Li et al. 2010; Ufana and Ashraf 2011; Arenas et al. 2013).

In this manuscript, effective PPy-TiO_2 composites have been fabricated by chemical oxidative polymerization method. The prepared PPy-TiO_2 nanocomposite was characterized by SEM, XRD, UV-DRS, FT-IR and PL studies. The photodegradation studies were carried out by the prepared PPy-TiO_2 nanocomposite. The prepared PPy-TiO_2 nanocomposite, a novel solar light driven catalyst PPy-TiO_2, shows high photocatalytic activity as compared to TiO_2.

Experimental

Chemicals and reagents

All these reagents are of AR grade and used without further purification. The list of chemicals is illustrated in Table 2.

Table 2 Chemicals

Chemicals	Source
Titanium dioxide (~21 nm)	Sigma-Aldrich
Ammonium persulphate	Thomas Baker
Pyrrole (98 + %) (AR)	Alfa Aesar
Methylene blue	Qualigens
Sulphuric acid	E Merck

Preparation of conductive PPy-TiO_2 nanocomposites

The preparation of conductive PPy-TiO_2 nanocomposite is illustrated in Scheme 1. Chemical oxidative polymerization of PPy was performed in the presence of negatively charged TiO_2 nanoparticles (different wt%, i.e. 0.5, 1.0, 1.5 and 2.0 wt%) using ammonium persulfate as an oxidant. A typical procedure is outlined: Pyrrole (5 mmol) was added to 500 ml of 0.1 M Sulphuric acid solution. Then, the TiO_2 nanoparticles (different wt%, i.e. 0.5, 1.0, 1.5 and 2.0) were added with continuous stirring. Then, it was kept under 5 °C cooled condition. Then, 20 ml of Ammonium persulfate solution was added drop by drop. The green coloured solution was obtained. After constant stirring, the precipitation occurred in dark colour. The obtained product was washed with distilled water to remove the remaining ammonia solution. Finally, the product was dried out at room temperature at overnight.

Instrumental analysis

UV–visible spectrophotometer (DRS) was recorded using "SHIMADZU" model: UV 2450, FT-IR spectrum was recorded using "SHIMADZU" (Model: 8400S). The crystallographic structures of the materials were determined by high-resolution powder diffractometer model—RICH SIEFRT & CO with Cu as the X-ray source ($\lambda = 1.5406 \times 10^{-10}$ m). The surface morphology of the sample was recorded using scanning electron microscopy (SEM-EDAX) (Model: FEG Quantum 250). The photoluminescence (LS 45) was recorded using a Perkin Elmer spectra. The thermal stability of the material was recorded using Thermogravimetric Analyser (Model : TG/DTA 6200) "SII nanotechnology".

Photodegradation studies

The photocatalytic activity of efficient PPy-TiO_2 nanocomposites was estimated by measuring the decolorization rates of MB. In our experiments, 50 ml of MB and 20 mg of catalysts were kept under dark room to get adsorption desorption equilibrium. Then, it suspended into sunlight irradiation. At given intervals of enlightenment, a sample of the catalyst particulate was collected and centrifuged. The supernatant was analysed by spectrophotometer (ELICO SL 207 MINI SPEC) at $\lambda_{max} = 665$ nm. The determined absorption was changed to concentration through the standard curve method. From that experiment, we work out the percentage removal of dye.

The amount of removal of the MB dye, in terms of percentage removal, has been calculated using the following relationship:

$$\text{Percentage removal } (\%R) = \left[\frac{C_i - C_f}{C_i}\right] \times 100 \quad (1)$$

where C_i and C_f are the initial and final concentrations of dye (ppm) at a given time.

The reaction was carried out the time between 12.00 and 1.30 p.m. The intensity of Sun light was measured by Lux Meter (TES 1332A digital Lux meter). In this work, we did not consider the water evaporation and consequent solution concentration. The processes will be included in our future studies.

Results and discussion

Surface morphology of PPy-TiO_2 nanocomposite (SEM-EDAX)

The morphology of the composite strongly depends upon the character as well as the method of preparation. The

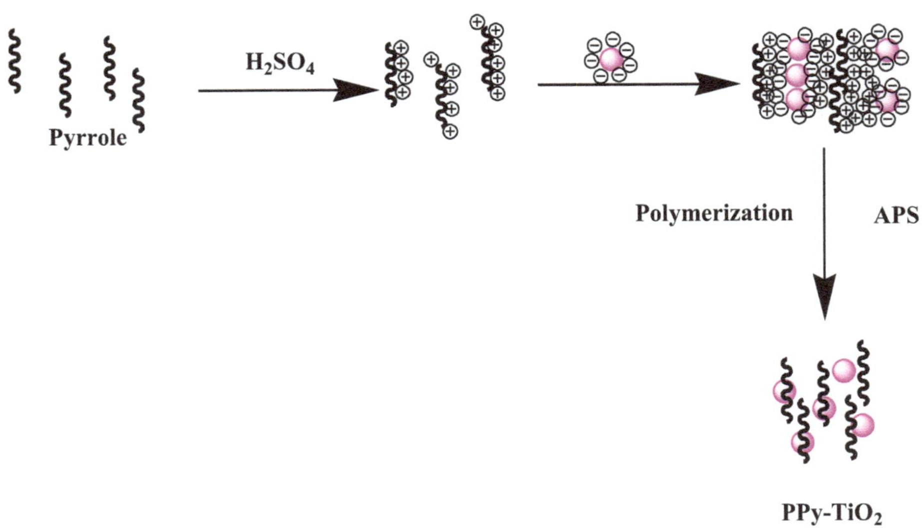

Scheme 1 Synthesis of PPy-TiO_2 nanocomposite by chemical oxidative polymerization method

surface morphology of the prepared PPy-TiO_2 nanocomposite was examined by SEM-EDAX. Figure 1a shows the surface morphology Polypyrrole having flower like structure. Figure 1b represents that the surface morphology of TiO_2 is a ball-like round shape deposited on the surface of polypyrrole. Based on these conclusions in PPy-TiO_2 nanocomposite, the TiO_2 particles deposited on the outside of polypyrrole (Hongyu et al. 2010; Guoquan et al. 2011).

FT-IR studies

The molecular structure and functional group of PPy and PPy-TiO nanocomposites (0.5, 1.0, 1.5, 2.0 wt%) were determined by FTIR spectrum. Figure 2 gives the typical FTIR spectra of PPy polymer, and the bands at 1552.15 and 1635.13 cm^{-1} are attributed to the stretching vibrations of C=C and C=N in the phenazine ring, respectively. The peaks at 1400.22 and 1100 cm^{-1} are attributed to the C–N stretching in the benzenoid and quinoid imine units. In addition, the bands at 520.78 cm^{-1} which are the description of C–H out-of plane bending vibrations of benzene nuclei of PPy. The FTIR spectra of the PPy-TiO_2 nanocomposites with different wt% TiO_2 represent the same characteristic peaks of PPy polymer. The peak in the region of 1100 cm^{-1} is related with vibrational modes of N=Q=N (Q refers to the pyrrole ring), representing that PPy is formed in the composite. But the stretching is decreased with the increase in the loading amount of TiO_2. The results of FTIR spectra of PPy agree well with the earlier reports (Xiaofang et al. 2006; Deivanayaki et al. 2012; Mi et al. 2009; Chougulea et al. 2012, 2013; Chatterjee et al. 2013; Etelino et al. 2013; Juraj et al. 2013).

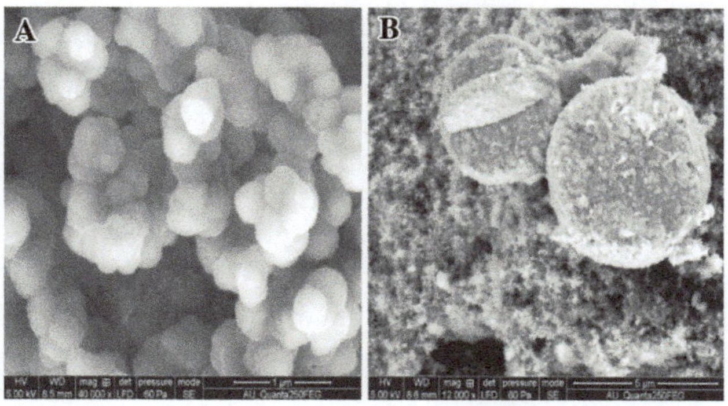

Fig. 1 SEM images of **a** pure polypyrrole, **b** 2.0 wt% TiO_2-PPy nanocomposite

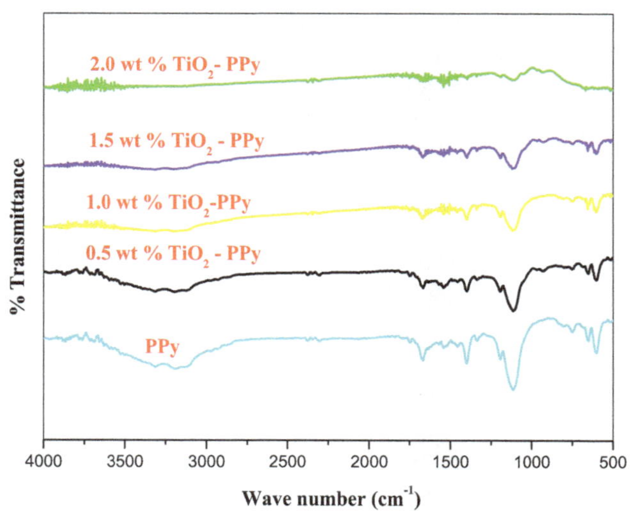

Fig. 2 FTIR spectra of PPy, 0.5 wt% TiO_2-PPy, 1.0 wt% TiO_2-PPy, 1.5 wt% TiO_2-PPy, 2.0 wt% TiO_2-PPy

UV–vis DRS studies of PPy-TiO_2 nanocomposite

The optical properties of PPy-TiO_2 nanocomposites investigated by UV-DRS spectroscopy are shown in Fig. 3a. The band gap energies of PPy-TiO_2 can be estimated from the wavelength values equivalent to the junction point of the perpendicular and parallel parts of the spectra.

From that spectra, the polymer expands the absorption range of the TiO_2 from 400 nm to over 550 nm. In fact, the spectra of PPy-TiO_2 were the spectra of polymer PPy adsorbed onto the surface of the powders. It is an essential condition for photo-sensitization that the photosensitizer be adsorbed onto the surface of semiconductor. PPy-TiO_2 seemed to be capable of responding to visible light.

The position of the fundamental absorption edge of PPy-TiO_2 is determined using the equation:

$$(\alpha h\upsilon)^2 = A(h\upsilon - E_g)^n \quad (2)$$

where α, h, υ, E_g and A are the absorption coefficients, plank constant, light frequency, band gap and a constant, respectively. The n value depends on the transition characteristics. The E_g value can be approximated by extrapolating the straight portion of the $(\alpha h\upsilon)^2 - (h\upsilon)$ plot shown in Fig. 3b. The obtained E_g value occurs between 2.22 and 2.23 e V for PPy-TiO_2, respectively. The band gap value of 2.0 wt% TiO_2-PPy nanocomposite was smaller when compared to other components as shown in Table 3. The band gap values estimated are somewhat smaller than the reported values which can be accredited to wide particle size distribution and surface defects (Rohit et al. 2013).

X-ray diffraction pattern of PPy-TiO_2 nanocomposite

The crystallographic construction of obtained PPy-TiO_2 nanocomposite was evidenced by XRD measurements. Figure 4a shows the XRD pattern of TiO_2 and Fig. 4b shows the XRD pattern of PPy/TiO_2 nanocomposites. The

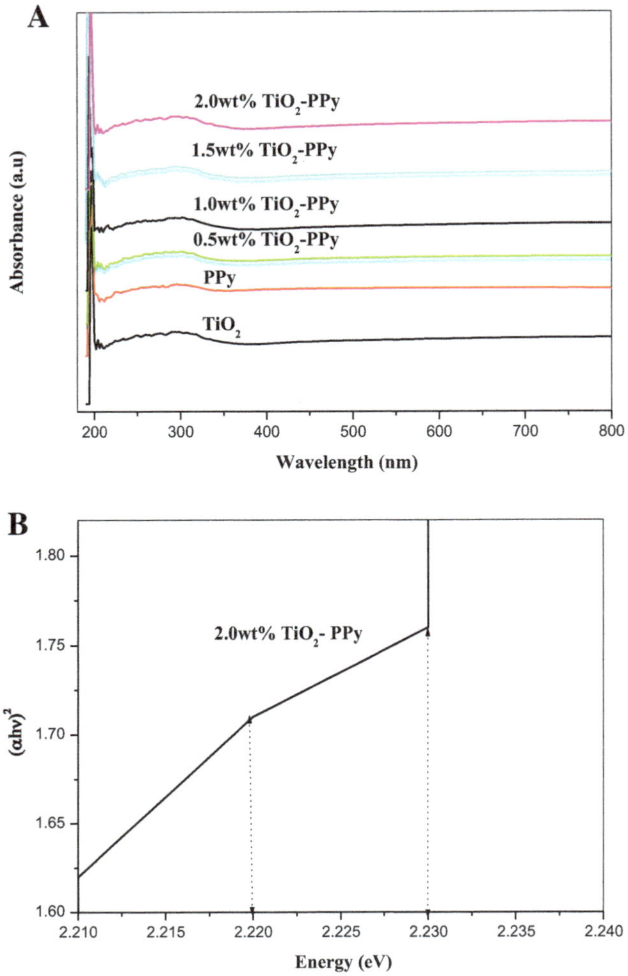

Fig. 3 **a** UV-DRS spectra of PPy-TiO_2 nanocomposite. **b** Plots of the $(\alpha h\upsilon)^2$ vs photon energy ($h\upsilon$) for PPy-TiO_2 nanocomposite

Table 3 Calculated band gap values for other components

Catalyst	Band gap (eV)
TiO_2	3.2
PPy	2.4
0.5 wt% TiO_2-PPy	3.0
1.0 wt% TiO_2-PPy	2.9
1.5 wt% TiO_2-PPy	2.6

XRD peaks denote the arrangement of PPy-TiO_2 nanocomposite. The respected peaks occur at 25.3°, 26.2°, 36.0°, 38.2°, 48.9°, 54.3°, 56.6°, 63.1°, 69.9°. In addition, it can be prominent due to the TiO_2 deposited on the surface of polypyrrole (Hasani and Hossein 2013; Yucheng et al. 2013).

The mean crystalline sizes of PPy-TiO_2 nanocomposites are designed using Scherrer's formula:

$$D\ Scherrer = k\lambda/\beta \cos \Theta$$

where λ is the wavelength of the X-ray radiation ($\lambda = 1.54 \times 10^{-9}$nm), and k is the Scherrer constant ($k = 0.89$)

Θ is the diffraction angle and β is the line width at half-maximum height of the most intense peak. Based on the XRD results, the crystalline range of PPy-TiO_2 nanocomposites was in the order of 10-20 nm. The results are in good agreement with the SEM images.

Photoluminescence spectra

To study the cause for the higher photocatalytic activity of PPy-TiO_2 nanocomposites, PL spectra of samples have been taken, as shown in Fig. 5. The rate of recombination of e^-/h^+ pairs may also be a key factor disturbing the photoactivity of TiO_2. Probably, the TiO_2 photocatalysts show a broad PL emission band. The peak intensities of the PL spectra of PPy-TiO_2, demonstrating that the electron—hole recombination rate of self—trapped excitation in TiO_2, are reduced by the introduction of PPy. The significant PL quenching of TiO_2 can be observed after combination with PPy, representing the efficient transfer of photo-generated electron from the TiO_2 to PPy, most important to the enhanced photocatalytic activity of the PPy/TiO_2 nanocomposites (Wang et al. 2010).

Thermogravimetric analysis

Figure 6 displays the thermogravimetric analysis of Pure PPy, 1.0 wt% TiO_2-PPy and 2.0 wt% TiO_2-PPy nanocomposites under N_2 atmosphere with the heating rate 20 °C/min. As observed from Fig. 6, it demonstrates that the thermal stability of 2.0 wt% TiO_2-PPy is much advanced than that of other composites in the temperature range of 30–800 °C. The first weight loss occurred at 100 °C due to the evaporation of remaining water molecule. The second weight loss occurs between 110 and 250 °C. This demonstrates the structural decay of PPy. Finally, the third weight loss occurs at 250–730 °C. This is due to the skeletal degradation of PPy. In contrast, 2.0 wt% TiO_2-PPy nanocomposites have greater thermal stability than other two nanocomposites. The thermogram shows that the TiO_2 nanoparticle lost about 11.9 % weight owing to the loss of polymer coatings on the surface of the nanoparticle.

Effect of concentration of dye

The outcome of initial concentration of Methylene Blue dye on the photocatalytic activity of photocatalysts (TiO_2 and PPy-TiO_2 nanocomposites) was carried out under solar light irradiation within the concentration range of

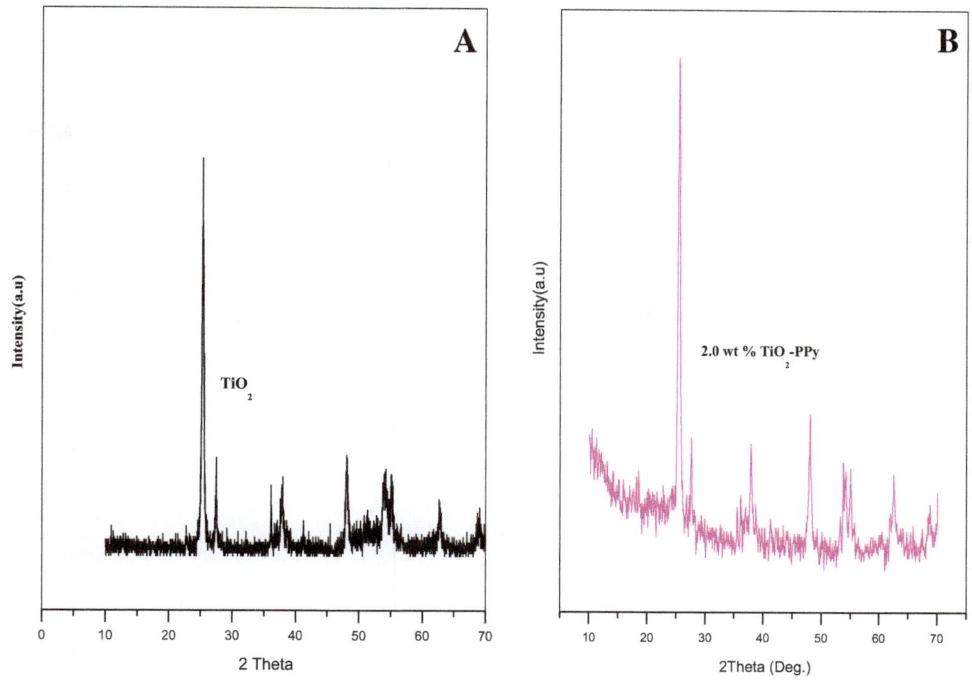

Fig. 4 XRD spectra of **a** TiO_2, **b** PPy-TiO_2 nanocomposite

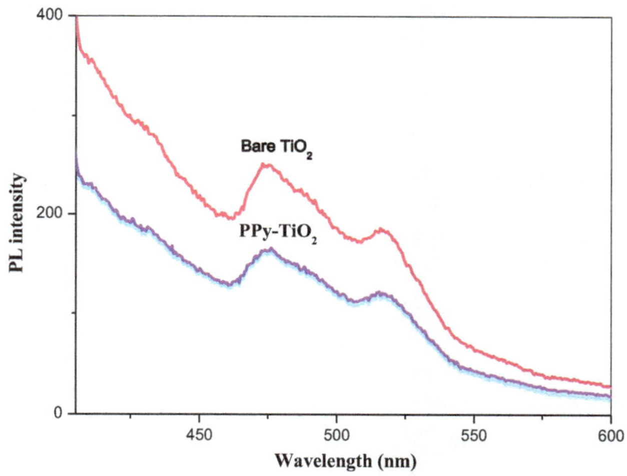

Fig. 5 PL spectra of **a** PPy-TiO_2 nanocomposite, **b** TiO_2

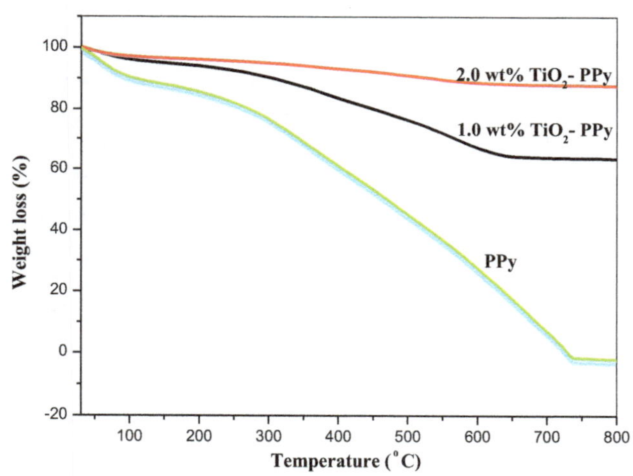

Fig. 6 TGA/DTA spectrum for 2.0 wt% TiO_2-PPy nanocomposite

10–60 ppm and constant amount of catalyst and irradiation time and the results are shown in Fig. 7. These results indicated that PPy-TiO_2 nanocomposites photocatalyst showed a high efficiency for the photocatalytic degradation of MB in the presence of solar light irradiation than TiO_2.

When increasing the dye concentration from 10 to 20 ppm, the percentage decoloration of MB dye was decreased at 90–5 %. Finally, the degradation quickly decreases for 60 ppm (25 %). This may be due to:

1. higher dye concentration might serve as an inner filter shunting the photons away from the catalyst surface;
2. non-availability of oxidative free radicals;
3. inverse effect is explained that if the dye concentration increases in the treating solution, large amount of dye molecules is adsorbed on catalyst surface active site. The surface active sites to contribute in the degradation reaction will help to enlarge the degradation efficiency. Due to the increase in dye concentration, there is no space for the formation of OH· radicals. So at last there is a drop in degradation efficiency. Hydroxyl radical is the strongest oxidizing agent promoting the degradation rate. According to the Beer–Lambert law, as the initial dye concentration increases, the pathway distance end to end of photon incoming the solution decreases consequently resulting

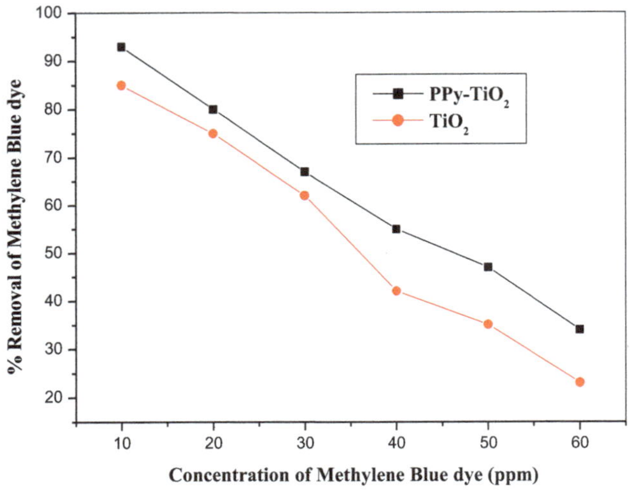

Fig. 7 Effect of initial concentration on the photodegradation of MB dye on PPy-TiO_2 nanocomposite and TiO_2

in lower photo degradation rate (Zhao et al. 2012; Wang et al. 2012; Selvam et al. 2007).

Effect of dose variation

The influence of dose on the degradation of Methylene Blue has been investigated by PPy-TiO_2 nanocomposite. This percentage removal of dye using PPy-TiO_2 nanocomposites was higher than TiO_2 (Fig. 8). The increase in the efficiency seems to be due to the increase in the whole surface area, namely the number of active sites obtainable for the photo catalytic reaction as the quantity of photo catalyst increased. When photocatalyst is added in excess, the degradation of dye molecules remains constant.

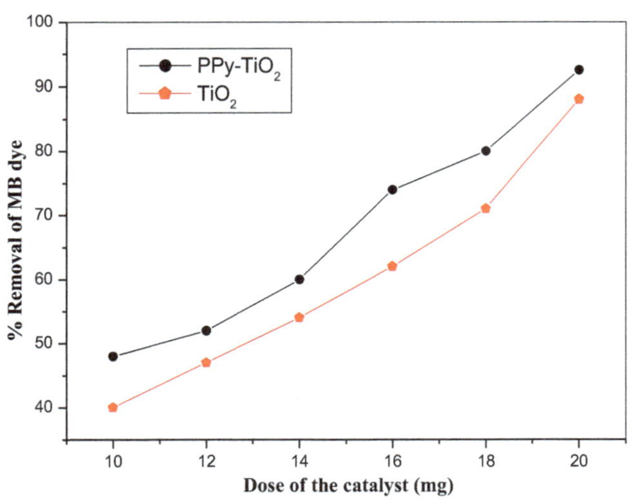

Fig. 8 Effect of dose of the catalysts on the photodegradation of MB dye on PPy-TiO_2 nanocomposite and TiO_2

The reason for this is the deceased light penetration, the increased light scattering and the slaughter in surface area occasioned by agglomeration at high solid concentration (Baruah et al. 2010).

Effect of time variation

The degradation of methylene blue was exponential with time. The degradation rate constant k for methylene blue was obtained from the plot in Fig. 9 using

$$\ln\left(\frac{C_O}{C_f}\right) = kt \tag{3}$$

where C_o and C_f are the initial and final concentrations of dye and t is the irradiation of time.

The degradation rate constants are similar for both the sample with decayed methylene blue fraction of 93 % leading to enlightenment for 90 min (Kansal et al. 2007).

UV–vis absorption spectrum of MB with different reaction time under solar light irradiation in the presence of PPy-TiO_2 nanocomposites is illustrated in Fig. 10. Solar light irradiation leads to decreases in absorbance in MB in the presence of PPy-TiO_2 nanocomposites and the decrease of the absorption intensities indicated that the dye has been degraded. As can be seen in figure, the departure of the distinguishing band of MB dye at 665 nm after 90 min under solar light irradiation indicates that Methylene Blue has been degraded completely.

The value of $\ln(C_0/C_t)$ is plotted against time (in min) and the plots are found to be linear. From the slope, the rate constants were considered for the degradation of MB dye in the presence and absence of PPy. The pseudo first-order rate constant (k min^{-1}) values of TiO_2 and PPy-TiO_2 system can be found out from the Fig. 9. The degradation efficiency of the MB dye PPy-TiO_2 nanocomposite is higher than that of TiO_2 as shown in Fig. 11.

Effect of pH variation

To study the effect of pH on the decolorization efficiency, experiments were carried out at various pH values, ranging from 3 to 11 for constant dye concentration (10 ppm) and catalysts loading (Fig. 12). It has been experimental that the decolorization efficiency increases with increasing pH exhibiting a maximum rate of degradation at pH 11 and 9. The zero point charge for TiO_2 is 6.25. The pH increases from 7 to 11. The plane of the catalyst will become negatively charged. So the cationic dye (MB) easily attached to the catalyst surface. Finally, the percentage removal of dye increased. In contrast, at acidic condition the degradation efficiency will be decreased (Madhusudhana et al. 2012).

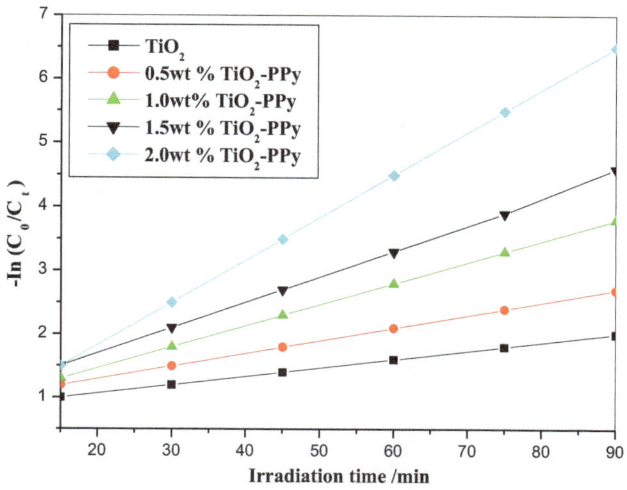

Fig. 9 Apparent first-order linear transforms—ln(C_0/C_t) vs. t of MB dye (dye concentration: 10 ppm) and (catalyst dose: 0.4 g/l^{-1})

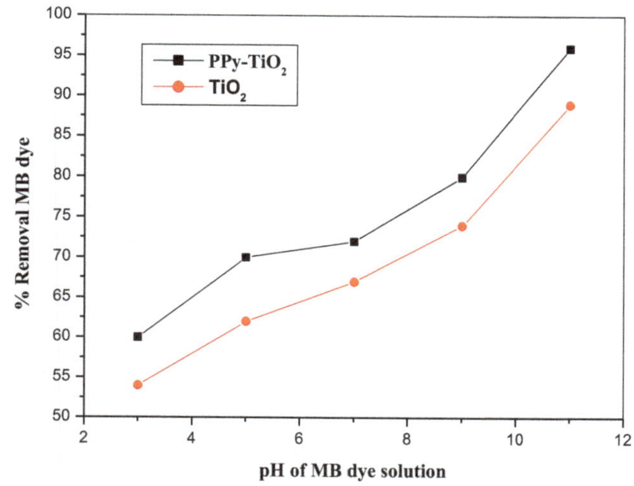

Fig. 12 Effect of pH variation on the photodegradation of MB dye on PPy-TiO_2 nanocomposite and TiO_2

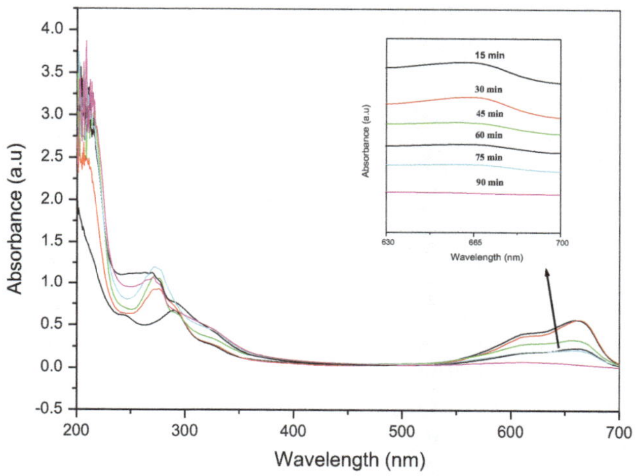

Fig. 10 UV–vis absorption spectra of MB dye degradation using PPy-TiO_2 nanocomposite under solar light with different intervals of time [15, 30, 45, 60, 75, 90]

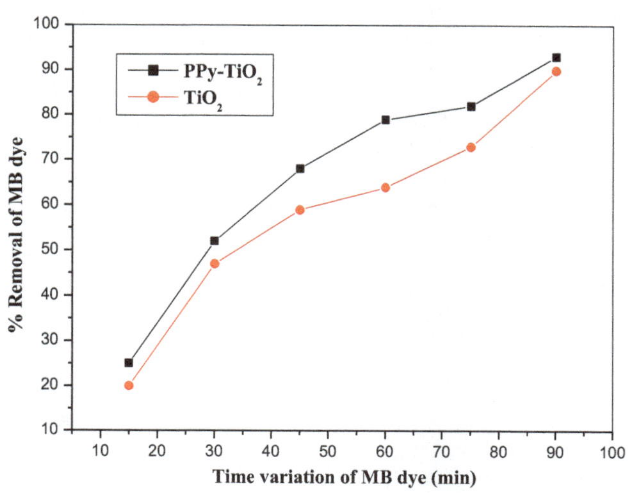

Fig. 11 Effect of time variation on the photodegradation of MB dye on PPy-TiO_2 nanocomposite and TiO_2

Mechanism

Photocatalytic mechanism of PPy-TiO_2 nanocomposites can be described as below. The photocurrent values obtained from PPy to TiO_2 nanocomoposite are higher than TiO_2. PPy is a p-type narrow band gap semiconductor when compared with other inorganic semiconductors (TiO_2, $SrTiO_3$, ZnO) having *n*-type narrow band gap semiconductor. The lowest vacant molecular orbital level of PPy is positioned in an actively higher location than the conduction band edge of TiO_2. Consequently, when the system was illuminated, the excited electrons of PPy in the highest engaged molecular orbital level could move to the LUMO level, then these electrons could be transferred into the conduction band of the TiO_2. Thus, more and extra photo energized electrons formed in TiO_2 and the combination of the semiconductors (*p*–*n* junction) should have a beneficial role in getting better charge separation. As a result, this scheme delivers important photo responses in this intersection.

The above-discussed mechanism is illustrated in Scheme 2.

Reuse of catalysts

The reproducibility of the photocatalytic degradation activity of a 0.02 g sample PPy-TiO_2 nanocomposite performed on a 10 ppm MB dye solution (50 ml) is expressed as the number of cycles, where the photocatalytic degradation of PPy-TiO_2 nanocomposite was reduced from 93 % on the first usage to 82, 29 and 20 % after the second, third and fourth cycles of reuse, respectively. This reduced dye photodegradation activity is in accord with the appearance of recycled PPy-TiO_2 nanocomposite, which was clearly

Scheme 2 Schematic illustration on the photodegradation mechanism of MB dye on PPy-TiO_2 nanocomposite under solar irradiation

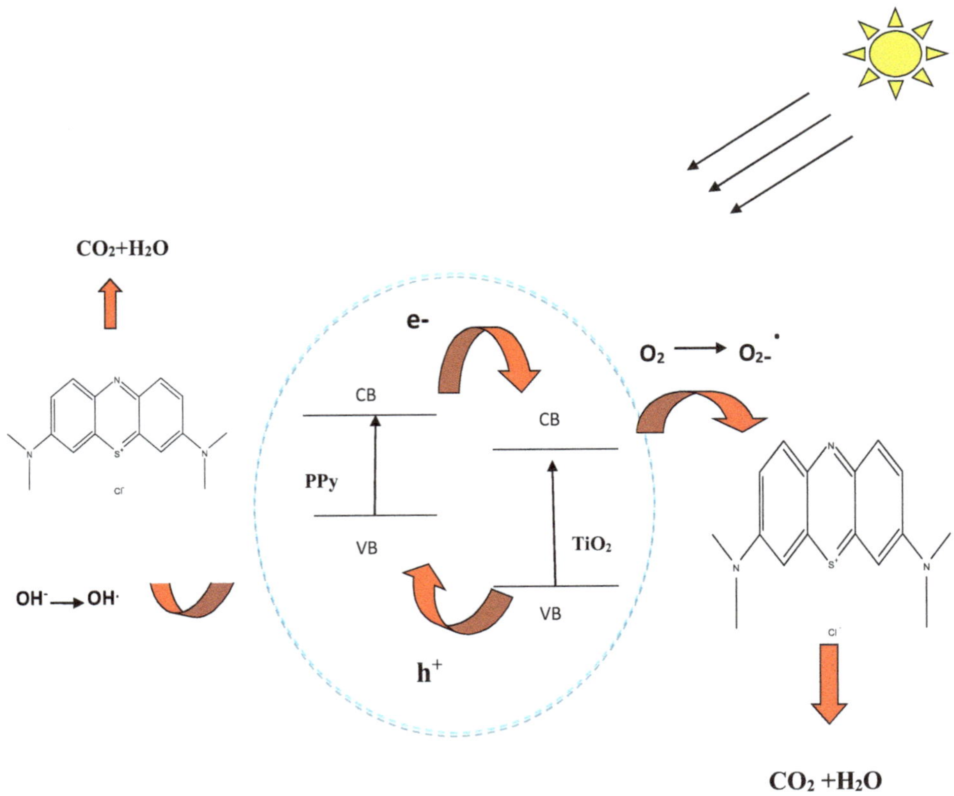

changed in the morphology of the PPy-TiO_2 nanocomposites due to the photodegradation.

Conclusions

PPy-TiO_2 nanocomposite was effectively prepared by the chemical oxidative polymerization method. The results from SEM analysis consistently specify that the TiO_2 has been deposited on Polypyrrole atmosphere. The FT-IR results show the major functional group present in PPy-TiO_2 nanocomposites. The UV–vis diffuse reflectance spectra confirm that the tailored catalyst immersed more photons under visible light irradiation. The TGA spectra indicate the thermal stability of PPy-TiO_2 nanocomposites. The MB dye was effectively degraded by PPy-TiO_2 nanocomposite under solar light irradiation. The fraction removal of dye degradation was higher (93 %) in PPy-TiO_2 nanocomposite than TiO_2.

Acknowledgments The authors thank the University Grants Commission, New Delhi for the Financial support in the form of Major Research Project (MRP) Grant. The authors also thank Ayya Nadar Janaki Ammal College, Sivakasi for providing laboratory facilities.

appropriate credit to the original author(s) and the source, provide a link to the Creative Commons license, and indicate if changes were made.

References

Arenas MC, Fernando Rodríguez-Núnez L, Domingo R, Omar Martínez A, Martínez-Alonso C, Castaño VM (2013) Simple one-step ultrasonic synthesis of anatase titania/polypyrrole nanocomposites. Ultrason Sonochem 20:777–784

Baruah S, Jaisai M, Imani R, Nazhad M, Dutta J (2010) Photocatalytic paper using zinc oxide nanorods. Sci Technol Adv Mater 11:1–7

Chatterjee S, Shit A, Nandi AK (2013) Nanochannel morphology of polypyrrole-ZnO nanocomposites towards dye sensitized solar cell application. J Mater Chem A (1):12302–12309

Chougulea MA, Dalavib DS, Mali S, Patilb PS, Moholkarb AV, Agawanec GL, Kimc JH, Shashwati S, Patila VB (2012) Novel method for fabrication of room temperature polypyrrole—ZnO nanocomposite NO_2 sensor. Measurement S0263-22410, 0193-5

Chougule MA, Khuspe GD, Shashwati S, Patil VB (2013) Polypyrrole–ZnO nanohybrids: effect of CSA doping on structure, morphology and optoelectronic properties. Appl Nanosci (3):423–429

Deivanayaki S, Ponnuswamy V, Jayamurugan P, Ashokan S (2012) The structure and properties of polypyrrole/titaniumdioxide nanospheres of various dopant percentages by Chemical oxidation method. Elixir Polymer 49B:10182–10185

Denga F, Li Y, Luob X, Lixia Y, Tu X (2012) Preparation of conductive polypyrrole/TiO_2 nanocomposite via surface molecular imprinting technique and its photocatalytic activity under simulated solar light irradiation. Colloid Surf A: Physicochem Eng Aspects 395:183–189

Eskizeybek V, Fahriyesari B, Gulce H, Gulce A, Ahmet A (2012) Preparation of the new polyaniline/ZnO nanocomposite and its photocatalytic activity for degradation of MB and MG dyes under UV and natural sunlights irradiations. Appl Catal B: Environ 119–120:197–206

Etelino F, Melo D, Kleber G, Alves B, Severinon A, Celso J, de Melo P (2013) Synthesis of fluorescent PVA/polypyrrole-ZnO nanofibers. J Mater Sci 48:3652–3658

Guoquan Z, Shuai W, Sha Z, Lei F, Guohua C, Fenglin Y (2011) Oxidative degradation of azo dye by hydrogenperoxide electrogenerated in situ on anthraquinonemonosulphonate/polypyrrole composite cathode with heterogeneous $CuO/\gamma\text{-}Al_2O_3$ catalyst. Appl Catal B: Environ 106:370–378

Hasani T, Hossein E (2013) Removal of Cd (II) by using polypyrrole and its nanocomposites. Synth Met 175:15–20

Hongyu M, Xiaogang Z, Youlong X, Fang X (2010) Synthesis, characterization and electrochemical behavior of polypyrrole/carbon nanotube composites using organometallic-functionalized carbon nanotubes. Appl Surf Sci 256:2284–2288

Hou J, Cao R, Jiao S, Zhu H, Kumar RV (2011) PANI/$Bi_{12}TiO_{20}$ complex architectures: controllable synthesis and enhanced visible-light photocatalytic activities. Appl Catal B: Environ 104:399–406

Ibrahim UG, Halim AA (2008) Heterogeneous photocatalytic degradation of organic contaminants over titanium dioxide: a review of fundamentals, progress and problems. J Photochem Photobiol C: Photochem Rev 9:1–12

Juraj D, Martin K, Gabriela B, Martin K, Iva M (2013) Electrochemical fabrication and characterization of porous silicon/polypyrrole composites and chemical sensing of organic vapors. Int J Electrochem Sci 8:1559–1572

Kandiel TA, Dillert R, Bahnemann DW (2009) Enhanced photocatalytic production of molecular hydrogen on TiO_2 modified with Pt: polypyrrole nanocomposites. J Photochem Photobiol Sci 8:683–690

Kangwansupamonkon W, Jitbumpot W, Sudakiatkamjornwong C (2010) Photocatalytic efficiency of TiO_2/poly[acrylamide-co-(acrylic acid)] composite for textile dye degradation. Polym Degrad Stab 95:1894–1902

Kansal SK, Singh M, Sud D (2007) Studies on photodegradation of two commercial dyesin aqueous phase using different photocatalyst. J Hazard Mater 141:581–590

Li XY, Wang DS, Cheng GX, Luo QZ, An J, Wang YH (2008) Preparation of polyaniline-modified TiO_2 nanoparticles and their photocatalytic activity under visible light. Appl Catal B: Environ 81:267–273

Li Q, Zhang C, Li J (2010) Photocatalysis and wave-absorbing properties of polyaniline/TiO_2 microbelts composite by in situ polymerization method. Appl Surf Sci 257:944–948

Madhusudhana N, Yogendra K, Mahadevan M (2012) A comparative study on photocatalytic degradation of violet GL2B azo dye using CaO and TiO_2 nanoparticles. Int J Eng Res Appl 2(5):1300–1307

Maeda K (2012) Photocatalytic water splitting using semiconductor particles: history and recent developments- Invited review. J Photochem Photobiol C: Photochem Rev 12:237–268

Mi O, Ru B, Yi X, Cheng Z, Chun MA, Man W, Hong-zheng C (2009) Fabrication of polypyrrole/TiO_2 nanocomposite via electrochemical process and its photoconductivity. Trans Nonferrous Met Soc China 19:1572–1577

Muthirulan P, Meenakshi Sundaram M, Kannan N (2012a) Beneficial role of ZnO photocatalyst supported with porous activated carbon for the mineralization of alizarin cyanin green dye in aqueous solution. J Adv Res. doi:10.1016/j.jare.2012.08.005

Muthirulan P, Nirmala Devi C, Meenakshi Sundaram M (2012b) Synchronous role of coupled adsorption and photocatalytic degradation on CAC-TiO2 composite generating excellent mineralization of alizarin cyanine green dye in aqueous solution. Arab J Chem 1:1–7

Muthirulan P, Kannan Nirmala Devi C, Meenakshi Sundaram M (2013) Facile synthesis of novel hierarchical TiO_2@Poly(o-phenylenediamine) core–shell structures with enhanced photocatalytic performance under solar light. J Environ Chem Eng 1:620–627

Ochiai T, Fujishima A (2012) Photoelectrochemical properties of TiO_2 photocatalyst and its applications for environmental purification-invited review. J Photochem Photobiol C: Photochem Rev 13:247–262

Park H, Park Y, Kim W, Choi W (2013) Surface modification of TiO_2 photocatalyst for environmental applications: invited review. J Photochem Photobiol C: Photochem Rev 15:1–20

Rohit B, Madhulika S, Bahadur D (2013) Visible light-driven nanocomposites ($BiVO_4/CuCr_2O_4$) for efficient degradation of organic dye+. Dalton Trans 42(19):6736–6744

Selvam K, Muruganandham M, Muthuvel I, Swaminathan M (2007) The influence of inorganic oxidants and metal ions on semiconductor sensitized photodegradation of 4-fluorophenol. J Chem Eng 128:51–57

Ufana R, Ashraf SM (2011) Semi-conducting poly(1-naphthylamine) nanotubes: a pH independent adsorbent of sulphonate dyes. Chem Eng J 174:546–555

Wang DS, Wang YH, Li XY, Luo QZ, An J, Yue JX (2008) Sunlight photocatalyticss activity of polypyrrole: TiO_2 nanocomposites prepared by 'insitu' method. Catal Commun 9(1162–1166):7

Wang D, Zhang J, Luo Q, Li X, Duan Y, An J (2009) Characterization, photocatalytic activity of poly(3-hexylthiophene)-modified TiO_2 for degradation of methyl orange under visible light. J Hazard Mater 169:546–550

Wang Y, Shi R, Lin J, Zhu Y (2010) Significant photocatalytic enhancement in methylene blue degradation of TiO_2 photocatalysts via grapheme-like carbon in situ hybridisation. Appl Catal B: Environ 100:179–183

Wang WS, Wang DH, Qu WG, Lu LQ, Xu AW (2012) Large ultrathin anatase TiO2 nanosheets with exposed 001 facets on grapheme for enhanced visible light photocatalytic activity. J Phys Chem C 116:19893–19901

Xiaofang L, Qidong Z, Xin LC, Dajun W, Wanjin Z, Wang C, Wei Y (2006) Prepartion and characterization of polypyrrole/TiO_2 coaxial nanocables. Macromol Rapid Commun 27:430–434

Xu S, Gu L, Yang KH, Song Y, Jiang L, Dan Y (2012) The influence of the oxidation degree of poly(3-hexylthiophene) on the photocatalytic activity of poly(3-hexylthiophene)/TiO_2 composites. Sol Energy Mater Sol Cells 96:286–291

Yucheng Y, Junwei W, Jianhong W, Xiong R, Shi J, Chun-Xu P (2013) Polypyrrole-decorated Ag-TiO_2 nanofibers exhibiting enhanced photocatalytic activity under visible light illumination. ACS Appl Mater Interfaces 5(13):6201–6207

Zhang L, Liu P, Su Z (2006) Preparation of PANI/TiO_2 nanocomposites and their solidphase photocatalytic degradation. Polym Degrad Stab 91:2213–2219

Zhao D, Sheng G, Chen C, Wang X (2012) Enhanced photocatalytic degradation in methylene blue under visible irradiation on grapheme@TiO_2 dyade structure. Appl Catal B: Environ 111:303–308

16

Impact of physicochemical parameters on phytoplankton compositions and abundances in Selameko Manmade Reservoir, Debre Tabor, South Gondar, Ethiopia

Tilahun Adugna Wassie[1] · Ayalew Wondie Melese[2]

Abstract Impact of physicochemical parameters on phytoplankton compositions and abundances in Selameko Reservoir, Debre Tabor, South Gondar from August 2009 to May 2010 was assessed. Water quality parameters, such as temperature, water transparency, water depth, dissolved oxygen, pH, total dissolved solids, phosphate, nitrate, and silicate were measured in situ from two sites (littoral and open water zone) of the reservoir. Phytoplankton compositions and abundances were analyzed in Tana fisheries and other aquatic organisms' research center. ANOVA result of the physicochemical parameters included chlorophyll-a showed the presence of significance difference among seasons and between sites ($P < 0.05$). A total of seven families, 36 genera from three groups (Diatom, Blue green algae and Green algae) of phytoplankton were identified during the study period. From all groups, diatoms were the most abundant at both sites and Blue green algae were the least abundant. ANOVA of all phytoplankton showed highly significant difference among seasons and between sites ($P < 0.05$). ANOVA of all phytoplankton showed highly significant difference among seasons and between sites ($P < 0.05$). Based on the stepwise regression, a total number of phytoplanktons had positive correlation with some of the physicochemical parameters (R2 = 0.99, $P < 0.001$, $N = 16$). The study concluded that some of physicochemical parameters (NO_3-N and PO_4-P) indicated the presence of reservoir water pollution. This is supported by the presence of pollution-resistant phytoplankton species such as *Melosira* and *Microcystis*. The reservoir water was eutrophic (productive) throughout the year. To avoid such pollution, basin and reservoir management are recommended.

Keywords Physico-chemical parameters · Phytoplankton composition · Abundance

Introduction

Phytoplankton are free-floating uni-cells and colonies that grow photoautotrophically in aquatic environments. Phytoplankton play a key role in the primary production and global nutrient cycles of the Earth (Daniel 2001) by making up the main producers in any given water body (Biddanda and Benner 1997). It colonizes the upper part of the water column, down to the limit of penetration of light. The structure and abundance of the phytoplankton populations are mainly controlled by inorganic nutrients such as nitrogen, phosphorus, and silica (Daniel 2001) and mainly available nitrogen as nitrate, nitrite and ammonia, phosphorus as soluble orthophosphate (USEPA 2000) and silicone as silicate forms.

Phytoplankton communities are sensitive to changes in their environment and therefore phytoplankton total biomass and many phytoplankton species are used as indicators of water quality (Reynolds et al. 2002; Brettum and Andersen 2005).

Wholly, the use of living organisms to determine the presence, amounts, changes in and effects of physical,

✉ Tilahun Adugna Wassie
tawamtmr@gmail.com; adugnawassie@gmail.com; tilahun46@yahoo.com

Ayalew Wondie Melese
ayaleww2001@yahoo.com

1 Biology Department, College of Natural and Computational Science, Mizan Tepi University, Tepi, Ethiopia

2 Department of Biology, Faculty of Natural Science, Bahir Dar University, Bahir Dar, Ethiopia

chemical, and biotic factors in the environment are termed biological monitoring (Baker 1976). Therefore, the water quality has to be monitored by measuring the phytoplankton content in order to come up with preventive measures such as aeration to prevent fish kills during the decay of the phytoplankton biomass, and the sustainable use of drinking water supply and recreational activities (Imhoff and Alberrecht 1975) and other uses of the water bodies. This work was aimed to determine the impacts of physicochemical parameters on the phytoplankton compositions and abundances in Selameko manmade reservoir, Debre Tabor, South Gondar, Ethiopia.

Materials and methods

Description of the study area

This study was conducted at Selameko manmade reservoir, south-west of Debre tabor town, Ethiopia. This manmade reservoir is found 2513 meter above sea level (masl), specifically located at 38°05′E and 11°53′24″N (Fig. 1). The reservoir was constructed in 2007 with a total of 11.6 hectares (ha) with 20 m depth to irrigate nearly 63 ha. The catchments of the reservoir are extended from 2513 to 2726 masl with 879.25 ha of total area.

Hydrology and climate

Based on the National Metrology Agency of Bahir Dar Branch Office, the mean annual temperature of the study area is 16.23 °C (ranges from 9.2 to 23.26 °C) (from 1997 to 2009 years) and its mean annual rainfall is 1371.2 mm (ranges from 1096.7 to 1645.7 mm). The climate of reservoir is characterized roughly by four seasons: (1) A main-rainy season (MRS) with heavy rains during July–September, (2) a post-rainy season (PORS) between October and November, (3) a dry season (DS) between December and April and (4) a pre-rainy season (PRS) from May to June (Tamiru 2006; Ayalew et al. 2007).

Measurements of physicochemical parameters

Samples of physicochemical parameters were collected from August 2009 to May 2010 in the four seasons, i.e., main rainy/MRS/, post-rainy/PORS/, dry/DS/, and pre-

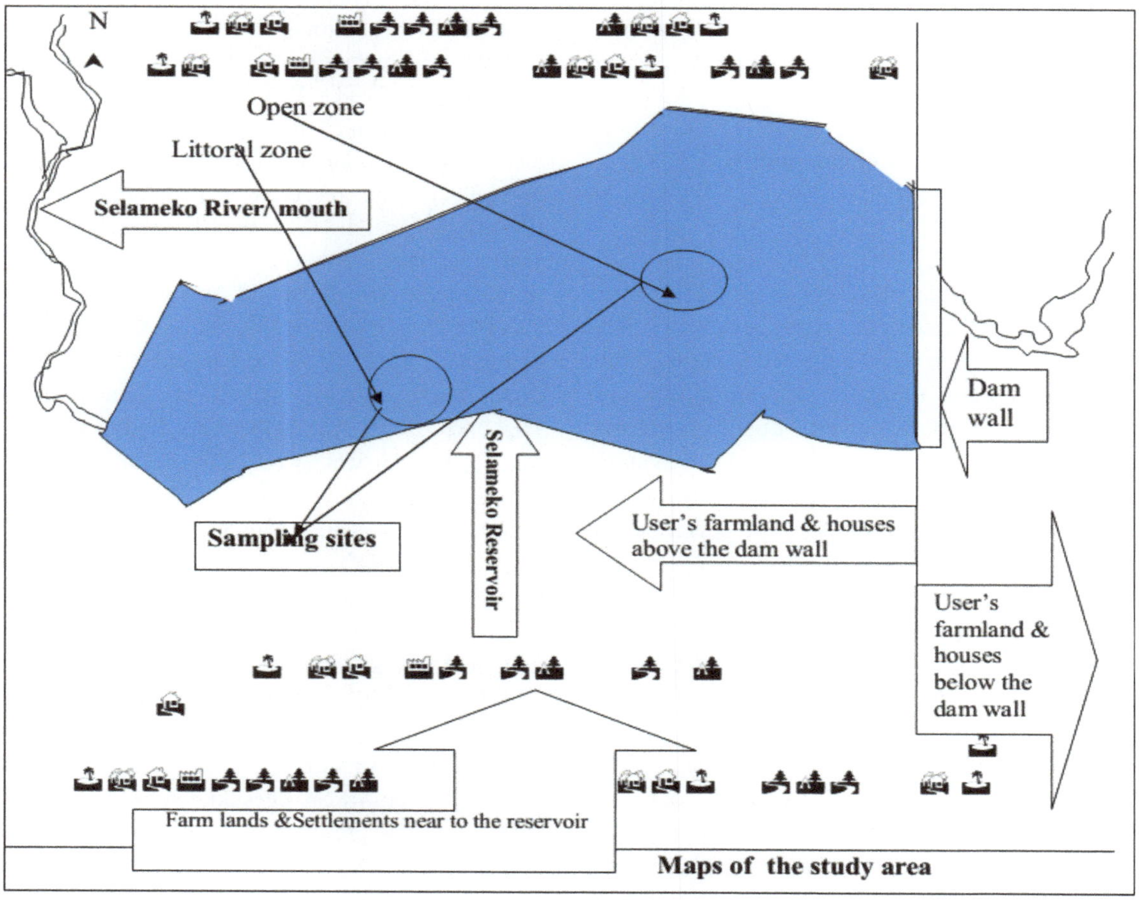

Fig. 1 Map and location of the study area (Selameko Manmade Reservoir)

rainy/PRS/seasons. The water samples were taken only from littoral (SI) and open water zone (SII) two times from each for each parameter (Fig. 1).

Water temperature and pH, dissolved oxygen (DO), total dissolved substances (TDSs), and water transparency were determined in situ. Water temperature and pH were measured with coupled pH/TDS/CON Meter (Model Tochpro II); DO was determined by portable oxygen analyzer (JPB-607); TDSs were measured by cond/TDS meter (Model CE 470 Cond. Meter 01189); water depth was measured by standardized meter (the calibrated meter that had metal tip part was dipped into the reservoirs water until it touched the bottom of the reservoir three times at the central and the average recorded depth was taken) and transparency was measured by standardized Secchi disk. The physico-chemical parameter measurements were taken after the probes dipping down from the surface water to 50 cm down in the reservoir water. Major nutrients, nitrate (NO_3-N), phosphate (PO_4-P), and silicate (SiO_2), were measured in situ immediately by using a portable water analyzer kit (Wagtech international, Palintest transmittance display photometer 5000, Palintest Ltd., and UK) (Palintest Ltd 1989). The collected water samples from the two sites were first filtrate by Whatman GF/C, 0.6–0.7 μm pore size membrane filter to avoid unnecessary large-sized particulate materials (debris) that cause further nutrient release before nutrient analysis was made.

Measurement of phytoplankton community

Phytoplankton

Integrated phytoplankton samples were collected two times for each from littoral (SI) and open water zone (SII) using Van Dorn water sampler. The collected samples were concentrated in 100 mL using 55 μm (mesh opening) phytoplankton net and preserved with Lugol's solutions. Again, 100 mL was allowed to settle in graduate cylinder overnight and the supernatant was siphoned off till 10 mL remained. Of this concentrated sample, 1 mL was used in a Sedgwick-Rafter Cell, of which 100 microscopic field were counted for major species according to Wetzel and Likens (2000). Identification and counting were made under Olympus (CH-2) compound microscope (200X) based on a key guideline of Yamaguchi and Gould (2007) and Blomqvist and Olsen (1981), and abundance of each species of phytoplankton was calculated based on Lind (1979). Chlorophyll-a (Chl-a) was calculated based on monochromatic methods (Lorenzen 1967 in Wetzel and Likens 2000).

Data analyses

Analysis of variance (ANOVA) was used to test significant differences between those like spatial and temporal variations of physico-chemical and phytoplankton ($P < 0.05$). Tukey (Honestly Significantly Differently Test) test was used to determine significance in mean catches and estimates. Stepwise regression was used to test the relationship between the physico-chemical parameters and biological variables; the observed (r) values were then compared to the table values at $P < 0.05$ level of significance, whereas abundance of phytoplankton was analyzed using percentage. Two-way ANOVA was applied for both physico-chemical and phytoplankton. In general, data were calculated and organized using appropriate statistical software, such as the SPSS (2007) version 16 and SAS (2003).

Results and discussion

Physicochemical measurements (in seasons and sites)

Most of the physicochemical parameters are very suitable for phytoplankton growth. The ANOVA result showed that there were highly significant differences in temperature, pH, DO, WTD, TDS, NO_3-N, PO_4-P, SiO_2 and chl-a among seasons and between sites (Table 1). The Tukey test also showed similar scenario (Table 1). Specifically, temperature of the reservoir water ranged from 18.7 °C (PORS, SI) to 24.2 °C (DS, SI and SII, PRS, SII). This registered temperature is good for planktons but not optimum for the growth of fish as suggested by Korai et al. (2008) which is between 22 and 31 °C. pH of the reservoir extended from 7.01 (SII) to 8.01 (SI). These pH values are good for aquatic life including fish and fall (Oso and Fagbuaro, 2008) within the EPA Redbook recommended range for fresh water (6.5–9.0) (Schmirz 1996) and recommended by others (Chapman 1996; Goldman and Horne 1983). DO was ranged from 5.0 (DS) to 6.15 mg/L (MRS). The obtained DO concentration satisfied the minimum recommended standard (>5 ppm) set by EPA Redbook and others (USEPA 2008; Yajurvedi 2008) and good for fishing and planktons. The transparency depth was ranged from 32 (MRS, SI) to 97 cm (PORS, SII) and is eutrophic (Horne and Goldman 1994). The lowest was due to the presence of high suspended matter (sands, silts) that increased water turbidity in the MRS (Meesukko et al. 2007; Rafique et al. 2002), and the presence of large numbers phytoplankton that favored by high nutrient loads in rain season (Musta-

Table 1 Physico-chemical parameters of Selameko reservoir from 2009 to 2010 respect to seasons and locations

Physico-chemical parameters	SI					SII				
	Seasons					Seasons				
	MRS	PROS	DS	PRS	Av ± SD	MRS	PROS	DS	PRS	Av ± SD
T °C	20.0^{d}	18.7^{e}	24.2^{a}	23.1^{b}	21.5 ± 1.28	19.7^{d}	20.8^{c}	24.2^{a}	24.1^{a}	22.2 ± 1.16
pH	$7.72^{a,b}$	8.01^{a}	$7.42^{b,c}$	7.59^{b}	7.69 ± 0.12	$7.42^{b,c}$	7.39^{bc}	7.01^{d}	$7.22^{c,d}$	7.26 ± 0.094
DO (mg/L)	$5.59^{b,c}$	$5.9^{a,b}$	5.1^{d}	$5.3^{c,d}$	5.47 ± 0.17	6.15^{a}	$5.9^{a,b}$	5.0^{d}	$5.2^{c,d}$	5.56 ± 0.27
WTD (cm)	32^{f}	67^{c}	$62^{c,d}$	40^{e}	50.25 ± 8.46	42^{e}	97^{a}	89^{b}	60^{d}	72 ± 12.72
TDSs (ppm)	98.4^{c}	67.1^{f}	74.3^{e}	87.4^{d}	81.8 ± 6.94	137.2^{a}	67.1^{f}	90.43^{d}	114.3^{b}	102.26 ± 15.11
NO_3-N (mg/L)	1.85^{b}	$0.21^{e,f}$	0.25^{e}	0.53^{d}	0.71 ± 0.34	2.0^{a}	0.22^{e}	0.10^{f}	0.71^{a}	0.758 ± 0.43
PO_4-P mg/L)	0.83^{a}	$0.12^{c,d}$	0.44^{b}	$0.16^{c,d}$	0.388 ± 0.26	0.45^{b}	0.24^{c}	0.24^{c}	0.08^{d}	0.253 ± 0.075
SiO_2 (mg/L)	22.5^{a}	6.81^{c}	2.24^{d}	0.66^{e}	8.05 ± 4.98	22.5^{a}	10.06^{b}	2.08^{d}	0.09^{f}	8.68 ± 5.08
pH	$7.72^{a,b}$	8.01^{a}	$7.42^{b,c}$	7.59^{b}	7.69 ± 0.12	$7.42^{b,c}$	$7.39^{b,c}$	7.01^{d}	$7.22^{c,d}$	7.26 ± 0.094

The table included the mean differences of (Tukey Test) "physico-chemical" parameters. Means of the two columns of a particular parameter followed by the same letter (s) are not significantly different from each other ($P < 0.05$, Tukey HSD). (average ± SD)

Av average, *SD* standard deviation, *WTD* water transparency depth

pha and Omotosho 2005). Similarly, the TDSs of the reservoir were between 67.1 (in PORS, SI, SII) and 137.2 ppm (in MRS, SII). This registered value is very conducive for the growth of aquatic organisms including fish (Mohamed et al. 2009) (Table 1).

The concentration of NO_3-N was between 0.1 and 2 mg/L, for PO_4-P was between 0.08 and 0.83 mg/L, and for silicate was between 0.09 and 22.5 mg/L. All nutrients, the highest values were recorded in MRS due to surface water inflow that brought nutrients from the surrounding agricultural areas (Meesukko et al. 2007; Stanley et al. 2003) and the leachates of municipal wastes from waste disposal sites and sanitary landfills (Bennett 1998). Additionally, PO_4-P concentration was due to washing and bathing with phosphate-based detergents and soaps (Davies et al. 2009), and silicate was due to the presence of high rate of rock and soil weathering in the water body (Little 2004). All these high concentrations of nutrients abnormally increased plankton abundances and composition in rainy time than other seasons. Generally, the obtained nitrate concentration is tolerable by fish and other aquatic organism and satisfies surface water quality standards (<5 mg/L) (PCD 1997 in Chattopadhyay and Banerjee 2007) and fulfills the minimum level of nitrate in lake to be productive (Yajurvedi 2008). The registered silicate concentration is good for plankton growth and satisfies freshwater ranges from 1 to 30 mg/L (Wetzel 2001 in Meesukko et al. 2007; Chapman 1996). However, the obtained concentration of PO_4-P is greater than other standards (0.005–0.020 mg/L PO_4-P, Chapman 1996; 0.01–0.03 ppm phosphorus, Yajurvedi 2008) and an indication of the presence of pollution.

Phytoplankton compositions and abundances (in season and sites)

A total of 36 genera from seven taxonomic families and three groups were recorded during the study period.

The ANOVA result showed that there were highly significant differences in zygnemaphyceae, bacillariophyceae, chlorophyceae, cryptophyceae, cyanophyceae, euglenenophyceae, and dinophyceae among seasons and between sites (Tables 2 and 3). According to stepwise regression, total number of phytoplankton had positive relationship with physicochemical parameters ($R2 = 0.99$, $P < 0.001$, $N = 16$). Grand total of phytoplanktons $= +2.6 \times e + 6$ NO_3-N, $+4.5 \times e + 6$ PO_4-P, $+1.8 \times e + 5$ SiO_2, $+1 \times e + 5$ chl-a.

From the total numbers of phytoplankton, the MRS was higher than the other seasons and PORS was the least abundance. The highest record was due to the presence of excess nutrients that come from human sources such as waste and agricultural runoffs (Osondu 2008) mainly attributed to the increase in nitrates, phosphates, and silicates (Kobbia et al. 1991) (Table 1). These alternative nutrients became food sources of various phytoplankton types. The lowest abundance in PORS was probably due to the presence of small amounts of nutrients like phosphates and nitrates. From all seasons, Bacillariophyceae was the most abundance family due to majorly the presence the high concentration and most favorite food alternative, silicate (Radwan 2005; Little 2004) and to some extant nitrate (Khenari 2007). Among Bacillariophyceae, *Melosira* and in Cyanophyceae, *Microcystis* were some of the dominant genera and were indicators of the presence of pollution

Table 2 Seasonal and temporal abundance (ind/L) and distribution of phytoplankton in Selameko manmade reservoir during the four seasons in two locations, South Gondar, Debre Tabor

Phytoplankton genera	MRS			PORS			DS			PRS					Total
	sSI	sSII	TTotal	sSI	sSII	TTotal	sSI	sSII	TTotal	sSI	sSII	TTotal	sSI	sSII	
Diatom															
Bacillariophyceae															
Cyclotella	11633	15863	27496	697	6502	7199	9288	3483	12771	13003	11378	24381	34621	37226	71847
Cymatpleura	–	–	–	–	–	–	–	464	464	10	–	10	10	464	474
Cymbella	–	–	–	–	10	–	10	10	–	–	–	–	10	20	30
Pinnularia	–	–	–	12	–	–	–	–	–	–	464	464	12	464	476
Fragilaria	–	–	–	–	10	–	10	10	–	–	–	–	10	20	30
Navicula	–	100	100	232	1393	1625	3947	1393	5340	1858	464	2322	6037	3350	9387
Melosira	1E+7	8.2E+6	1.8E+7	59792	345049	404841	12074	10217	22291	20666	16951	37617	1.0E+6	8.5E+6	1.9E+7
Synedra	–	100	100	–	–	–	–	–	–	10	–	10	10	100	110
Suririella	–	–	–	10	–	10	–	–	–	–	–	10	10	10	20
Nitzstcha.	–	–	–	–	20434	20434	3483	–	3483	1161	1393	2554	4644	21827	26471
Gyrosigma	–	–	–	–	–	–	232	–	232	–	–	–	232	–	232
Tabularia	–	–	–	–	–	–	232	–	232	–	–	–	232	–	–
Total	1E+7	8.2E+6	18452519	60743	373398	434141	29276	15577	44853	36708	30660	67368	1E+7	8.6E+6	1.9+7
% Bacillariophyceae													94.8	93.1	94
Green algae's															
Zygnemaphyceae															
Cosmarium.	100	–	100	17531	61301	78832	145357	91022	236379	2090	1625	3715	165078	153948	3.2E+5
Staurastrum	–	–		12	1393	1405	232	232	464	–	–		244	1625	1869
Total	100	–	100	17543	62694	80237	145589	91254	236843	2090	1625	3715	165322	155573	3.2E+5
% Zygnemaphyceae													1.5	17	1.6
Chlorophyceae															
Ankistrodesmus	–	–		1277	1161	2438	1858	–	1858	10	–	10	3145	1161	4306
Piiiastrum	–	–		232	232	464	6734	7430	14164	1161	3947	5108	8127	11609	19736
Sieimiasmui	2215	5288	7503	348	3483	3831	5341	10449	15790	9056	7895	16951	16960	27115	44075
Oocytis	25380	22208	47588	104	1393	1497	3019	2786	5805	10	1858	1868	28513	28245	56758
Closterium	–	3173		348	232	–	232	–	–	10	464	474	590	3869	4459
Chlamydomonas	–	–		2786	2089	4875	5108	4179	9287	–	–	–	7894	6268	14162
Pandorina	–	–		–	–	–	–	–	–	10	–	10	10	–	10
Kirchnelriella	–	–		348	1858	2206	–	464	464	10	–	10	358	2322	2680
Staurastrum	–	–		5	5	10	1393	10	1403	–	–	–	1398	15	1413
Chodatella	–	–		–	10		78716	18344	–	929	464	1393	79645	18818	98463
Crucigenia	28553	25380	53933	1393	6037	7430	13003	13932	26935	22059	20201	42260	65008	65550	130558

Table 2 continued

Phytoplankton genera	MRS			PORS			DS			PRS					Total
	sSI	sSII	TTotal	sSI	sSII	TTotal	sSI	sSII	TTotal	sSI	sSII	TTotal	sSI	sSII	
Sphaerocysis	7403	15863	–	–	–	–	–	–	–	10	10	20	7413	15873	23286
Monoraphidium	–	–		–	1161	1161	3715	1161	4876	–	10	10	3715	2332	6047
Coelastrum	20093	14805	34898	813	464	1277	–	10	10	–	–	–	20906	15279	36185
Total	83644	867I7	170361	7654	18125	25779	119119	58765	177884	33265	34849	68114	243682	198456	442138
% Chlorophyceae													2.2	2.1	2.2
Cryptophyceae															
Cryptomonas	–	–	–	–	2089		3483	7663					3483	9752	13235
Euglenophyceae															
Euglena	100	100	–	–	–	–	–	–	–	–	–	–	100	100	200
Phacus	200	2115		10	232	–	–	–	–	–	–	–	210	2347	2557
Total	300	2215		10	232	–	–	–	–	–	–	–	310	2447	2757
% Euglenophyceae	–	–	–	–	–	–	–	–	–	–	–	–	0.01	0.01	
Dinophyceae															
Peredinium	100	–		464	4644		7663	3367		64319	94041	158360	72546	1E+5	174597
% Dinophyceae													0.7	1.1	0.9
Blue green algae															
Cyanophyceae															
Oscillatoria	–	–	–	–	–	–	–	–	–	10	–	10	10	–	10
Anabaena	–	–	–	8127	17415	25542	10	1393	1403	464	697	1161	8601	19505	28106
Microcystis	46530	92003	13853333	2322	4876	7198	11610	36455	48065	14861	16951	31812	75323	1502855	225608
Planktosphaneria	–	–	–	464	–	–	–	–	–	–	–	–	464	–	464
Total	46530	92003	138533	10913	22291	33204	11620	37848	49468	15335	17648	32983	84398	1697900	254188
% Cyanophyceae													0.8	1.8	1.3
Grand total (ind/1)	1E+7	8.4E+7	1.9e+7	97327	483473	580800	316750	214474	531224	151717	178823	330540	10956334 334	9250357	2E+7
T/no of genera	12	13		22	25		24	22		21	18				
% abundance	94.84	90.5		0.89	5.2		2.89	2.32		1.38	1.93		100	100	100

SI site one (littoral), *SII* site two (open); Most abundance > 201,000 Less abundance = between 200,000 and 50,000; Common = between 49,999 and 2000; Rare < 1999; dash represents implied absence

Table 3 The mean differences of (Tukey Test) biological parameters (phytoplankton) of Selameko manmade reservoir, South Gondar (Debre Tabor) from 2009 to 2010

Biological parameters	Seasons	Locations	
		SI	SII
Zygnemaphyceae	Main rainy	100.0[g]	0.0[f]
	Post-rainy	17543[d]	62,694.0[c]
	Dry	1.5e + 5[a]	91,254.0[b]
	Pre-rainy	2090.0[e]	1625[f]
Bacillariophyceae	Main rainy	1.0 + e+7[a]	8192653[b]
	Post-rainy	60743[d]	373398[c]
	Dry	29276[g]	15577[h]
	Pre-rainy	36708[e]	30660[f]
Chlorophyceae	Main rainy	83644[c]	86717[b]
	Post-rainy	7654[h]	18125[g]
	Dry	1.2e + 5[a]	58765[d]
	Pre-rainy	33265[f]	34849[e]
Cryptophyceae	Main rainy	0.0[d]	0.0[d]
	Post-rainy	0.0[d]	2089[c]
	Dry	3483[b]	7663[a]
	Pre-rainy	0.0[d]	0.0[d]
Cyanophyceae	Main rainy	46530[b]	92003[a]
	Post-rainy	10913[h]	22291[d]
	Dry	11620[g]	37848[c]
	Pre-rainy	1.5e + 5[f]	17648[e]
Euglenenophyceae	Main rainy	300.0[b]	2215[a]
	Post-rainy	10.0[d]	232[c]
	Dry	0.0[d]	0.0[d]
	Pre-rainy	0.0[d]	0.0[d]
Dinophyceae	Main rainy	100[g]	0.0[g]
	Post-rainy	464[f]	4644[d]
	Dry	7663[c]	3367[e]
	Pre-rainy	6.4e + 5[b]	94041[a]
Chlorophyll-a	Main rainy	39.29[a]	31.23[b]
	Post-rainy	11.02[e]	23.03[c]
	Dry	21.33[c]	15.65[d,e]
	Pre-rainy	17.88[c,d]	12.43[e]

Means of the two columns of a particular parameter followed by the small letter(s) are not significantly different from each other ($P = 0.05$, Tukey HSD)

(Heiskary and Markus 2001) and the alkaline nature of the reservoir water (Onyema 2007). Totally, very large numbers of phytoplankton were recorded. The total numbers of phytoplankton had positive correlation with NO_3-N, PO_4-P, and SiO_2 as well as chl-a. When the concentrations of nutrients increased, the phytoplankton abundances increased and vices verse (Onyema 2007). Being these, very large concentrations of chl-a (39.29 μg/L) were recorded in MRS than any other seasons (Meesukko et al. 2007; Radwan 2005). Based on this registered chl-a concentration, reservoir water is productive (eutrophic) because its value is between 10 and 40 μg/L (ADEQ 2008), and 3.0–78.0 μg/L (Wetzel 1983).

Conclusions

Based on the present observation, Selameko manmade reservoir is rich in species diversity and composition and the nutrient status is high enough to support the plankton community. However, the high concentration of PO_4-P indicates the presence of strong anthropogenic pressure. To solve such big pressure, both basin and reservoir management are recommended to solve such acute problems.

References

ADEQ (2008) Narrative nutrient standard implementation procedures for lakes and reservoirs. p 11

Ayalew W, Seyoum M, Vijverberg J, Eshetie D (2007) Seasonal variation in primary production of a large high altitude lakes (Lake Tana, Ethiopia): effects of nutrient availability and water transparency. Springer, Netherlands

Baker JM (1976) Biological monitoring—principles, methods and difficulties. In: Baker JM (ed) Marine ecology and oil pollution. The Institute of Petroleum, Great Britain, pp 41–54

Bennett BL (1998) Land use influences on benthic invertebrate assemblages in southern Appalachian agricultural streams. Virginia Polytechnic Institute and State University Blacksburg, Virginia, pp 38–45

Biddanda B, Benner R (1997) Carbon, nitrogen and carbohydrate fluxes during the production of particulate and dissolved organic matter by marine phytoplankton. Limnol Oceanogr 42:506–518

Blomqvist P, Olsen P (1981) *VAXTPLANKTONKOMPENDIUM*. UPPSALA Publisher, Germany **(German)**

Brettum P, Andersen T (2005) The use of phytoplankton as indicators of water quality. NIVA report SNO 4818-2004

Chapman D (ed) (1996) Water quality assessments—a guide to use of biota, sediments and water in environmental monitoring, 2nd ed. E & FN Spon, an imprint of Chapman & Hall. Great Britain at the University Press, Cambridge

Chattopadhyay C, Banerjee TC (2007) Temporal changes in environmental characteristics and diversity of net phytoplankton in a freshwater lake. Turk J Bot 31:287–296

Daniel V (2001) Phytoplankton. Encyclopedia of life sciences. Macmillan Publishers Ltd, Nature Publishing Group, New York, pp 1–5

Davies OA, Abowei JFN, Otene BB (2009) Seasonal abundance and distribution of plankton of Minichinda Stream, Niger Delta, Nigeria. Am J Sci Res ISSN 1450-223X. Euro-Journals Publishing, Inc. 2:20–30

Goldman CR, Horne AJ (1983) Limnology. McGraw-Hill Inc., New York

Heiskary S, Markus H (2001) Establishing the relationships among nutrient concentration, phytoplankton abundance, and biological oxygen demand in USA, and Minnesota, Rivers. J Lake Reserv Manag 17(4):251–262

Horne AJ, Goldman CR (1994) Understanding lake ecology, 2nd edn. McGraw-Hill, New York

Imhoff KR, Alberrecht D (1975) Oxygen dynamics in the impounded lower Ruh River Programme. Water Technol 7(314):505–518

Khenari AG (2007) Distribution, abundance and biomass of phytoplankton in the southern parts of Caspian Sea. In: Iranian waters aters. Masters thesis, Universiti Sains Malaysia, USM Plankton Research Laboratory

Kobbia IA, Hassan SKM, Shoulkamy MA (1991) Dynamics of phytoplankton succession in the River of Nile at Minia (Upper Egypt): as influenced by agricultural runoff. Department of Botany, Faculty of Science, Cairo and Minia Universities, Egypt. J Islam Acad Sci 4(3):234–241

Korai AL, Sahato GA, Lashari KH, Arbani SN (2008) Biodiversity in Relation to Physicochemical Properties of Keenjhar Lake, Thatta District, Sindh, Pakistan. Turk J Fish Aquat Sci 8:259–268

Lind OT (1979) Handbook of common methods in limnology. The CV Mosby Company, St. Louis

Little JCL (2004) Influences of Artificial Desertification on Limnology Processes in Lake Samsonvale (North Pine Dam) Greenland, Australia. Griffin university Australian school of Environmental Studies Faculty of Environmental Science, Greenland

Ltd Palintest (1989) Photometer systems for water analysis: Wagtech international, Palintest transmittance display photometer 5000. Palintest Ltd, UK

Meesukko C, Gajaseni C, Peeraponpisal Y, Voinova A (2007) Relationships between seasonal variation and phytoplankton dynamics in Kaeng Krachan reservoir. Phetchaburi Province, Thailand. Nat Hist J Chulalongkorn Univ 7(2):131–143

Mohamed AS, Thirumaran G, Arumugam R, Kannan RR, Anantharaman P (2009) Studies of phytoplankton diversity from Agnitheertham and Kothandaramar Koil Coastal Waters, Southeast Coast of India. Aristotle University of Thessaloniki, School of Biology, Laboratory of Ichthyology, GR-54124,Thessaloniki, Greece; Ul. Politechniczna, PL-32020 Wieliczka, Poland. Biologia, Bratislava. J Environ Res. 59(2):119–124

Mustapha MK, Omotosho JS (2005) An assessment the physicochemical properties of Moro Lake, Kwara State Nigeria. Department of Natural Sciences, university of Ilorin, Ilorin Nigeria. Afr J Appl Zool Environ Biol 7:73–77

Onyema IC (2007) The Phytoplankton composition, abundance and temporal variation of a polluted estuarine creek in Lagos, Nigeria. Department of Marine Sciences, University of Lagos, Akoka, Lagos Nigeria. Turk J Fish Aquat Sci 7:89–96

Oso JA, Fagbuaro O (2008) An assessment of the physicochemical properties of a tropical reservoir, Southwestern Nigeria. Department of Zoology, University of AdoEkiti, Ado Ekiti Nigeria, ISSN: 1871-3381. J Fish Int 3(2):42–45

Osondu CA (2008) Phytoplankton and nutrient dynamics of a tropical estuarine system, Imo River Estuary, Nigeria. Department of Biology, Bahir Dar University, Bahir Dar, Ethiopia. Int MultiDiscip J Index Afr J 2(2):253–264

Radwan AM (2005) Some factors affecting the primary production of phytoplankton in Lake Burullus. Egypt J Aqua Res 31(2):72–88

Rafique RM, Mahboob S, Ahmad M Saleem S (2002) Seasonal limnological variations in Mangla Reservoir at Sukhian, Mirpur (Azad Kashmir). Int J Agric Biol 1560–8530/2002/04–2–223–226. Faisalabad, Pakistan

Reynolds C, Huszar V, Kruk C, Naselli-Flores L, Melo S (2002) Towards a functional classification of the freshwater phytoplankton. J Plankton Res 24:417–428

SAS (Statistical Analysis System) 2003. Statistical Analysis System. User's Guide SAS/STA-t version, 8th edition, SAS, Institute, Inc. Cary

Schmirz RJ (1996) Introduction to water pollution biology. Gulf Publishing Company, Houston

Schultz GA (1991) Hydrology of Manmade Lakes. Hydrology of manmade lakes (Proceedings of the Vienna symposium) LAHS Publ. No. 206. Institute of hydrology, water resources and environmental techniques, Ruhr University Bochum, Germany

SPSS (2007) SPSS survival manual: a step by step guide to data analysis using SPSS for Windows. 3rd edn. Version 16.0; SPSS Inc., Chicago

Stanley CD, Clarke RA, McNeal BL, Macleod BW (2003) Impact of agricultural land use on nitrate levels in Lake Manatee, Florida. Soil and Water Science Department, Florida Cooperative Extension Service, IFAS, University of Florida. pp 1–9

Tamiru A (2006) Groundwater occurrence in Ethiopia. UNESCO. Addis Ababa Universities, Ethiopia

USEPA (2008) Nutrient criteria technical guidance manual wetlands. EPA document: EPA-822-B-08-001

USEPA (United State Environmental Protection Agency) (2000) Limnology, water quality parameters, conditions, and ecoregions. pp 1–3

Wetzel RG (1983) Limnology, 2nd edn. W.B. Saunders Company, Philadelphia

Wetzel RG, Likens GE (2000) Limnological analyses, 3rd edn. Springer, New Delhi

Yajurvedi HN (2008) A study of growth on co-efficient and relative condition of factor of the major carp (Catla catla) in two lakes differing in water quality. Department of Zoology, University of Mysore, Mysore

Yamaguchi E, Gould A (2007) Phytoplankton identification guide. The University of Georgia Marine Education Center and Aquarium, Georgia

17

Morphometric evaluation of Swarnrekha watershed, Madhya Pradesh, India: an integrated GIS-based approach

Abhishek Banerjee[1] · Prafull Singh[1] · Kamleshwar Pratap[2]

Abstract The quantitative analysis of the watershedis vital to understand the hydrological setup of any terrain. The present study deals with quantitative evaluation of Swarnrekha Watershed, Madhya Pradesh, India based on IRS satellite data and SRTM DEM. Morphometric parameters of the watershed were evaluated by computations of linear and areal aspect using standard methodology in GIS environment. ARC GIS software was utilized for morphometric component analysis and delineation of the watershed using SRTM digital elevation model (DEM). The watershed is drained by a fifth-order river and shown a dendritic drainage pattern, which is a sign of the homogeneity in texture and lack of structural control. The drainage density in the area has been found to be low which indicates that the area possesses highly permeable soils and low relief. The bifurcation ratio varies from 3.00 to 5.60 and elongation ratio is 0.518 which reveals that the basin belongs to the elongated shape basin and has the potential for water management. The main objective of the paper is to extract the morphometric parameters of the watershed and their relevance in water resource evaluation management. The results observed from this work would be useful in categorization of watershed for future water management and selection recharge structure in the area.

Keywords Morphometric analysis · GIS · Swarnrekha watershed · Madhya Pradesh · India

✉ Prafull Singh
Pks.jiwaji@gmail.com; psingh17@amity.edu

1 Amity Institute of Geo-Informatics and Remote Sensing, Amity University, Sector 125, Noida 201303, India

2 WAPCOS, Ministry of Water Resources, Government of India, New Delhi 110001, India

Introduction

One of the most concerning issues currently being faced by the society is the increase in world's population and its impact on the availability of fresh water. Presently high inhabitant's expansion, fast urbanization and climate change along with the irregular frequency and intensity of rainfall make difficulty for appropriate water management and storage plans. Therefore, there is an urgent need for the evaluation of water resources because they play a primary role in the sustainability of livelihood and regional economics throughout the world (Singh et al. 2011, 2013). Recent studies show an increase in urban sprawl and population growth in the developing countries like India which leads to enormous stress on the water resources of the country. This excess need puts a strain on the available surface and groundwater supplies, which might result in an acute water shortage.

The drainage morphometric analyses are prerequisite for selñection of water recharge site, watershed modeling, runoff modeling, watershed delineation, groundwater prospect mapping and geotechnical investigation (Magesh et al. 2011; Thomas et al. 2012). Quantitative morphometric analysis of watershed can provide information about the hydrological nature of the rocks exposed within the watershed. A drainage map of basin provides a reliable index of permeability of rocks and their relationship between rock type, structures and their hydrological status. Watershed characterization and management require detailed information for topography, drainage network, water divide, channel length, geomorphologic and geological setup of the area for proper watershed management and implementation plan for water conservation measures (Sreedevi et al. 2013).

The over exploration of groundwater and variation in rainfall are the major concerns for depletion of groundwater resource and their degradation in terms drying of potential aquifer in the area (Singh et al. 2013, Ellis and Revitt 2010; Rao 2008; Eyquem 2007).

Morphometric analysis for evaluation of drainage basin, their characterization and classification has been performed from the early nineties by the pioneer workers for linear, aerial and relief aspect and interpretation of their interrelationship with reference to land and water management. (Horton 1945; Smith 1950; Strahler 1957). Knowledge of satellite data and relative GIS methods has also been used along with traditional methods of drainage basin analyses. Information regarding the hydrological nature of the exposed rocks in a drainage basin can be extracted by quantitative morphometric analysis of drainage basin (Subramanyan 1974; Rao and Babu 1995; Pakhmode et al. 2003; Sreedevi et al. 2005).

Further advancement in computer-based technology precise work has been accomplished using advanced techniques like geographical information system and related tools for automated determination of the basin parameters, which results in time-efficient, accurate and reliable results (Grohmann 2004; Gangalakunta 2004; Grohmann et al. 2007; Hlaing et al. 2008; Javed et al. 2009). Remote sensing and GIS-based drainage basin evaluation has been carried out by number of researchers for different terrains and it is proved to be a very scientific tool for generation of precise and updated information for characterization of drainage basin parameters (Thomas et al. 2012; Pankaj and Kumar 2009; Singh and Singh 2009; Das and Mukherjee 2005).

Digital elevation models (DEMs), such as from the Shuttle Radar Topography Mission (SRTM), or the ASTER GDEM product (USGS, Denver, Colorado, USA), have been used to extract different geomorphological parameters of drainage basins, including drainage networks, catchment divides, slope gradient and aspect, and upstream flow contributing areas (e.g. Mark 1984, Tarboton 1997). GIS-based watershed evaluation using Shuttle Radar Topographic Mission (SRTM) data has given a precise, fast, and an inexpensive way for analyzing hydrological systems (Farr and Kobrick 2000; Grohmann et al. 2007). Recently remote sensing data and digital elevation models were utilized to extract the catchment hydrological parameters for management of water resources and they concluded that DEMs are very accurate tool for morphometric parameter evaluation and watershed delineation for watershed management (El Bastawesy et al. 2013).

Very recently number of case studies based on remote sensing, GIS and satellite-borne digital elevation data were used for morphometric analysis and watershed prioritization for different geological and hydrological setups for water resource evaluation and management (Singh et al. 2014; Gajbhiye et al. 2014; Kaliraj et al. 2014; Rai et al. 2014). The findings of all the studies show that integrated application of remote sensing and GIS is the potential tool for water resource evaluation and extraction of hydrological parameters of any terrain.

In the present study, an integrated approach of GIS, remote sensing and DEM has been used for generation and interpretation of drainage parameters for water resource management of Swarnrekha Watershed. The watershed selected for the present work is the important source of water for agriculture and domestics purpose in the area.

Study area

The Swarnrekha watershed is the major part of urban and suburban area of Gwalior and Morena district and lies between 26°06′N–26°26′N latitude and 78°01′E–78°13′E longitude. The elevation ranges from 160 to 350 m average mean sea level (AMSL) and the study area is located in the northeastern part of the Gwalior district in the central India with an area about 292 km^2 (Fig. 1). It is covered in the Survey of India (SOI) toposheet numbers 54J/3 and J/4 on a 1:50,000 scale. The climate of the basin is extreme with hot summers and cold winters. It experiences southwestern monsoon rains in June–September with an annual rainfall of 600 mm. May is the hottest month with a maximum temperature of 45 °C and minimum temperature 4 °C in the month of December. Gwalior is one of the erstwhile princely cities with population of over 1.2 million having semi-arid climate marked by extreme temperatures and erratic rainfall patterns. The overall climate of the region is dry, except during the south-west monsoon season (June–October). Groundwater is the major source of water supply for drinking, agricultural, and other activities in the area and groundwater level depletion over a time is the major concern. The incessant water scarcity and the ever increasing demand for agriculture and industries have put immense stress on the limited groundwater resources in the area and have pronounced the need for locating additional sources. The river Swarnrekha, which is the tributary of Yamuna River Basin flowing through Gwalior urban and suburban area, is the major source of water supply in the area.

Hydrogeological and geological setup of the watershed

The Gwalior basin is located on the northwestern fringe of the Bundelkhand massif and is intracratonic in nature, and the complete lithological unit lies on the Bundelkhand granite which comprises basal arenaceous Par formation which is overlaid by a sequence of volcanic sediments of the Morar formation. The Gwalior group (which dips in due north) shows continuity and similarity with the Bijawar group (Singh et al. 2011). The basic structure is quite

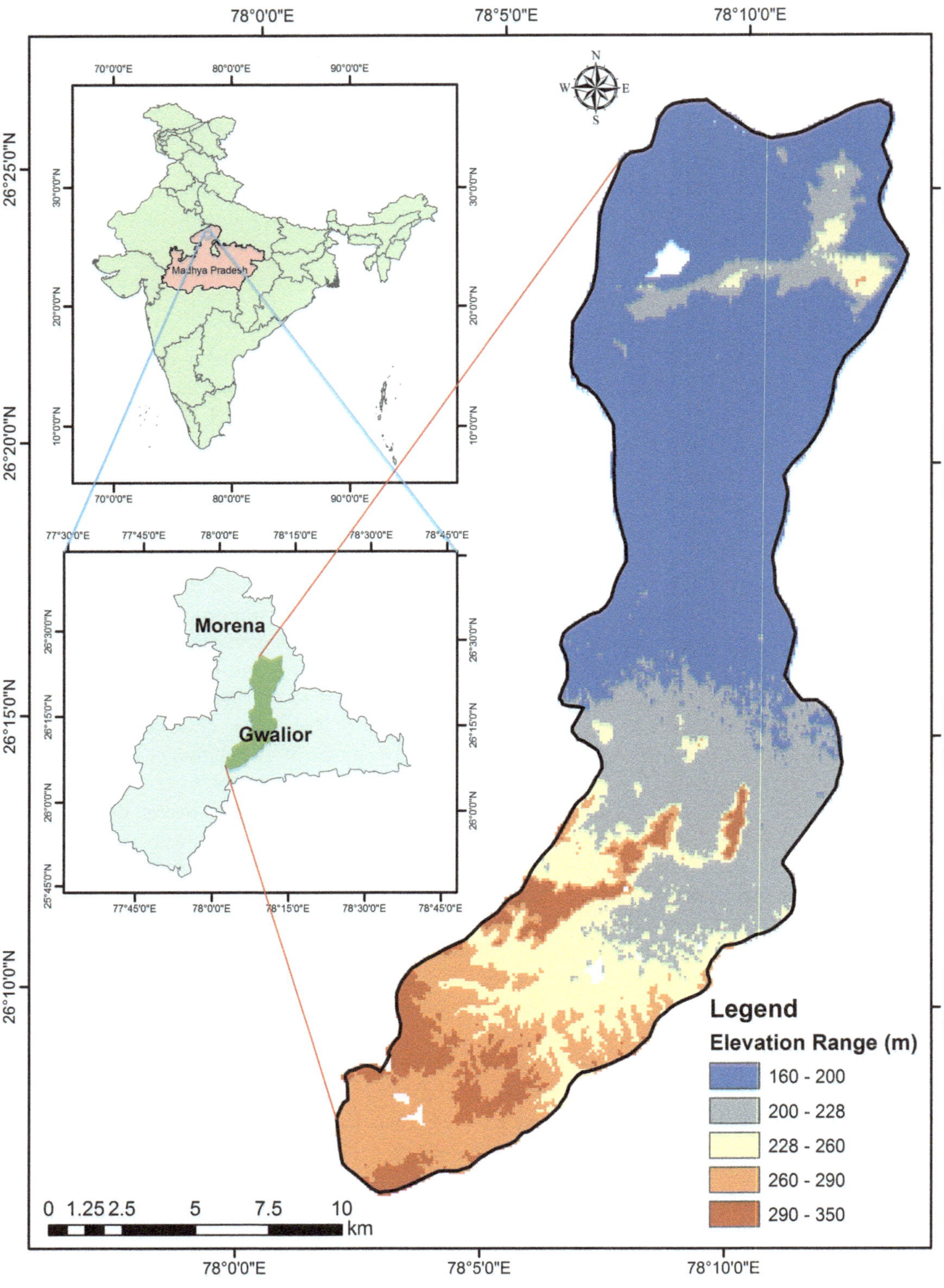

Fig. 1 Location map showing elevation range of the study area

similar as the base is made up of quartzite; the shallow water conditions are characteristics of ferruginous shale with chert and jasper; and transgression follows after the rapid development of the basin formation. In few parts of the basin, the Gwalior group shows fine or coarse doleritic sills and dykes. The hard rocks of the region comprise sandstone and shale as they belong to the Archaean group to Precambrian group; however, the doleritic dykes are

Table 1 The data used for the present study

Type of data	Details	Sources
SRTM digital satellite images	Path 145, row 042, year 2000	University of Maryland, USA
Toposheets	54J/3, 54J/4, scale 1:50,000	Survey of India, Dehradun, India
Geological quadrangle map	Scale 1:250,000	Geological Survey of India, Hyderabad, India

recent alluvium deposits where granite is the base rock of the study area. By the regional geological map of the area which has been prepared based on the satellite images and geological maps, the Gwalior basin shows much of the Morar formation shale with lower region of Par formation of sandstone. Surrounding the region is the Kaimur group sandstone and Bundelkhand granite. The chief groundwater deposits in the region are formed by the weathered shale zones, alluvial deposits and sandstone forming a coarse- or medium-grained unconsolidated aggregation. The thickness of aquifer formation increases to the northern portion of the watershed with three major water bearing formation at 100 m depth. The fracturing, openness, size of the fracture and the interconnection nature between the fractures dominate the water yielding capacity of the rocks. The Swarnrekha watershed is mostly covered with alluvium which varies in thickness and constitute the major aquife system of the area. The intracratonic Gwalior basin is present on the northwestern fringes of granitic Bundelkhand massif and Gwalior litho-units group rest over Bundelkhand granite. These comprise basal arenaceous Par formation overlain by volcano-sedimentary sequences of Morar formation consisting of ferruginous shale with bands of chert, jasper and limestone. The predominant rocks comprise sandstone, shale, quartzite, doleritic dykes and alluvium. The thickness of this weathered zone is generally greater in the northern portion. The Gwalior Group of rocks overlies the Vindhyans group and is the youngest rock in the region.

Database generation and detail methodology

The quantitative analysis of drainage basin provides background information about the hydrological conditions and nature of rock formation exposed within the watershed. Morphometric analysis of a basin provides an indication about permeability and storage capacity of the rocks, and gives an indication of the yield of the basin. In the present paper, satellite-borne SRTM DEM and Survey of India topographical sheets were utilized for generation of database and extraction of various drainage parameters. The watershed boundary of the study area was automatically delineated using SRTM data which are freely available on the website (http://glcf.umd.edu). Details of data used are shown in Table 1.

The morphometric parameters such as number of stream segments, stream order, drainage pattern, subbasin length, perimeter and area were delineated from the SRTM DEM (90 m) data and verified with Survey of India topographical map (1:50,000). The observed parameters were used for computation the morphometric characteristics such as drainage density, stream frequency, bifurcation ratio, form factor, circulation ratio, relief ratio and overland flow using mathematical equations (Table 2) in the GIS environment. Moreover, the slope and relief of the area were extracted from the same data source using spatial analyst tool in ArcGIS 10 software. The following procedure was followed for morphometric analysis of Swarnrekha watershed.

1. Firstly, SRTM DEM was downloaded from USGS site and primary processing was performed.
2. The Geocoded SOI toposheets were correlated with SRTM data for delineation of watershed.
3. Watershed area of Swarnrekha is delineated from SRTM data along with the SOI toposheets of the study area using Erdas Imagine Software.
4. All the extracted parameters from SRTM and Survey of India topographical sheets, like the number and lengths of streams of each different order, drainage area, basin perimeter and total basin length, and width were calculated using ARC GIS software; drainage density, drainage frequency, shape, form factor, circulatory ratio, and elongation ratio, were calculated from these parameters. The methodologies adopted for the computation of morphometric parameters are given in Table 2 and results are presented in Table 3.

Results and discussion

Quantitative evaluations of drainage basin are the primary-level water resource management at watershed level in hard and soft rock areas. Quantitative description of drainage network and basin characteristics has been carried out for the Swarnarekha Watershed. Hydrological parameters such as number and lengths of streams of different order, drainage area, basin perimeter and maximum basin lengths were calculated after cleaning and then topology building of the drainage layer in GIS software; from these

Table 2 The parameters adopted for morphometric parameters computation

S. no.	Morphometric parameters	Calculational Formulae	References
1.	Stream order	Hierarchical rank	Strahler (1964)
2.	Stream length	The length of the stream	Horton (1945)
3.	Mean stream length (L_{sm})	$L_{sm} = L_u/N_u$; L_u = total stream length of order u, N_u = total no. of stream segments of order u	Strahler (1964)
4.	Stream length ratio (R_L)	$R_L = L_u/(L_u - 1)$; L_u = total stream length of order u, $(L_u - 1)$ = total stream length of the next lower order	Horton (1945)
5.	Stream frequency (F_s)	$F_s = N_u/A$; N_u = total no. of streams of all orders, A = area of the basin	Horton (1932)
6.	Drainage density (D)	$D = L_u/A$; L_u = total stream length of all orders, A = area of basin	Horton (1932)
7.	Drainage texture (R_t)	$R_t = N_u/P$; N_u = total no. of streams of all orders, P = perimeter of the basin	Horton (1945)
8.	Bifurcation ratio (R_b)	$R_b = N_u/(N_u + 1)$; N_u = total no. of stream segments of order u, $(N_u + 1)$ = number of segments of next higher order	Schumn (1956)
9.	Form factor (R_f)	$R_f = A/L_b^2$; A = area of basin, L_b^2 = square of the basin length	Horton (1932)
10.	Elongation ratio (R_e)	$R_e = 2\sqrt{(A/\pi)}/L_b$; A = area of the basin, π = 3.14, L_b = basin length	Schumn (1956)
11.	Circularity ratio (R_c)	$R_c = 4\pi(A/P^2)$; A = area of the basin, π = 3.14, P = perimeter of the basin	Miller (1953)
12.	Relief ratio (R_h)	$R_h = H/L_b$; H = total relief of the basin, L_b = basin length	Schumn (1956)
13.	Length of overland flow (L_g)	$L_g = 1/(2D)$; D = drainage density of the basin	Horton (1945)
14.	Relative relief (R_r)	$R_r = 100H/P$; H = total relief of the basin, P = perimeter of the basin	Schumn (1956)

Table 3 Linear aspect of the watershed

Stream order	No. of streams	Bifurcation ratio	Mean bifurcation ratio	Total length of streams (km)	Mean length of the streams (km)	Average mean length of the streams (km)	Length ratio
1st	460		4.69	315.93	0.69	0.95	
2nd	126	3.65		151.41	1.20		0.479
3rd	28	4.5		76.29	2.72		0.504
4th	5	5.6		41.12	8.22		0.539
5th	1	5		5.55	5.55		0.135

parameters, various drainage characteristics such as bifurcation ratio (Rb), drainage density (Dd), stream frequency (Fs), circulatory ratio (Rc), and elongation ratios (Re) are calculated for evaluation of watershed. All the important watershed parameters were discussed under following important sections: linear, aerial and relief aspects. Linear aspects of the watershed refer to the analysis of stream order, stream number, bifurcation ratio and stream length ratio. Stream ordering is the first step of quantitative analysis of the drainage basin. In the present study, stream ordering has been carried out using Strahler's system of stream calculation under GIS environment. The adopted mathematical equations for computation of drainage parameters have been summarized in Table 2.

The watershed contains a dendritic drainage pattern which indicates homogenous subsurface strata of the study area. The stream ordering has been ranked based on a method proposed (Strahler 1964) from the streams extracted from SRTM DEM and updated from Survey of India toposheets. The order-wise stream numbers and their linear characteristics are shown in Table 3. The drainage pattern analysis of the watershed indicated that the area is

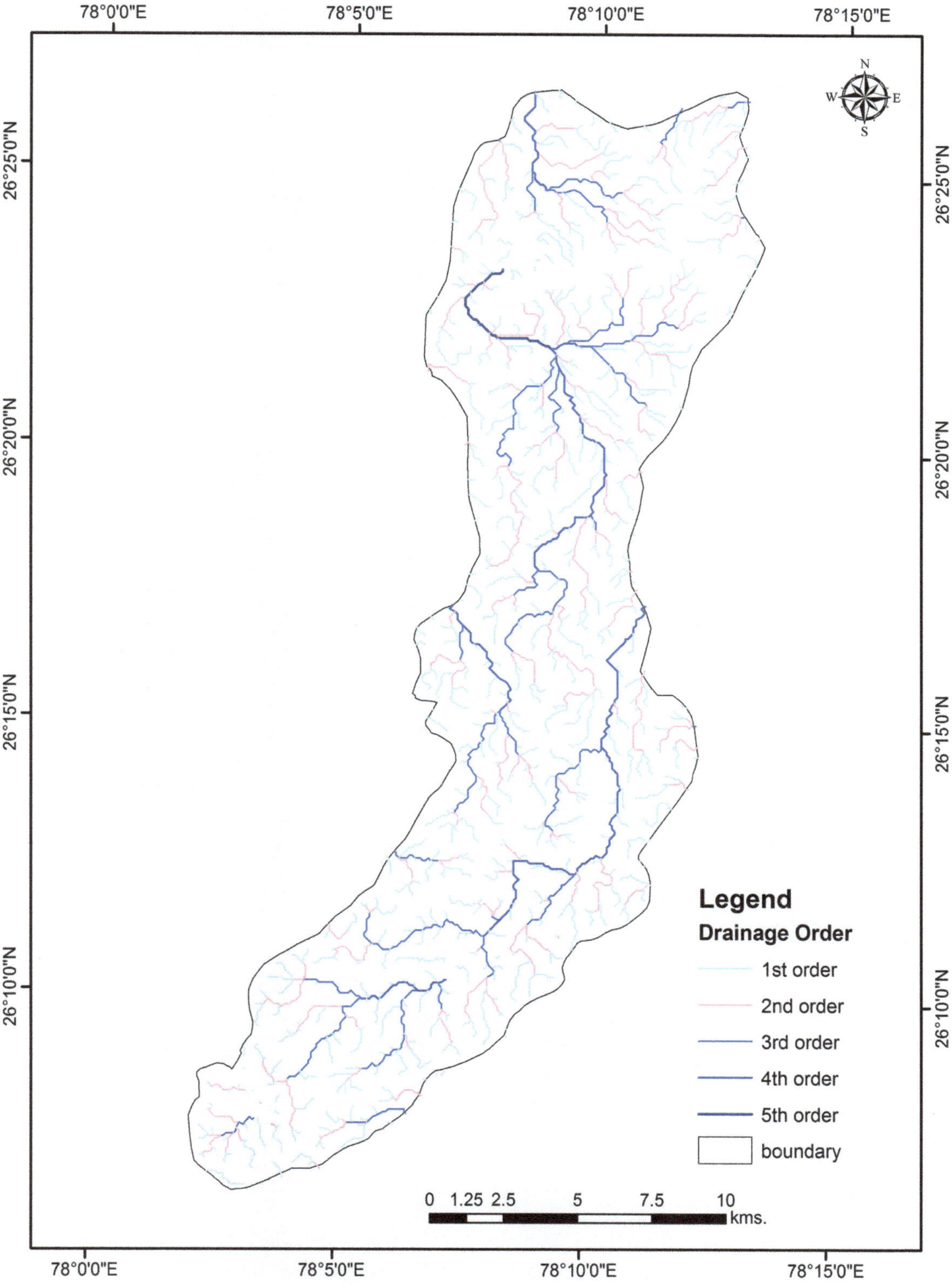

Fig. 2 The drainage stream ordering of Swarnrekha river watershed

having a lake of structural tectonic control. Maximum number of stream was found in the first order and as the stream order increases with a decrease in stream number. The drainage map of watershed with stream order is shown in Fig. 2. Hydrologically dendritic drainage pattern is a symbolic of homogeneity in texture and lack of structural control. The number of streams in each order gives a fair idea of the distribution of the water channels of the smallest

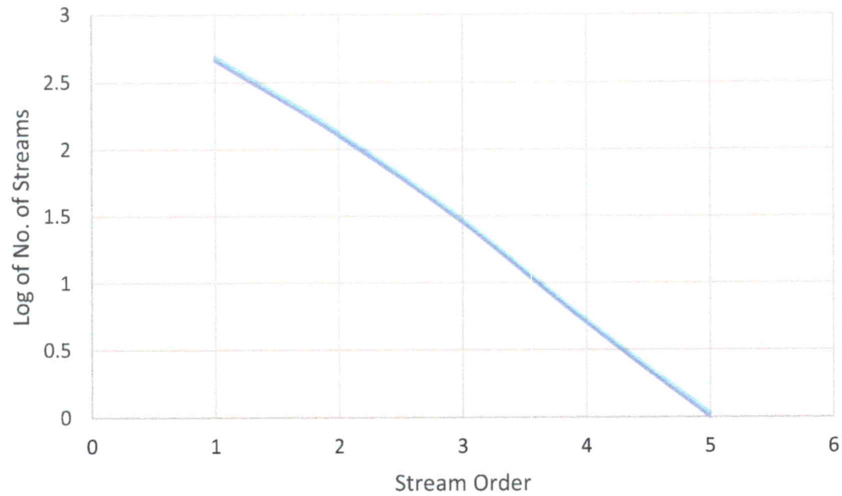

Fig. 3 Plot of inverse linear relationship between drainage networks

Table 4 Areal aspects of the watershed

Basin area (km^2)	Perimeter (km)	Width (km)	Form factor	Elongation ratio	Circularity ratio	Drainage density	Stream frequency	Drainage texture	Relief ratio
292.49	103.36	8.16	0.211	0.518	0.344	2.02	2.12	5.99	0.005
Texture ratio	Length of overland flow								
4.45	0.25								

to largest level in the region and how well the area is covered in the water flow network. These stream segments were plotted against stream order and show an inverse geometric sequence; mostly the drainage networks show an almost linear relationship (Fig. 3).

The relationship between the bifurcation ratio and the stream length ratio is determined by hydrogeologic, physiographic, and geological characteristics. Value of total length, mean length and length ratio of different stream orders of the watershed are shown in Table 3. The mean bifurcation ratio of watershed is 4.69, indicating that the drainage pattern is not much affected by tectonic and structural disturbances. Stream frequency (Fs) or channel frequency is the total number of stream segments of all orders per unit area and values indicate positive correlation with the drainage density of the basin, suggesting an increase in stream population with respect to increase in drainage density (Horton, 1932). The observed stream frequency (Fs) is 2.12 for the basin and exhibit a positive correlation with the drainage density value of the area, indicating the increase in stream population with respect to increase in drainage density (Table 3). Horton (1932) has introduced the drainage density as an expression to indicate the closeness of spacing of channels. The drainage density of the watershed is 2.02 (Table 4). The value of Dd suggested that area have low drainage density in the northern portion of the watershed. The low drainage density is likely to result in the area of highly resistant permeable subsoil material under dense vegetation and low relief. High drainage density is the resultant of weak impermeable subsurface material, sparse vegetation and mountainous relief. Multi-spectral satellite images of the study area also suggest that the area has good vegetation in terms of shrub's and agricultural lands in addition to reserve forests and dense forests.

Drainage texture (Rt) is one of the important concepts of geomorphology which means the relative spacing of drainage lines. According to Horton (1945), Rt is the total number of stream segments of all orders per perimeter of that area. The observed Rt is 5.99 which indicates the moderate drainage texture. The moderate drainage texture and lower value of drainage density (2.02) indicate the presence of highly resistant permeable material with low relief (Table 4). Low drainage density and moderately drainage texture indicate the presence of permeable soil strata within the watershed. Texture ratio depends on the underlying lithology of the area, the infiltration capacity and relief aspect of the watershed terrain. The present area has a texture ratio of 4.45 and hence shows a moderate nature. Form factor (Horton 1932) is a ratio of the basin area to square of the basin length. The area, perimeter and length of the watershed are basic measurements which are useful in generating dimensionless ratios for indexing the watershed shape. The R_f value for the watershed is 0.211

which suggests its elongated shape. A watershed with low form factor and an elongated shape denotes its flatter peak of flow for long period. A flood flow for such an elongated shape will be easier to manage in due course of time (Table 4). Elongation ratio is defined as the ratio between the diameter of the circle of the similar area as of the drainage basin with a maximum length equal to that of the basin (Schumn 1956). Depending on the climatic and geological conditions of the area, the value of the R_e varies considerably between 0.6 and 1.0, those with value close to the highest, i.e. 1.0 are categorized as typical low relief and with value of 0.6 and nearby are associated with high relief and steep slope (Strahler 1964). The values are grouped in all as three categories: circular (with values >0.9), oval ($0.8 \leq$ value ≤ 0.9) and elongated (value <0.8). The elongation ratio of the current study area is 0.518 which places it in the category of elongated basin (Table 4). Miller (1953) defined a dimensionless circularity ratio (Rc) as the ratio of the basin area to the area of circle having the same perimeter as the basin. Rc was influenced by the length and frequency of streams, geological structures, land use/land cover, climate, relief and slope of the basin. Circularity ratios range 0.4 to 0.5 which indicate strongly elongated and highly permeable homogenous geology, and the observed circularity ratio of the basin is 0.344 (Table 3) which indicates that the basin is elongated in shape, low discharge of runoff and highly permeability of the subsoil condition.

The elevation difference between the highest and the lowest points on the valley floor of a basin is known as the total relief of that basin. The relief ratio (Rh) is termed as the ratio of maximum relief to horizontal distance along the longest dimension of the basin parallel to the principal drainage line (Schumn 1956). It measures the overall steepness of a drainage basin and is an indicator of the intensity of the erosion process operation on the slope of the basin. In the present study, the Rh value of the basin 0.005 shows the major portion of the basin having gentle slope (Table 4). As per Horton (1945), the length of overland flow is half of reciprocal of the drainage density. It is defined as the water length above ground before concentrating into definite stream channels. The present study has L_g equal to 0.25 indicating a very young topography.

Conclusions

GIS-based morphometric analysis using remote sensing and GIS quantitative morphometric have been proved as an efficient tool for hydrological evaluation and drainage network analysis of watershed. The computation of linear, areal and relief parameters of the Swarnrekha watershed confirms that there is a positive correlation between hydrological behavior of the basin and their associated landforms which is the potential for water management activities. The average slope of the watershed is towards south to north which justifies the surface and subsurface water movement within the watershed. The drainage network of the area is mainly dendritic type indicating homogeneity in texture and lack of structural control which could thereby help in understanding the various terrain parameters viz. bedrock nature, infiltration capacity, runoff, drainage density, etc. A low stream frequency and drainage density indicate a high rate of permeability of the subsurface formation. The morphological characteristics and hydrological parameters of Swarnrekha watershed, such as bifurcation ratio, circularity ratioand relief ratio, have special significance in deciding the overall pattern and changes in the subbasin, shape and drainage pattern of the area. The results observed from the analysis could be helpful in prioritizing watershed with respect to soil and water conservation and management. Thus, concluding remote sensing information and GIS tools together form a powerful, effective and efficient method to study the hydrologic and morphometric characteristics of a drainage basin. This work shall prove useful and beneficial to the planners and decision makers for proper natural resource management even at micro-level of the Swarnrekha watershed.

Acknowledgement The authors expresses his gratefulness to the Vice Chancellor and Director Amity Institute of Geoinformatics and Remote Sensing, Amity University, Noida for constant encouragement. Thanks are also due to the anonymous reviewers for their many helpful suggestions.

References

Das AK, Mukherjee S (2005) Drainage morphometry using satellite data and GIS in Raigad district, Maharashtra. J GeolSoc India 65:577–586

ElBastawesy M, White KH, Gabr S (2013) Hydrology and geomorphology of the Upper White Nile Lakes and their relevance for water resources management in the Nile basin. Hydrol Process 27:196–205

Ellis JB, Revitt D (2010) The management of urban surface water drainage in England and Wales. Water Environ J 24:1–8

Eyquem J (2007) Using fluvial geomorphology to inform integrated river basin management. Water Environ J 21:54–60

Farr TG, Kobrick M (2000) Shuttle radar topography mission produces a wealth of data. Am Geophys Union Eos 81:583–585

Gajbhiye S, Mishra SK, Pandey A (2014) Prioritizing erosion prone area through morphometric analysis: an RS and GIS perspective. Appl Water Sci Springer 4:51–61

Gangalakunta P (2004) Drainage morphometry and its influence on landform characteristics in a basaltic terrain, central India: a remote sensing and GIS approach. Int J Appl Earth ObsGeoinf 6:1–16

Grohmann CH (2004) Morphometric analysis in geographic information systems: applications of free softwares. ComputGeosci 30:1055–1067

Grohmann CH, Riccomini C, Alves FM (2007) SRTM based morphotectonic analysis of the Pocos De Caldas alkaline massif southeastern Brazil. ComputGoesci 33:10–19

Hlaing TK, Haruyama S, Aye MM (2008) Using GIS based distributed soil loss modelling and morphometric analysis to prioritize watershed for soil conservation in Bago river basin of lower Myanmar. Front Earth Sci China 2:465–478

Horton RE (1932) Drainage basin characteristics. Trans AmerGeophys Union 13:350–361

Horton RE (1945) Erosional development of streams and their drainage basins; hydro physical approach to quantitative morphology. Geol Soc Am Bull 56(3):275–370

Javed A, Khanday MY, Ahmed R (2009) Prioritization of sub watershed based on morphometric and land use analysis using remote sensing and GIS techniques. J Indian Soc Remote Sens 37:261–274

Kaliraj S, Chandrasekar N, Magesh NS (2014) Morphometric analysis of the river Thamirabarani sub basin in Kanyakumari district, south west coast of Tamil Nadu, India, using remote sensing and GIS. Environ Earth Sci Springer. doi:10.1007/s12665-014-3914-1

Magesh N, Chandrasekar N, Soundranayagam J (2011) Morphometric evaluation of Papanasam and Manimuthar watersheds, parts of Western Ghats, Tirunelveli district, Tamil Nadu, India: a GIS approach. Environ Earth Sci 64(2):373–381

Mark DM (1984) Automatic detection of drainage networks from digital elevation models. Cartographica 21:168–178

Miller VC (1953) A quantitative geomorphologic study of drainage basin characteristics in the Clinch mountain area, Virginia and Tennessee, project NR 389042, technical report 3, Columbia University, Department of Geology, ONR, Geography branch, New York

Pakhmode V, Himanshu K, Deolankar SB (2003) Hydrological drainage analysis in watershed programme planning: a case from the Deccan basalt, India. Hydrogeol J 11:595–604

Pankaj A, Kumar P (2009) GIS based morphometric analysis of five major sub watershed of Song river, Dehradun district, Uttarakhand with special reference to landslide incidences. J Indian Soc Remote Sens 37:157–166

Rai PK, Mohan K, Mishra S, Ahmad A, Mishra V (2014) A GIS based approach in drainage morphometric analysis of Kanhar River basin, India. Appl Water Sci Springer. doi:10.1007/s13201-014-0238-y

Rao NS (2008) A numerical scheme for groundwater development in a watershed basin of basement terrain: a case study from India. Hydrogeol J 17:379–396

Rao JU, Babu VRRM (1995) A quantitative morphometric analysis of Gundalakamma river basin, Andhra Pradesh. Indian J Earth Sci 22:63–74

Schumn SA (1956) Evaluation of drainage system and slopes in Badlands at Perth Amboy, New Jersy. Natl Geol Soc Am Bull 67:597–646

Singh PK, Singh UC (2009) Water resource evaluation and management for Morar river basin, Gwalior district, Madhya Pradesh, using GIS. E-J Earth Sci India 2:174–186

Singh P, Kumar S, Singh U (2011) Groundwater resource evaluation in the Gwalior area, India, using satellite data: an integrated geomorphological and geophysical approach. Hydrogeol J 19:1421–1429

Singh P, Thakur JK, Singh UC (2013) Morphometric analysis of Morar river basin, Madhya Pradesh, India, using remote sensing and GIS techniques. Environ Earth Sci Springer 68:1967–1977

Singh P, Gupta A, Singh M (2014) Hydrological inferences from watershed analysis for water resource management using remote sensing and GIS techniques. Egypt J Remote Sens Space Sci 17:111–121

Smith KG (1950) Standard for grading textures of erosional topography. Am J Sci 248(9):655–668

Sreedevi PD, Subrahmanyam K, Ahmed S (2005) The significance of morphometric analysis for obtaining groundwater potential zones in a structurally controlled terrain. Environ Geol 47:412–420

Sreedevi PD, Sreekanth PD, Khan HH, Ahmed S (2013) Drainage morphometry and its influence on hydrology in an semi-arid region: using SRTM data and GIS. Environ Earth Sci 70(2):839–848

Strahler AN (1957) Watershed geomorphology. Trans Am Geophys Union 38(6):913–920

Strahler AN (1964) Quantitative geomorphology of drainage basins and channel networks. In: Chow VT (ed) Handbook of applied hydrogeology. McGraw Hill, New York, pp 4–76

Subramanyan V (1974) A quantitative analysis of two drainage basins around Sagar, M.P. Indian Sci Acad 40:76–99

Tarboton DG (1997) A new method for the determination of flow directions and contributing areas in grid digital elevation models. Water Resour Res 33:309–319

Thomas J, Joseph S, Thrivikramji K, Abe G, Kannan N (2012) Morphometrical analysis of two tropical mountain river basins of contrasting environmental settings, the southern Western Ghats, India. Environ Earth Sci. 66(8):2353–2366

18

Assessment of Groundwater quality in Krishnagiri and Vellore Districts in Tamil Nadu, India

A. Shanmugasundharam[1] · G. Kalpana[1] · S. R. Mahapatra[1] · E. R. Sudharson[1] · M. Jayaprakash[1]

Abstract Groundwater quality is important as it is the main factor determining its suitability for drinking, domestic, agricultural and industrial purposes. The suitability of groundwater for drinking and irrigation has been assessed in north and eastern part of Krishnagiri district, South-western part of Vellore district and contiguous with Andhra Pradesh states, India. A total of 31 groundwater samples were collected in the study area. The groundwater quality assessment has been carried out by evaluating the physicochemical parameters such as pH, EC, TDS, HCO_3^-, Cl^-, SO_4^{2-}, Ca^{2+}, Mg^{2+}, Na^+ and K^+. The dominant cations are in the order of $Na^+ > K^+ > Ca^{2+} > Mg^{2+}$ while the dominant anions have the trends of $Cl^- > HCO_3^- > SO_4^{2-} > CO_3$. The quality of the water is evaluated using Wilcox diagram and the results reveals that most of the samples are found to be suitable for irrigation. Based on these parameters, groundwater has been assessed in favor of its suitability for drinking and irrigation purpose.

Keywords Groundwater · Major ions · Quality assessment

Introduction

Groundwater, being the largest freshwater resource after glaciers and polar ice in the world, plays an important role in socio-economic life of the people in Tamil Nadu for domestic, horticultural, agricultural, and hydropower generation purposes. The suitability of groundwater for different purposes depends upon its intrinsic quality which reflects inputs from the atmosphere, soil and rock weathering, as well as from anthropogenic activities. Public ignorance of environment and related considerations, indiscriminate disposal of increasing anthropogenic wastes, unplanned application of agrochemicals, and discharges of improperly treated sewage have resulted in the deterioration of surface and subsurface water (Singh and Hasnain 1998; Mitra et al. 2007; Kumar et al. 2008; Ishaku 2011; Ewusi et al. 2013; Kalpana and Elango 2013). The value of groundwater lies not only in its widespread occurrence and availability but also in its consistent good quality (Rajmohan et al. 2000; UNESCO 2000).

It has been estimated that once pollution enters the subsurface environment, it may remain concealed for many years, becoming dispersed over wide areas of groundwater aquifer and rendering groundwater supplies unsuitable for consumption and other uses (Nagarajan et al. 2010). Water chemistry differs depending on the source of water, the degree to which it has been evaporated, the types of rock and mineral it has encountered, and the time it has been in contact with reactive minerals (Plummer et al. 2003; Arshid et al. 2011). Assessment of water quality is very important for knowing the suitability for various purposes (Ifatimehin and Musa 2008; Arshid et al. 2011).

The urban aquifers are the only natural resource for drinking water supply, they are often perceived as of lesser relevance for the drinking water supply, leading to crisis in terms of drinking water scarcity, becoming increasingly polluted thereby decreasing their potability (Tiwari et al. 2012). The knowledge of hydro-chemistry is important to assess the ground water quality in any area in which the ground water is used for both irrigation and drinking needs (Srinivas et al. 2013). The water quality assessment may

✉ M. Jayaprakash
emjaypee@gmail.com

1 Department of Applied Geology, Global and Environmental Change Research Group, University of Madras, Guindy Campus, Tamil Nadu, India

Table 1 Comparison of analytical results with international and national standards

Parameter	Max	Min	Mean	WHO Guideline Value (2004)	BIS Standards (2000)
EC (μS/cm)	2306.25	471.5	927.0	1500	–
pH	8.4	7.4	7.9	6.5–8.5	6.5–8.5
TDS (mg/l)	1476	301.8	593.3	1500	500–2000
Ca (mg/l)	54	10.0	21.8	200	75–200
Mg (mg/l)	42	3.6	26.9	150	–
Na (mg/l)	558	67.0	156.7	200	200–400
K (mg/l)	103	2.0	20.2	12	–
CO_3 (mg/l)	15	0.0	2.5	–	–
HCO_3 (mg/l)	94	30.5	69.3	500	–
Cl (mg/l)	710	96.0	254.2	600	250
SO_4 (mg/l)	59	35.3	42.2	250	200–400

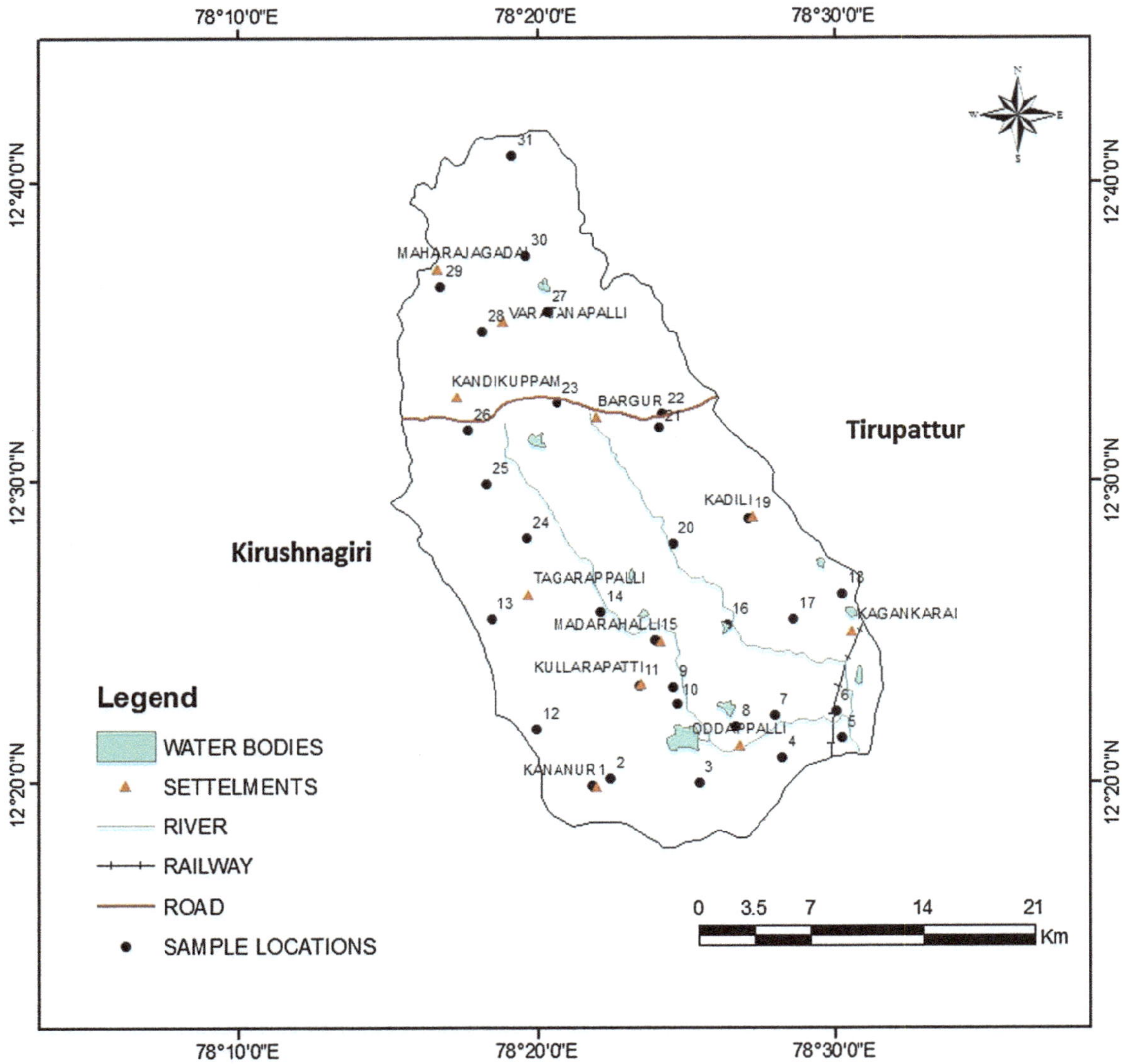

Fig. 1 Study area map

give clear information about the subsurface geologic environments in which the water presents (Raju et al. 2011).

Most studies on water quality have been carried out by various researchers in many places in India (Raju et al. 2011; Srinivasamoorthy et al. 2011; Subramanian 2011; Gnanachandrasamy et al. 2013; Annapoorna and Janardhanab 2015; Nagaraju et al. 2014; Sajil Kumar et al. 2013). Hydrogeochemical investigation of groundwater has been carried out in the coastal aquifers of southern Tamil Nadu, India (Chandrasekar et al. 2014). Ashwani and Abhay (2014) have studied groundwater chemistry of Pratapgarh district in Uttar Pradesh.

Since water is a precious natural resource, for sustaining all life on the earth and due to its multiple benefits and the problems created by its excesses, shortage and quality deterioration, water as a resource requires special attention. So the aim of the study is to assess the quality of groundwater and to assess the spatial distribution of various hydrogeochemical parameters for suitability of groundwater resources in the study area as it is densely populated area and they mostly demand on the groundwater resources.

Study area

The study area is located in the northern part of Tamil Nadu State in India and is situated between north eastern part of Krishnagiri district and south western part of Vellore district. It lies between latitudes 12°17′40″ and 12°41′53″ and longitudes 78°14′56″ and 78°31′38″ (Fig. 1). The study area is drained by Bargur and Mathur rivers. These two rivers merge at the southeast corner, where the Pambar River originates and finally joins the river Ponnaiyar. The study area covers an area of 781 sq. km in the Survey of India toposheet numbers 57L/6, 57L/7 and 57L/11 on a scale of 1:50,000. Topography of the area is highly and full of massive rock shoots with fracture zone. A wide array of litho unit ranging from alkali syenites and ultramafics complexes and younger dolerite like intrusive are exposed in the study area. Also, an intrusive igneous complex of Proterozoic age, younger dykes and recent alluvial cover along the Bargur and Mattur river course cover up rest of the geology. The area has a sub-tropical climate without any sharp variations. Temperatures vary from 40 °C in summer to around 20 °C in the winter season. The average rainfall is 857 mm/year. Public Work Department Report (2004).

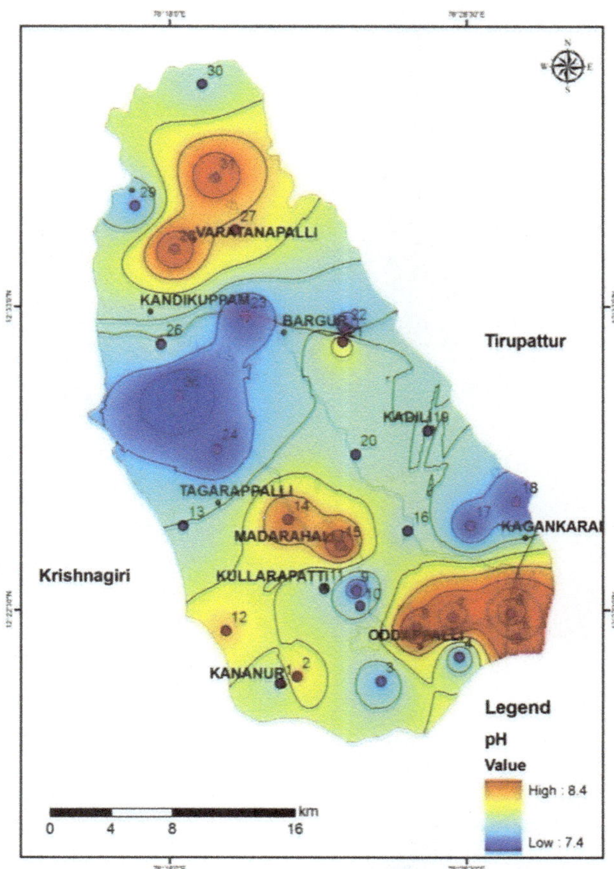

Fig. 2 Spatial variation of distribution of pH in study area

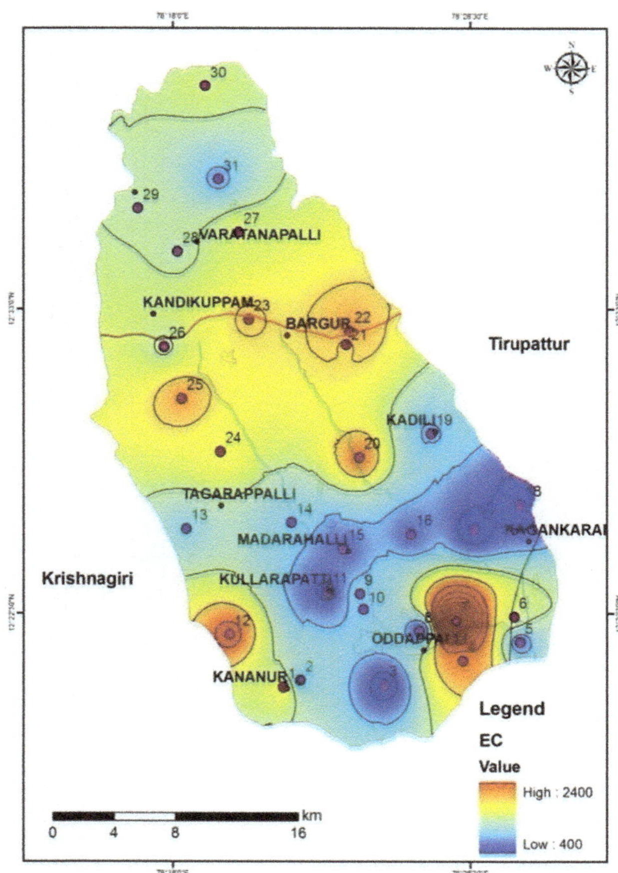

Fig. 3 Spatial variation of distribution of EC in study area

In the present research work, the study area consists of part of two districts namely, Vellore and Krishnagiri. Both the districts have large industrials profiles such as textiles, leather tanneries and small-scale dying industries. The effluents of the leather industries, usage of the chemical fertilizers for agriculture and small-scale dying industries fall heavily on the quality of the drinking water. The impact is felt very much on the drinking water sources which are available for the people, settled on the banks of the river. (Dhiviyaa Pranavam et al. 2011).

Hydrogeology

Weathered and fractured Archaean crystalline rocks construct the major aquifer systems in the study area. The thickness of weathered zones ranges from less than a meter to more than 15 m. Generally, the groundwater occurs under phreatic conditions in the weathered mantle and under semi-confined conditions in the fractured zones at deeper levels. The occurrence and movement of ground water are controlled by various factors such as physiography, climate, geology and structural features (CGWBoard 2009).

Methodology

The Groundwater samples were collected from 31 well locations which fall in 28 mini watersheds in both Krishnagiri and Vellore districts. Among the 31 well locations, most of the wells are situated near or surroundings of Bargur and Mattur river, which are the mini tributaries Ponnaiyar river. The water samples were collected in definite intervals in a grid pattern. The samples were stored in plastic bottles, which pre-cleaned with 1 N hydrochloric acid and rinsed 3–4 times with distilled water. Water analysis was done using standard methods for the examinations of water and waste water (APHA 1999). EC and pH were measured in the field using calibrated thermometer with a resolution of 0.1 and Elico portable water quality analyser, respectively. Total dissolved solids (TDS) were computed by multiplying the electrical conductivity (EC) by a factor (0.64). Total hardness (TH) as $CaCO_3$ and calcium (Ca) were analyzed titrimetrically, using standard EDTA. Magnesium (Mg) was calculated by taking the differential value between total hardness (TH) and calcium (Ca) concentrations. Chloride (Cl)) was determined titrimetrically by standard $AgNO_3$ titration. The content of sodium (Na) and potassium (K) in

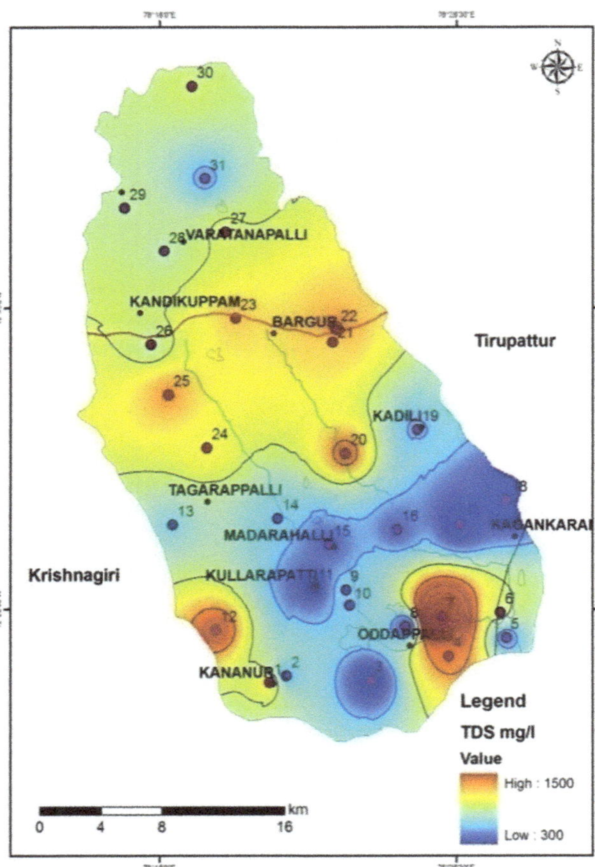

Fig. 4 Spatial variation of distribution of TDS in study area

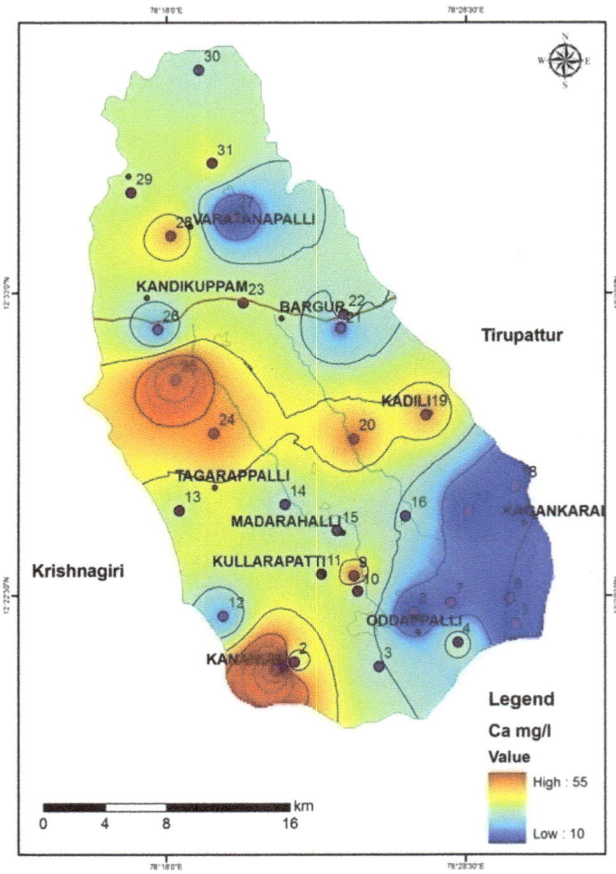

Fig. 5 Spatial variation of distribution of Ca in study area

groundwater was estimated using EEL flame photometer with proper Air-LPG flame, sodium and potassium sulfate standards of appropriate concentrations were used.

All parameters are expressed in milligrams per liter (mg/l), except pH (no units) and electrical conductivity (EC). The electrical conductivity (EC) is expressed in microsiemens/cm (μS/cm) at 25 °C.

Result and discussion

The various physicochemical parameters of groundwater samples were analyzed and the descriptive statistics of the analyzed parameters are given in Table 1. The results are compared to the World Health Organization recommended maximum permissible limits and BIS standards.

The data in Table 1 showed that pH of the groundwater ranged from 7.4 to 8.4 and average value is 7.9 indicating slight alkalinity. The lowest pH (7.4) was found in the station 25 and the highest pH (8.4) was found in the sample station 6. Spatial distribution of pH is shown in Fig. 2. The permissible range of pH for drinking and agricultural purposes is 6.5–8.5 (IS: 10500-1991). Most of the samples fall in within the permissible range.

The electrical conductivity in groundwater ranged from 471.5 to 2306 μS/cm and the mean value is 927 μS/cm. The higher EC values show at station 7. Figure 3 shows the spatial distribution of EC values and that is increasing north to south. It is indicating the flow direction. The large variations in EC are mainly attributed to anthropogenic activities and to geochemical processes prevailing in this region. EC generally increases along a groundwater flow path because of the combined effects of evaporation, ion-exchange, and topographic conditions (Toth 1999). The spatial distribution diagram (Fig. 4) of TDS also showing similar type variation like EC, and the values range from 301 to 1476 mg/l and mean 593 mg/l the TDS concentration basically depends on what are the ions present in water. The calcium and magnesium in waters are generally used to classify the suitability of water. Calcium and magnesium are directly related to hardness of the water and these ions are the most abundant elements in the

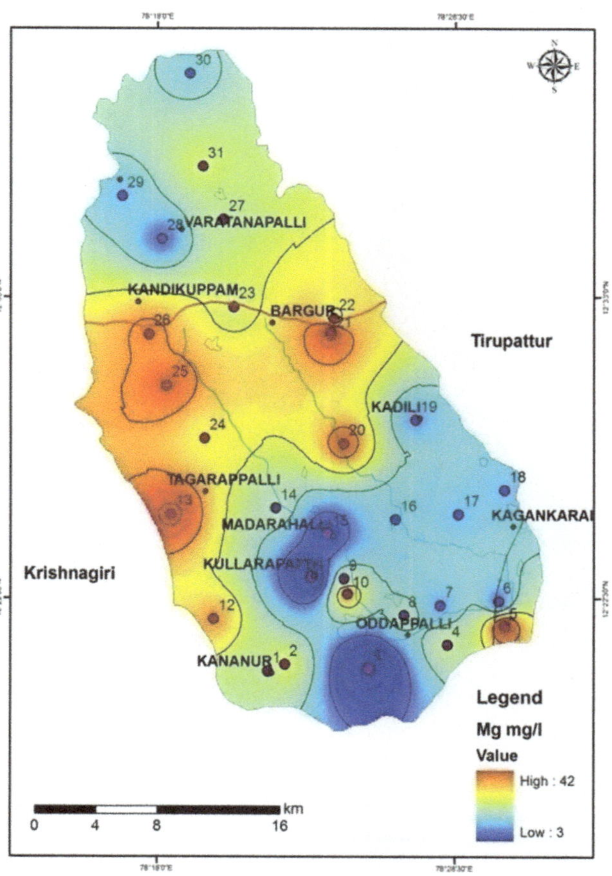

Fig. 6 Spatial variation of distribution of Mg in study area

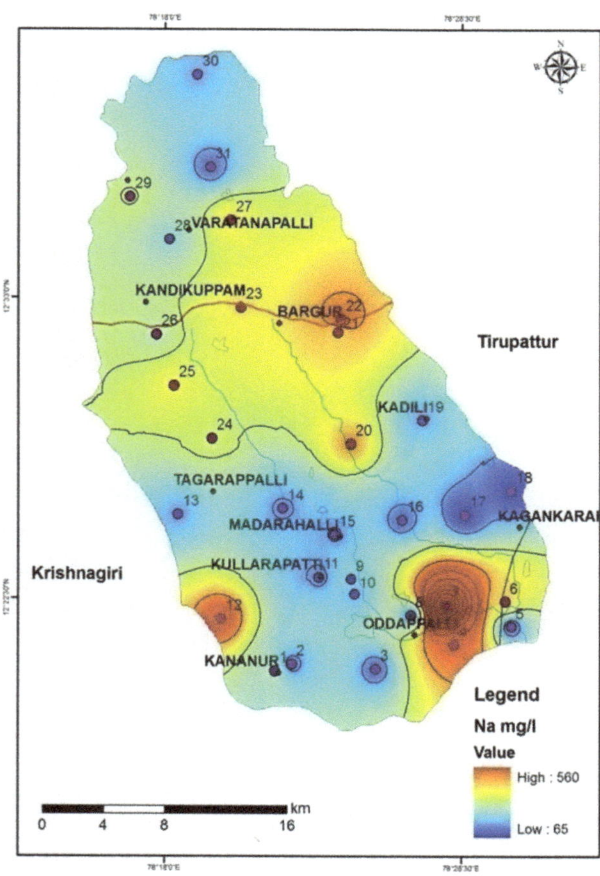

Fig. 7 Spatial variation of distribution of Na in study area

surface and groundwater and exist mainly as bicarbonates and to a lesser degree in the form of sulfate and chloride (Krishna Kumar et al. 2014). Calcium in the study area varied from 10 to 54 mg/l. Mg varies from 3.6 to 42 mg/l. Ca and Mg concentrations are within the permissible limits. The order of abundance of cations is $Na^+ > K^+ > Ca^{2+} > Mg^{2+}$. Higher concentration of Na observed in station 7 may be due to contribution from silicate weathering process and agricultural sources. K is higher due to weathering of feldspar and clay minerals from the surrounding substance to the aquifer. The anion chemistry shows that Cl^- and HCO_3^- are the dominant anions followed by SO_4^{2-}, and CO_3^- (Table 1). The chloride ion is the most predominant natural form of the element chlorine and is extremely stable in water. The chloride in groundwater may be from diverse sources such as weathering, leaching of sedimentary rocks and soil, domestic and municipal effluents (Sarath Prasanth et al. 2012). Cl was higher due to leaching from upper soil layers derived from industrial and domestic activities and dry climates (Srinivasamoorthy et al. 2008). The higher concentration of HCO_3 was may be due to the effect dissolution of silicates and rock weathering. Higher concentration of chloride in water is often found in conjunction with higher sodium concentration. ICMR and BIS have prescribed 250 mg/l as the maximum permissible value. If the chlorine value exceeds 300 mg/l and the presence of a major cation is sodium, then the water becomes salty and similar observation is seen in the present study area (Ravisankar and Poongothai 2008). The spatial distribution of all the anions and cations is shown in the Figs. 5, 6, 7, 8, 9, 10, 11, 12. The spatial distributions of major ions of groundwater samples are clearly showing the flow of ground water northwest to southeast direction.

Hydrochemical facies

The conventional techniques such as trilinear plots, statistical techniques are widely accepted methods to evaluate water quality for various uses. The problems of groundwater quality are more acute in areas that are coastal, densely populated and thickly industrialized and have shallow groundwater tube wells (Krishna Kumar et al. 2011; Padmalal et al. 2012; Bagyaraj et al. 2013; Selvam et al. 2013). The chemical processes and the evolution of

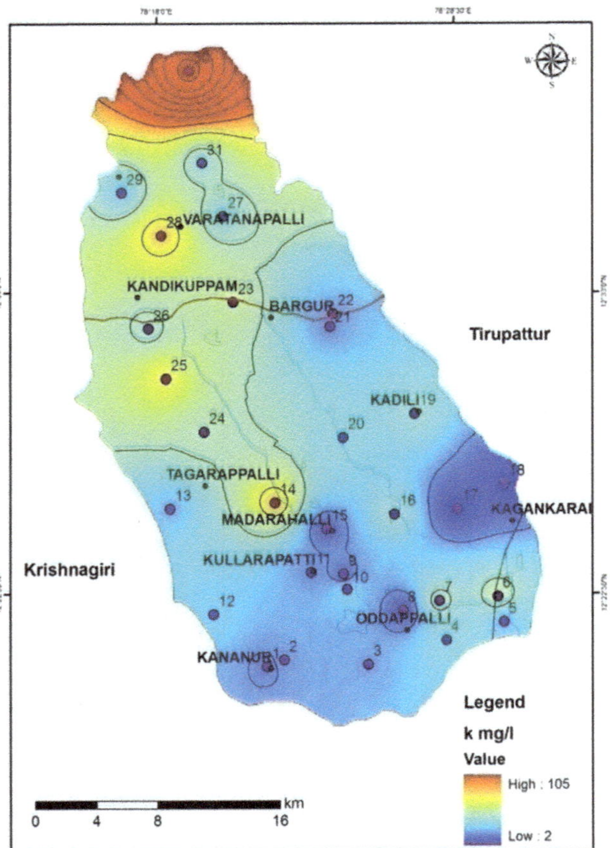

Fig. 8 Spatial variation of distribution of K in study area

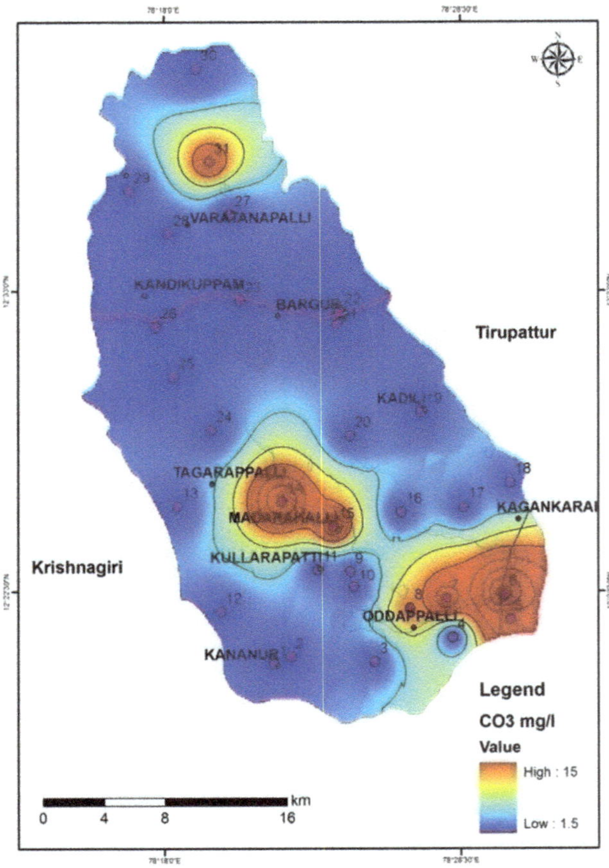

Fig. 9 Spatial variation of distribution of CO_3 in study area

the groundwater in the aquifers due to the residence and the flow may be evaluated using the hydrochemical facies. This can be well interpreted by drawing the Piper diagram (Piper 1953). The elevation of hydrochemical parameters of ground water can be understood by plotting the concentration of major cations and anions in the piper diagram (Fig. 13). The plots show that most of the water falls in the field of NaCl with one stations showing mixed CaMgCl. From the plot, it is observed that alkalis (Na and K) exceed the alkaline earth (Ca and Mg) and Cl exceeds other anions.

Wilcox diagram

According to the sodium % and specific conductance in evaluating the suitability of the water samples are varying from excellent to doubtful classes for irrigation. The results show that (Fig. 14) about 52 % of the samples fall in the very good to good region; 6 % samples are in the good to permissible region indicating that these waters are very much suitable for irrigation. 32 % of the samples fall in the permissible to doubtful region and only one sample fall in the doubtful to unsuitable region (Table 2). This station demonstrates high TDS value with higher concentration of Na. Since excess sodium affects the plants, the water belongs to this station (Station 7) and is not suitable for irrigation.

Statistical analysis

The correlation coefficient is commonly used to measure the relationship between two variables. It is a measure to exhibit how well one variable predicts the behavior of the other (Lee et al. 2003). The results show that (Table 2) the Na has highly significant correlation with Cl and SO_4. Similarly, Mg and K have positive correlation with Cl. The hydrochemical character of alkalies explains the correlation of sodium with chlorine ion. CO_3 showing high significant correlation with pH, indicating that changes in the hydrogen ion concentration will directly correlation of carbonate in the water. The Ca/Na ratio is estimated to be 0.14 which is slightly below the value deduced by Galy and France-Lanord (1999). Among the cations, no significant correlation is observed highly competitive relationship but in the case of anions Cl has significant correlation with SO_4. HCO_3 and CO_3 have low positive correlation with SO_4, low correlation exists between K and Cl; Na and CO_3. Cl is the dominant ion in anions, and its higher concentration is observed along the course of the river,

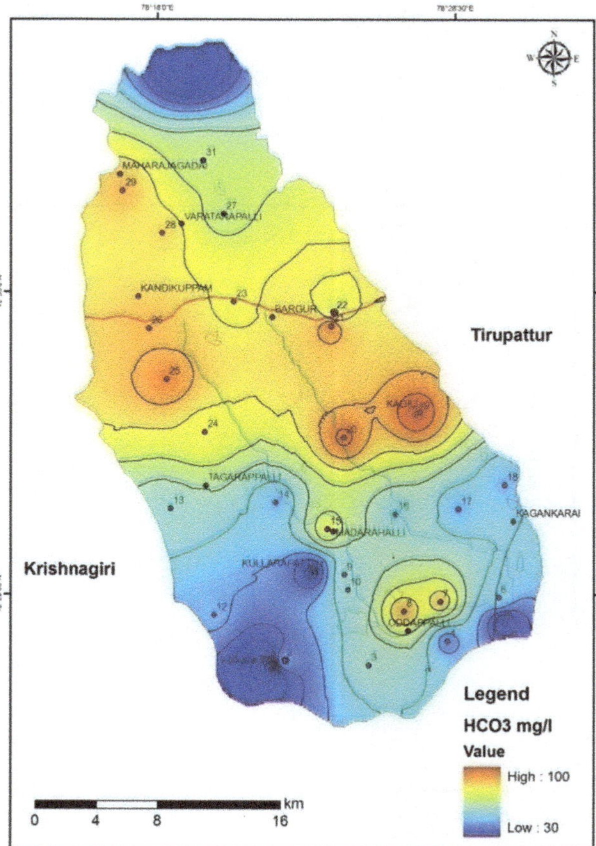

Fig. 10 Spatial variation of distribution of HCO_3 in study area

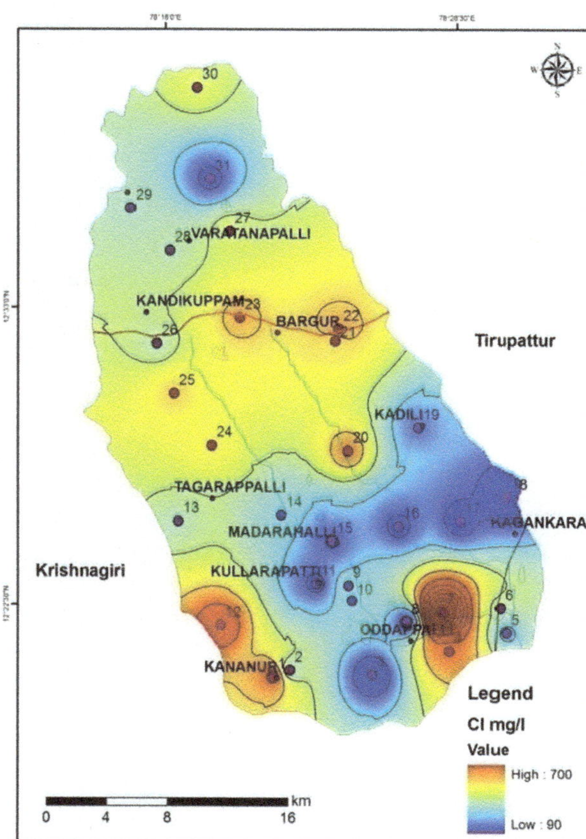

Fig. 11 Spatial variation of distribution of Cl in study area

may be due to sewage infiltration or domestic effluents (Ranjana and Naverathna 2011).

Cluster analysis

This is a group of multivariate techniques which primarily classify (Massart and Kaufmann 1983) variables or cases (observation or samples) into the cluster with high homogeneity level within the class and high heterogeneity level between classes The spatial variability of groundwater was determined by the CA. CA was first performed to group all sample sites to classify them into a cluster to minimize their number. We use CA to link sample site in the configuration of a tree with different branches (Dendogram) which provide a visual summary of the clustering process, presenting a picture of the group and their proximity. Branches that have linkage closer to each other indicate a stronger relationship between sample/variables or cluster of sampling site/variable. In this present study, CA was applied for the grouping of parameters using the ward's linkage method (Ward 1963).

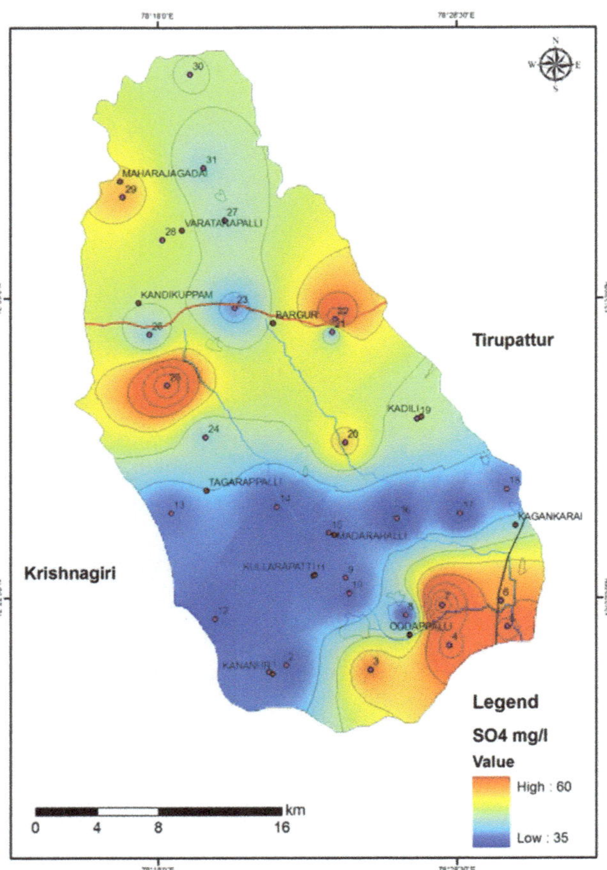

Fig. 12 Spatial variation of distribution of SO_4 in study area

Cluster 1 comprised of EC and TDS showing close similarities which can be interpreted as natural mineralization and is controlled by cation exchange (Fig. 15). The cluster also indicates Ca–Mg–Cl facies, which resulted from the reverse cation exchange. The presence of TDS in this cluster is an indication that the cations and anions influence TDS and thus increases the water's electrical conductivity (EC). At high TDS concentration, water becomes saline (Shahbazi and Esmaeili-Sari 2009).

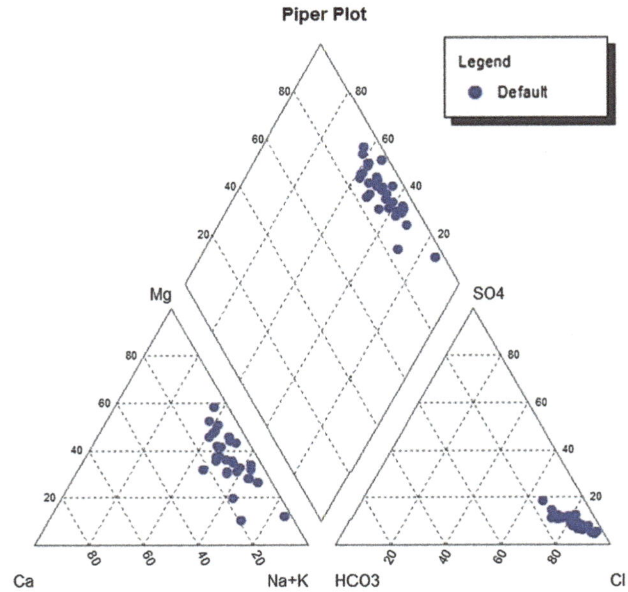

Fig. 13 Piper diagram

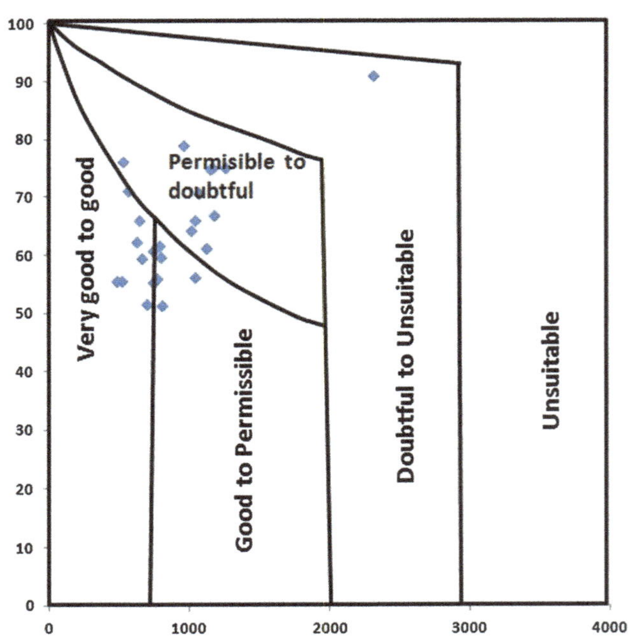

Fig. 14 Wilcox diagram

Table 2 Correlation coefficient for groundwater quality parameters of Krishnagiri district and Vellore district

	EC	pH	TDS	Ca	Mg	Na	K	CO_3	HCO_3	Cl	SO_4
EC	1.00										
pH	0.16	1.00									
TDS	1.00	0.16	1.00								
Ca	0.11	−0.38	0.11	1.00							
Mg	0.24	−0.07	0.24	0.07	1.00						
Na	**0.97***	0.20	**0.97***	−0.05	0.09	1.00					
K	0.18	0.04	0.18	0.08	−0.01	0.04	1.00				
CO_3	0.12	**0.75**	0.12	−0.31	−0.12	0.17	0.03	1.00			
HCO_3	0.22	0.00	0.22	−0.08	0.16	0.26	−0.24	−0.04	1.00		
Cl	**0.98***	0.12	**0.98***	0.16	0.28	**0.93***	0.17	0.06	0.10	1.00	
SO_4	0.59	0.01	**0.59***	0.05	0.02	**0.60***	0.14	0.18	0.21	0.51	1.00

* Correlation significant at 0.5 level

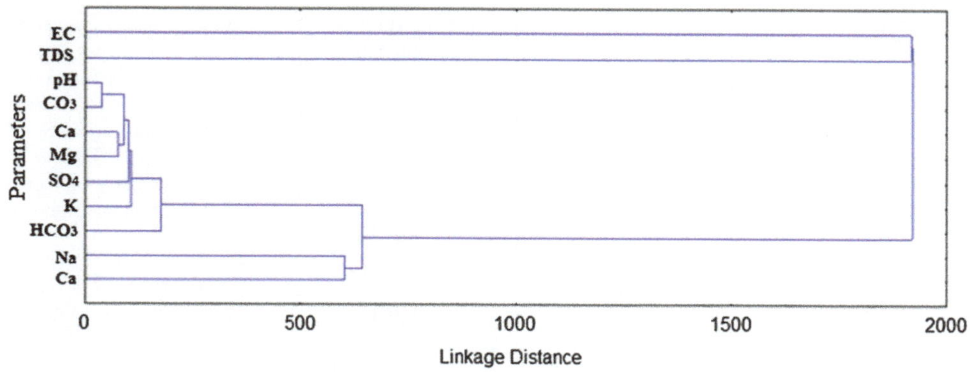

Fig. 15 Parameter wise analysis using cluster for groundwater quality of Krishnagiri and Vellore District

Conclusion

The cation chemistry of the groundwater shows excessive presence of Na except in few locations. The order of abundance is $Na^+ > Mg^{2+} > Ca^{2+} > K^+$. The anion chemistry of the groundwater shows excessive presence of chloride except in few samples. The order of abundance is $Cl^- >$- $> SO_4^{2-} > CO_3^{2-}$. The result of hydrochemical facies reveals the type of water as NaCl with mixed CaMgCl. In statistical analysis, the results show that the Na has significant correlation with Cl similarly correlated with SO_4. Rock dissolution, Agriculture, domestic and other industrial effluence in the study area is responsible for the Cl and Na increase. Wilcox plot illustrates that 58 % samples are suitable for irrigation. When comparing the analytical results with the World Health Organization's permissible limits and BIS standards that 40 % of the sample is not potable water. The key source of all the hydrogeochemical process in the study area is due to lithology which is the parent control as well as anthropogenic influence such as effluents released from the nearby tanneries which govern the hydro geochemistry of the groundwater and make it unsuitable for drinking purpose. This inference will be much helpful for water resource managers to solve environmental problems in the society.

References

Annapoorna H, Janardhanab MR (2015) Assessment of groundwater quality for drinking purpose in rural areas surrounding a defunct copper mine. Aquatic Procedia 4:685–692. doi:10.1016/j.aqpro.02.088

APHA (1999) Standard methods for the examination of water and wastewater, 20th edn. American Public Health Association, Washington DC

Arshid J, Aasimah T, Yousuf AR, Akbar M, Aabid HN (2011) Geochemistry and irrigation quality of groundwater along River Jhelum in South Kashmir, India. Recent Res Sci Technol 3(6):57–63

Ashwani KT, Abhay KS (2014) Hydrogeochemical investigation and groundwater quality assessment of Pratapgarh District, Uttar Pradesh. J Geol Soc India 83:329–343

Bagyaraj M, Ramkumar T, Venkatramanan S, Gurugnanam B (2013) Application of remote sensing and GIS analysis for identifying groundwater potential zone in parts of Kodaikanal Taluk, South India. Front Earth Sci 7(1):65–75

BIS (2000) Drinking water specification. Bureau of Indian Standards, New Delhi

Central Ground Water Board (2009) Report. South Eastern Coastal Region, Chennai

Chandrasekar N, Selvakumar S, Srinivas Y, John Wilson JS, Simon Peter T, Magesh NS (2014) Hydrogeochemical assessment of groundwater quality along the coastal aquifers of southern Tamil Nadu, India. Environ Earth Sci 71:4739–4750

Dhiviyaa Pranavam TS, Venkatesa Rao T, Punithavathi L, Karunanithi S, Bhaskaran A (2011) Groundwater pollution in the Palar Riverbed near Vellore, Tamil Nadu, India. Indian J Sci Technol 4:1

Ewusi A, Obiri-yeboah S, Voigt HS (2013) Groundwater quality assessment for drinking and irrigation purposes in Obuasi Municipality of Ghana: a preliminary study. J Environ Earth Sci 5(1):6–17

Galy A, France-Lanord C (1999) Weathering process in the Ganges-Brahmaputra basin and the riverine alkalinity budget. Chem Geol 159:31–60

Gnanachandrasamy G, Ramkumar T, Venkatramanan S, Vasudevan S, Chung SY, Bagyaraj M (2013) Accessing groundwater quality in lower part of Nagapattinam district, Southern India: using hydrogeochemistry and GIS interpolation techniques. Appl Water Sci. doi:10.1007/s13201-014-0172-z

Ifatimehin OO, Musa SD (2008) The prospects of sustainable management of domestic 48 water supply and sanitation in Kogi strata. J Environ Policy Issues 4(1–2):33–44

Ishaku JM (2011) Assessment of groundwater quality index for Jimeta-Yola area, northeastern Nigeria. J Geol Mining Res 3(9):219–231

Kalpana L, Elango L (2013) Assessment of groundwater quality for drinking and irrigation purposes in Pambar river sub-basin, Tamil Nadu. Indian J Environ Protection 33(1):1–8

Krishna Kumar S, Logeshkumaran A, Magesh NS, Prince SG, Chandrasekar N (2014) Hydro-geochemistry and application of water quality index (WQI) for groundwater quality assessment, Anna Nagar, part of Chennai City, Tamil Nadu, India. Appl Water Sci. doi:10.1007/s13201-014-0196-4

Krishna Kumar S, Chandrasekar N, Seralathan P, Godson PS, Magesh NS (2011) Hydrogeochemical study of shallow carbonate aquifers, Rameswaram Island, India. Environ Monit Assess 184(7):4127–4139

Kumar SK, Rammohan V, Sahayam JD, Jeevanandam M (2008) Assessment of groundwater quality and hydrogeo-chemistry of Manimuktha River basin, Tamil Nadu, India. Environ Monit Assess. doi:10.1007/s10661-008-0633-7

Lee SM, Min KD, Woo NC, Kim YJ, Ahn CH (2003) Statistical models for the assessment of nitrate contamination in urban groundwater using GIS. Environ Geol 44:210–221

Massar DL, Kaufman L (1983) The interpretation of chemical data by the use of cluster analysis. Wiley, New York

Mitra BK, Sasaki K, Enari N, Matsuyama N, Fujita M (2007) Suitability assessment of shallow ground water for agriculture in sand dune area of northwest Honshu Island, Japan. Appl Ecol Environ Res 5(1):177–188

Nagarajan R, Rajmohan N, Mahendran U, Senthamilkumar S (2010) Evaluation of groundwater quality and its suitability for drinking and agriculture use in Thanjavur city, Tamil Nadu, India. Environ Monit Assess 171:289–308

Nagaraju A, Sunil Kumar K, Thejaswi A (2014) Assessment of groundwater quality for irrigation: a case study from Bandalamottu lead mining area, Guntur District, Andhra Pradesh, South India. Appl Water Sci 4:385–396. doi:10.1007/s13201-014-0154-1

Padmalal D, Maya K, Narendra Babu K, Baiju RS, Babura B (2012) Hydro chemical characterization and water quality assessment of the coastal springs of southern Kerala, India. J Appl Geochem 14(4):466–481

Piper AM (1953) A graphic procedure in the chemical interpretation of water analysis, US Geological Survey Groundwater Note 12

Plummer LN, Bexfield LM, Anderholm SK (2003) How ground-water chemistry helps us understand the aquifer. In: Bartolino JR, Cole JC (eds) U.S. Geological Survey Circular, p 1222

Public Work Department Report (2004)

Rajmohan N, Elango L, Ramachandran S, Natarajan M (2000) Major ion correlation in groundwater of Kancheepuram region, south India. Indian J Environ Protection 20(3):188–193

Raju NJ, Shukla UK, Ram P (2011) Hydrogeochemistry for the assessment of groundwater quality in Varanasi: a fast-urbanizing center in Uttar Pradesh, India. Environ Monit Assess 173:279–300

Ranjana UK, Naverathna MPC (2011) River sand mining in southern Sri-Lanka and its effect on environment. In: 11th International River symposium on "A Future of extremes" Brisbane, Australia

Ravisankar N, Poongothai S (2008) A study of groundwater quality in Tsunami affected areas of Sirkazhi taluk, Nagapattinam district, Tamil Nadu, India. Sci Tsunami Hazards 27(1):47–55

Sajil Kumar PJ, Elango L, James EJ (2013) Assessment of hydrochemistry and groundwater quality in the coastal area of South Chennai, India. Arab J Geosci. doi:10.1007/s12517-013-0940-3

Sarath Prasanth SV, Magesh NS, Jitheshlal KV, Chandrasekar N (2012) Evaluation of groundwater quality and its suitability for drinking and agricultural use in the coastal stretch of Alappuzha District, Kerala, India. Appl Water Sci 2(3):165–175

Selvam S, Manimaran G, Sivasubramanian P (2013) Hydrochemical characteristics and GIS-based assessment of groundwater quality in the coastal aquifers of Tuticorin corporation, Tamil Nadu, India. Appl Water Sci 3:145–159

Shahbazi A, Esmaeili-Sari A (2009) Groundwater quality assessment in North of Iran: a case study of the Mazandaran Province. World Appl Sci J 5:92–97

SinghA K, Hasnain SI (1998) Major ion chemistry and weathering control in a high altitude basin: Alaknanda river, Garhwal Himalaya, India. Hydrol Sci 43(6):825–843

Srinivas Y, Hudson Oliver D, Stanley Raj A, Chandrasekar N (2013) Evaluation of groundwater quality in and around Nagercoil town, Tamil Nadu, India: an integrated geochemical and GIS approach. Appl Water Sci 3:631–651

Srinivasamoorthy K, Chidambaram S, Prasanna MV, Vasanthaviha M, Peter John, Anandhan P (2008) Identification of major sources controlling groundwater chemistry from a hard rock terrain: a case study from Mettur taluk, Salem district, Tamil Nadu, India. J Earth Syst Sci 117(1):49–58

Srinivasamoorthy K, Nanthakumar C, Vasanthavigar M, Vijayaraghavan K, Rajivgandhi R, Chidambaram S, Anandhan P, Manivannan R, Vasudevan S (2011) Groundwater quality assessment from a hard rock terrain, Salem District of Tamil Nadu, India. Arab J Geosci 4:91–102

Subramanian A (2011) Ground water quality assessment of Nagercoil Town (Hand pumps). J Environ Earth Sci 1(1):1–5

Tiwari KK, Prasad RN, Ram Chandra, Mondal NC (2012) Geochemical parameters for assessment of groundwater quality around urban and suburban areas of Dausa city in Rajasthan, India. J Appl Geochem 14(2):184–193

Toth J (1999) Groundwater as a geologic agent: an overview of the causes, processes and manifestations. Hydrogeol J 7:1–14

UNESCO (2000) Groundwater pollution. International Hydrological Programme

Ward JH Jr (1963) Hierarchical grouping to optimize an objective function. J Am Stat Assoc 58:236–244

WHO (2004) WHO Guidelines for Drinking-water Quality 4th edn. World Health Organization

19

An evaluation method of the sustainability of water resource in karst region: a case study of Zunyi, China

Bo Li[1,2] **· Ganlu Wang**[2] **· Hanghang Ding**[2] **· Yulong Chen**[3]

Abstract Water resource is of great significance to the survival and development of human. However, the water resource system in karst regions is sensitive to external interference owing to the special geological processes which cause soil impoverishment, severe rocky desertification and large topographic height difference. Therefore, evaluating the sustainability of the water resource in karst regions is beneficial to reasonably use and protect water resource. This paper puts forward to evaluate the water resource from four aspects, including water resources system, water requirement system, ecosystem and social economic system. Moreover, on this basis, 18 evaluation indexes were selected to construct the sustainability evaluation index system and method. This method was used to evaluate the sustainability of the water resource in the typical karst region—Zunyi, Guizhou province, China, and was verified according to the actual situation in the research area. All these provide reference for the evaluation of the sustainability of the water resource in similar regions.

✉ Bo Li
libo1512@163.com

Ganlu Wang
47850901@qq.com

Hanghang Ding
1679760386@qq.com

Yulong Chen
673054399@qq.com

1 Key Laboratory of Karst Environment and Geohazard Prevention, Ministry of Education, Guizhou University, Guiyang 550000, Guizhou, China

2 College of Resource and Environmental Engineering, Guizhou University, Guiyang 550000, Guizhou, China

3 Department of Civil Engineering, University of Tokyo, Tokyo 113-8656, Japan

Keywords Karst region · Water resource · Evaluation index system · Sustainability · Evaluation

Introduction

The water resource is the basic resources for controlling the ecological environment and the material base for social development. The *World Water Resources Comprehensive Assessment (1997.1)* of the United Nations reported that water crisis would seriously restrict global society and economy development in the twenty-first century, and might lead to conflicts among countries (Simonovic et al. 1995; Loucks 2000; Giacomoni and Zechman 2011). Currently, the water crisis in all the countries has not been solved fundamentally and even become more serious. Accordingly, performing relative studies on the sustainability of the water resource is beneficial to the reasonable utilization and protection of water resource (Rijsberman et al. 2000).

Currently, there are a lot of researchers studying on water resources sustainability; the Fuzzy UTASTAR multicriteria analysis method (based on linear programing) has been used for sustainability evaluation by Haider et al. (2015). The groundwater resource carrying capacity of the irrigation areas is evaluated using the improved catastrophe evaluation method by Xiao et al. (2011). To evaluate the long-term sustainability of water withdrawals in the United States, the magnitudes of human withdrawals from surface water and ground water sources and the stored water requirements during the warmest months of the year were evaluated by Roy, SB (Roy et al. 2015). Kuzdas et al.

(2014) presented a criteria-based and participatory sustainability appraisal of water governance in a region with such characteristics—the dry tropics of NW Costa Rica.

Previous researches on the evaluation of the sustainability of the water resource mainly focus on non-karst regions, while rarely evaluate the sustainability of that in karst regions. With large topographic height difference and developed karst channels, karst regions are generally covered with thin overlying soil (Chu et al. 2008; Zhang et al. 2004). These characteristics lead to these regions which have more abundant water resource underground than on the ground; moreover, the water resource system responds more sensitively to the changes in the surrounding environment. For the sustainability of the water resource, except for the influence of the fixed factors of water, it is also related to the water requirement, ecological environment and social economy in the region. Considering this, studying the sustainability of water resource merely from one aspect cannot reflect the whole situation. Therefore, in view of water resource system, water requirement system, ecosystem and social economic system, the evaluation index system for the sustainability of the water resource in the karst region is constructed. Based on the system, the typical karst mountainous area, that is, Zunyi, in Guizhou, China, is served as an example to evaluate the sustainability of water resource.

Evaluation index system

The evaluation index system for the sustainability of the water resource in the karst region is complicated, thus attention has to be paid to two points in the selection of evaluation indexes. Firstly, the evaluation indexes are supposed to present the key factors affecting the sustainability of the water resource; secondly, the selected indexes are operable (Bossel 1999). Therefore, from four aspects including water resources system, water requirement system, ecosystem and social economic system, this paper selected corresponding indexes to establish the evaluation index system.

Water resource system

The water resource system refers to the factors influencing the water resource quantity in a natural state, mainly including four ones. First is the annual mean precipitation. As the main source of water, precipitation can reflect the water resource quantity in the research area. Second is the annual mean evaporation which means the amount of moisture evaporated to air in a certain period. Thirdly, water production modulus presents the ratio of total water resource to the total area of the region. Fourth is a drought index, which mirrors the degree of drought and is generally defined as the ratio of annual evaporation to annual precipitation.

Water requirement system

The water requirement system presents the requirement degree of water resource for social development. It is primarily influenced by the following five factors: firstly, it is affected by the exploitation and utilization rate of water resource, which shows the percentage of annual water consumption in the total water resource. The second one is the development difficulty coefficient of water resource. It is an important index for measuring the development difficulty of the water resource. The third influencing factor is water consumption per capita, which is the ratio of water consumption to total population in the research area. Fourth is the proportion of industrial water. It presents the ratio of industrial water consumption to total water consumption in a certain region. The fifth is the proportion of agricultural water, which shows the ratio of water used for agricultural irrigation to total water used in a certain region.

Ecosystem

Aiming at the present ecological problems caused by the shortage and irrational utilization of water in the karst region, the relation between the ecological environment and water resource has to be studied. The main factors influencing the sustainability evaluation of the water in the ecosystem are as follows. First is the average standard-reaching percentage of water quality. It is the percentage of the quantity of water reaching the quality standard in the total detected water. Second is the intensity coefficient of karst development which determines the capacity of groundwater for receiving rainfall and the degree of being polluted. Thirdly, rocky desertification is the ratio of rocky desertification area to the total area of this region. Rocky desertification results in the reduction capacity of soil for water conservation. Fourth is the overburden thickness which is the thickness of surface soil. Different overburden thicknesses correspond to the different capacities of natural degradation of pollutant composition and pollution infiltration. The fifth influencing factor is the vegetation coverage, which presents the ratio of forest cover to the total area of the research region.

Social economic system

Social economy is closely related to the sustainability of the water resource. It is mainly influenced by the following factors: firstly, natural growth rate of population, which is an important index for reflecting the development speed of

Table 1 Evaluation index system for the sustainability of water resource in the karst region

Target layer *A*	Evaluation of the sustainability of water resource in the karst region and its influencing factors (*A*)			
Criteria layer *B*	Water resources system (B_1)	Water requirement system (B_2)	Ecosystem (B_3)	Social economic system (B_4)
Index layer *C*	Annual mean precipitation (C_1)	Exploitation and utilization rate of water resource (C_5)	Average standard-reaching rate of water quality (C_{10})	Natural growth rate of population (C_{15})
	Annual mean evaporation (C_2)	Development difficulty coefficient of water resource (C_6)	Rocky desertification rate (C_{11})	Water consumption per unit GDP (C_{16})
	Water production modulus (C_3)	Water consumption per capita (C_7)	Intensity coefficient of karst development (C_{12})	Perfect degree of regulations related to water resource (C_{17})
	Drought index (C_4)	Proportion of industrial water (C_8)	Overburden thickness (C_{13})	Mean life expectancy (C_{18})
		Proportion of agricultural water (C_9)	Vegetation coverage (C_{14})	

population and formulating population plan. Secondly, water consumption per unit gross domestic product (GDP), which is the ratio of the total water consumption to the urban GDP. Thirdly, the perfect degree of regulations related to water resource, which mirrors the perfect degree of laws and regulations related to water resource in this region. Fourthly, average life expectancy, which presents the average lifetime of people in the research region.

Based on the above analysis, the research object is divided into three layers. As the ultimate propose, the evaluation of the sustainability of water resource in the karst region is acted as the target layer (*A* layer) of the model; though water resources system, water requirement system, ecosystem and social economic system determine the possibility of sustainability, their effects have to be reflected by concrete indexes, thus serving as the criteria layer (*B* layer); the concrete main controlling factors compose the index layer (*C* layer) of the model (Table 1).

Methods for evaluating the sustainability of the water resource in the karst region

Calculation for the comprehensive indexes of the sustainability of water resource

The indexes in the sustainability evaluation system of water resource present the sustainability of the water resource from different aspects, so the influencing effects of each index have to be comprehensively considered. In this paper, using the multi-index linear weighted function, the evaluation model is as follows:

$$A_i = \sum C_{ik} \times W_{ik} \tag{1}$$

where *A* presents the comprehensive evaluation value of the sustainability of water resource, C_{ik} shows the index value of the *k*th factor in the *i*th subsystem, while W_{ik} is the weight of the *k*th index in the *i*th subsystem.

The index values are quantized based on the standard in Table 2. While according to the effect of index values on the system, directional indexes are divided into positive and inverse ones. Then, suppose that $a_1 < a_2 < a_3 < a_4$ and $b_1 < b_2 < b_3 < b_4$ are the grading scores of the positive and reverse indexes, respectively.

Based on the scoring standard of the indexes in Table 2, linear index quantization is used to quantize the indexes. Suppose *P* is the score of the indexes and *x* presents the actual values of the indexes, the following index quantization formula is obtained.

1. I-grade index
   For the positive index, when $x \le a_1$, $p = 1$.
   For the reverse index, $p = 1$ when $x \ge b_4$.
2. II-grade index
   $p = 1 + (x - a_1)\frac{1}{a_2 - a_1}$, when *x* is positive index;
   $p = 1 + (b_4 - x)\frac{1}{b_4 - b_3}$, when *x* is reverse index.
3. III-grade index
   $p = 2 + (x - a_2)\frac{1}{a_3 - a_2}$, when *x* is positive index;
   $p = 2 + (b_3 - x)\frac{1}{b_3 - b_2}$, when *x* is reverse index.
4. IV-grade index
   $p = 3 + (x - a_3)\frac{1}{a_4 - a_3}$, when *x* is positive index;
   $p = 3 + (b_2 - x)\frac{1}{b_2 - b_1}$, when *x* is reverse index.
5. V-grade index
   For the positive indexes, when $x \ge a_4$, $p = 4$.
   For the reverse indexes, $p = 4$ when $x \le b_1$.

Methods for determining the weights of indexes

This paper employs the analysis hierarchy process (AHP) to determine the weights of indexes for factors. Proposed by American operation researcher and professor Saaty in

Table 2 The marking standard of the sustainability evaluation index of water resource in karst regions

Category	Range	Weaker sustainability I grade	Weak sustainability II grade	Moderate sustainability III grade	Strong sustainability IV grade	Stronger sustainability V grade
Positive indexes	Value range	$\le a_1$	a_1–a_2	a_1–a_2	a_1–a_2	$\ge a_4$
	Score range	1	1–2	2–3	3–4	4
Reverse indexes	Value range	$\ge b_4$	b_3–b_4	b_2–b_3	b_1–b_2	$\le b_1$
	Score range	1	1–2	2–3	3–4	4

the 1980s, this method is a practical decision-making method for multi-schemes or multi-objectives. The concrete procedures are as follows (Saaty 1977, 1980): The hierarchical structure of the system is firstly established by analyzing the relation of factors. Next, the 1–9 scale method is utilized to construct the judgment matrix through pairwise comparison. Then, single ordering and consistency check are conducted to the hierarchies. Finally, hierarchy general ranking and consistency check are performed.

Application example

Research situation

1. Position

Located in the southwest of China and the north of Guizhou province, namely, 105°36′ −108°13′E and 27°8′ −29°12′N, Zunyi is situated in the slope zone where the terrain changes from Yun-Gui plateau to Hunan hill and Sichuan basin. With large topographic relief and complex landforms, Zunyi, at an altitude of 1000–1500 m, is distributed with extensive corrosion landform (karst landform) which accounts for 75 % of total land area (Fig. 1).

2. Meteorology and hydrology

Situated in the subtropical humid monsoon climate zone, Zunyi shows an annual mean precipitation of 1167.3 mm. Moreover, the maximal and minimal annual precipitations are 1498.4 and 871.0 mm separately. The precipitation here generally occurs from May to September, which occupies 59–74 % of the annual precipitation. The rivers in the research region belong to the Yangtze River with the basin area of 30,763 square kilometers.

3. Hydrogeology

In the research region, the outcropping strata include those from Sinian system to Quaternary system, except for Silurian system, Devonian system, Carboniferous system, upper Jurassic series and Tertiary system. The groundwater is mainly carbonate rock karst water, which is mainly supplied from the atmospheric precipitation. Afterwards, under the influence of surface hydrographic net and lithology as well as tectonics, the groundwater flows out as springs or underground rivers through the cracks and pipelines developed in the rock mass.

4. Current situations of social economy and exploration of water resource

At the end of 2014, there were 6.1549 million permanent residents in the research area. The regional gross product and total financial income were 187.436 billion yuan and 41.266 billion yuan, respectively. Urban and rural per capita disposable incomes were 22,728 and 8365 yuan separately. The total surface water amount and groundwater quantity in the research area were 21.48 billion cubic meters and 4.872 billion cubic meters, respectively. While the total water used in this year was 2.15 billion cubic meters in which 1.962 billion cubic meters were surface water.

5. Ecological environment

As a mountainous area characterized by bare karsts, this region has widely distributed carbonate rocks and intensely developed karsts. In addition, karst forms including sinkhole, doline, depression, etc. are densely distributed. The groundwater is accumulated and flows through cracks, pipelines, karst carves, etc. The forest coverage in Zunyi city is about 53.08 %; besides, the karst area accounts for 75 % of total area of Zunyi and the rocky desertification area takes up 18.79 % of the karst area.

Quantization for indexes

The indexes are quantized using the method proposed in the third section, and then the results are graded according to the standard with five grades in Table 2. The obtained quantization results of the indexes in the research region are presented in Table 3.

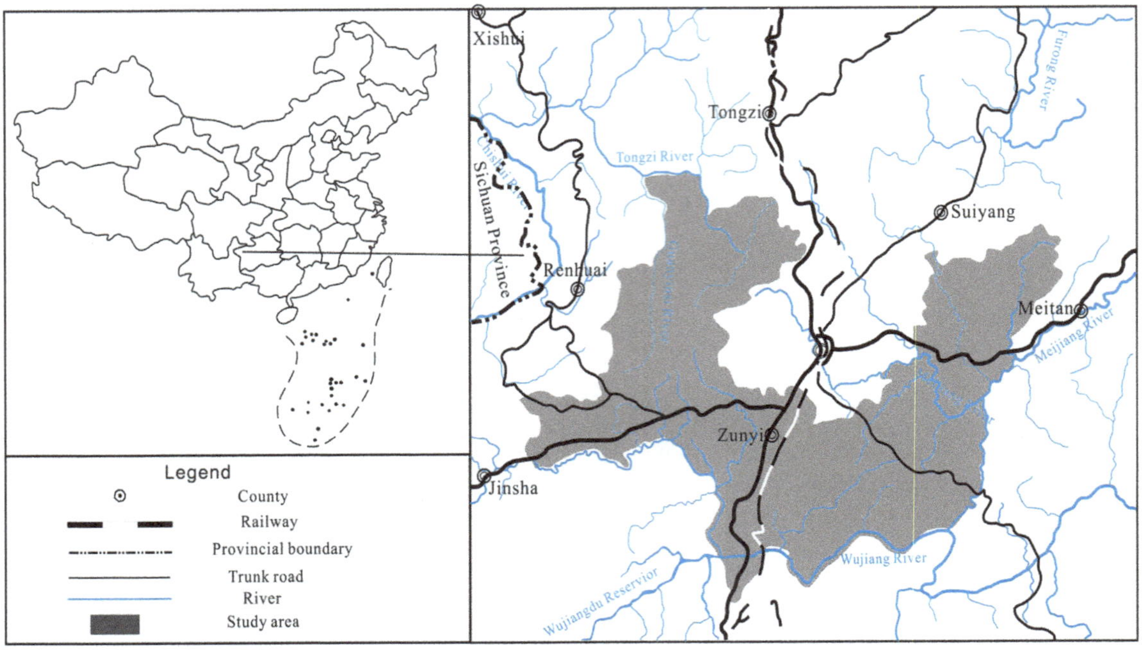

Fig. 1 Site location

Table 3 The quantization results of the indexes

Indexes	Quantized values	Indexes	Quantized values
Annual mean precipitation (C_1)	2.00	Average standard-reaching percentage of water quality (C_{13})	3.10
Annual mean evaporation (C_2)	1.59	Rocky desertification (C_{10})	1.16
Water production modulus (C_3)	2.11	Intensity coefficient of karst development (C_{14})	1.26
Drought index (C_4)	1.55	Overburden thickness (C_{12})	1.05
Exploitation and utilization rate of water resource (C_5)	2.10	Vegetation coverage (C_{11})	2.55
Development difficulty coefficient of water resource (C_8)	1.20	Population growth rate (C_{15})	2.40
Water consumption per capita (C_7)	1.52	Water consumption per unit GDP(C_{16})	1.40
Proportion of industrial water (C_8)	2.00	Perfect degree of regulations related to water resource (C_{17})	1.20
Proportion of agricultural water (C_9)	1.92	Mean life expectancy (C_{18})	2.63

Determination of the weights of the indexes

According to the analysis on the main controlling factors on the vulnerability of water resource, the relative importance of each factor is evaluated by experts in the field through expert evaluation method, thus an evaluation set of the influencing factors is obtained. On this basis, the judgment matrix for evaluating the sustainability of the water resource based on the AHP method in the karst region is established. The weights of the indexes in the

Table 4 Judgment matrix $A \sim B_i$ ($i = 1$–4)

A	B_1	B_2	B_3	B_4	W (A/B_4)
B_1	1	2	2	2	0.39521
B_2	½	1	2	2	0.27807
B_3	½	1/2	1	1	0.16336
B_4	½	1/2	1	1	0.16336

$\lambda_{max} = 4$, CI 0.0202, CR $= 0.02246 < 0.1$

Table 5 Calculation results of the weights of the indexes (i = 1–18)

Evaluation indexes	Weights	Evaluation indexes	Weights
Annual mean precipitation (C_1)	0.1337	Comprehensive standard-reaching percentage of water quality (C_{10})	0.026
Annual mean evaporation (C_2)	0.0941	Rocky desertification (C_{11})	0.0597
Water production modulus (C_3)	0.0553	Intensity coefficient of karst development (C_{12})	0.0452
Drought index (C_4)	0.0553	Overburden thickness (C_{13})	0.0343
Exploitation and utilization rate of water resource (C_5)	0.0816	Vegetation coverage (C_{14})	0.0394
Development difficulty coefficient of water resource (C_8)	0.0408	Population growth rate (C_{15})	0.0282
Water consumption per capita (C_7)	0.0816	Water consumption per unit GDP (C_{16})	0.0564
Proportion of industrial water (C_8)	0.0361	Perfect degree of regulations related to water resource (C_{17})	0.0564
Proportion of agricultural water (C_9)	0.0479	Mean life expectancy (C_{18})	0.0282

criteria layer are firstly determined. Then, for the four evaluation criteria, the factors are pairwise compared using the expert evaluation method. Consequently, the weighted complementary judgment matrix is acquired, as illustrated in Table 4.

As can be seen, the CR value computed based on the matrix is less than 0.1, which implies that the consistency check is passed. Likewise, using the same method for determining the weights in the criteria layer, the weights of indexes are obtained in Table 5 using the AHP.

Comprehensive evaluation of the sustainability of the water resource in the research region

By combining the calculated weights and the quantization results of the indexes in the research region, the comprehensive evaluation value of the sustainability is acquired based on the Formula (1) as $C = \Sigma A_{ik} \times W_{ik} = 1.76$. According to the standard in Table 2, the water resource in the research region exhibits weak sustainability and the main reasons are as follows. With the low forest coverage and thin soil, Zunyi presents serious rocky desertification. Moreover, as there are many mountainous areas, the topographic height difference here is large, thus causing extreme difficulty in exploring the water resource. Furthermore, both per unit GDP and per capita water consumption are large, and meanwhile the regulations related to water resource remain to be improved.

Conclusions

The exploration and utilization of the water resource in the karst region present special characteristics as uneven distribution of water resource, difficult exploration and sensitivity of the water resource system to the external interferences. Based on the analysis of other evaluation methods, this paper proposes that the sustainability of the water resource in the karst region is supposed to be evaluated from four aspects. They include water resource system, water requirement system, ecosystem and social economic system. Then, from these aspects, the evaluation index system for the sustainability of the water resource in the karst region is established. Meanwhile, the corresponding evaluation method is put forward by combining the AHP method. Afterwards, this method is employed to evaluate the sustainability of the water resource of the typical karst region—Zunyi of Guizhou province, China. Furthermore, favorable effect is obtained in the verification based on the regional situation. Therefore, this evaluation method provides a certain basis and reference for evaluating the sustainability of the water resource in karst regions.

Acknowledgments This research was financially supported by China National Natural Science Foundation (Grant Nos. 41430318 and 41272276), the China National Scientific and Technical Support Program (Grant Nos. 201105060-06 and 2012BAB12B03), Guizhou Province Science and Technology Agency Foundation (qian ke he LH zi[2014]7617), Guizhou University Introducing Talents Research Foundation (2014-61), Guizhou Province Geological Exploration Fund (Guizhou karst groundwater system functions sustainable utilization).

References

Bossel H (1999) Indicators for sustainable development: theory, method, applications. International Institute for Sustainable Development, Winnipeg

Chu WH, Yan GQ, Su WC (2008) Study on water resources sustainable utilization of karst river basins. Shui Li Xue Bao 39(6):753–757

Giacomoni MH, Zechman EM (2011) Assessing sustainability of integrated urban water resources systems through a complex adaptive systems approach. In: World Environmental and Water Resources Congress, ASCE, 2844–2851

Haider H, Singh P, Ali W, Tesfamariam S, Sadiq R (2015) Sustainability evaluation of surface water quality management options in developing countries: multicriteria analysis using fuzzy UTASTAR method. Water Resour Manag 29(8):2987–3013

Kuzdas C, Wiek A, Warner B, Vignola R, Morataya R (2014) Sustainability appraisal of water governance regimes: the case of Guanacaste, Costa Rica. Environ Manag 54(2):205–222

Loucks DP (2000) Sustainable water resources management. Water Int 25(1):3–10

Rijsberman MA, De Ven V, Frans HM (2000) Different approaches to assessment of design and management of sustainable urban water system. Environ Impact Assess Rev 129(3):333–345

Roy SB, Ricci PF, Summers KV, Chung CF, Goldstein RA (2015) Evaluation of the sustainability of water withdrawals in the United States, 1995–2025. J Am Water Resour Assoc 41(5):1091–1108

Saaty TL (1977) A scaling method for priorities in hierarchical structures. J Math Psychol 15(3):234–281

Saaty TL (1980) The analytic hierarchy process. McGraw-Hill, New York

Simonovic SP, Kundzewicz Z, Rosbjerg D, Takeuchi K (1995) Modeling and management of sustainable basin-scale water resource systems. IAHS Press, Wallingford

Xiao L, Han XJ, Dong Z (2011) The sustainability evaluation model of irrigation district groundwater resources carrying capacity based on catastrophe theory. In: 5th international symposium on integrated water resources management, Nanjing, China, IWRM 2010 and the 3rd international symposium on methodology in hydrology, pp 241–247

Zhang LP, Zhu ZL, Deng LJ (2004) Present situation of and problems in water resource evaluation indicator systems in China. Resour Surv Assess 21(4):5–8

Amino-functionalized mesoporous MCM-41 silica as an efficient adsorbent for water treatment: batch and fixed-bed column adsorption of the nitrate anion

Mehdi Ebrahimi-Gatkash[1] · Habibollah Younesi[1] · Afsaneh Shahbazi[2] · Ava Heidari[3]

Abstract In the present study, amino-functionalized Mobil Composite Material No. 41 (MCM-41) was used as an adsorbent to remove nitrate anions from aqueous solutions. Mono-, di- and tri-amino functioned silicas (N-MCM-41, NN-MCM-41 and NNN-MCM-41) were prepared by post-synthesis grafting method. The samples were characterized by means of X-ray powder diffraction, FTIR spectroscopy, thermogravimetric analysis, scanning electron microscopy and nitrogen adsorption–desorption. The effects of pH, initial concentration of anions, and adsorbent loading were examined in batch adsorption system. Results of adsorption experiments showed that the adsorption capacity increased with increasing adsorbent loading and initial anion concentration. It was found that the Langmuir mathematical model indicated better fit to the experimental data than the Freundlich. According to the constants of the Langmuir equation, the maximum adsorption capacity for nitrate anion by N-MCM-41, NN-MCM-41 and NNN-MCM-41 was found to be 31.68, 38.58 and 36.81 mg/g, respectively. The adsorption kinetics were investigated with pseudo-first-order and pseudo-second-order model. Adsorption followed the pseudo-second-order rate kinetics. The coefficients of determination for pseudo-second-order kinetic model are >0.99. For continuous adsorption experiments, NNN-MCM-41 adsorbent was used for the removal of nitrate anion from solutions. Breakthrough curves were investigated at different bed heights, flow rates and initial nitrate anion concentrations. The Thomas and Yan models were utilized to calculate the kinetic parameters and to predict the breakthrough curves of different bed height. Results from this study illustrated the potential utility of these adsorbents for nitrate removal from water solution.

Keywords MCM-41 · Nitrate · Amine functionalized · Isotherm · Breakthrough analysis

Introduction

Water has a vital role in life on earth. High level of nitrate anion in aquatic environments is a serious environmental worldwide problem, because this anion is implicated in the eutrophication of receiving surface waters and also the presence of nitrate anions in drinking water causes many public health problems (Kapoor and Viraraghavan 1997). Agricultural and urban wastewater, improper disposal of municipal and industrial wastes, leakage in septic systems, landfill leachate, animal manure are sources of nitrate anion contamination (Bhatnagar et al. 2010). While nitrate is relatively nontoxic to adults, because of nitrate excretion from the kidneys, concentrations >10 mg/l $N–NO_3^-$ are fatal for children under 6 months. A sort of "blue baby syndrome" is caused by the reduction of nitrate to nitrite, resulting in the decrement of O_2 carrying capacity of hemoglobin in the blood and finally death. US EPA has set a most extreme contaminant level of 10 mg/l of $N–NO_3^-$ in drinking water (EPA 2010). Eutrophication is the abundance of aquatic plants, the growth of algae, and

✉ Habibollah Younesi
hunesi@modares.ac.ir; hunesi@yahoo.com

1 Department of Environmental Science, Faculty of Natural Resources, Tarbiat Modares University, P.O. Box 46414-356, Noor, Iran

2 Environmental Sciences Research Institute, Shahid Beheshti University, GC, Tehran, Iran

3 Natural Resources and Environment College, Ferdowsi University of Mashhad, Mashhad, Iran

depletion of dissolved oxygen (Levine and Schindler 1989).

Several processes were investigated for the removal of nitrate in water and wastewaters which include chemical denitrification using zero-valent iron (Fe^0) (Cheng et al. 1997; Huang and Zhang 2004), ionic exchange (Broséus et al. 2009), reverse osmosis (Schoeman and Steyn 2003), electrodialysis (Hell et al. 1998), catalytic denitrification (Pintar et al. 2001) and biological denitrification (Soares 2000). Though, current available technologies for water treatment of nitrate are found to be expensive, inefficient and generate additional by-products. Among them, adsorption systems are favorable, because they permit straightforward and economical operation, bringing about less sludge generation and disposal issues. So that adsorbents have been developed to remove various pollutants of aquatic environments.

Mesoporous silica materials have gotten extensive consideration due to their high specific surface area, well-defined pore size and pore shape (Beck et al. 1992). Mostly, the functionalizations of mesoporous silica have been investigated. For example, Yoshitake et al. (2002) investigated the amino-functionalized mesoporous silicas for removal of arsenate from an aqueous solution. Heidari et al. (2009) prepared an adsorbent from mesoporous silica Mobil Composite Material No. 41 (MCM-41) by grafting organic amine groups for removal of heavy metal pollution. Hamoudi et al. (2007b) evaluated the potential of removing nitrate and phosphate from aqueous solutions using ammonium-functionalized mesoporous silica MCM-48. However, that is necessary to provide treatment method for removing nitrate anion by other mesoporous silica materials. In this study, mesoporous MCM-41 silica functionalized with amine organic groups was used for nitrate anion adsorption. Amino-functionalized mesoporous MCM-41 silica was synthesized with post-synthesis Grafting method. The grafting method consists in reacting a suitable organosilane with surface silanol groups belonging to a previously synthesized mesoporous material using an appropriate solvent under reflux conditions (Lim and Stein 1999).

In the present study, the adsorption properties of adsorbents for nitrate anion adsorption were investigated using batch and fixed-bed column systems. The Freundlich and Langmuir isotherms were utilized to demonstrate the equilibrium adsorption data. The kinetic data of adsorption and the evaluation of equilibrium parameters were described by the pseudo-second-order equations. Also, the investigation was performed to remove nitrate anion from aqueous solution by amine functionalized mesoporous silica MCM-41 in a fixed-bed column and also to investigate the impacts of operating conditions, for example, bed height, flow rate and initial concentration on the column dynamics. A breakthrough analysis for the adsorption of nitrate on NNN-MCM-41 sieve was performed. The adsorption capacity was calculated based on the breakthrough analysis and Thomas and Yan models.

Experimental section

Chemicals

3-Aminopropyltriethoxysilane (99 %), cetyltrimethylammonium bromide (CTAB, 99 %), hydrochloric acid (HCl, 37 %), dry toluene (99.9 %) and sodium nitrate ($NaNO_3$) were purchased from Merck (Germany). Tetraethylorthosilicate (TEOS, 98 %) and sodium hydroxide (NaOH) were obtained from Sigma-Aldrich (USA). [3-(2-Aminoethylamino)propyl]trimethoxysilane (99 %) and 3-[2-(2-Aminoethylamino)ethylamino] propyltrimethoxysilane (99 %) were purchased from Alfa Aesar.

Adsorbents synthesis

Synthesis of MCM-41

The source of silicon was TEOS and the structure directing factor was CTAB. MCM-41 was prepared by slowly adding 20.41 g of TEOS into a solution containing 3.64 g of CTAB and 1.2 g of NaOH in 108 ml of deionized water (Chen and Wang 2002). After mixing for 1 h at room temperature, the subsequent homogeneous blend was solidified under static hydrothermal conditions at 110 °C in a Teflon bottle for 96 h. The molar composition of the starting gel mixture was 1.0:0.1:0.3:60 TEOS/CTAB/NaOH/H_2O. The final product was acquired by filtration, washed with deionized water, ethanol reflux for 48 h, and then dried at 100 °C overnight, and calcined 6 h at 540 °C (heated from room temperature to 540 °C with a heating rate of 3 °C/min).

Synthesis of amine functionalized silica MCM-41 by grafting

The mesoporous materials MCM-41 with amino (i.e., monoamine, diamine and triamine) were prepared according to the method described in the literature (McKittrick and Jones 2003). Prior to the functionalization with organic grapes, calcined MCM-41 was dried in a vacuum oven for 4 h at approximately 150 °C. To graft the silica surface, 1 g of so-obtained MCM-41 materials was suspended in dried toluene (25 ml) and then 6 mmol (1.381 ml) of 3-aminopropyltriethoxysilane precursor was slowly added to the mixture. This mixture was refluxed for 24 h. The functionalization of MCM-41 with diamine and triamine groups was conducted by the same aforementioned procedure, but 1.268 ml (6 mmol) of [3-(2-aminoethylamino)propyl]trimethoxysilane and 1.592 ml (6 mmol) of 3-[2-(2-aminoethylamino)ethylamino]propyltrimethoxysilane was added in each run instead. Solid materials were collected by filtration, washed with dried toluene, ethanol reflux for 24 h and then dried at 70 °C overnight. Amine-grafted silica materials were labeled N-MCM-41, NN-MCM-41 and NNN-MCM-41 compared to organic chains containing one, two and three amino-functional groups, as shown in Fig. 1.

Characterizations

X-ray powder diffraction (XRD) experiments were carried out with Cu radiation (MMA007, manufactured by GBC Australia). The scans were made in the 2θ range 0.5°–10° (low angle diffraction). The morphology of the mesoporous adsorbents was acquired by electron scanning microscopy (SEM, LEO 1455VP, Cambridge, UK). Nitrogen adsorption–desorption isotherms were evaluated at −196 °C using a volumetric adsorption analyzer (BEL sorp-mini II, Japan) to ascertain the textural properties. Before the analysis, approximately 0.1 g of all MCM-41 samples was degassed at 100 °C for 3 h under argon gas flow. The specific surface area was assessed utilizing the Brunauer–Emmett–Teller (BET) technique, and the pore size distribution was estimated from desorption branches of nitrogen isotherms using the Barrett–Joyner–Halenda (BJH) system. FT-IR measurements were accomplished on a TENSOR27 (manufactured by Bruker, Germany) in the range of 400–4000 cm^{-1}. A thermogravimetric analysis (TGA) of sample was also carried out on a Pyris Diamond Thermogravimetric analyzer (PerkinElmer, UK). The sample was heated under nitrogen atmosphere from an ambient temperature to 840 °C at 10 °C/min.

Adsorption experiments in batch system

The adsorption capability was investigated of the synthesized adsorbent materials toward nitrate anions, using aqueous solutions of $NaNO_3$. Adsorption was performed batchwise in Erlenmeyer flasks on a stirrer at 150 rpm

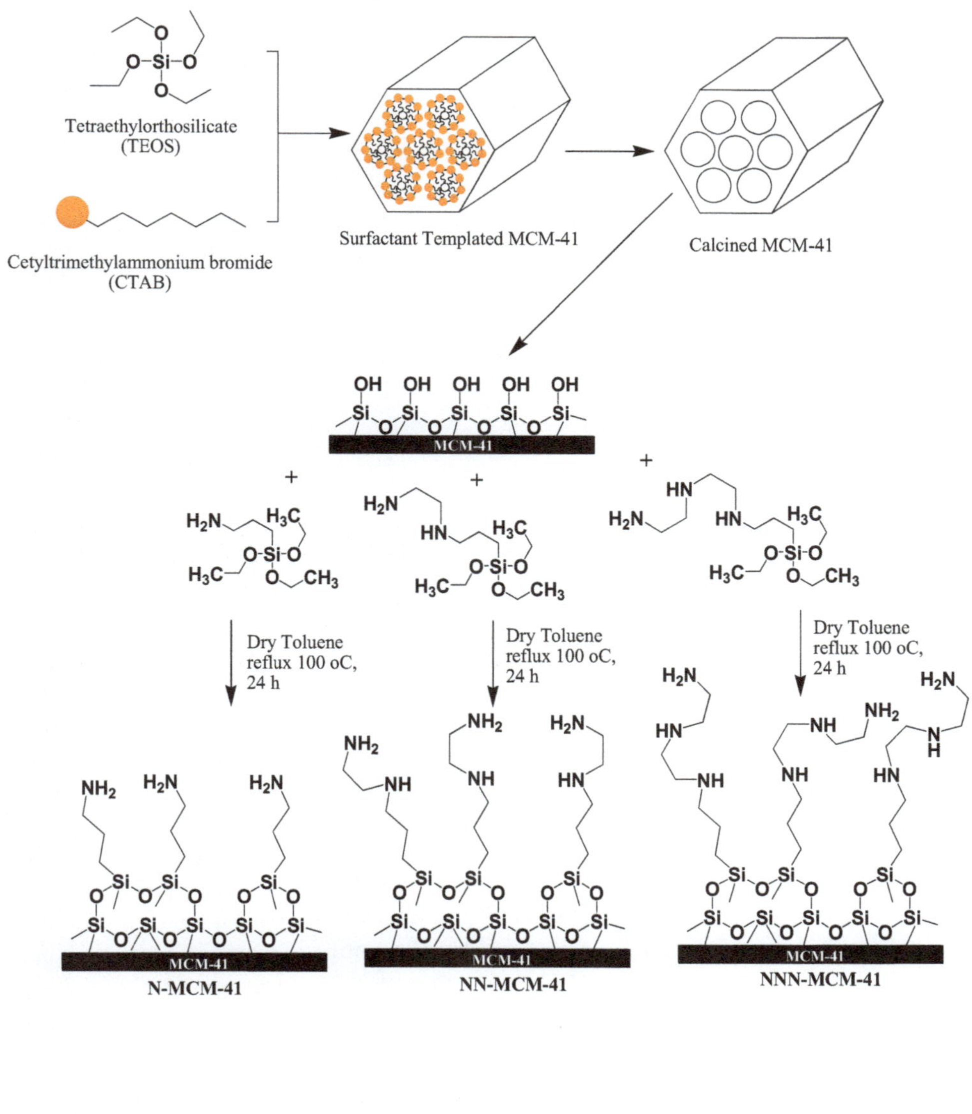

Fig. 1 Synthesis of amino-functionalized mesoporous silica materials via post-synthesis grafting method

under the following conditions: adsorbent loading, 0.5–15 g/l; anion initial concentration, 30–250 mg/l; and pH 4–8. Preliminary adsorption tests with sampling intervals of 5 min showed that equilibrium was attained within 10–15 min. Thus, for all the adsorption tests presented thereafter, the stirring was continued for 120 min before aliquots of the solutions were withdrawn, centrifuged (236 HK, Hermle, German) for 10 min at 12,000 rpm and 25 °C and subsequently filtered with filter paper (0.45 μm) to remove the adsorbent. The concentration of nitrate anion was measured using a UV visible spectrophotometer set at 210 nm (Lambda 25, Perkin-Elmer, USA).

The percent of removed nitrate anions by the adsorbents was calculated by the following equation:

$$R = \frac{C_0 - C_t}{C_0} \times 100 \tag{1}$$

where R is the removal efficiency of the nitrate anions, C_0 is the initial concentration and C_t is the concentration of the nitrate anions at t time in mg/l. The amount of nitrate adsorbed (q_e in mg/g) was calculated as follows:

$$q_e = \frac{(C_o - C_e)V}{m} \tag{2}$$

where C_e is the equilibrium concentration of nitrate anion in the aqueous solution (mg/l), V is the volume of solution (l) and m is the quantity of the adsorbent (g).

The equilibrium adsorption data of nitrate anions were fitted into both the Freundlich and Langmuir isotherm equations. In the Langmuir model, it is presumed that the adsorption surface places have the same energy and each adsorbate molecule has been situated on a solitary site and thus this model anticipates the formation of monolayer of an adsorbate on the adsorbent surface (Heidari et al. 2009; Vasconcelos et al. 2008). The Langmuir isotherm is given by the following equation:

$$q_e = \frac{q_m b C_e}{1 + b C_e} \tag{3}$$

where q_e is the equilibrium sorption capacity of sorbent in mg anion/g sorbent, C_e is the equilibrium concentration of nitrate anion in mg/l, q_m is the maximum amount of anion sorbed in mg nitrate/g sorbent, and b is the factor that alludes to the bonding energy of adsorption in l/mg. Nonlinear regression analysis was performed in SigmaPlot software (SigmaPlot 12.0, SPSS Inc., USA) to estimate the values of q_m and b parameters.

The Freundlich model depicts a reversible heterogeneous adsorption, since it does not limit itself to a monolayer of adsorbent covering (Ghorbani et al. 2008). Actually, the Freundlich isotherm predicts that the concentration of nitrate anion on the adsorbent will rise until there is an increment of the anion concentration in the liquid solution (Vaghetti et al. 2009). The Freundlich isotherm is given by the following equation:

$$q_e = K_f C_e^{\frac{1}{n}} \tag{4}$$

where q_e is the adsorption capacity at equilibrium time in mg anion/g adsorbent, C_e is the equilibrium concentration of nitrate anion in mg/l, K_f and n are the parameters identified with the adsorption capacity in mg nitrate/g adsorbent, and the adsorption intensity of the sorbent, respectively. Nonlinear regression analysis was done by SigmaPlot software (SigmaPlot 12.0, SPSS Inc., USA) to predict the K_f and n parameters.

Two kinetic models (pseudo-first-order and pseudo-second-order) are used to test the experimental data for showing the controlling mechanisms of adsorption processes. The rate constant of adsorption is assigned from the subsequent pseudo-first-order rate expression given by Lagergreen (Soner Altundoğan et al. 2000; Wu et al. 2001).

$$\log(q_e - q_t) = \log q_e - \frac{k_1 t}{2.303} \tag{5}$$

where q_e and q_t are the quantity of nitrate adsorbed (mg/g) at the equilibrium and t times (min), respectively, and k_1 (l/min) is the rate constant of first-order adsorption (Sivaraj et al. 2001). A pseudo-second-order equation based on equilibrium adsorption capacity may be represented as:

$$\frac{t}{q_t} = \frac{1}{k_2 q_e^2} + \frac{t}{q_e} \tag{6}$$

where k_2 g/(min mg) is the rate constant of second-order adsorption.

Adsorption experiments in a fixed-bed column

The adsorption efficiency of NNN-MCM-41 for removal of nitrate anion was investigated using fixed-bed glass column. A glass column of 24 cm length and 1 cm internal diameter was used to fixed bed column adsorption. The column was shut by glass wool, which guarantees good liquid distribution. A solution containing the nitrate anion was pumped upward to the column. A peristaltic pump was used to control the flow rate at the inlet and outlet (pump drive 5101, Heidolph, Germany). Samples of the effluent were collected periodically and analyzed for the remaining nitrate anion concentration using a spectrophotometer. For improving the column dynamics, adsorption analyses were done at various bed heights (2, 4 and 6 cm but the same inner diameter 6 mm) of NNN-MCM-41, various influent flow rates (0.8, 1.6 and 2.4 ml/min) and various initial concentrations (70, 110 and 150 mg/l) at pH 7 and ambient temperature.

The breakthrough curves show the loading behavior of NNNMCM-41 for the removal of nitrate from aqueous solution in a fixed-bed column. The area under the breakthrough curve obtained by integrating the adsorbed concentration of nitrate anion (mg/l) versus time (min) plot can be used to find the total adsorbed quantity of nitrate anion (q_{total}, mg) and for a certain inlet concentration and fluid flow rate (ml/min) and it can be calculated from Eq. (7) (Aksu et al. 2002; Aksu and Gönen 2004):

$$q_{total} = \frac{QC_0}{1000}\int_0^t \left(1 - \frac{C_e}{C_0}\right)dt \tag{7}$$

where C_0 is the initial concentration of nitrate anion, t_e is the exhaustion time, $(1 - C_e/C_0)$ is the concentration of anion adsorbed (mg/l), Q is the flow rate (ml/min) that can be computed by dividing the effluent volume (V_e, ml) to the total time (t_{total}, min). The total mass of nitrate anion adsorbed to column (m_{total}) can be found from Eq. (8) (Vijayaraghavan et al. 2005a):

$$m_{total} = \frac{QC_0t_e}{1000} \tag{8}$$

Total removal percent of nitrate anion (column performance) with respect to flow time can also be found from the ratio of the total adsorbed amount of nitrate (q_{total}) to the total of a quantity of nitrate anion sent to column (m_{total}) Eq. (9):

$$R = \frac{q_{total}}{m_{total}} \times 100 \tag{9}$$

The equilibrium adsorption capacity, q_e (mg/g), can be calculated using the following equation:

$$q_e = \frac{q_{total}}{m} \tag{10}$$

where m is the mass of the adsorbent (g).

Equilibrium studies on adsorption provide information about the adsorbent capacity or the quantity required to eliminate a unit mass of pollutants under the system. The column flow data were evaluated using various models. Two mathematical models including the Thomas model and Yan model were used. The Thomas model is expressed in the following form:

$$\frac{C_e}{C_0} = \frac{1}{1 + \exp[\frac{k_{Th}}{Q}(q_0m - C_0V)]} \tag{11}$$

where C_e is the effluent concentration (mg/l), C_0 is the influent concentration (mg/l), Q is the flow rate (ml/min), V is the effluent volume (ml), m is the mass of the adsorbent (g), t is the time (min) k_{Th} is the Thomas rate constant (ml/min mg) and q_0 is the maximum adsorption capacity (mg/g). The Thomas model parameters (k_{Th}, q_{Th}) can be assigned from the plot of C_e/C_0 against t using nonlinear regression analysis (Thomas 1944). Yan model proposed an empirical equation which could overcome the drawback in the Thomas model, especially its serious deficiency in predicting the effluent concentration at time zero (Yan and Viraraghavan 2001):

$$\frac{C}{C_0} = 1 - \frac{1}{1 + (\frac{QC_o}{q_Ym}t)^a} \tag{12}$$

where C_0 is the initial solution concentration (mg/l), Q is the flow rate (ml/min), q_Y is the maximum adsorption capacity (mg/g) of anion nitrate estimated by Yan model, and m is the amount of adsorbent (g).

Results and discussion

Characterization of adsorbents

Figure 2 gives the X-ray diffraction patterns of mesoporous MCM-41 silica before and after functionalization with amino groups. In the pattern of mesoporous MCM-41 silica as shown in Fig. 2a, a dominant (100) peak with a small (110), (200) and (210) reflections is attributed to the hexagonal structure (Yoshitake et al. 2002). The prominent peak at 2θ ranging between 2° and 2.5° corresponds to (100) plane, which is indicative of standard MCM-41 (Parida and Rath 2009). The XRD peak intensities of the MCM-41 mesoporous silica grafted by amino functionalized groups are lower than that of MCM-41 (Fig. 2b–d). The decrease in peak intensity provides evidence that amino-functionalized groups occurred mainly inside the mesopore channels.

The estimated hexagonal pore diameters from d_{100} values are listed in Table 1. The results of N_2 sorption

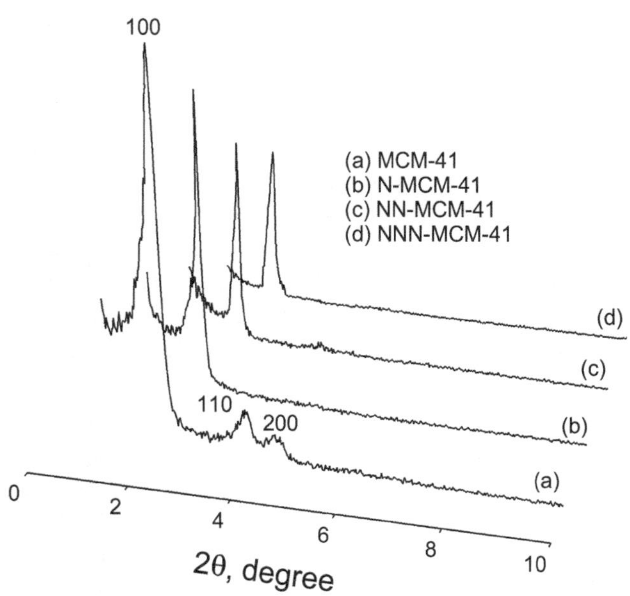

Fig. 2 XRD patterns of (*a*) MCM-41, (*b*) N-MCM-41, (*c*) NN-MCM-41and (*d*) NNN-MCM-41

Table 1 Physical properties of products mesoporous materials obtained by BET and XRD

Sample	Specific surface area (m^2/g)	Total pore volume (cm^3/g)	d_{100} (nm)	Wall thickness (nm)	Average pore diameter[a] (nm)	Average pore diameter[b] (nm)
MCM-41	1002	0.703	2.26	1.21	3.61	3.81
N-MCM-41	864	0.372	2.24	1.21	2.58	2.39
NN-MCM-41	742	0.341	2.22	1.21	2.56	2.18
NNN-MCM-41	733	0.336	2.14	1.21	2.28	1.86

[a] Unit cell diameter that calculated from XRD data ($a^\circ = 2d_{100}/3^{1/2}$)

[b] Unit cell diameter that calculated from BET data

Fig. 3 SEM figures of mesoporous MCM-41 silica

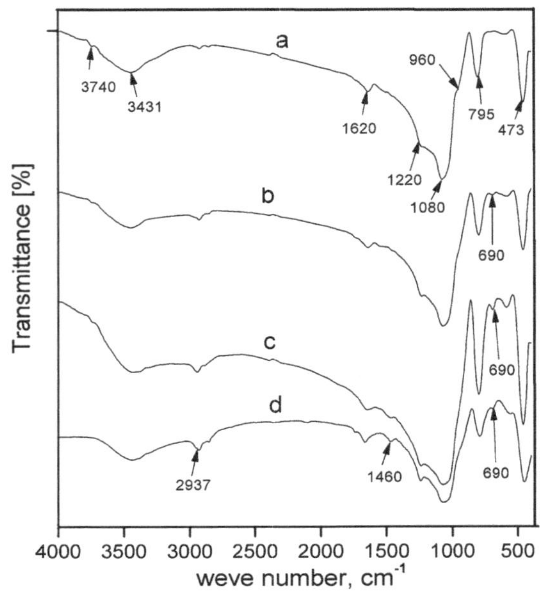

Fig. 4 FT-IR spectra of (*a*) MCM-41, (*b*) N-MCM-41, (*c*) NN-MCM-41 and (*d*) NNN-MCM-41

experiments are listed in Table 1. The specific surface area, pore size distribution and total pore volume of the mesoporous adsorbents were determined by the BET equation. MCM-41 was a porous material with a surface area of 1001 m^2/g and mean pore diameter of 2.8 nm. The BET surface area, total pore volume, and pore diameter decreased with increasing number of nitrogen atoms in amino-functionalized MCM-41.

The morphology of the MCM-41 was analyzed by a scanning electron microscope. A typical image is presented in Fig. 3, which shows that many amorphous and spherical shapes with the sizes of the particles range from 400 to 900 nm, which is in good agreement with the MCM-41 morphology presented in previous reports (Chen and Wang 2002; Luan et al. 1995).

Figure 4 shows the infrared spectra of MCM-41 prior to and following functionalized with amino groups. In all the materials, the bands around 473, 795, 1080 and 1220 cm^{-1} are assigned to links of (Si–O–Si), which is associated with the formation of a condensed silica network (Wang et al. 2005). A very wide absorption band focused at 3431 cm^{-1} determined to hydrogen-bonded Si–OH groups was intruded by physically adsorbed water, while a sharp absorption is perceived at 3740 cm^{-1} ascribed to free Si–OH groups. After amino-functionalization, the band at 3740 cm^{-1} for free Si–OH groups was evacuated (Peng et al. 2004). The presence of bands in 690, 1460 and 2937 cm^{-1} is assigned to N–H bending vibration, N–H asymmetric bending vibration and C–H link, respectively (Li et al. 2008).

The organic content of the surface modified mesoporous silica MCM-41, assessed by TGA is shown in Fig. 5. The weight loss of MCM-41 is relatively low due to the removal of absorbed water. The amino-functionalized MCM-41 showed a higher weight loss. In all materials, the weight loss below 150 °C corresponds to the desorption of water molecules. The curve corresponding to the MCM-41 (Fig. 5a) shows that the weight loss at higher temperatures (above 150 °C) is assigned to dehydroxylation of silanol groups. This is because the surfactant well was removed in the final stages of synthesis (Budi Hartono et al. 2009; Maria Chong and Zhao 2003). In curves corresponding to amino-functionalized MCM-41, the weight loss at above 150 °C is due to the decomposition of amine anchored to the pore walls (Parida and Rath 2009).

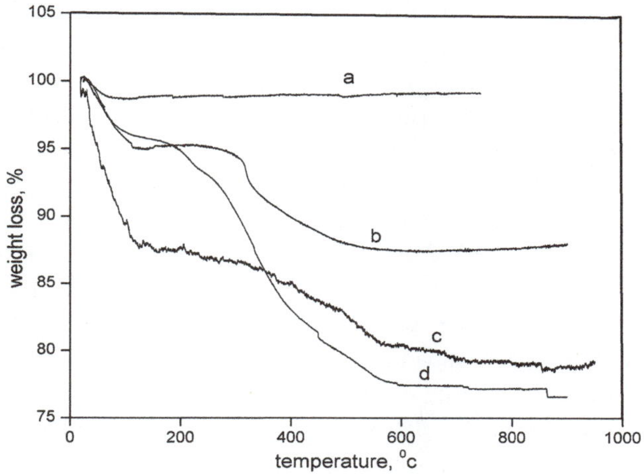

Fig. 5 TGA of (*a*) MCM-41, (*b*) N-MCM-41, (*c*) NN-MCM-41 and (*d*) NNN-MCM-41

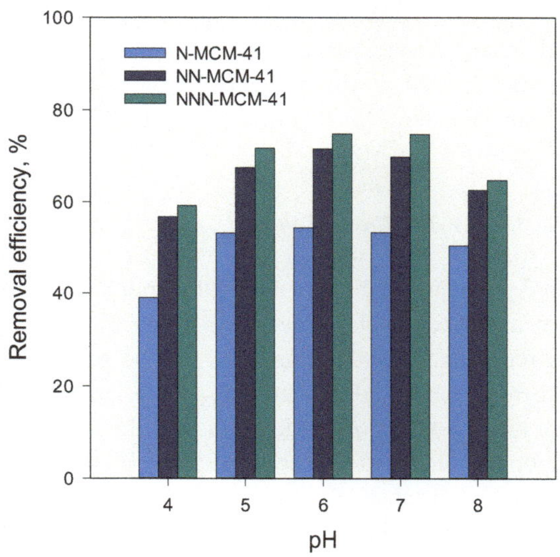

Fig. 6 Effect of pH on nitrate anion removal by N-MCM-41, NN-MCM-41and NNN-MCM-41: operated at anion concentration of 110 mg/l, adsorption dose of 5 g/l

Adsorption of nitrate anion in batch system

Effect of pH

The pH of the solution is a significant factor in the investigated adsorption process. So that the chemical characteristics of adsorbent and anion changed by changing the pH. The effect of the solution pH on the sorption of nitrate anions by amino-functionalized MCM-41 is shown in Fig. 6. For all adsorbents, the results of the adsorption tests showed that removal of nitrate anion increased with increase in pH up to 6, thereafter no significant change was observed (pH 6-7), but further increasing the pH from 7 to 8 resulted to a decrease in the removal percentage of nitrate

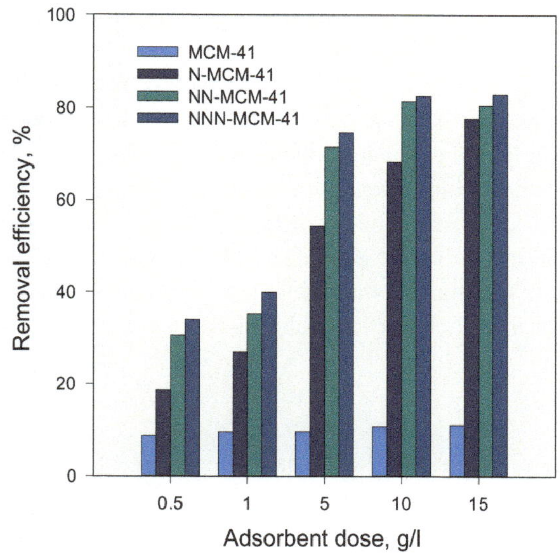

Fig. 7 Effect of adsorbent dose on the nitrate anion removal efficiency by N-MCM-41, NN-MCM-41 and NNN-MCM-41: operated at anion concentration of 110 mg/l, pH 7

anion. This may be due to the competition for the active sites by OH^- ions and the electrostatic repulsion of anionic nitrate by the negatively charged mesoporous surface by adsorption of nitrate anion at higher pH (Bhatnagar et al. 2010). Although the highest nitrate removal was achieved at pH 6, the adsorption experiments were performed at the pH 7 (neutral conditions), since only about 2 % of adsorbed nitrate was increased at pH 6. Hamoudi et al. (2007b) investigated the adsorptive removal of phosphate and nitrate anions using ammonium-functionalized mesoporous silica. Their results showed that the nitrate adsorption performance depends on pH and the obtained optimum pH was 6 that is in agreement with this study.

Effect of adsorbent dose

The amount of adsorbent determines the adsorption capacity in a given initial concentration of nitrate concentration. The adsorbent concentration was varied between 0.5 and 15 g/l in this set of experiments. The effect of the adsorbent doses on the removal efficiency of nitrate anions is shown in Fig. 7. It was found that the adsorption of nitrate anion in aqueous solution enhanced with an increment of adsorbent dose. It is due to the rising of interference between activating sites of adsorbent at higher sorbent dosages and also increase in the number of adsorptive site and surface area (Phuengprasop et al. 2011; Rome and Gadd 1987). The highest amount of nitrate anion removal in different adsorbents was observed in the amount of 15 g/l adsorbent dose. As percents of nitrate anion removal for N-MCM-41, NN-MCM-41 and NNN-MCM-41 obtained were 77.7, 80.5 and 82.9 %, respectively. On the other hand, the adsorption capacity reduced by increasing the adsorbent dose. At low dose of amine-modified MCM-41, the available binding sites on the adsorbent surface were occupied by nitrate ion, since the nitrate ion concentration raised the binding site yielding high equilibrium adsorbent nitrate concentration. However, at high dose of amine-modified MCM-41, a large number of binding sites were available on the adsorbent surface and only a fraction of those binding sites was occupied by nitrate ions yielding low solid-phase concentration

Adsorption isotherms

There are numerous mathematical equations for examining experimental adsorption equilibrium data. The parameters of these equilibrium models regularly give some information about the adsorption mechanism, the surface features and affinity of the adsorbent. The nonlinear Langmuir and Freundlich adsorption isotherms of nitrate anions obtained at the adsorbent dose of 5 g/l and temperature of 25 °C are shown in Fig. 8. The b, q_m, n, K_f values and the nonlinear regression coefficient of determination (R^2) for Langmuir and Freundlich isotherms are given in Table 2. The coefficient of determination indicates that adsorption was fitted better by the Langmuir ($R^2 = 0.98$–0.99) than the Freundlich model ($R^2 = 0.94$–0.97). Therefore, the adsorption process can be described by the formation of a monolayer coverage of the adsorbate on the adsorbent surface (Myers 2002). The amount of b in NNN-MCM-41 adsorbent was more than the other two adsorbents, indicating that bonding energy of sorption in this adsorbent was more from other two adsorbents. The maximum adsorption capacity (q_m) by Langmuir model was determined as 31.68, 38.58 and 38.81 mg/g for N-MCM-41, NN-MCM-41 and NNN-MCM-41, respectively. The sample NNN-MCM-41 exhibited a higher adsorption capacity than N- and NN-MCM-41. It is apparent that the adsorption capacities of nitrate improved by changing the amine type. By altering amine type from diamine to triamine, the number of adsorption places on the surface of amine functionalized MCM-41 increases, leading to a rise in the quantity of adsorbed nitrate anion. This trend was observed in the study conducted by Bui (2013). They functionalized SBA-16 with amine groups (mono-, di- and tri-ammonium silane) via post-synthesis grafting and utilized as potential adsorbent for nitrate adsorption from water. Their results show that the adsorption capacity values increase from 52.75 to 58.5, then 83.25 for g N-, NN-, NNN-SBA-16, respectively.

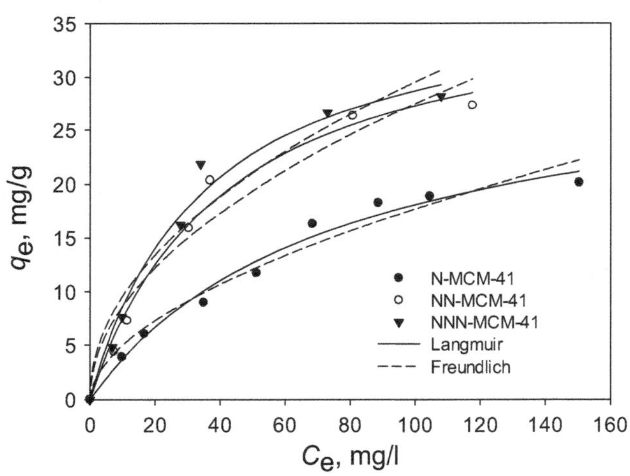

Fig. 8 Adsorption isotherm of nitrate anion on N-MCM-41, NN-MCM-41 and NNN-MCM-41: Langmuir and Freundlich at the adsorbent dose of 5 g/l, pH 7.0 and temperature of 25 °C

Compassion with other adsorbent

Table 3 compares the maximum adsorption capacity for nitrate anion removal with various silica material adsorbents reported in the literature. As can be seen, different adsorbents exhibit different adsorption capacities for nitrate anion adsorption, which indicates the role of a kind and property of a grafted functional group on the surface of sorbents. For instance, Saad et al. (2008) synthesized three different ammonium-functionalized mesoporous materials (MCM-41, MCM-48 and SBA-15) through post-synthesis grafting procedures and investigated their efficiency to remove nitrate anions in aqueous solutions. The adsorbents presented great adsorption capacities reaching 46.5 mg/g for N-MCM-48 and 36 mg/g for the MCM-41 and SBA-15, under the operating conditions explored. Also, Hamoudi et al. (2007a) modified MCM-48 with ammonium groups and used as an adsorbent for removing nitrate ion from aqueous solution. The adsorbent was effective for the removal of nitrate and the maximum adsorption capacity of

Table 2 Langmuir and Freundlich parameters for adsorption of nitrate anion onto N-MCM-41, NN-MCM-41 and NNN-MCM-41

Adsorbent	Langmuir model			Freundlich model		
	q_m, mg/g	b, l/mg	R^2	K_f	n	R^2
N-MCM-41	31.7	0.013	0.984	1.39	1.8	0.9669
NN-MCM-41	38.6	0.024	0.985	2.70	2.0	0.9482
NNN-MCM-41	38.8	0.028	0.982	3.10	2.1	0.9443

Table 3 Comparative evaluation adsorption capacities of mesoporous silica material with other adsorbents for nitrate removal

Adsorbent	Amount adsorbed (mg/g)	Initial concentration (mg/l)	Contact time (h)	Temperature (°C)	pH	Ref.
N-MCM-41	35.6 ± 3.8	130	3	25	<8	Saad et al. (2008)
N-MCM-48	46.5 ± 0.4					
N-SBA-15	35.9 ± 1.0					
N-MCM-48-N	51.8	100–700	1	5	–	Hamoudi et al. (2007b) and Saad et al. (2007)
N-SBA-16	70.46	250	1	5	3	Bui (2013)
PASBA-15	55.24	100–700	1	25	–	Dioum and Hamoudi (2014)
PTMA-SBA-15	62.50					
N-MCM-41	31.7	30–250	2	25	7	This Study
NN-MCM-41	38.6					
NNN-MCM-41	38.8					

51.8 mg/g was obtained at 5 °C. Besides, Dioum and Hamoudi (2014) compared the adsorptive performances of mono and quaternary ammonium-functionalized SBA-15 through post-synthesis grafting method for the removal of nitrate ion from aqueous solution. The results showed that adsorption capacity was affected by the organic functional group nature, thus the quaternary ammonium-functionalized SBA-15 has remarkably high adsorption capacity (63 mg/g) compared to mono one (55 mg/g). A similar tendency was observed in this study between N-MCM-41 (31.7 mg/g) and NNN-MCM-41 (38.8 mg/g). Moreover, the adsorption of nitrate ion from water onto functionalized amine SBA-16 mesoporous silica was investigated by Bui (2013). The adsorbents were synthesized by post-synthesis grafting method using mono-, di-, tri-ammonium functional groups. Based on the results, the maximum adsorption capacity was found to be 70.46 mg/g at a temperature of 5 °C. The quantity of adsorbed nitrate increased by changing the amine type from mono-, di-amine to tri-amine, which was in agreement with the results of this study. Comparison was made between the results of this study and the others from the above-mentioned studies of the nitrate adsorption on the amine-modified mesoporous silica according to the adsorption capacity and removal efficiency. The result showed that the adsorption capacity and removal efficiency NNN-MCM-41 synthesized in this study were higher than those mono-amine functionalized MCM-41 in the literature, while presented smaller adsorption capacity compared to MCM-48 and SBA-15 materials mentioned above (Table 3).

Adsorption kinetics

The adsorption kinetic study has an important role in the treatment of wastewater as it gives significant information on the pathways and mechanism of adsorption reactions. The kinetic rates of the nitrate anion adsorption process were investigated by the pseudo first-order and the pseudo second-order models, but the pseudo-second-order model were found to explain the experimental adsorption kinetic data, as shown in Fig. 9. The kinetic parameters (R^2, q_{e2} and K_2) of the pseudo-second-order sorption kinetics model were determined from the plots at different concentrations (Fig. 9). The values of R^2 and K_2 are presented in Table 4. As can be seen from Table 4, the coefficient of determination values of the pseudo-second-order model was >0.99 and the estimated adsorption capacity values acquired from pseudo-second-order model were more compatible with the experimental amounts of adsorption capacity. This suggested that the pseudo-second-order model represented better the sorption kinetics of nitrate anion. As seen in Fig. 9, the rate of nitrate adsorption increased rapidly for all the adsorbents, during the first 30 min of the experiment, then reduced gradually and reached steady state, 2 h after the beginning of the experiment. Since pseudo-second-order kinetic model assumes that the adsorption occurred via chemisorption (Bouhamed et al. 2012; Ho et al. 2003), so here, it implied that nitrate anion adsorption onto amino-functionalized MCM-41 may occur through a chemical procedure including valence forces through the sharing or trade of electrons. A good correlation coefficient of the kinetics data can show the mechanism of the nitrate adsorption to the solid phase (Ho and McKay 2000), whereas the pseudo-second-order kinetic model presumes that chemisorption may be the rate-limiting step (Low et al. 2000).

Desorption studies

Recovery and regeneration of the adsorbent and adsorbate have obtained economic, scientific and environmental motivations in recent years (Hamoudi et al. 2007a). For example, desorption study is helpful for elucidating the mechanism of adsorption, to regenerate the adsorbent, to

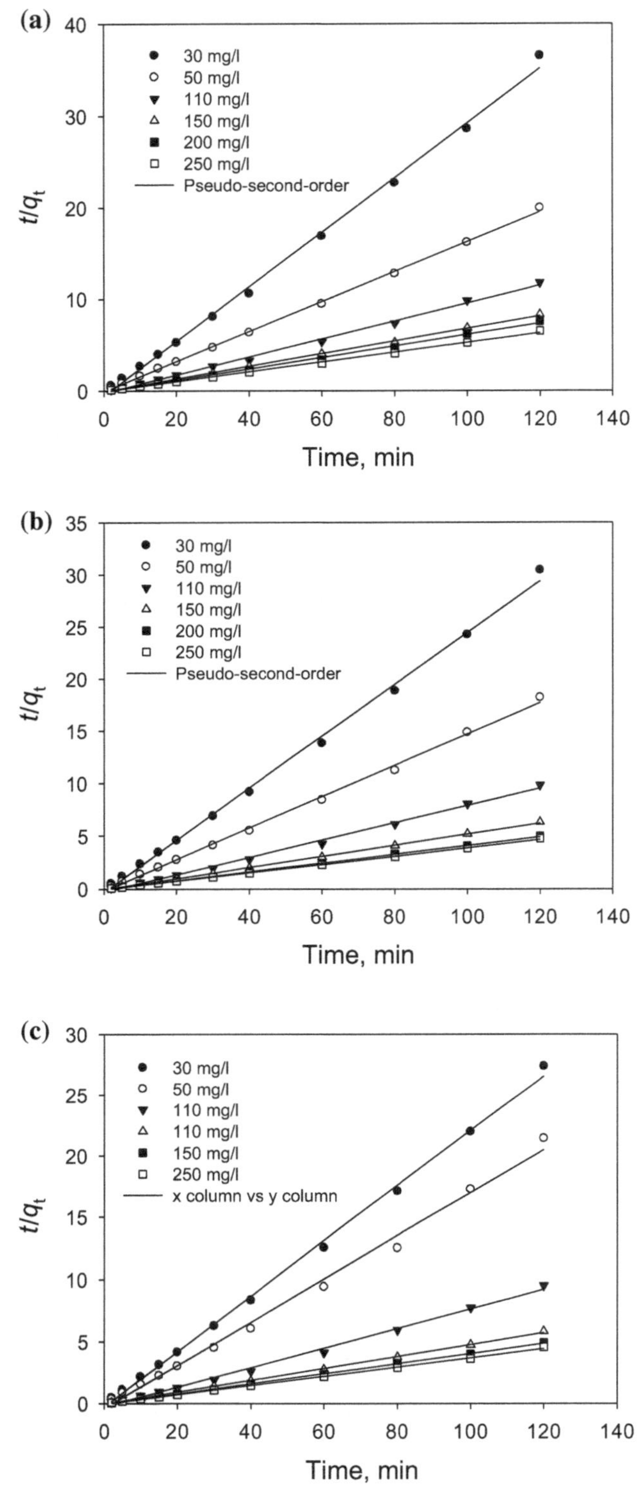

Fig. 9 Pseudo-second-order kinetic for adsorption of nitrate anion onto N-MCM-41 (**a**), NN-MCM-41 (**b**) and NNN-MCM-41 (**c**)

recover nitrate from the adsorbent and to protect the environment from the solid waste disposal problem. For these purposes and for examining desorption process, adsorbent was entered into the NaOH solution with concentration of 0.01 M. Afterward for the sake of activation, the adsorbent surface was entered into HCl solution with a concentration of 0.01 M. The desorption kinetics of nitrate anion in amino-functionalized MCM-41 in NaOH solutions is depicted (figure not shown). As observed, the desorption was rapid and complete for nitrate anions within 10 min. A maximum desorption of above 95 % N-MCM-41, NN-MCM-41 and NNN-MCM-41 was achieved by contact with NaOH (0.01 M), pH 7 and an adsorbent dose of 5 g/l for 10 min. Due to this, the adsorption of this anion with amino-functionalized adsorbents is completely reversible. The adsorbent was reused in four successive adsorption–desorption cycles. The adsorption capacity of the nitrate anion after three cycles indicated a loss in the adsorption capacity of only 5.5 % by N-MCM-41, 5.1 % by NN-MCM-41 and 5.7 % by NNN-MCM-41 compared to the initial cycle, indicating a good regeneration capacity of the adsorbent. Therefore, amino-functionalized MCM-41 can be used for the removal of nitrate anion from polluted waters.

Table 4 Kinetic adsorption parameters obtained using pseudo-second-order models

Adsorbent	Anion concentration, mg/l	Pseudo-second-order		
		k_2 (g/mg min)	q_{e2} (mg/g)	R^2
N-MCM-41	30	7.536	3.36	0.9970
	50	0.957	6.10	0.9990
	110	1.831	10.19	0.9960
	150	0.854	14.51	0.9980
	200	0.61	16.00	0.9970
	250	0.490	18.80	0.9980
NN-MCM-41	30	1.689	4.03	0.9970
	50	0.571	6.71	0.9970
	110	4.987	12.21	0.9950
	150	3.683	19.12	0.9990
	200	1.372	24.21	0.9990
	250	0.339	25.64	0.9990
NNN-MCM-41	30	1.642	4.49	0.9990
	50	0.365	5.75	0.9930
	110	4.971	12.66	0.9950
	150	3.872	20.75	0.9990
	200	1.116	24.33	0.9990
	250	0.447	26.67	0.9980

Adsorption of nitrate anion in fixed-bed column

For continuous adsorption experiments, NNN-MCM-41 adsorbent was used for the removal of nitrate anion from aqueous solutions due to high adsorption performance which obtained from batch experiments.

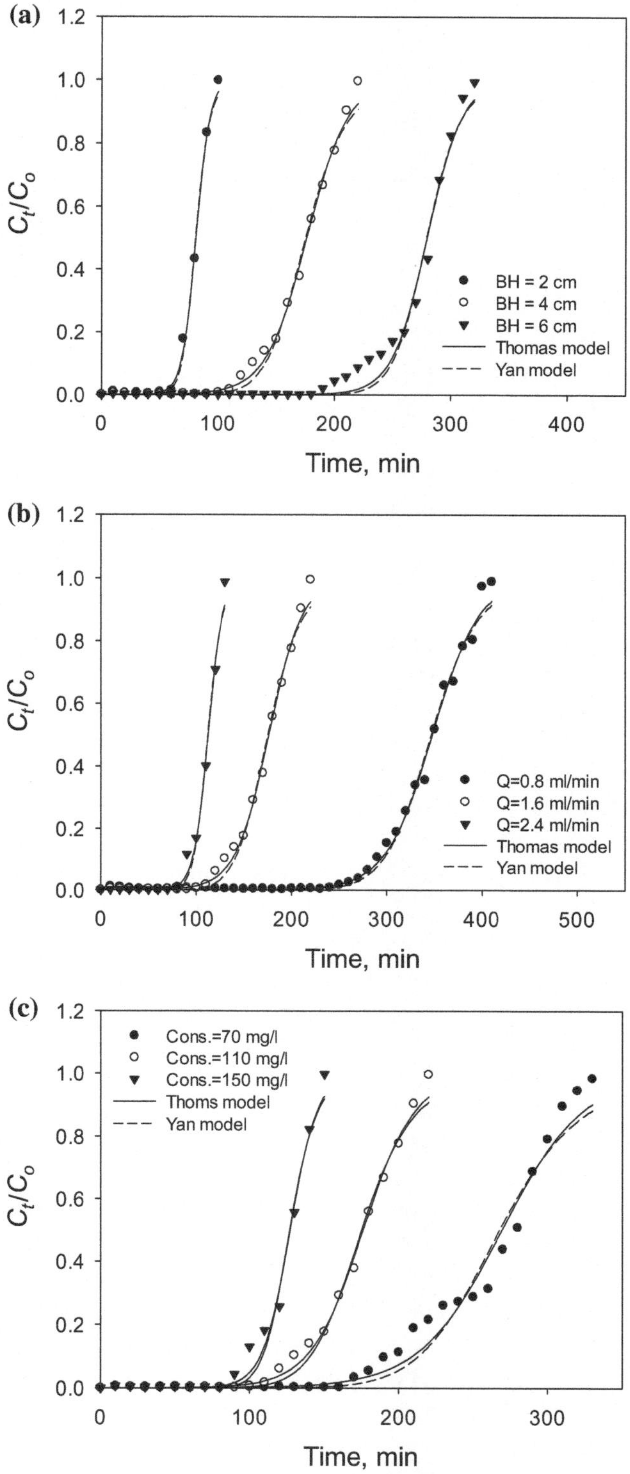

Fig. 10 Experimental breakthrough curve **a** at 1.6 ml/min flow rate, 110 mg/l concentration and different bed heights and **b** at 110 mg/l concentration, 4 cm bed height and different flow rate and **c** at 1.6 ml/min flow rate, 4 cm bed height and different concentration for nitrate anion removal by the NNN-MCM-41

Effect of bed height

Accumulation of nitrate in the packed bed column is mostly determined by the amount of adsorbent inside the column. The effect of NNN-MCM-41 bed height on nitrate anion adsorption in fixed-bed column was investigated using three different bed heights of 2, 4, and 6 cm at a constant flow rate of 1.6 ml/min and influent concentration of 110 mg/l. The adsorption breakthrough curves at different bed heights for studied anion are shown in Fig. 10a and the breakthrough analysis is summarized in Table 5. From this figure, it can be observed that the slope of breakthrough curves decreased by increasing the bed height. Both breakthrough and exhaust times were found to be increased with increasing the bed height from 2 to 6 cm, which resulted in a higher removal percentage of nitrate anion in the column because of the more prominent number of binding sites for adsorption process and also because the transfer zone requires more time to reach the column end for the nitrate anion to have more time to be in contact with NNN-MCM-41. Besides, it is well noted that the exhaustion time was 100, 220 and 320 min at the bed depth of 2, 4 and 6, respectively (Table 5). Further, the breakthrough analysis showed the increased maximum adsorption capacity and the equilibrium adsorption capacity of nitrate anion in the solution with increasing bed height. In this study, the equilibrium adsorption capacity of the nitrate anion increased from 20.13 to 23.01 mg/g with increasing bed height from 2 to 6 cm. It means that high uptake was observed at the highest bed height (6 cm). The reason for this is that the adsorption efficiency is dependent on the number of available adsorption sites for nitrate adsorption resulting in a higher surface area and hence higher capacity for adsorption (Baral et al. 2009; Vijayaraghavan et al. 2005a).

Effect of flow rate

The flow rate is one of the significant properties in assessing adsorbents for continuous treatment of nitrate effluents on an industrial scale. So, in the second stage of removal studies in the fixed-bed column with NNN-MCM-41, the impact of flow rate was examined by changing the flow rate from 0.8 to 2.4 ml/min while the inlet nitrate anion concentration in the feed was held constant at 110 mg/l and 4 cm bed height. The breakthrough curves of nitrate anion concentration versus time at different flow rates are given in Fig. 10b and the breakthrough analysis is summarized in Table 5. The results indicated that the

Table 5 The breakthrough analysis and parameters of Thomas and Yan models of NNN-MCM-41 for nitrate anion removal of various column conditions

Influent concentration, mg/l	Bed height, cm	Flow rate, ml/min	Breakthrough analysis					Thomas model			Yan model		
			q_e, mg/g	t_b, min	t_e, min	R, %	MTZ, cm	k_{Th}, ml/h/g	q_{Th}, mg/g	R^2	a	q_y, mg/g	R^2
110	4	0.8	21.62	274.4	410	83.90	1.32	0.371	21.85	0.994	14.06	21.78	0.992
110	4	1.6	21.87	117.5	220	79.06	1.86	0.521	22.15	0.995	10.04	22.02	0.992
110	4	2.4	20.91	83.8	130	85.29	1.42	1.209	21.18	0.991	15.06	21.13	0.988
110	2	1.6	20.13	62.2	100	80.07	0.76	1.534	20.34	0.997	13.69	20.27	0.995
110	6	1.6	23.01	204	320	85.79	2.23	0.612	23.43	0.986	19.25	23.41	0.982
70	4	1.6	21.08	178	330	79.84	1.84	0.522	21.54	0.978	9.67	21.42	0.970
150	4	1.6	21.39	91	150	83.16	1.57	0.736	21.79	0.988	14.30	21.74	0.983

adsorption of nitrate on the NNN-MCM-41 was strongly influenced by the flow rate. All the breakthrough curves had a comparable shape. In general, with the increasing flow rate, the percentage removal and slope of breakthrough curves increased. As indicated in Fig. 10b, at the lowest flow rate of 0.8 ml/min, relatively higher nitrate removal values were observed for nitrate adsorption to NNN-MCM-41. In contrast to bed height results, both exhaustion time and adsorption capacity of nitrate anion decreased with the increasing flow rate. Earlier breakthrough time and saturation time were appeared for higher flow rate, resulting in low uptake and minimum percent removal. This behavior can be explained that nitrate anion adsorption by NNN-MCM-41 is affected by insufficient residence time of the solute in the column, due to the reduced contact time between adsorbent and the nitrate anion in the column (Akar et al. 2009; Qaiser et al. 2009; Vijayaraghavan et al. 2005b). It is also observed that maximum nitrate uptake and nitrate removal obtained were 21.87 mg/g and 85.29 % at 2.4 and 1.6 ml/min, respectively (Table 5). Therefore, this should be emphasized that the fixed-bed column system has a good performance at the lower flow rate.

Effect of influent nitrate anion concentration

The adsorption performance of NNN-MCM-41 was tested at various nitrate anion inlet concentrations. The sorption breakthrough curves obtained by changing the inlet nitrate anion concentration from 70 to 150 mg/l at 1.6 ml/min flow rate and 4 cm bed height are given in Fig. 10c. As seen from Fig. 10c, a decreased inlet nitrate concentration gave an extended breakthrough curve and the time needed for the saturation of NNN-MCM-41 was longer at lower concentrations of nitrate in the influent. At the most nitrate concentration (150 mg/l), the NNN-MCM-41 was saturated rapidly, leading to earlier column stoppage and steeper breakthrough and exhaustion time. The breakthrough parameters of different nitrate anion concentrations are listed in Table 5. A maximum adsorption capacity (21.87 mg/g) was obtained with initial nitrate concentration 110 mg/l and at 1.6 ml/min flow rate and 4 cm bed height. The difference between the concentration of nitrate anions on the adsorbent and that in the aqueous solution is considered as the driving force (Aksu and Gönen 2004). Thus, the high driving force due to the high nitrate concentration resulted in better nitrate uptake. However, with increasing initial nitrate concentration, the removal efficiency decreased. This is because the functional groups in the adsorbent are not probably enough for adsorbed all the adsorbate molecules. Hence, with the increment of nitrate concentration, the number of available places becomes less for the adsorption of nitrate anion moles in solution.

Mathematical modeling of breakthrough curves

All the breakthrough curves acquired with various bed heights, flow rates and initial nitrate concentrations were fitted well to sigmoidal-type mathematical models, namely, Thomas and Yan and models (Fig. 10) and the adsorption capacity was calculated based on the fitting results (Table 5). The Thomas model is one of the most general and broadly utilized models for depicting column breakthrough data. The Thomas model, which accepts Langmuir adsorption and desorption kinetics and no axial dispersion, is determined with the adsorption that the rate driving force complies second-order kinetics (Aksu and Gönen 2004; Vijayaraghavan et al. 2005a). The fixed-bed column data were fitted to the Thomas model to determine maximum solid-phase concentration (q_{Th}). Fig. 10 shows the comparison of experimentally determined and the Thomas model predicted breakthrough curves of nitrate adsorption by NNN-MCM-41 at a flow rate of 1.6 ml/min, influent concentration of 110 mg/l and different bed height. The Thomas rate constant (k_{Th}) and the maximum adsorption capacity (q_{Th}) along with the coefficient of determination, under different conditions, are given in Table 5. It is clear from Fig. 10 that the model gave a good fit of the experimental data obtained from all flow rates, bed heights and initial nitrate anion concentration with a correlation coefficient, r^2, ranging from 0.978 to 0.995 for NNN-MCM-41, which agrees well with the breakthrough experiments. As can be seen from Table 5, the estimations of k_{Th} and q_{Th} were impacted by different conditions. As the bed height increased, the values of q_{Th} increased and the values of k_{Th} decreased. The maximum amount of q_{Th} was 23.43 mg/g at bed height 6 cm. The capacity q_{Th} diminished and Thomas constant k_{Th} expanded with expanding flow rate. In contrast, q_{Th} improved and k_{Th} reduced with increasing initial nitrate concentration.

Although the high coefficient of determination was acquired utilizing the Thomas model, Yan model allows us to predict with great precision, the breakthrough curve, in the whole range of effluent concentration and at time zero and also this model resolves the error resulting from the use of the Thomas model, (Yan and Chen 2001; Pokhrel and Viraraghavan 2008). The breakthrough curves predicted by the Yan model under different conditions are shown in Fig. 10. The model constants (R^2, k_y, q_Y) obtained by the nonlinear fit of the data are presented in Table 5. At all conditions examined, the breakthrough behavior of the nitrate fed column packed with NNN-MCM-41 was fitted well with the Yan model (based on the R^2 value). Also the maximum adsorption capacity (q_Y) predicted by the Yan model showed the same trend predicted by the Thomas model. Due to the high regression coefficient of determination and close to q_e of experimental value and q_y in each of the three column heights, it can be concluded that Yan model can be a well-described nitrate anion adsorption with mesoporous NNN-MCM-41 in the continuous system. The quantity of nitrate uptake predicted by this model corresponding to 2, 4 and 6 cm bed height of the column was 20.27, 21.13 and 23.41 mg/g, respectively.

Conclusions

The adsorption of nitrate anion with amino-functionalized mesoporous silica was investigated in batch and fixed-bed column systems. Due to their high surface areas, their open pore structure and their appropriate surface functionalization, these adsorbents proved to be high capacity for nitrate anions in water. Batch experiments were conducted at different pH, initial concentration of anions, and adsorbent dose. The results show that the highest removal of nitrate was achieved at pH 7 and adsorption capacity increased with the adsorbent loading. The experimental equilibrium data of nitrate anion fitted with the Langmuir isotherm better than the Freundlich isotherm, with a higher coefficient of determination. The maximum uptake capacity for nitrate anion was 31.68, 38.58 and 38.81 mg/g for N-MCM-41, NN-MCM-41 and NNN-MCM-41, respectively. The kinetics of nitrate anion adsorption onto amino-functionalized MCM-41 revealed that adsorption satisfactorily follows the pseudo-second-order equation. Desorption was rapid and it was completely achieved within 10 min for nitrate anion using NaOH at 0.01 M. Continuous experiments in the fixed-bed column were performed to investigate the impact of bed height, flow rate and initial nitrate concentration. On the basis of the experimental results, the increase of influent concentration and flow rate decreased the breakthrough and exhausted times. In contrast, both breakthrough and exhaust times were found to be increased with increasing the bed height. The Thomas and Yan models are used to describe column data obtained at different conditions. Both models were able to predict breakthrough curves with a coefficient of determination ranging from 0.97 to 0.995.

Acknowledgments The authors wish to thank the Iranian Nano Technology Initiative Council and Tarbiat Modares University for financing this research work.

References

Akar T, Kaynak Z, Ulusoy S, Yuvaci D, Ozsari G, Akar ST (2009) Enhanced biosorption of nickel(II) ions by silica-gel-immobilized waste biomass: biosorption characteristics in batch and dynamic flow mode. J Hazard Mater 163(2–3):1134–1141

Aksu Z, Gönen F (2004) Biosorption of phenol by immobilized activated sludge in a continuous packed bed: prediction of breakthrough curves. Process Biochem 39(5):599–613

Aksu Z, Gönen F, Demircan Z (2002) Biosorption of chromium(VI) ions by Mowital®B30H resin immobilized activated sludge in a packed bed: comparison with granular activated carbon. Process Biochem 38(2):175–186

Baral SS, Das N, Ramulu TS, Sahoo SK, Das SN, Chaudhury GR (2009) Removal of Cr(VI) by thermally activated weed *Salvinia cucullata* in a fixed-bed column. J Hazard Mater 161(2–3):1427–1435

Beck JS, Vartuli JC, Roth WJ, Leonowicz ME, Kresge CT, Schmitt KD et al (1992) A new family of mesoporous molecular sieves prepared with liquid crystal templates. J Am Chem Soc 114(27):10834–10843

Bhatnagar A, Kumar E, Sillanpää M (2010) Nitrate removal from water by nano-alumina: characterization and sorption studies. Chem Eng J 163(3):317–323

Bouhamed F, Elouear Z, Bouzid J (2012) Adsorptive removal of copper(II) from aqueous solutions on activated carbon prepared from Tunisian date stones: equilibrium, kinetics and thermodynamics. J Taiwan Inst Chem Eng 43(5):741–749

Broséus R, Cigana J, Barbeau B, Daines-Martinez C, Suty H (2009) Removal of total dissolved solids, nitrates and ammonium ions from drinking water using charge-barrier capacitive deionisation. Desalination 249(1):217–223

Budi Hartono S, Qiao S, Jack K, Ladewig BP, Hao Z, Lu G (2009) Improving adsorbent properties of cage-like ordered amine functionalized mesoporous silica with very large pores for bioadsorption. Langmuir 25(11):6413–6424

Bui THL (2013) Removal of nitrate from water and wastewater by ammonium-functionalized SBA-16 mesoporous silica. Laval University, Quebec City

Chen H, Wang Y (2002) Preparation of MCM-41 with high thermal stability and complementary textural porosity. Ceram Int 28(5):541–547

Cheng IF, Muftikian R, Fernando Q, Korte N (1997) Reduction of nitrate to ammonia by zero-valent iron. Chemosphere 35(11):2689–2695

Dioum A, Hamoudi S (2014) Mono- and quaternary-ammonium functionalized mesoporous silica materials for nitrate adsorptive removal from water and wastewaters. J Porous Mater 21(5):685–690

EPA (2010) Basic information about nitrate in drinking water. US EPA, Washington, DC

Ghorbani F, Younesi H, Ghasempouri SM, Zinatizadeh AA, Amini M, Daneshi A (2008) Application of response surface methodology for optimization of cadmium biosorption in an aqueous solution by *Saccharomyces cerevisiae*. Chem Eng J 145(2):267–275

Hamoudi S, Saad R, Belkacemi K (2007a) Adsorptive removal of phosphate and nitrate anions from aqueous solutions using ammonium-functionalized mesoporous silica. Ind Eng Chem Res 46(25):8806–8812

Hamoudi S, Saad R, Belkacemi K (2007b) Adsorptive removal of phosphate and nitrate anions from aqueous solutions using ammonium-functionalized mesoporous silica. Ind Eng Chem Res 46(25):8806–8812

Heidari A, Younesi H, Mehraban Z (2009) Removal of Ni(II), Cd(II), and Pb(II) from a ternary aqueous solution by amino functionalized mesoporous and nano mesoporous silica. Chem Eng J 153(1–3):70–79

Hell F, Lahnsteiner J, Frischherz H, Baumgartner G (1998) Experience with full-scale electrodialysis for nitrate and hardness removal. Desalination 117(1–3):173–180

Ho YS, McKay G (2000) The kinetics of sorption of divalent metal ions onto sphagnum moss peat. Water Res 34(3):735–742

Ho KY, McKay G, Yeung KL (2003) Selective adsorbents from ordered mesoporous silica. Langmuir 19(7):3019–3024

Huang YH, Zhang TC (2004) Effects of low pH on nitrate reduction by iron powder. Water Res 38(11):2631–2642

Kapoor A, Viraraghavan T (1997) Nitrate removal from drinking water: review. J Environ Eng 123(4):371–380

Levine SN, Schindler DW (1989) Phosphorus, nitrogen, and carbon dynamics of experimental lake 303 during recovery from eutrophication. Can J Fish Aquat Sci 46(1):2–10

Li J, Miao X, Hao Y, Zhao J, Sun X, Wang L (2008) Synthesis, amino-functionalization of mesoporous silica and its adsorption of Cr(VI). J Colloid Interface Sci 318(2):309–314

Lim MH, Stein A (1999) Comparative studies of grafting and direct syntheses of inorganic-organic hybrid mesoporous materials. Chem Mater 11(11):3285–3295

Low KS, Lee CK, Liew SC (2000) Sorption of cadmium and lead from aqueous solutions by spent grain. Process Biochem 36(1–2):59–64

Luan Z, Cheng C-F, Zhou W, Klinowski J (1995) Mesopore molecular sieve MCM-41 containing framework aluminum. J Phys Chem 99(3):1018–1024

Maria Chong AS, Zhao XS (2003) Functionalization of SBA-15 with APTES and characterization of functionalized materials. J Phys Chem B 107(46):12650–12657

McKittrick MW, Jones CW (2003) Toward single-site functional materialspreparation of amine-functionalized surfaces exhibiting site-isolated behavior. Chem Mater 15(5):1132–1139

Myers D (2002) Frontmatter and index. In: Surfaces, interfaces, and colloids: principles and Applications, 2nd edn. John Wiley & Sons, Inc., New York

Parida KM, Rath D (2009) Amine functionalized MCM-41: an active and reusable catalyst for Knoevenagel condensation reaction. J Mol Catal A Chem 310(1–2):93–100

Peng Q, Yang Y, Yuan Y (2004) Immobilization of rhodium complexes ligated with triphenyphosphine analogs on amino-functionalized MCM-41 and MCM-48 for 1-hexene hydroformylation. J Mol Catal A Chem 219(1):175–181

Phuengprasop T, Sittiwong J, Unob F (2011) Removal of heavy metal ions by iron oxide coated sewage sludge. J Hazard Mater 186(1):502–507

Pintar A, Batista J, Levec J (2001) Catalytic denitrification: direct and indirect removal of nitrates from potable water. Catal Today 66(2–4):503–510

Pokhrel D, Viraraghavan T (2008) Arsenic removal in an iron oxide-coated fungal biomass column: analysis of breakthrough curves. Bioresour Technol 99(6):2067–2071

Qaiser S, Saleemi AR, Umar M (2009) Biosorption of lead from aqueous solution by Ficus religiosa leaves: batch and column study. J Hazard Mater 166(2–3):998–1005

Rome L, Gadd GM (1987) Copper adsorption by *Rhizopus arrhizus*, *Cladosporium resinae* and *Penicillium italicum*. Appl Microbiol Biotechnol 26(1):84–90

Saad R, Belkacemi K, Hamoudi S (2007) Adsorption of phosphate and nitrate anions on ammonium-functionalized MCM-48: effects of experimental conditions. J Colloid Interface Sci 311(2):375–381

Saad R, Hamoudi S, Belkacemi K (2008) Adsorption of phosphate and nitrate anions on ammonium-functionnalized mesoporous silicas. J Porous Mater 15(3):315–323

Schoeman JJ, Steyn A (2003) Nitrate removal with reverse osmosis in a rural area in South Africa. Desalination 155(1):15–26

Sivaraj R, Namasivayam C, Kadirvelu K (2001) Orange peel as an adsorbent in the removal of Acid violet 17 (acid dye) from aqueous solutions. Waste Manag 21(1):105–110

Soares MIM (2000) Biological denitrification of groundwater. Water Air Soil Pollut 123(1):183–193

Soner Altundoğan H, Altundoğan S, Tümen F, Bildik M (2000) Arsenic removal from aqueous solutions by adsorption on red mud. Waste Manag 20(8):761–767

Thomas HC (1944) Heterogeneous ion exchange in a flowing system. J Am Chem Soc 66(10):1664–1666

Vaghetti JCP, Lima EC, Royer B, da Cunha BM, Cardoso NF, Brasil JL et al (2009) Pecan nutshell as biosorbent to remove Cu(II), Mn(II) and Pb(II) from aqueous solutions. J Hazard Mater 162(1):270–280

Vasconcelos HL, Camargo TP, Gonçalves NS, Neves A, Laranjeira MCM, Fávere VT (2008) Chitosan crosslinked with a metal complexing agent: synthesis, characterization and copper(II) ions adsorption. React Funct Polym 68(2):572–579

Vijayaraghavan K, Jegan J, Palanivelu K, Velan M (2005a) Batch and column removal of copper from aqueous solution using a brown marine alga Turbinaria ornata. Chem Eng J 106(2):177–184

Vijayaraghavan K, Jegan J, Palanivelu K, Velan M (2005b) Biosorption of cobalt(II) and nickel(II) by seaweeds: batch and column studies. Sep Purif Technol 44(1):53–59

Wang X, Lin KSK, Chan JCC, Cheng S (2005) Direct synthesis and catalytic applications of ordered large pore aminopropyl-functionalized SBA-15 mesoporous materials. J Phys Chem B 109(5):1763–1769

Wu F-C, Tseng R-L, Juang R-S (2001) Kinetic modeling of liquid-phase adsorption of reactive dyes and metal ions on chitosan. Water Res 35(3):613–618

Yan G, Viraraghavan T, Chen M (2001) A new model for heavy metal removal in a biosorption column. Adsorpt Sci Technol 19(1):25–43

Yan G, Viraraghavan T (2001) Heavy metal removal in a biosorption column by immobilized M. rouxii biomass. Bioresour Technol 78(3):243–249

Yoshitake H, Yokoi T, Tatsumi T (2002) Adsorption of chromate and arsenate by amino-functionalized mcm-41 and SBA-1. Chem Mater 14(11):4603–4610

Hydrogeochemical characteristics of groundwater in the coal-bearing aquifer of the Wugou coal mine, northern Anhui Province, China

Song Chen[1,2,3] · **Herong Gui**[1,2]

Abstract Major ions, trace elements, and isotope concentrations were measured in 11 representative groundwater samples that were collected from a series of aquifers in the Wugou coal mine, Anhui Province. The geochemical characteristics of the groundwater samples were examined using conventional graphical and multivariate statistical approaches, and the results showed that almost all of the groundwater samples collected from the coal-bearing aquifer were the Na-SO_4 type, whereas the samples from the Quaternary and the limestone aquifer were the Ca-SO_4 and Na-Cl types, respectively. The groundwater in the study area is not suitable for drinking without treatment because of the higher values of total dissolved solids and other parameters, whereas the lower value of the sodium adsorption ratio indicates that it can be used for irrigation. The total rare earth element concentrations ranged from 0.0398 to 0.1874 mg/L, and had an average of 0.075 mg/L. There were negative cerium and positive europium anomalies in the groundwater. The δD and δ^{18}O values in groundwater ranged from −9.01 to −8.81 ‰, and from −74.7 to −71.4 ‰, respectively. Meteoric water with variable degrees of evaporation is the main source of the groundwater in the coal-bearing aquifer.

✉ Song Chen
szxychensong@163.com

[1] School of Resources and Civil Engineering, Suzhou University, Suzhou 234000, China

[2] National Engineering Research Center of Coal Mine Water Hazard Controlling, Suzhou 234000, China

[3] School of Earth and Space Science, University of Sciences and Technology of China, Hefei 230000, China

Keywords Hydro-geochemistry · Groundwater · Wugou coal mine · REE · Isotopes

Introduction

Deep groundwater systems not only form an important consideration in coal exploitation, but also serve as major storage zones for fresh water resources. Many studies have examined the hydrochemical characteristics, water–rock interactions, sources, and quality of groundwater in coal mining areas (Gui 2014; Nakano 2014; Chen et al. 2014). These studies provide supporting information to efficient groundwater development schemes and help reduce coal mining hazards as well as water quality deterioration. Groundwater under different hydrogeological conditions has different hydrochemical characteristics due to the influence of surrounding rocks, water flow, and stratum. Therefore, studies that provide information on the hydrochemistry of groundwater from coal-bearing aquifers of individual coal mining regions are needed. To date, few studies have provided information about groundwater conditions in the Wugou coal mine. In particular, no studies exist on either the trace (and rare earth) elements or isotopic contents of groundwater in this coal mine.

Previous studies have shown that signatures of rare earth elements (REE) in groundwater, such as total REE concentrations, anomalies of cerium (Ce), europium (Eu), or yttrium (Y), and normalized REE patterns, vary in different aquifers (Gui et al. 2011; Sun et al. 2011); The varying pattern of these constituents could also provide vital information about groundwater source. In this study, 11 representative groundwater samples were collected from different aquifers in the Wugou coal mine. To derive information about the hydrochemical characteristics of deep groundwater in this coal mine, major ions, trace elements, and isotope signatures (δD and $\delta^{18}O$) were measured. The main objectives of the study were (1) to define the geochemical characteristics of groundwater collected from different aquifers, and (2) to discuss the characteristics of trace elements in groundwater samples.

Materials and methods

The Wugou coal mine is located in northern Anhui Province, China (Fig. 1). The area has a marine-continental climate, and annual average temperature of 14.9 °C. Previous studies have shown that the deep groundwater system in the Wugou coal mine comprises three main aquifers, the Quaternary aquifer (QA), coal-bearing aquifer (CA), and a limestone aquifer (LA), all of which have influence on coal exploration (Gui and Chen 2007). The QA ranges in depth between 280 and 300 m, and composed of mainly yellow mudstone, sandstone, and conglomerate. The CA is characterized by mudstone, siltstone, and sandstone, occurring at a depth range between 300 and 700 m. The LA belongs to the Taiyuan and Ordovician Formations, and composed of mainly limestone (Fig. 2).

Nine representative samples of groundwater were collected from the main CA in the Wugou coal mine and two groundwater samples were collected from the QA (WG6) and LA (WG9) to permit comparison. Groundwater samples were collected via drainage holes in alleys, and then filtered into pre-cleaned and sterilized polyethylene bottles that had been cleaned following trace element cleaning procedures through a 0.45-µm membrane. All groundwater samples were analyzed for major ions and trace elements. Four samples were also analyzed for δD and $\delta^{18}O$ concentrations. Major ions were determined in the analysis testing laboratory of the Department of Coal Geology, Anhui Province, China. The K^{+}+Na^{+}, SO_4^{2-} and Cl^{-}, Ca^{2+} and Mg^{2+}, and alkaline concentrations were determined by atomic absorption spectrometry, ion chromatography, EDTA titration, and acid-based titration, respectively. After liquid–liquid extraction, trace element concentrations were determined by inductively coupled plasma-mass spectrometry (POEMS III) in the State Key Laboratory of Geological Processes and Mineral Resources, China University of Geosciences (Wuhan). The analytical precision for all trace elements was 10 % of the relative standard deviation or better. The isotopic compositions were determined in the laboratory of the Institute of Karst Geology of the Chinese Academy of Geological Sciences. The isotopic data are reported relative to standard mean ocean water, and the overall precisions of $\delta^{18}O$ and δD were 0.2 and 2 ‰, respectively.

Results and discussion

Major ion chemistry

The analysis results of groundwater samples collected from the Wugou coal mine are presented in Table 1. The ionic balance between anions and cations was checked by AqQa software, and the relative deviation was less than 2.0 %. In general, the pH values of groundwater ranged from 7.7 to 8.5, and had a mean value of 8.1, indicating that all the groundwater samples were alkaline in nature. The anionic concentrations (Cl^{-}, SO_4^{2-}, HCO_3^{-}, and CO_3^{2-}) in the groundwater ranged from 142.9 to 954.2, 244.1 to 1940.7, 242.4 to 715.5, and 0 to 52.2 mg/L, with average values of 303.1, 941.1, 344.5, and 11.1 mg/L, respectively. The cationic concentrations (K^{+}+Na^{+}, Ca^{2+}, and Mg^{2+}) ranged from 293.3 to 800.2, 7.1 to 401.4, and 3.8 to 195.9 mg/L, with average values of 458.0, 147.3, and 84.0, respectively. The groundwater samples were mainly the Na-SO_4 and Na-Cl types, indicating that Na^{+}, SO_4^{2-}, and Cl^{-} were

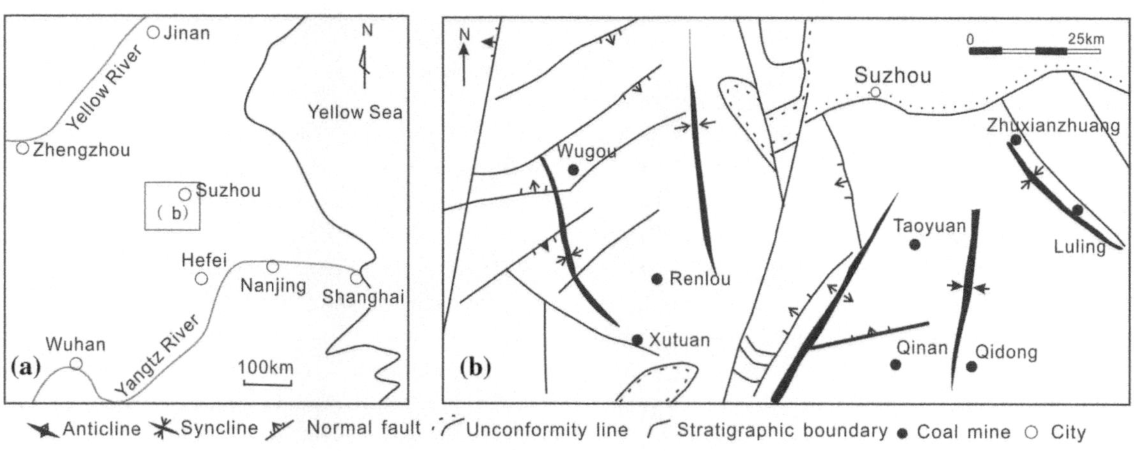

Fig. 1 Geographic location (**a**) and geological map (**b**) of the study area in northern Anhui Province, China

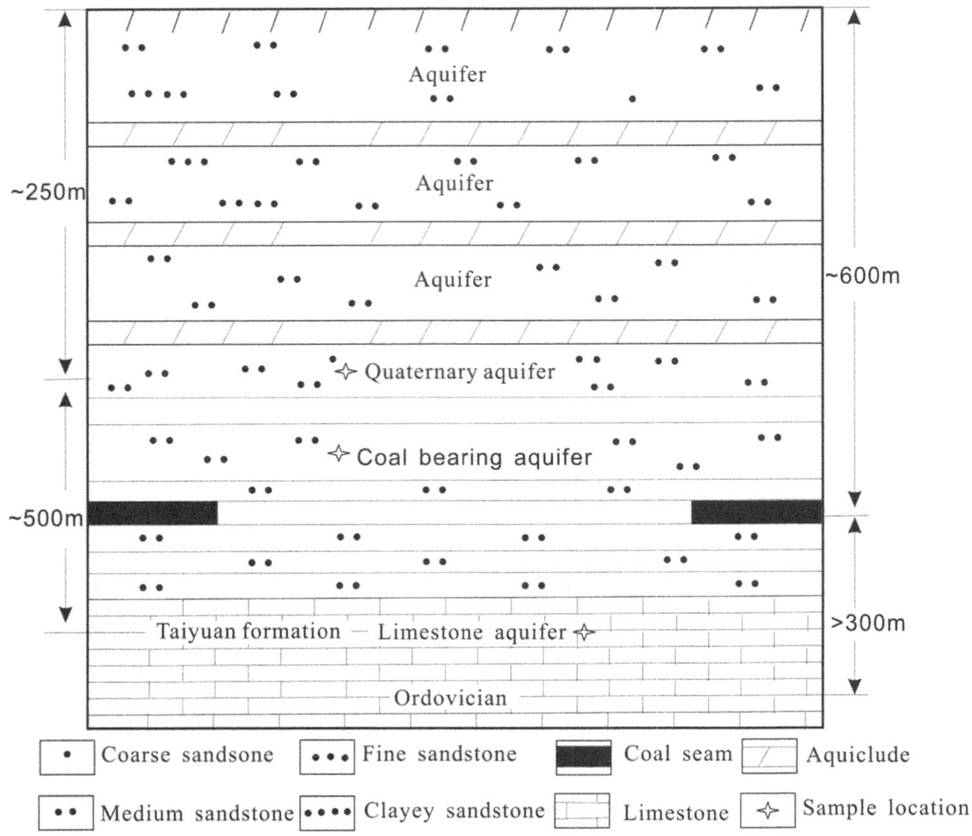

Fig. 2 Hydrogeological profile and sampling points in the study area

the dominant ions in the groundwater. Sample WG6 from the QA was the Ca-SO_4 water type, and had high concentrations of Ca^{2+} and Mg^{2+}. Almost all of the groundwater samples from the CA were described as the Na-SO_4 type, while the sample from LA was the Na-Cl type.

To further understand the hydrochemical characteristics of the groundwater from the different aquifers, the ionic concentrations were plotted on a Piper diagram (Fig. 3). This diagram shows that, with the exception of samples WG6 and WG5, almost all the alkali elements (K^+ and Na^+) exceeded the alkaline earth elements (Ca^{2+} and Mg^{2+}) in the groundwater samples, and strong acids (SO_4^{2-} and Cl^-) exceeded the weak acids (HCO_3^- and CO_3^{2-}) in the groundwater samples.

To evaluate the suitability of the groundwater in the study area for various uses, the analytical results of the groundwater samples were compared with the World Health Organization (WHO) standards (World Health Organization (WHO) 1993). This comparison indicated that the magnesium (84.0 mg/L) and calcium (147.3 mg/L) concentrations, as well as the pH values were within the WHO standards. Other parameters were found to exceed threshold limits. Concentrations of total dissolved solids (TDS) in the groundwater were higher than 1000 mg/L, indicating that the groundwater in the study area is not suitable for drinking before treatment. The higher concentrations of Na^+ in the groundwater could influence crop growth if used for irrigation. The suitability of groundwater for irrigation is generally evaluated using various indices. The sodium adsorption ratio (SAR) (United States Salinity Laboratory (USSL) 1954), which can be calculated using the formula: $SAR = Na+/SQRT((Ca^{2+}-Mg^{2+})/2)$, is frequently used. SAR values can be divided into four categories: excellent (<10), good (10–18), permissable (18–26), and unsuitable (>26). The mean value of the calculated SAR results for the study area was 10.7, indicating that the groundwater was suitable for irrigation use.

Trace elements

The REE concentrations of groundwater samples are listed in Table 1 and the Post Archean Average Shale (PAAS) normalized REE patterns are presented in Fig. 4 (Taylor and McLennan 1985). The total REE concentrations ($\sum$REE) ranged from 0.0398 to 0.1874 mg/L, with an average value of 0.075 mg/L. The normalized neodymium/ytterbium (Nd_{SN}/Yb_{SN}) ratios (SN indicates PAAS normalization) ranged from 0.033 to 0.357, with an average of 0.103. This ratio, when combined with the features of Fig. 4, indicates that, when normalized to PAAS, the groundwater samples are more enriched in heavy REEs

Table 1 Major ions (mg/L), trace elements (μg/L), isotopes, and physico-chemical properties of groundwater in the study area

Parameter	WG1	WG2	WG3	WG4	WG5	WG6	WG7	WG8	WG9	WG10	WG11
$K^{+}+Na^{+}$	666.0	539.2	311.1	293.3	367.5	334.7	337.9	588.6	800.2	366.9	433.0
Ca^{2+}	137.0	159.1	78.4	125.9	190.8	401.4	109.3	7.1	190.8	114.8	105.3
Mg^{2+}	43.2	72.0	65.3	78.7	165.2	195.9	68.7	3.8	94.1	76.8	60.5
Cl^{-}	206.4	195.8	231.1	352.8	142.9	220.5	294.6	273.4	954.2	231.1	231.1
SO_4^{2-}	1291.6	1260.7	587.8	447.4	1441.4	1940.7	572.5	244.1	1012.5	756.9	796.9
HCO_3^{-}	403.9	309.3	247.0	380.8	242.4	247.0	339.3	715.5	247.0	294.7	362.4
CO_3^{2-}	0	13.62	0	0	31.78	0	0	52.22	0	24.97	0
PH	7.91	8.36	8.16	7.91	8.31	7.71	7.85	8.5	8.05	8.38	8.16
TDS	2557	2405	1405	1497	2471	3227	1563	1536	3186	1731	1819
Aquifer	CA	CA	CA	CA	CA	QA	CA	CA	LA	CA	CA
Water type	Na-SO_4	Na-SO_4	Na-SO_4	Na-Cl	Na-SO_4	Ca-SO_4	Na-SO_4	Na-HCO_3	Na-Cl	Na-SO_4	Na-SO_4
SAR	12.7	8.9	6.28	5.05	4.7	3.42	6.23	44.1	11.8	6.5	8.32
La	0.0067	0.0036	0.0022	0.0039	0.0032	0.003	0.0023	0.0022	0.0031	0.0021	0.0026
Ce	0.032	0.0079	0.0049	0.0039	0.0078	0.0059	0.004	0.0062	0.0051	0.003	0.0025
Pr	0.0013	0.0015	0.0009	0.001	0.0014	0.0023	0.0006	0.0013	0.0009	0.0012	0.0017
Nd	0.009	0.01	0.0034	0.0047	0.0073	0.0052	0.005	0.0087	0.006	0.005	0.015
Sm	0.0054	0.0069	0.0081	0.0049	0.01	0.007	0.006	0.0097	0.0056	0.0053	0.0084
Eu	0.0023	0.0013	0.0013	0.0029	0.0024	0.004	0.0015	0.011	0.002	0.0023	0.0043
Gd	0.0043	0.0042	0.0075	0.006	0.0024	0.0023	0.0041	0.0073	0.0044	0.0032	0.011
Tb	0.0015	0.0015	0.0008	0.0011	0.0013	0.0018	0.0007	0.0022	0.0014	0.0011	0.0027
Dy	0.0051	0.0044	0.0062	0.0021	0.0015	0.0054	0.0058	0.012	0.0056	0.0029	0.011
Ho	0.0018	0.0016	0.0017	0.0011	0.003	0.0009	0.001	0.0032	0.0021	0.0017	0.0084
Er	0.04	0.013	0.0037	0.0071	0.0048	0.0047	0.0032	0.015	0.0071	0.0046	0.07
Tm	0.0019	0.0014	0.001	0.0015	0.0002	0.0011	0.0014	0.0034	0.0005	0.0007	0.02
Yb	0.02	0.006	0.0034	0.0046	0.0017	0.0044	0.0054	0.022	0.0051	0.0054	0.24
Lu	0.006	0.0006	0.0008	0.0007	0.0009	0.0011	0.0013	0.0057	0.0014	0.0013	0.058
Sr	12,036	11,660	7147	7052	7752	12,735	7608	516	19,537	6140	6917
Y	0.079	0.065	0.039	0.035	0.046	0.064	0.035	0.083	0.1	0.027	0.062
Y/Ho	43.89	40.63	22.94	31.82	15.33	71.11	35.00	25.94	47.62	15.88	7.38
Y/Dy	15.49	14.77	6.29	16.67	30.67	11.85	6.03	6.92	17.86	9.31	5.64
∑REE	0.1373	0.0639	0.0459	0.0455	0.0479	0.0491	0.0423	0.1099	0.0503	0.0398	0.1874
LREE	0.0567	0.0312	0.0208	0.0213	0.0321	0.0274	0.0194	0.0391	0.0227	0.0189	0.0345
HREE	0.0806	0.0327	0.0251	0.0242	0.0158	0.0217	0.0229	0.0708	0.0276	0.0209	0.1529
Ce/Ce*	2.49	0.75	0.77	0.46	0.81	0.44	0.78	0.76	0.70	0.40	0.241
Eu/Eu*	2.25	1.12	0.78	2.47	1.92	4.22	1.42	6.15	1.90	2.59	2.06
Nd_{SN}/Yb_{SN}	0.037	0.139	0.083	0.085	0.357	0.098	0.077	0.033	0.098	0.077	0.052
Y*	1.79	1.67	0.84	1.48	0.90	2.04	1.02	0.94	1.98	0.76	0.38
δD	−74.7	−71.4	−72.8		−73.5						
$\delta^{18}O$	−8.84	−8.81	−8.95		−9.01						

than light REEs. In particular, the total REE concentrations in groundwater samples from the QA (WG6) and LA (WG9) were lower than those from the CA.

Dissolved strontium (Sr) concentrations in the groundwater in the study area ranged from 516 to 19,537 μg/L, with an average value of 9009.1 μg/L (Table 1). The Sr concentrations varied between the different aquifers, with higher concentrations in the QA and the LA, and lower concentrations in the CA. The dissolved Sr concentrations in both the QA and LA were 12,735 and 19,537 μg/L, respectively. Sr and Ca are both alkaline earth elements, and therefore their geochemical behaviors are very similar. Previous studies (Nakano 2014) have reported that the Sr concentration in water is approximately 1 % that of Ca, the dominant element, and that the concentrations of the two elements are strongly correlated. Therefore, we would expect to see positive correlations between Ca and Sr in the groundwater samples. From Fig. 5, we observed that Ca

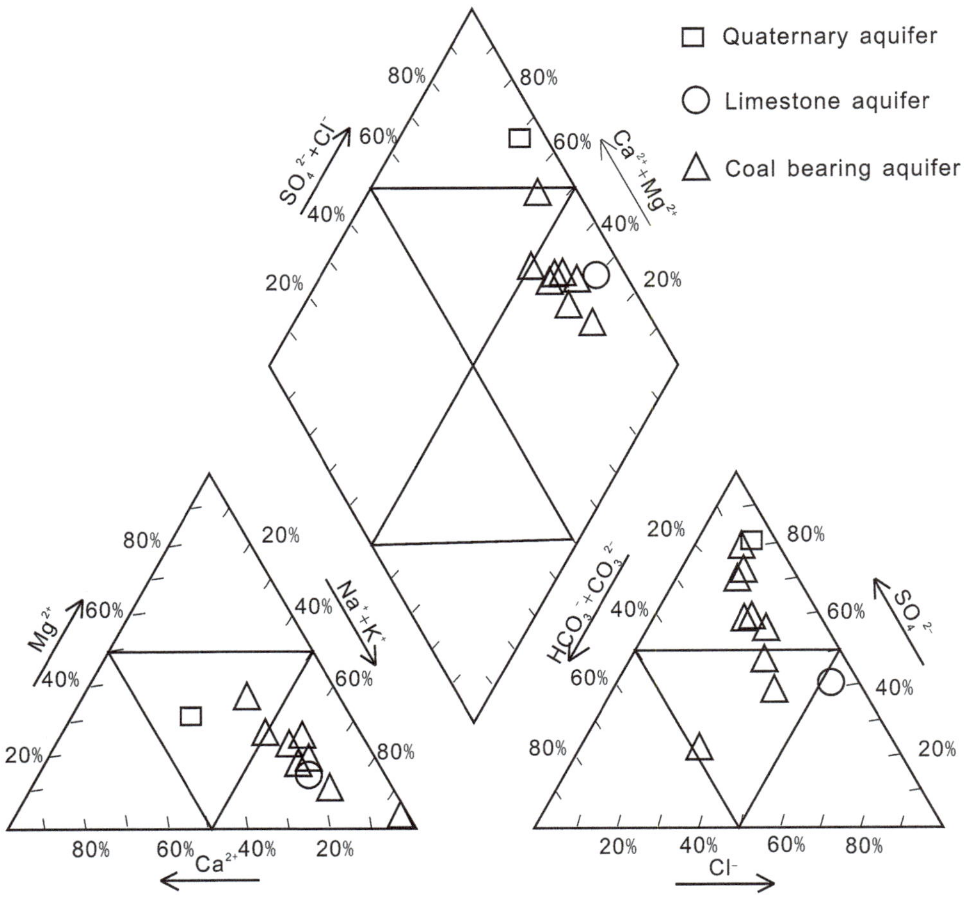

Fig. 3 Piper diagram of groundwater from the coal-bearing aquifer in the study area

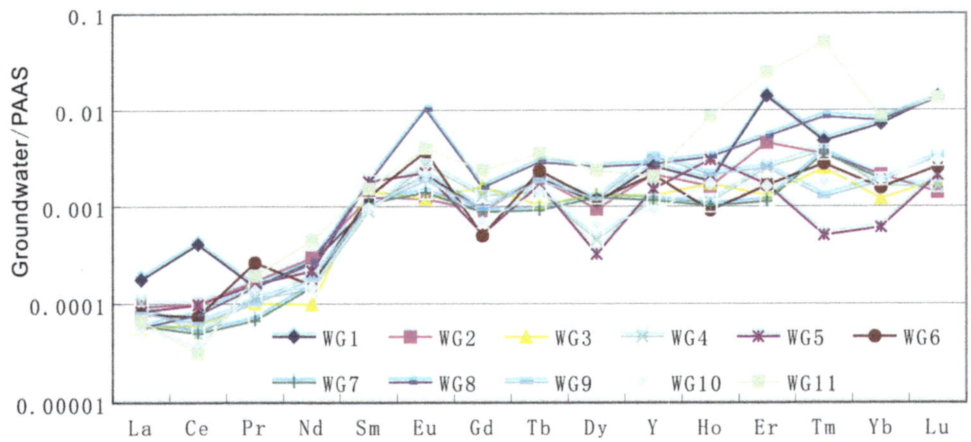

Fig. 4 PAAS-normalized REE patterns in groundwater from the coal-bearing aquifer in the study area

and Sr are positively correlated, which indicates that the Sr in the groundwater is released by weathering (dissolution) of Ca-containing minerals, and that there are no other sources such as human activity.

In addition, the groundwater samples demonstrated negative Ce anomalies when normalized by PAAS; the average Ce/Ce* value was 0.78 when calculated using the formula $Ce/Ce^* = 2^*Ce_{SN}/(La_{SN} + Pr_{SN})$. Eu, however, demonstrated positive anomalies; the average Eu/Eu* value was 2.44 when calculated by the formula $Eu/Eu^* = 2^*Eu_{SN}/(Sm_{SN} + Gd_{SN})$. The negative Ce anomalies in groundwater samples may reflect the oxidative conditions of the aquifer, as Ce^{3+} is oxidized to Ce^{4+}, thereby reducing the concentrations of soluble Ce (Leybourne et al. 2000). However, there is no obvious proof that this occurs in the study area, as the groundwater came from the coal-bearing aquifer, which is regarded as a deoxidized environment. Alternatively, the Ce anomalies in the

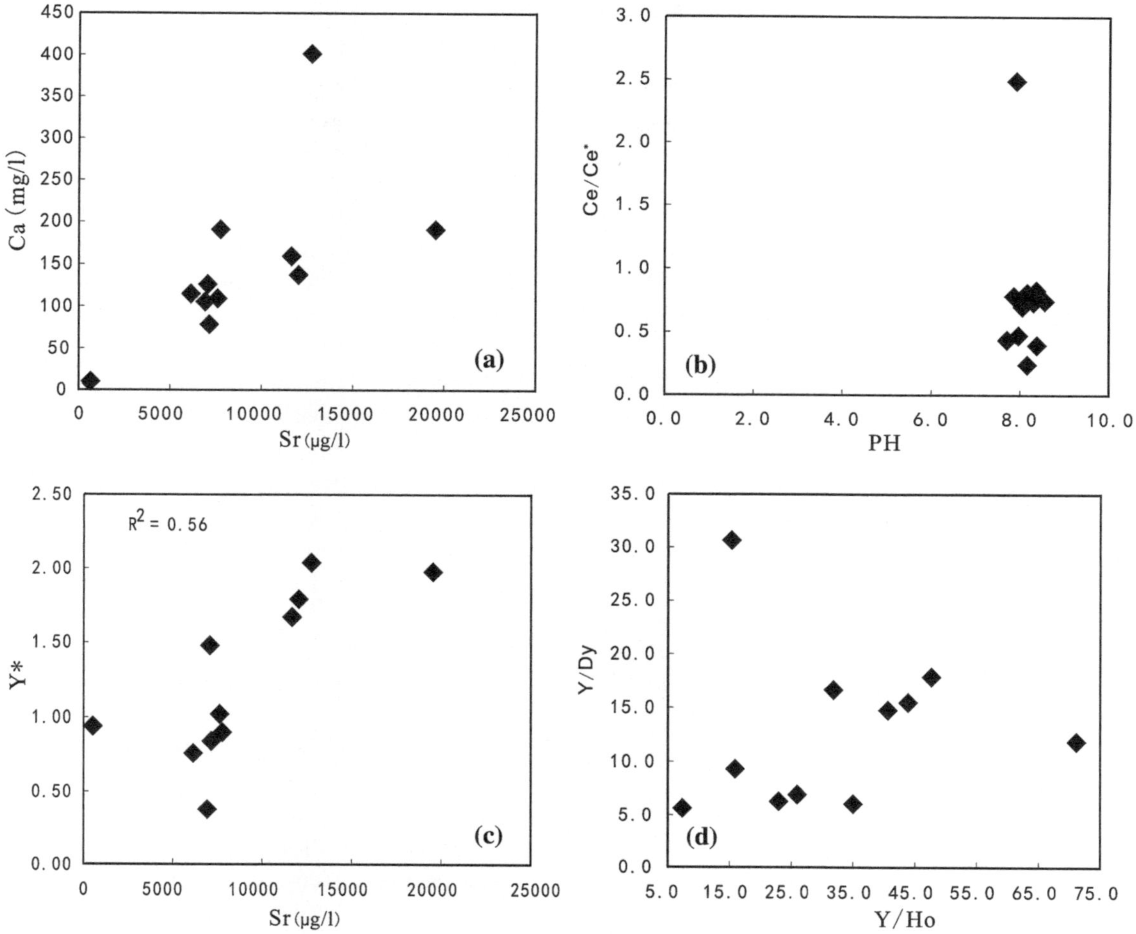

Fig. 5 Scatter diagrams among Ca-Sr(a), pH-Ce/Ce*(b), Sr-Y*(c) and Y/Ho-Y/Dy(d) in groundwater samples from the coal-bearing aquifer in the study area

groundwater reflect differences in the solubility of Ce redox species related to pH conditions in the aquifer system. Previous studies have shown that the degree of negative Ce anomalies varies as the pH increases, as Ce^{3+} is more stable in low pH conditions (Johannesson et al. 2005). The plot of Ce anomalies and pH shows that the magnitude of the Ce anomalies varied gradually as the pH values changed (Fig. 5), suggesting that this mechanism was not the main reason for Ce depletion. The Ce negative anomalies in the groundwater are therefore the result of water–rock interactions.

The oxidative environment and water–rock interactions may have caused the positive Eu anomalies. Positive Eu anomalies in groundwater may indicate an oxidative environment (Jeong 2001), whereas the aquifers in the study area have been confirmed as deoxidative. Water–rock interactions were, however, confirmed as the reason for the positive Eu anomalies, especially for the minerals with higher Eu contents such as feldspar. By coincidence, the surrounding rock is mainly composed of feldspar minerals. Further, the positive Eu anomalies in deep groundwater may be partly attributable to exchange reactions between Eu^{2+} and Sr^{2+} in reducing conditions, as Eu^{2+} and Sr^{2+} have similar chemical properties (Yan et al. 2013). Therefore, if the Sr^{2+} concentrations in groundwater are high, exchange reactions may occur between Eu^{2+} and Sr^{2+}. Consequently, Sr^{2+} is readily precipitated, whereas the concentration of Eu^{2+} remains high in groundwater. As mentioned above, the concentrations of Sr are higher in groundwater samples, so this mechanism possibly also contributes to the positive Eu anomalies in groundwater in this study.

The geochemical behavior of the elements Y and holmium (Ho) in the geological environment are similar because of their nearly ionic radius; the Y/Ho ratio in minerals is generally 27. The Y/Ho ratio is also 27 in silicates, clastic rocks, and fresh river water, whereas the Y/Ho ratio in seawater is generally higher than 27, and may reach 80 (Nozaki et al. 1997). This can be explained by the higher removal rate of Ho, which is nearly double that of Y in seawater. Thus, seawater and limestone formed in the sea frequently present positive Y anomalies. The Y/Ho ratios of groundwater from the study area varied from 7.4 to 71.1, and had an average value of 32.5; the Y/Ho ratios in the QA and the LA reached 71.11 and 47.62, respectively. When the two samples from the QA and the LA are excluded, the average Y/Ho value in the CA was only 26.5. This suggests that the Y/Ho ratios in the groundwater samples from the QA and the LA are influenced by carbonates such as limestone, which are abundant in the QA and the LA. However, there are no carbonates in the CA. The positive correlation between Y/Dysprosium (Dy) and Y/Ho (Fig. 5) also supports this perspective. When calculated using the equation $Y^* = 2^*Y_{SN}/(Dy_{SN} + Ho_{SN})$, the Y* values showed positive anomalies, varied from 0.38 to 2.0, and had an average of 1.25. The values of Y* in groundwater from the QA and the LA were higher than the average, and were 2.04 and 1.98, respectively. The higher values may reflect water–rock interactions over a long time, as the surrounding rocks in the QA and the LA comprise calcite or limestone, whereas the CA is mainly composed of feldspar. Therefore, if this hypothesis is correct, there would be a positive relationship between Y* and Sr, as is supported by Fig. 5.

Isotopic composition

The results of δD and δ^{18}O analysis of groundwater from the coal-bearing aquifer are plotted in Fig. 6. The values of δ^{18}O in groundwater varied from −9.01 to −8.81 ‰, and had an average value of −8.90 ‰, while δD varied from −74.7 to −71.4 ‰, and had an average value of −73.1 ‰, respectively. To obtain more information about the isotopic features of groundwater samples, additional data are required. Thus, we used the global meteoric water line (GMWL), the local meteoric water line (LMWL), and the local surface water line (LSWL) for δD and δ^{18}O. The GMWL is described by $\delta D = 8^*\delta^{18}O + 10.56$, as defined by Craig (1961); the LMWL was characterized as $\delta D = 7.9^*\delta^{18}O + 8.2$, which was summarized from the measured data of the stable isotopes (Zhang, 1989), and the LSWL is described by the formula $\delta D = 6.74^*\delta^{18}O - 3.33$ (Gui et al. 2005). All the lines and the isotopic signatures of the groundwater samples are plotted in Fig. 6.

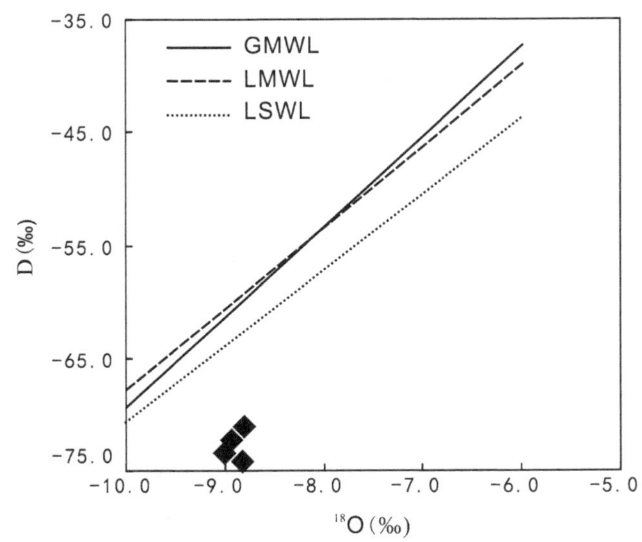

Fig. 6 Plot of the δ^{18}O and δD values in groundwater from the Wugou coal mine, Anhui Province, China

The four groundwater samples plotted below the LSWL, LMWL, and the GMWL, indicating that the source of groundwater from the CA was meteoric water that had varying degrees of evaporation. In particular, the values of δD showed clear variation, whereas the values of δ^{18}O were stable. The upward movement in the vertical or approximately vertical direction in the plot clearly indicates variability in the δD values.

In general, the variability in δ^{18}O and δD may be the result of factors such as evaporation, reservoir temperature, residence time, and water–rock interactions (Truesdell and Hulston 1980). These groundwater samples (WG1, WG2, WG3, and WG5) are all categorized as the Na-SO_4 type (Table 1), with Na^+ and SO_4^{2-} as the dominant ions. This indicates that hydrochemical processes in the groundwater may be influenced by weathering of feldspar minerals, and that Na^+ may be released into the groundwater by sulfuric acid reactions. The information in Fig. 6 suggests that the CA is a reducing environment, where the exchange reactions of δD between groundwater and alkyl are more obvious, whereas δ^{18}O exchanges are in equilibrium, as also reported by a previous study (Gui et al. 2005).

Conclusions

Nine groundwater samples were collected from the CA in the Wugou coal mine. Two samples were also collected from the QA and the LA in the same area to permit comparison. The 11 groundwater samples were analyzed for major ion chemistry and trace elements, and four samples were analyzed for their δD and δ^{18}O concentrations, and hydrochemical characteristics.

This study demonstrates that almost all the groundwater samples collected from the CA can be described as the Na-SO_4 type, whereas the samples from the QA and the LA were the Ca-SO_4 and Na-Cl types, respectively. The groundwater in the study area is not suitable for drinking before treatment because of the higher concentrations of TDS and other parameters relative to the WHO standards, whereas the lower average SAR value shows that the groundwater can be used for irrigation.

The total REE concentrations ($\sum$REE) varied from 0.0398 to 0.1874 mg/L, with an average of 0.075 mg/L, and groundwater was more enriched with heavy REEs than light REEs when normalized to PAAS. The higher dissolved Sr concentrations (average 9009.1 μg/L) in groundwater were attributable to release from weathering of Ca-containing minerals.

Ce and Eu showed negative and positive anomalies in groundwater, with average values of 0.78 and 2.44 for Ce/Ce* and Eu/Eu*, respectively. These results reflect water–rock interactions, with only minor influence from pH and redox actions. The calcite content, such as limestone and carbonate, of the surrounding rocks was determined by the Y/Ho and Y* values, the values of which were significantly higher in the QA and the LA than in the CA. However, the values of Y/Ho and Y* were lower in groundwater collected from the CA, which is surrounded by silicate rocks such as feldspar.

The δD and $\delta^{18}O$ values in groundwater samples varied from −9.01 to −8.81 ‰ and from −74.7 to −71.4 ‰, with an average of −8.90 and −73.1 ‰, respectively. The groundwater in the CA is mainly from meteoric water, with a variable degree of evaporation. The drift in δD may be the result of detritus and exchange reactions between groundwater and alkyl.

Acknowledgments The study was supported by the National Nature Science Foundation of China (41,373,095), the Program for Innovative Research Team in Suzhou University (2013kjtd01).

References

Chen S, Gui HR, Sun LH (2014) Quality evaluation and its controlling factors of groundwater from Wolonghu Mining area, northern Anhui province China. Nat Environ Pollut Technol 13(3):577–582

Craig H (1961) Isotopic variation in meteoric water. Science 133:1702–1703

Gui HR (2014) Geochemical characteristics of deep groundwater from sandstone aquifer in Qianyingzi mine, northern Anhui province China. Water Prac Technol 9(1):95–103

Gui HR, Chen LW (2007) Hydro-geochemistry evolution and discrimination of groundwater in mining district. Geological Publishing House, Beijing **(in Chinese)**

Gui HR, Chen LW, Song XM (2005) Drift features of oxygen and hydrogen stable isotopes in deep groundwater in mining area of northern Anhui. J Harbin inst Technol 37:111–114 **(in Chinese)**

Gui HR, Sun LH, Chen LW, Chen S (2011) Rare earth element geochemistry of groundwater from a deep seated sandstone aquifer, northern Anhui province, China. Min Sci Technol 21:477–482

Jeong CH (2001) Mineral-water interaction and hydro-geochemistry in the Samkwang mine area, Korea. Geochem J 35:1–12

Johannesson KH, Cortes A, Leal JAR, Ramirez AG, Durazo J (2005) Rare earth elements in groundwater flow systems. Springer, Netherlands, pp 188–222

Leybourne MI, Goodfellow WD, Boyle DR, Hall GM (2000) Rapid development of negative Ce anomalies in surface waters and contrasting REE patterns in groundwaters associated with Zn-Pb massive sulphide deposits. Appl Geochem 15:695–723

Nakano T (2014) Use of water quality analysis for groundwater traceability. In groundwater as a key for adaptation to changing climate and society. Springer, Japan, pp 45–67

Nozaki Y, Zhang J, Amakawa H (1997) The fractionation between Y and Ho in the marine environment. Earth Planet Sci Lett 148:329–340

Sun LH, Gui HR, Chen LW (2011) Geochemistry of rare earth elements in groundwater from deep seated limestone aquifer in Renlou Coal Mine, Anhui Province, China. J Cent South Univ 18:1646–1653

Taylor SR, McLennan SM (1985) The continental crust: its composition and evolution. Blackwell Scientific Publications, Oxford

Truesdell AH, Hulston JR (1980) Isotopic evidence of environments of geothermal systems. In: Fritz P, Fontes JCh (eds) Handbook of environmental Isotope geochemistry, 1: The terrestrial Environment. Elsevier, Amsterdam, pp 179–226

United States Salinity Laboratory (USSL) (1954) Diagnosis and improvement of saline and alkaline soils. US Department of Agriculture, Washington

World Health Organization (WHO) (1993) Guidelines for drinking water quality, recommendation, 2nd edn. WHO, Geneva

Yan Z, Liu G, Sun R, Tang Q, Wu D, Wu B, Zhou C (2013) Geochemistry of rare earth elements in groundwater from the Taiyuan Formation limestone aquifer in the Wolonghu Coal Mine, Anhui Province, China. J Geochem Explor 135:54–62

Zhang HP (1989) Study on the background value of stable isotopes in precipitations of China. Site Investig Sci Technol 1:6–13 **(in Chinese)**

22

Contemplating the feasibility of vermiculate blended chitosan for heavy metal removal from simulated industrial wastewater

N. Prakash[1] · M. Soundarrajan[2] · S. Arungalai Vendan[3] · P. N. Sudha[4] · N. G. Renganathan[5]

Abstract Wastewater contaminated by heavy metals pose great challenges as they are non biodegradable, toxic and carcinogenic to the soil and aquifers. Vermiculite blended with chitosan have been used to remove Cr(VI) and Cd(II) from the industrial wastewater. The results indicate that the vermiculite blended with chitosan adsorb Cr(VI) and Cd(II) from industrial waste water. Batch adsorption experiments were performed as a function of pH 5.0 and 5.5 respectively for chromium and cadmium. The adsorption rate was observed to be 72 and 71 % of chromium and cadmium respectively. The initial optimum contact time for Cr(VI) was 300 min with 59.2 % adsorption and 300 min for Cd(II) with 71.5 % adsorption. Whereas, at 4–6 there is saturation, increasing the solid to liquid ratio for chitosan biopolymers increases the number of active sites available for adsorption. The optimum pH required for maximum adsorption was found to be 5.0 and 5.5 for chromium and cadmium respectively. The experimental equilibrium adsorption data were fitted using Langmuir and Freundlich equations. It was observed that adsorption kinetics of both the metal ions on vermiculite blended chitosan is well be analyzed with pseudo-second-order model. The negative free energy change of adsorption indicates that the process was spontaneous and vermiculite blended chitosan was a favourable adsorbent for both the metals.

Keywords Chromium(VI) · Cadmium(II) · Chitosan mixed vermiculite · Adsorption isotherms · Kinetics

Introduction

Chromium and cadmium are highly toxic heavy metals and these have to be removed from the water sources. These metals are from various industrial effluents such as tanneries, electroplating and paints. The chromium toxicity is mainly induced from its hexavalent form, Cr(VI) which is readily soluble in water. Its concentration should not exceed 0.05 mg l^{-1} in drinking water and become more toxic with potential carcinogenic effects (Baird and Cann 2005). Cadmium belongs to the hazardous metal group. It is fairly mobile in soil and is primarily present as an organically bound, exchangeable and water soluble species (Holm et al. 1995; Chlopecka 1996). The removal of cadmium from the wastewater by various techniques such as chemical precipitation, electro deposition, electro coagulation process, ion exchange and emulsion liquid membrane (Vasudevan et al. 2011; Elkady et al. 2011; Ahmad et al. 2012). These techniques are expensive and ineffective at low concentration of metal ions. Adsorption method is reported to be a favourable method owing to its low cost and high efficiency for high and low concentration of metal ions (Ngah and Hanafiah 2008).

Novel adsorbents prepared from orange peel and Fe_2O_3 nano particles have been used (Gupta et al. 2012) to

✉ N. Prakash
prakash.nagapillai@gmail.com

1 Department of Chemistry, Thangavelu Engineering College, Chennai 600 097, India

2 Department of Chemistry, Valliammai Engineering College, Kattankulathur 603 203, India

3 School of Electrical Engineering, VIT University, Vellore, India

4 Department of Chemistry, D.K.M. College for Women, Vellore 632 001, India

5 Department of Chemistry, Vel Tech University, Avadi, Chennai, India

remove cadmium from aqueous solutions. The removal of cadmium from the electroplating industry effluent is reported and is shown to have desorption and reusability without loss of efficiency. Material from the low cost fertilizer industry viz., carbon slurry, has been chemically treated, activated and used as adsorbent to remove hexavalent chromium from aqueous solutions (Gupta et al. 2010) and the kinetics of adsorption follows pseudo second order rate equation based on batch experiments. The removal of lead and chromium from aqueous solutions, has been reported using inexpensive red mud, an aluminium industry waste and bagasse fly ash. The blast furnace waste generated in steel plants has been used for the removal of lead and chromium (Srivastava et al. 1997) and the authors concluded that the efficiency is higher for lead than that of chromium.

Low cost adsorbents have been investigated from agricultural and fishery wastes such as rice hull and saw dust (Asadi et al. 2008), peanut husk (Li et al. 2007), pumpkin waste (Horsfall et al. 2006), Nipah palm (*Nypa fruticans* Wurmb), shoot biomass (Wankasi et al. 2006), prawn shell (Chu 2002), crab shell (Dahiya et al. 2008), chitin (Jayakumar et al. 2009) and chitosan (Tao et al. 2009) to remove metal ions from wastewater. Chitosan biopolymer (Kumari et al. 2009; ShahinHydari et al. 2012; Prakash et al. 2011) and vermiculite clay (Malandrino et al. 2006; Maria Rosaria Panuccio et al. 2009) are being explored recently for the removal of heavy metals from wastewater.

Chitosan, poly (β(1-4)-2-amino-2-deoxy-D-glucose is obtained from partial *N*-deacetylation of chitin with a strong alkali solution such as sodium hydroxide. It is characterized by a high content of nitrogen, present as amine groups capable of binding the metal ion through several mechanisms, including chemical interactions, such as chelation, electrostatic interactions (or) ion-exchange. The interaction depends (Ng et al. 2003; Dzul Erosa et al. 2001) on the metal ions and initial pH of the medium. The chemical stability of chitosan is enhanced by several techniques such as cross linking (Du et al. 2009) carboxymethylation (Xu et al. 2009) grafting (Morimoto et al. 2002) blending (Wang and Kuo 2008; Ngah et al. 2004) coating (Popuri et al. 2009) and sulphonation (Holme and Perlin 1997). Among all the techniques cross linking is mainly focused because of its simple procedure and there are enormous opportunities to form macromolecular super structures for various specific applications (Crini 2005). Several studies have emphasized that chitosan and its modified form (cross linking coating and blending) is a very efficient adsorbent to remove various toxic and strategic metals (Gerente et al. 2007; Chang et al. 2007).

Vermiculite refers to a class of hydrated ferro-magnesium–aluminum silicates that contains various mineral groups. It looks like mica and has a lattice layer of Mg^{2+} (or) K^{+} with water molecules (McBride 1994). Vermiculite can adsorb metals via two mechanisms: (1) cation exchange at the planar sites, resulting from the interaction between metal ions and negative permanent charge (outer-sphere complexes) and (2) formation of inner-sphere complexes through Si–O- and Al–O-groups at the clay particle edges. Removal of toxic metals such as Cd(II), Pb(II) and Cu(II) by vermiculite (Allan et al. 2007; Prakash et al. 2012) has also been reported.

Multiwall carbon nanotubes with the magnetic properties of iron oxide have been used as a composite adsorbent for the adsorption of Cr(III) and it has been shown (Gupta et al. 2011) that the adsorption of Cr(III) is strongly dependent on contact time, agitation speed and pH in the batch mode and on flow rate and the bed thickness in the fixed bed mode. Chemical treatment technologies for wastewater recycling has been reviewed in (Gupta et al. 2012) and a brief guideline for the selection of the appropriate technologies for specific applications has been neatly evaluated. The carbon occurs adsorbent prepared from carbon slurry is reported (Jain et al. 2003) to be efficient and can be used for the removal of dyes from solution.

Bottom ash and despoiled soya have been used as adsorbents for the removal of a hazardous azodye and reported in Mittal et al. (2008). A hazardous textile dye, safranin-T, has been successfully photo degraded using Tio_2 as catalyst and reported in Gupta et al. (2007). The adsorption performance of the materials namely bottom ash, a power plant water and de-oiled soya waste has been appreciable while employing batch and column methods as reported (Mittal et al. 2010). The hexavalent chromium, Cr(VI) has been successfully bioadsorbed on raw and acid treated *Oedogonium hatei* from aqueous solutions (Gupta et al. 2010c).

From the available literature reports, it is observed that the investigations on ternary blends with the presence of cross linking agents likely to enhance the thermal and chemical stability for the removal of heavy metals are scanty and inadequate. Hence, in the present study an attempt has been made to investigate the possibility of the vermiculite blended chitosan in the presence of glutaraldehyde as the cross linking agent for the adsorption of Cr(VI) and Cd(II) ions from the aqueous solution. Vermiculite is mixed with chitosan was prepared in ratios of (1:1), (3:1) in the presence and absence of cross linking agent glutaraldehyde. The blend was characterized by using FT-IR, XRD, TGA and DSC techniques. The kinetic adsorption studies of these blends with chromium and cadmium in aqueous solution were evaluated using different isotherms.

Materials and methods

Instrumentation

FT-IR studies

Fourier Transform Infra Red spectra of all the adsorbents of various combinations were recorded in the frequency range of 400–4000 cm^{-1} using Thermo Nicolet AVATAR 330 spectrophotometer. The samples were pressed into pellets with KBr.

X-ray diffraction studies

X-ray diffractograms of powdered samples were obtained using a X-ray powder diffractometer (XRD-Shimadzu XD-D1) with Ni-filter, Cu $K\alpha$, radiation source. The relative intensity was recorded in the scattering range 2θ of 10°–90°.

Thermogravimetric analysis

Thermo gravimetric analysis of all the adsorbents with various combinations was conducted by SOT Q600. V8. This analysis is to find out the thermal stability of the synthesized polymer and the heating range is 20–50 °C and the heating rate is 20 °C per min.

Differential scanning calorimetric analysis

The thermal behaviour of the adsorbents was studied using NET 2 SCH DSC thermal analyzer. The sample was inserted into the Al pan and DSC scan was made from 30 to 300 °C in a nitrogen atmosphere at a heating rate of 20 °C min^{-1}. The results were recorded and analyzed.

Stock solution

Stock solutions of chromium(VI) ions and cadmium ions were prepared (Dean 1995). All other concentration values, varying between 40 and 670 mg l^{-1} of Cd(II) were prepared by dilution from the stock solution. The initial pH of the working solution was obtained by adding small volumes of 0.5 N HNO_3 (for pH 2.0 and 3.0), or acetate buffer (CH_3COOH/CH_3COONa), for $3.0 < pH < 6.0$.

Preparation of adsorbent: chitosan mixed vermiculite

The chitosan polymer solution was prepared by dissolving 1 g of chitosan in 2 % acetic acid and mixed with a known weight of the vermiculite. The mixture was then stirred thoroughly in the presence of glutaraldehyde (10 ml) as a cross linking agent at room temperature for 1 h. Then the above prepared mixture were poured into a petri dish and dried in vacuum for 10 h to remove the solvent completely.

Methodology for batch process of adsorption studies

For the study of equilibrium isotherms, the adsorption studies were done at varying concentrations of the metal, i.e., 30, 60, 90, 120, 150, 180, and 210 mg l^{-1} of chromium and cadmium and keeping the amount of chitosan (2.0 g l^{-1}) constant at pH 5.0. Polymer blended solutions (30 ml) were added to the reactor vessel. Solutions were stirred magnetically throughout the experiments. Samples (100 μl) were taken from the feed cell into which metal was adsorbed and the solution is estimated for chromium and cadmium using the usual EDTA method (Prakash et al. 2012). Chromium and cadmium (30 mg l^{-1}) were added to the reactor vessel. The reactor vessel is stirred magnetically throughout the experiments. Following the chromium and cadmium concentrations were estimated, the equilibrium isotherm was prepared by plotting uptake in mg g^{-1} versus the equilibrium concentration of chromium and cadmium in mg l^{-1}.

Removal of chromium and cadmium

Synthetic solutions of Cr(VI) ions and Cd(II) ions were taken in stopper bottles and agitated with the prepared adsorbents separately at 30 °C in orbit shaker at fixed speed of 160 rpm. The extent of heavy metal removal was investigated separately by changing the amount of adsorbent, contact time and different pH of the solution. After attaining the equilibrium the adsorbent was separated by treating it with suitable reagents after leaching the metals from the adsorbents. Analysis of the concentration of the metals has been done using atomic adsorption spectrophotometer (Varian AAA 220 FS). Chromium has been determined from the analysis by fixing the wave length at 357.9 nm and cadmium has been obtained by fixing the wavelength at 228.8 nm.

Results and discussion

Characterization

Vermiculite blended chitosan were characterized by IR spectroscopy, X-ray diffraction, thermal gravimetric analysis (TGA/DTA), Differential Scanning Calorimetry (DSC). Further, kinetic analysis have also been reported and discussed.

FT-IR spectra analysis

The FT-IR spectrum of chitosan gave a characteristic band at 3450 cm^{-1} which is attributed to $-NH_2$ and –OH stretching vibration and the band for amide I at 1652 cm^{-1} is observed in the infrared spectrum of chitosan. The characteristic carbonyl stretching of chitosan observed at 1733 cm^{-1} and the broader peak at 1635 cm^{-1} corresponding to the chitosan NH_2 band. In 3:1 ratio the peaks observed at 3421 cm^{-1} are due to OH stretching, –NH stretching and intermolecular hydrogen bonding. The peaks at 2935, 1640 and 1579 cm^{-1} indicate asymmetric C–H stretching, C=O stretching and N–H bending respectively. The peaks at 1386, 1254, 1029 and 452 cm^{-1} arises due to C–H bending, O–H bending, Si–O stretching, Al–O stretching vibrations respectively. The IR spectrum of chitosan:vermiculite (1:1) shows the peak at 3430 cm^{-1} corresponding to the –OH stretching, –NH stretching, which may be due to intermolecular hydrogen bonding and polymeric association. The peaks at 2927, 1605 cm^{-1} correspond to the asymmetric C–H stretching and N–H bending. The peaks at 1383, 1021 and 460 cm^{-1} are due to the C–H bending, C–OH was stretching, Si–O and Al–O stretching vibrations.

The spectrum of chitosan:vermiculite+Glu (1:1), blend has a peak at 3000–3600 cm^{-1}, which is due to OH stretching, NH stretching and intermolecular hydrogen bonding. The peak positions are slightly modified in the presence of cross linker. The peaks at 2938, 1713, 1583 and 1381 cm^{-1} correspond to the asymmetric C–H stretching, C=O stretching, N–H bending and C–H bending vibrations. The peak at 1020, 681, 499 and 450 cm^{-1} indicate the C–O stretching, N–H deformation and Al–O & Si–O stretching vibration respectively. The IR spectrum for the ratio of chitosan:vermiculite+Glu (3:1) has stretching and due to intermolecular hydrogen bonding. The peaks at 2927 cm^{-1} to aldehydic asymmetric and symmetric CH stretching, the peaks at 1716, 1383 cm^{-1} the corresponds to C=O stretching, N–H bending and C–H bending vibrations respectively. The peaks at 1022 and 449 cm^{-1} correspond to C–OH stretching and Al–O, Si–O stretching vibrations respectively.

The sharp peak observed at 1610 cm^{-1} may be due to C=N (amide I band) of the Schiff's base formed between the free amino groups from the chitosan and aldehyde groups from glutaraldehyde which acts as a cross linking agent. In addition, the observed decrease in the intensity of the broad peak around may be attributed to the cross linking via acetalization reaction the glucosamine units of chitosan and aldehyde groups of glutaraldehyde. Similar observations were reported by Ruiz et al. (2000). An intensity of the sharp peak was observed at 1624 cm^{-1} which may be due to the N–H bending (amide II band). The increase is the chitosan composition in blend leads to more number of –NH functional groups. When chitosan was mixed with vermiculite the absorption peaks were observed at lower frequencies, due to the intermolecular hydrogen bonding of chitosan with vermiculite (Xie and Wang 2009).

When chitosan was blended with vermiculite in the presence of glutaraldehyde as cross linking agent an intensity of the sharp peak was observed at 1610 cm^{-1} which may be due to the C=N (amide I band) of the formed Schiff's base between the free amino groups that are present in the chitosan and aldehyde groups of the cross-linker glutaraldehyde (Ruiz et al. 2000).

X-ray diffraction studies

Figure 1a, b show the XRD spectrum of chitosan:vermiculite (1:1) and (3:1). From the XRD spectrum, the appearance of short sharp peaks at $2\theta = 19°$, 33° and 36° for 1:1 and three sharp peaks at $2\theta = 11°$, 15° and 25° for 3:1 indicates high crystalline nature of blends.

In the presence of glutaraldehyde, the XRD spectrum of chitosan:vermiculite for ratio (1:1) and (3:1) are shown in Fig. 1c, d. The XRD peaks are observed at $2\theta = 19°$, 26°, 34° for 1:1 and at $2\theta = 25°$, 39° for 3:1. From the XRD data it is clear that the nature of the sample is semi crystalline in nature for 1:1 ratio, where as in the case of 3:1, the observation of two long sharp peaks at $2\theta = 25°$ indicate crystalline nature of the blend.

Thermo gravimetric analysis (TGA)

Figure 2a, b represents the thermogram of chitosan:vermiculite mixture for (1:1) and (3:1) ratios. The maximum weight loss takes place at the temperature range from 250 to 500 °C. Around 54.57 % of the sample had disintegrated at the temperature of approximately 731.25 °C. At the end of the experiment, 45.43 % of the sample was obtained as a residue. 12 % of the sample had disintegrated at the temperature range of 510–860 °C. It may be due to loss of water. In the case of 3:1 ratio, the maximum weight loss takes place at the temperature range from 200 to 663 °C. Around 63.10 % of the sample had disintegrated at the temperature of about 657.12 °C. At the end of the experiment 36.90 % of the sample remained as a residue. 15 % of the sample had disintegrated at a temperature range of 450–200 °C.

Figure 2c, d represents the thermogram chitosan:vermiculite (1:1)+Glu (10 ml) ratio. The maximum weight loss takes place at the temperature range from 150 to 375 °C. Around 75.311 % of the sample had disintegrated at the end of the experiment. 24.68 % of the sample remained as a residue. 5 % of the sample had disintegrated at the temperature range of 500–150 °C. The weight loss

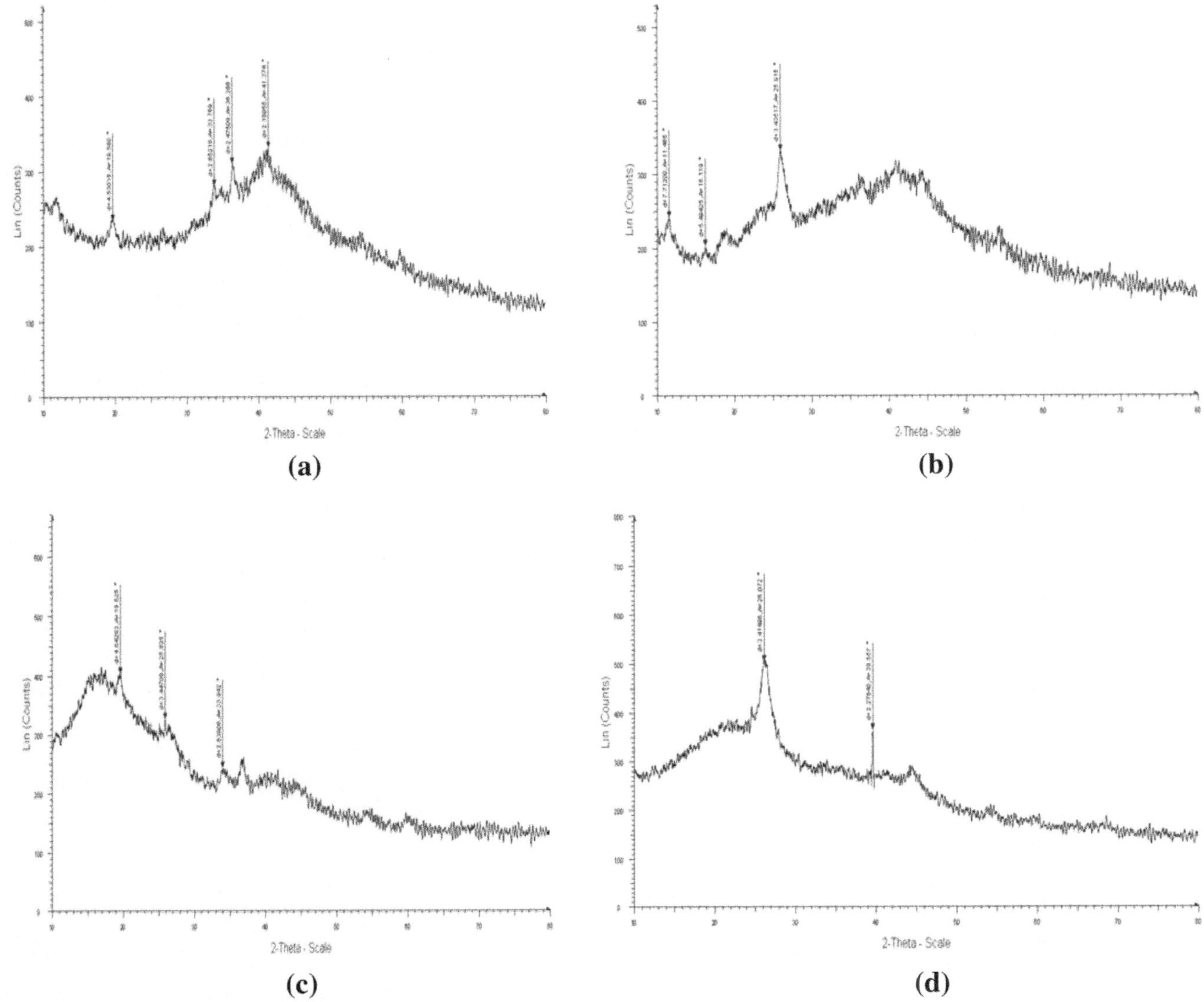

Fig. 1 **a** XRD diffractogram of chitosan mixed vermiculite (1:1), **b** XRD diffractogram of chitosan mixed vermiculite (3:1), **c** XRD diffractogram of chitosan mixed with vermiculite (1:1)+Glu, **d** XRD diffractogram of chitosan mixed with vermiculite (3:1)+Glu

between 50 and 150 °C may be due to water loss. The maximum weight loss occurs at 150 and 375 °C. The weight loss may be due to the decomposition of the –CH_2-linkages, H-bonding and other interaction between the two polymers. After 400 °C there was a sudden rise in the decomposition of the pyranose ring of the polysaccharides. When we compare the thermal stability with or without cross linking agent, the cross linked chitosan was thermally more stable.

Based on Fig. 2d, the thermogram of chitosan:vermiculite+Glu (3:1), the maximum weight loss takes place at the temperature range from 200 to 400 °C. Around 63.552 % of the sample had disintegrated at the temperature range of 700 °C. At the end of the experiment, 36.448 % of the sample remained as a residue. The maximum weight loss that occurred at 200–400 °C may be due to the decomposition of the –CH_2-linkage, H-bonding and other interaction between the polymers.

Adsorption experiments

Experimental procedure

Series of experiments were performed to evaluate the influence of the following variables: initial pH value of the solution, contact time, and adsorbent dosage. The batch experiments were performed by adding chitosan mixed vermiculite to a 25 ml solution containing Cr(VI) ions and Cd(II) ions under intermittent stirring, at room temperature (25 ± 0.5 °C) and, with the exception of only

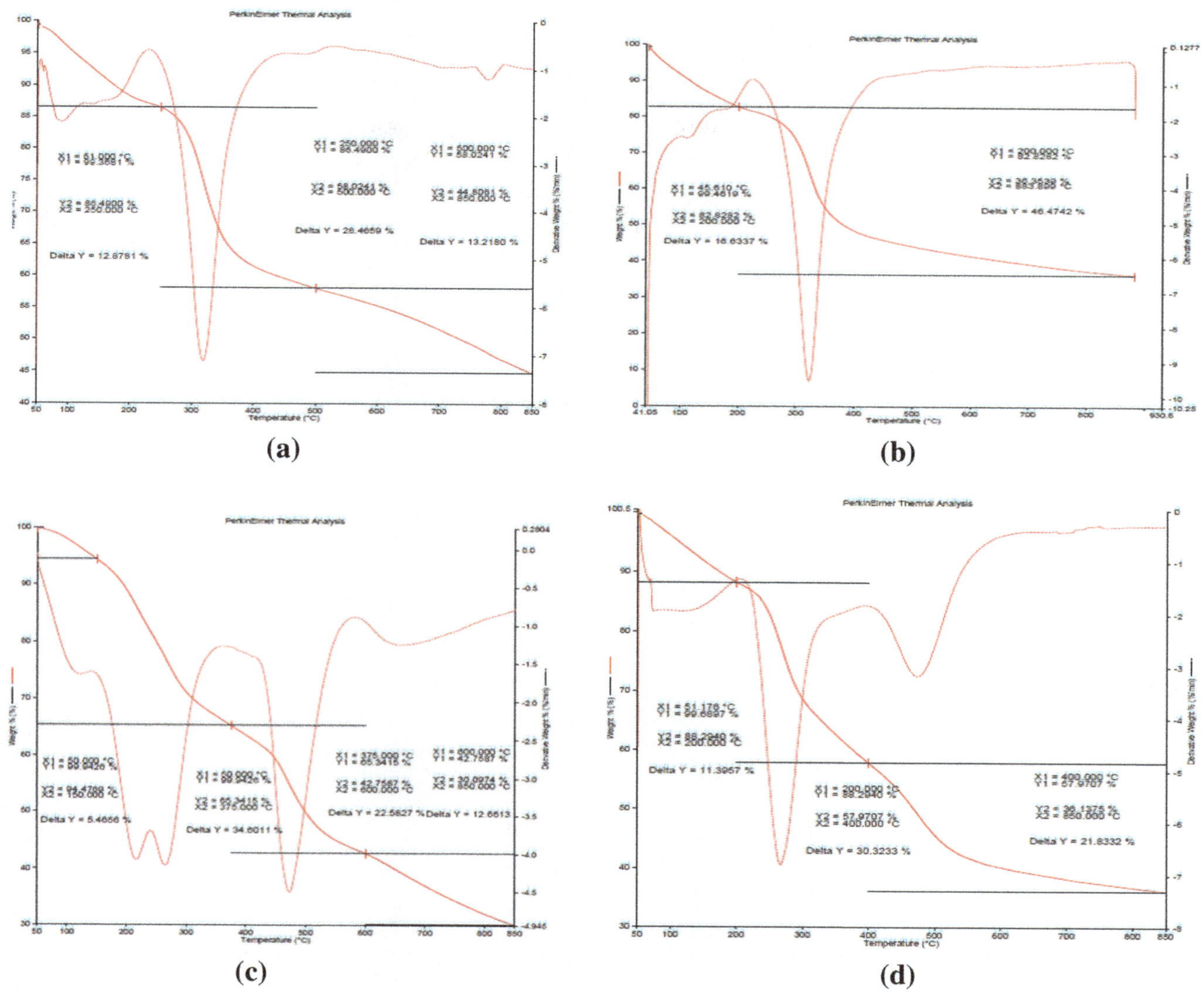

Fig. 2 **a** TGA of chitosan mixed with vermiculite (1:1), **b** TGA of chitosan mixed with vermiculite (3:1), **c** TGA of chitosan mixed with vermiculite (1:1)+Glu, **d** TGA of chitosan mixed with vermiculite (3:1)+Glu

one series, at a constant contact time of 24 h. The influence of the initial solution pH on Cr(VI) and Cd(II) ions capacity of adsorption onto chitosan mixed vermiculite was investigated over the 4.0–8.0 pH range, at a constant dosage of chitosan mixed vermiculite. To evaluate the effect of adsorbent dosage on Cr(VI) and Cd(II) adsorption, the metal ion solutions of a constant concentration (71.3 and 74.3 mg l^{-1}) was treated with variable dosages of chitosan mixed vermiculite. (Resulting in adsorbent concentrations in the mixture from 5 to 40 g l^{-1}, at pH 5.0 of the initial solution). The influence of contact time of both the metal ions concentration on chitosan mixed vermiculite adsorption capacity was studied over the 30–360 min range, at pH 5.0 of the initial solution and constant dosage of chitosan mixed vermiculite (5 g l^{-1} concentration in the reaction volume). It is seen from Fig. 3, that cadmium recovery is higher when compared to chromium recovery at all times. It is seen from the graph that the complexes of Cr^{6+} and Cd^{2+} behave in the same manner and cadmium and chromium complexes are stable during the studies. This may also be due to the stability of the complexes, cadmium complex is more stable compared to chromium complex (Prakash et al. 2012). The percentage of removal of cadmium is always higher compared to chromium at all contact time Fig. 3. Since cadmium is easily polarizable and charge to size ratio is less for cadmium compared to chromium the recovery of cadmium is more at all time compared to chromium. ($Cr^{6+} = 11.6 \times 10^{-24}$ and $Cd^{2+} = 7.2 \times 10^{-24}$) Fig. 3 also shows the effect of contact time on the removal of chromium and cadmium metals in the artificial wastewater.

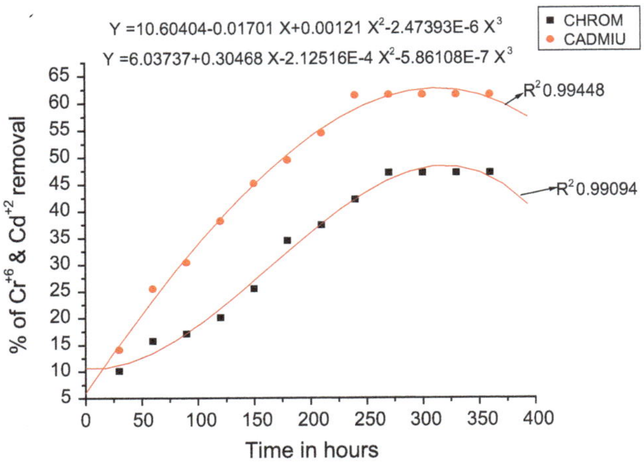

Fig. 3 Percentage of metal removal as a function of time

Percentage of metal removal as a function of pH

The adsorption process is strongly affected by the pH of the solution. The effect of pH change on adsorption was studied for both the metals by changing the pH of the contents from 4 to 8, using dilute solutions of HCl and NaOH. The observation shows that the maximum adsorption of both the metals took place in acidic media and this is evident from Fig. 4. The maximum adsorption for Cr(VI) was observed at pH 5 (72 % removal) and for cadmium at pH 5.5 (71 % removal) (Prakash et al. 2012). In general, at very high and very low pH values, the surface of the adsorption is surrounded mainly by H^+ and OH^- ions respectively. These positively and negatively charged ions may compete with the metal ions and as a result, adsorption decreases. That's why metal ions show low adsorption at very high and low pH values (Saeed and Iqbal 2003). Influence of pH on adsorption phenomenon is also related with the functional groups present on the bio-adsorbent.

The percentage removal of chromium and cadmium using chitosan mixed with vermiculite at different adsorbent dosage

Various operational parameters like solid/liquid ratio, pH, contact time, metal ion concentration and temperature were employed to determine the adsorption characteristics of chitosan biopolymers for heavy metal ion adsorption from aqueous solutions. The effect of the solid/liquid ratio for chromium adsorption on chitosan biopolymer was examined. In these experiments, operational parameters were kept constant $T = 298$ K, $C_0 = 200$ mg l^{-1} and the pH was varied initially for various time intervals. Figure 5 indicates that the percentage of adsorption of chromium and cadmium increases steadily with, solid to liquid ratio of 1–3 %. Whereas, at 4–6 there is saturation, increasing

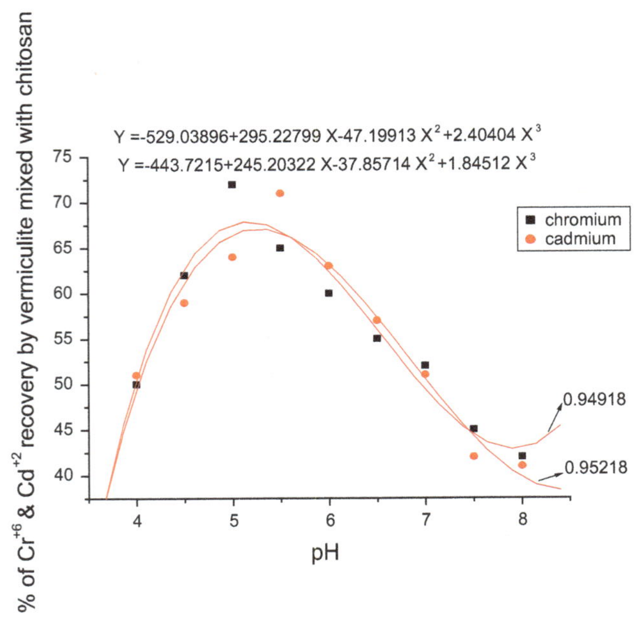

Fig. 4 Percentage of Cr(VI) and Cd(II) recovery by chitosan mixed with vermiculite at different pH

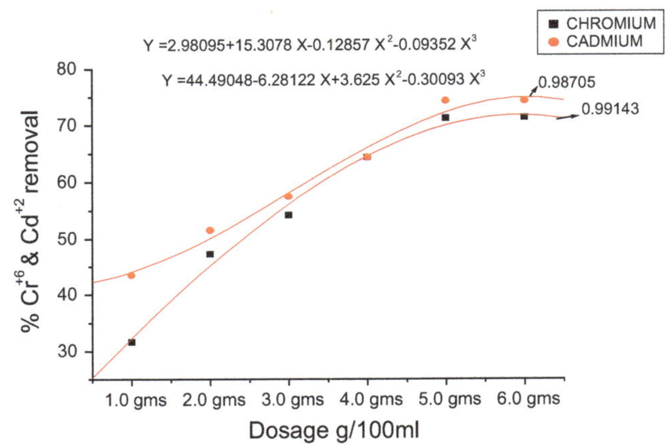

Fig. 5 Percentage removal of chromium and cadmium using chitosan mixed with vermiculite at different adsorbent dosage

the solid to liquid ratio for chitosan biopolymers increases the number of active sites available for adsorption.

Effect of contact time

It has been observed that at a constant concentration of metal ions and fixed amount of adsorbent, the adsorption efficiency increases with increase in the contact time up to a certain level. Figure 5 shows that the adsorption rate first increased rapidly, but after reaching the optimum time value, the removal efficiency slightly decreased with an increase in the contact time. The effect may be due to the saturation of adsorption sites with metal ions on the solid particle. The initial optimum contact time for Cr(VI) was

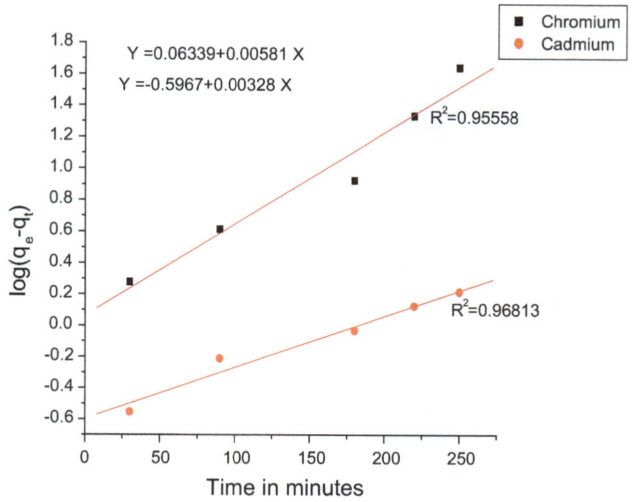

Fig. 6 Plot of ln $(q_e/q_e - q)$ versus time in chromium and cadmium

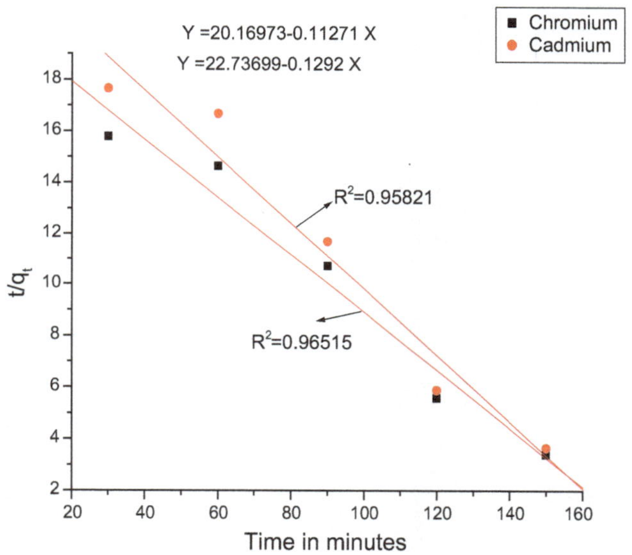

Fig. 7 Plot of t/q versus time in chromium and cadmium

300 min with 59.2 % adsorption and 300 min for Cd(II) with 71.5 % adsorption. The slight decrease in adsorption after the optimum contact time may be due to the breakage of newly formed weak adsorption bonds due to the constant shaking.

Kinetics of adsorption

Pseudo first and second order kinetics of adsorption

The sorption data of Cr(VI) and Cd(II) uptake by chitosan:vermiculite adsorbent was fitted using Lagergren pseudo first and second order model and these are given in Figs. 6 and 7. The linearized form of the first order Lagergren equation is given as Eq. 1.

$$\log(q_e - q_t) = \log q_e - \frac{K_{1,\text{ads}}}{2.303}t \quad (1)$$

The pseudo-second-order Eq. 2

$$\frac{t}{q} = \frac{1}{k_{2,\text{ads}}q_e^2} + \frac{t}{q_t} \quad (2)$$

where q_e is the mass of metal adsorbed at equilibrium (mg g^{-1}), q_t the mass of metal at time (min), k_1, ads the pseudo first order reaction rate of adsorption (per min), k_2, ads the pseudo second order rate constant of adsorption mg g^{-1} min^{-1}. A comparison between the two kinetic models suggested is presented in Table 1. These facts suggest that the obtained Cr(VI) and Cd(II) adsorption onto chitosan mixed vermiculite kinetic data followed the pseudo second order kinetic model which describes the biosorption as the rate limiting step (Hanif et al. 2007).

Adsorption isotherm models

Effect of adsorbent mass

Adsorption to a maximum of 5 g/100 ml (using 50 ppm metal solution) for Cr(VI) and 5 g/100 ml (10 ppm metal solution) for Cd(II). By increasing the quality of adsorption beyond the optimal mass, there is no significant change in the removal efficiency. These results indicate that the removal efficiency is directly related to the number of available adsorption sites. Once equilibrium is attained there is no effect on adsorption efficiency. The optimum adsorbent mass is 5 g/100 ml for both metals with 71.3 % removal of Cr(VI) and 74.3 % removal of Cd(II).

Adsorption isotherms

Langmuir adsorption isotherms The Langmuir is used to describe the experimental adsorption isotherm. The Langmuir isotherm equation may be formulated as

$$\frac{C_{eq}}{C_{ads}} = \left(\frac{1}{Qb}\right) + \left(\frac{C_{eq}}{Q}\right) \quad (3)$$

where non linearity is indicated in Figs. 8 and 9. From this it is not possible to correlate to the isotherm with the experimental data and it is clear that the experimental data is not following Langmuir isotherm. Another essential dimensionless equilibrium constant of Langmuir isotherm is R_L which is applied in many operations and is expressed as:

$$R_L = \frac{1}{1 + bC_i} \quad (4)$$

Table 1 Comparison between Lagergren pseudo first and second order kinetic of adsorption

Metal ion	Pseudo first order kinetic model			Experimental value	Pseudo second order kinetic model		
	q_e (mg g^{-1})	k_1 (min^{-1})	R^2	q_e (mg g^{-1})	q_e (mg g^{-1})	k_2 (g mg^{-1} min^{-1})	R^2
Cr(VI)	1071.86	0.001962	0.9555	124	360.73	0.005037	0.9582
Cd(II)	563.09	0.003804	0.96813	132	169.72	0.005231	0.9651

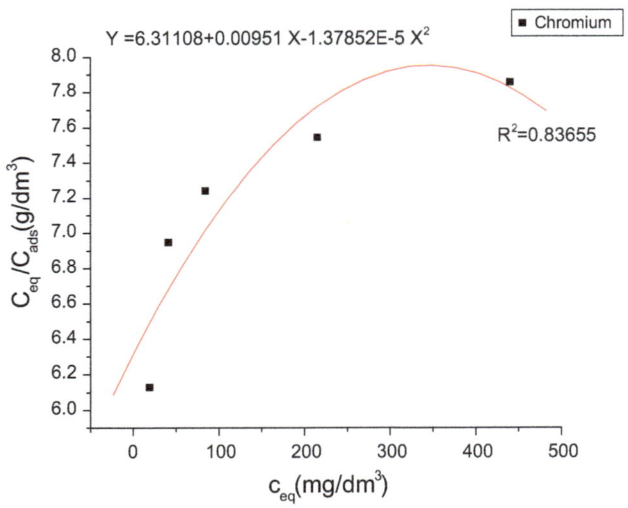

Fig. 8 Langmuir isotherm plot for chromium

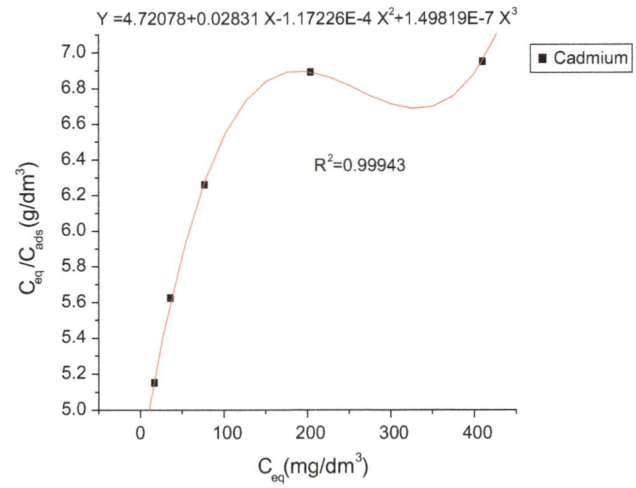

Fig. 9 Langmuir isotherm plot for cadmium

where, C_i the initial concentration is the analyst in the solution. If $R_L > 1$, then it is considered unfavourable. $R = 1$ corresponds to linear and $0 < R < 1$ is favourable (Table 2).

Freundlich adsorption isotherms The Freundlich model is used in essential of adsorption concentration of the adsorbate on the adsorbent surface (Freundlich 1926) and expressed as:

$$C_{ads} = KC_{eq}^{\frac{1}{n}} \quad (5)$$

The linearized (Figs. 10, 11) form of this equation is expressed as

$$\log C_{ads} = \frac{1}{n}\log C_{eq} + \log K \quad (6)$$

whereas, $1/n$ is Freundlich characteristic constant related to energy, and K is the maximum adsorption capacity. Both n and K are temperature dependant. By plotting log C_{ads} against log C_{eq}, if a straight line is obtained, then the value of $1/n$ and K can be calculated from the slope and intercept. At the regions of moderate coverage Freundlich isotherm is similar to the Langmuir isotherm. Values of $1/n$ indicate the steepness or flatness of the slopes. $1/n$ close to 1 indicates the steep slope and high adsorption capacity and equilibrium concentrations which diminishes at low equilibrium concentrations. On the other hand $1/n \ll 1$ indicates a flat slope and adsorption capacity is slightly reduced at low equilibrium concentration (Faust and Aly 1987).

Table 2 Adsorption isotherm constant, C_{max} and correlation coefficients

Metal ions	Langmuir constants			
	KL (dm^3 g^{-1})	b (dm^3 mg)$^{-1}$	C_{max} (mg g^{-1})	R^2
Cr(VI)	6.642	0.003139	2115.96	0.8365
Cd(II)	5.571	0.004061	1371.83	0.9994

Adsorption isotherms results Langmuir and Freundlich The adsorption isotherm is fundamental in describing the interactive behaviour between solutes and adsorbent (Chiou and Li 2003; Ofomoja and Ho 2006). According to Ofomaja and Ho (2006), the isotherm yields certain constant values, which expresses the surface properties and affinity of the adsorbent. It also plays an important role in the design of an adsorption system.

The parameter indicates the effect of separation factor on isotherm shape according to Table 3. The values of R_L

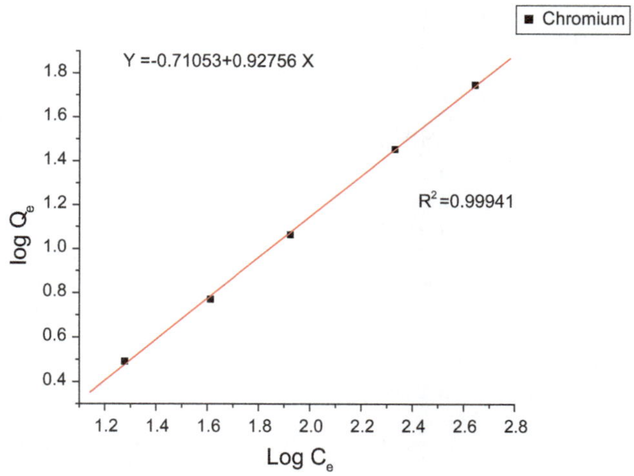

Fig. 10 Freundlich isotherm plot for chromium

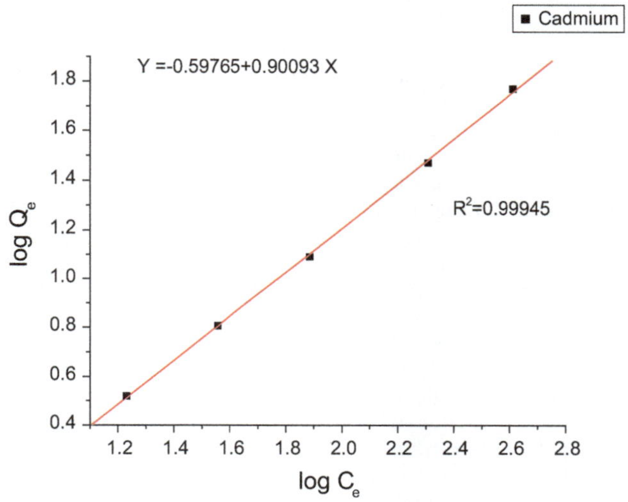

Fig. 11 Freundlich isotherm plot for cadmium

Table 3 Adsorption isotherm constant Freundlich

Metal ions	Freundlich constants		
	K	*n*	R^2
Cr(VI)	0.1948	1.0780	0.9994
Cd(II)	0.2526	1.1100	0.9994

calculated for different initial metal ion concentration are given in Table 4. If the R_L values are in the range of $0 < R_L < 1$, it indicates that the adsorption of Cr(IV) and Cd(II) onto chitosan beads are favourable. Thus, chitosan beads are favourable adsorbents. The mechanism of ion adsorption on porous adsorbents may involve three steps: (1) diffusion of the ions to the external surface of adsorbent; (2) diffusion of ions into the pores of adsorbents; (3) adsorption of the ions on the internal surface of adsorbent.

Table 4 R_L values based on Langmuir adsorption

Metal ions	Initial concentration C_0 (mg dm^{-3})	Final concentration C_f (mg dm^{-3})	R_L values
Cr(VI) ion	1000	440	0.4200
	500	215	0.5970
	200	84	0.7913
	100	41	0.8860
	50	19	0.9437
Cd(II) ion	1000	410	0.3752
	500	204	0.5469
	200	77	0.7618

Conclusion

Chitosan mixed with vermiculite has been successfully used as the adsorbent for the removal of Cr(VI) and Cd(II) from wastewater. Cadmium recovery is slightly higher at all times than chromium. The maximum adsorption for Cr(VI) has been obtained at pH 5.0 and for Cd(II) at pH 5.5. The percentage of adsorption of chromium and cadmium increases steadily with solid to liquid ratio of 1–3 % whereas at 4–6 saturation is attained. The adsorption of Cr(VI) and Cd(II) is found to be 59.2 and 71.5 % respectively with the contact time of 300 min. Kinetic data of adsorption studies indicate that adsorption follows pseudo second order model and bio adsorption is found to have rate limiting step. The optimum adsorption mass is 5 g/100 ml for both metals with 71.3 % removal of Cr(VI) and 74.3 % removal of Cd(II). The cross linker being glutaraldehyde enhances the thermal and chemical stability. Further, cross linking increases intra molecular hydrogen bonding enabling greater adsorption. The feasibility of using chitosan with vermiculite added with glutaraldehyde in the ratio 3:1 is established and this composition yields greater capacity for adsorption

References

Ahmad AL, Kusumastuti A, Derek CJC, Ooi BS (2012) Emulsion liquid membrane for cadmium removal: studies on emulsion diameter and stability. Desalination 287:30–34

Asadi F, Shariatmadari H, Mirghaffari N (2008) Modification of rice hull and sawdust sportive characteristics for removes heavy metals from synthetic solutions and wastewater. J Hazard Mater 154:451–458

Baird C, Cann M (2005) Environment chemistry, vol 1, 3rd edn. WH Freeman and Company, San Francisco, CA, p 1555

Chang YH, Huang CF, Hsu WJ, Chang FC (2007) Removal of Hg^{2+} from aqueous solution using alginate gel containing chitosan. J Appl Polym Sci 104:2896–2905

Chao AC, Shyu SS, Lin YC, Mi FL (2004) Enzymatic grafting of carboxy group on to chitosan to confer on chitosan the property of a cationic dye adsorbent. Bioresour Technol 91:157–162

Chiou MS, Li HY (2003) Adsorption behavior of reactive dye in aqueous solution on chemical cross-linked chitosan beads. Chemosphere 50:1095

Chlopecka A (1996) Assessment of form of Cd, Zn and Pb in contaminated calcareous and gleyed soils in Southwest Poland. Sci Total Environ 188:253–262

Chu KH (2002) Removal of copper from aqueous solution by chitosan in prawn shell: adsorption equilibrium and kinetics. J Hazard Mater B 90:77–95

Crini G (2005) Recent developments in polysaccharide-based materials used as adsorbents in wastewater treatment. Prog Polym Sci 30:38–70

Dahiya S, Tripathi RM, Hegde AG (2008) Biosorption of lead and copper from aqueous solutions by pre-treated crab and area shell biomass. Bioresour Technol 99:179–187

dos Santos ACV, Masini J (2007) Evaluating the removal of Cd(II), Pb(II) and Cu(II) from a wastewater sample of a coating industry by adsorption onto vermiculite. Appl Clay Sci 37:167–174

Du WL, Niu SS, Xu ZR, Xu YL (2009) Preparation, characterization, and adsorption properties of chitosan microspheres crosslinked by formaldehyde for copper (II) from aqueous solution. J Appl Polym Sci 111:2881–2885

Elkady MF, Abu-Saied MA, Abdel Rahman AM, Soliman EA, Elzatahry AA, ElsayedYossef M, MohyEldin MS (2011) Nano-sulphonatedpoly(glycidyl methacrylate) cations exchanger for cadmium ions removal: effects of operating parameters. Desalination 279:152–162

Erosa MSD, Medina TIS, Mendoza RN, Rodriguez MA, Guibal E (2001) Cadmium sorption on chitosan sorbents: kinetic and equilibrium studies. Hydrometallurgy 61:157–167

Gerente C, Lee VKC, Lee P, McKay G (2007) Application of chitosan for the removal of metals from wastewaters by adsorption—mechanisms and models review. Crit Rev Environ Sci Technol 37:41–127. doi:10.1080/10643380600729089

Gupta VK, Ali I (2004) Removal of lead and chromium from wastewater using bagasse flyash—a sugar industry waste. J Colloid Interface Sci 271(2):321–328

Gupta VK, Nayak A (2012) Cadmium removal and recovery from aqueous solutions by novel adsorbents prepared from orange peel and Fe_2O_3 nanoparticles. Chem Eng J 180:81–90

Gupta VK, Rastogi A (2010) Biosorption of nickel onto treated alga (*Oedogonium hatei*): application of isotherm and kinetic models. J Colloid Interface Sci 342(2):533–539

Gupta VK, Jain R, Mittal A, Mathur M, Sikarwar S (2007) Photochemical degradation of the hazardous dye Safranin-T using TiO_2 catalyst. J Colloid Interface Sci 309(2):464–469

Gupta VK, Rastogi A, Nayak A (2010) Adsorption studies on the removal of hexavalent chromium from aqueous solution using a low cost fertilizer industry waste material. J Colloid Interface Sci 342(1):135–141

Gupta VK, Agarwal S, Saleh TA (2011) Chromium removal by combining the magnetic properties of iron oxide with adsorption properties of carbon nanotubes. Water Res 45(6):2207–2212

Gupta VK, Ali I, Saleh TA, Nayak A, Agarwal S (2012) Chemical treatment technologies for wastewater recycling—an overview. RSC Adv 2:6380–6388

Hanif MA, Nadeema R, Zafar MN, Akhtar K, Bhatti HN (2007) Kinetic studies for Ni(II) adsorption from industrial wastewater by *Cassia fistula* (Golden Shower) biomass. J Hazard Mater 145:501–505

Ho YS, Wase DAJ, Forster CF (1996) Removal of lead ions from aqueous solution using sphagnum moss peat as adsorbent. Water SA 22:219–224

Ho YS, Huang CT, Huang HW (2002) Equilibrium sorption isotherm for metal ions on tree fern. Process Biochem 37:1421

Holm PE, Christensen TH, Tjell JC, McGrath SP (1995) Speciation of cadmium and zinc with application to soil solutions. J Environ Qual 24:183–190

Holme KR, Perlin AS (1997) Chitosan N-sulfate. A water-soluble polyelectrolyte. Carbohydr Res 302:7–12

Horsfall MJ, Spiff AI (2005) Sorption of lead, cadmium and zinc on sulfur-containing chemically modified wastes of fluted pumpkin. Chem Biodivers 2:373–385

Jain AK, Gupta VK, Bhatnagar A, Suhas (2003) A comparative study of adsorbents prepared from industrial wastes for removal of dyes. Sep Sci Technol 38(2):463–481

Jayakumar R, Rajkumar M, Freitas H, Selvamurugan N, Nair SV, Furuike TH, Tamura H (2009) Preparation, characterization, bioactive and metal uptake studies of alginate/phosphorylated chitin blend films. Int J Biol Macromol 44:107–111

Li Q, Zhai J, Zhang M, Wang J, Zhou J (2007) Kinetic studies of adsorption of Pb(II), Cr(III) and Cu(II) from aqueous solution by sawdust and modified peanut husk. J Hazard Mater 141:163–167

Malandrino M, Abollino O, Giacomino A, Aceto M, Mentasti E (2006) Adsorption of heavy metals on vermiculite: influence of pH and organic ligands. J Colloid Interface Sci 299:537–546

McBride M (1994) Environment chemistry of soils. Oxford University Press, New York

Mittal A, Gipta VK, Malviya A, Mittal J (2008) Process development for the batch and bulk removal and recovery of a hazardous, water-soluble azo dye (Metanil Yellow) by adsorption over waste materials (Bottom Ash and De-Oiled Soya). J Hazard Mater 151(2–3):821–832

Mittal A, Mittal J, Malviya A, Kaur D, Gipta VK (2010) Adsorption of hazardous dye crystal violet from wastewater by waste materials. J Colloid Interface Sci 343(2):463–473

Morimoto M, Saimoto H, Shigemasa Y (2002) Control of functions of chitin and chitosan by chemical modifications. Trends Glycosci Glycotechnol 14:205–222

Ngah WSW, Hanafiah MAKM (2008) Biosorption of copper ions from dilute aqueous solutions on base treated rubber (*Hevea brasiliensis*) leaves powder: kinetics, isotherm, and biosorption mechanism. J Environ Sci 20:1168–1176

Ngah WSW, Kamari A, Koay YJ (2004) Equilibrium and kinetic studies of adsorption of copper(II) on chitosan and chitosan/PVA beads. Int J Biol Macromol 3:155–161

Ofomoja AE, Ho YS (2006) Equilibrium sorption of anionic dye from aqueous by palm kernel fibre as sorbent. Dyes Pigments 12:249–257

Popuri SR, Vijaya Y, Boddu VM, Abburi K (2009) Adsorptive removal of copper and nickel ions from water using chitosan coated PVC beads. Bioresour Technol 100:194–199

Prakash N, Sudha PN, Renganathan NG (2011) Copper and cadmium removal from synthetic industrial wastewater using chitosan and nylon 6. Environ Sci Pollut Res 19(7):2930–2941
Prakash N, Latha S, Sudha PN, Renganathan NG (2012) Influence of clay on the adsorption of heavy metals like copper and cadmium on chitosan. Environ Sci Pollut Res 20(2):925–938
Saeed A, Iqbal M (2003) Bioremoval of cadmium from aqueous solution by black gram husk (*Cicer arientinum*). Water Resour 37(14):3472–3480
Srivastava SK, Gupta VK, Mohan Dinesh (1997) Removal of lead and chromium by activated slag—a blast-furnace waste. J Environ Eng 123(5):46–468
Takiguchi M, Yoshihara S (2006) New aspects of cadmium as endocrine disruptor. Environ Sci 13(2):107–116
Tao Y, Ye L, Pan J, Wang Y, Tang B (2009) Removal of Pb(II) from aqueous solution on chitosan/TiO_2 hybrid film. J Hazard Mater 161:718–722
Vasudevan S, Lakshmi J, Sozhan G (2011) Effects of alternating and direct current in electrocoagulation process on the removal of cadmium from water. J Hazard Mater 192:26–34
Wang JW, Kuo YM (2008) Preparation and adsorption properties of chitosan-poly (acrylic acid) nanoparticles for the removal of nickel ions. J Appl Polym Sci 07:2333–2342
Wankasi D, Horsfall HJ, Spiff AI (2006) Sorption kinetics of Pb^{2+} and Cu^{2+} ions from aqueous solution by Nipah palm (*Nypa fruticans* Wurmb) shoot biomass. Electron J Biotechnol 9:587–592
Xu X, Li L, He J (2009) The performances of carboxymethyl chitosan in wash-off reactive dyeing. Carbohydr Polym 75:203–207

23

Point source pollution and variability of nitrate concentrations in water from shallow aquifers

Jasna Nemčić-Jurec[1] · Anamarija Jazbec[2]

Abstract Agriculture is one of the several major sources of nitrate pollution, and therefore the EU Nitrate Directive, designed to decrease pollution, has been implemented. Point sources like septic systems and broken sewage systems also contribute to water pollution. Pollution of groundwater by nitrate from 19 shallow wells was studied in a typical agricultural region, middle Podravina, in northwest Croatia. The concentration of nitrate ranged from <0.1 to 367 mg/l in water from wells, and 29.8 % of 253 total samples were above maximum acceptable value of 50 mg/l (MAV). Among regions R1–R6, there was no statistically significant difference in nitrate concentrations ($F = 1.98$; $p = 0.15$) during the years 2002–2007. Average concentrations of nitrate in all 19 wells for all the analyzed years were between recommended limit value of 25 mg/l (RLV) and MAV except in 2002 (concentration was under RLV). The results of the repeated measures ANOVA showed statistically significant differences between the wells at the point source distance (proximity) of <10 m, compared to the wells at the point source distance of >20 m ($F = 10.6$; $p < 0.001$). Average annual concentrations of nitrate during the years studied are not statistically different, but interaction between proximity and years is statistically significant ($F = 2.07$; $p = 0.04$). Results of k-means clustering confirmed division into four clusters according to the pollution. Principal component analysis showed that there is only one significant factor, proximity, which explains 91.6 % of the total variability of nitrate. Differences in water quality were found as a result of different environmental factors. These results will contribute to the implementation of the Nitrate Directive in Croatia and the EU.

Keywords Nitrate · Pollution · Point source · Water · Shallow well · Croatia

Introduction

The most important and abundant source of drinking water in the European Union (EU) is groundwater constituting over 80 % of the supply (Council Directive 2000). Pollution of groundwater is caused by different pollutants, among which nitrate plays the most important role. Nitrate pollution from agriculture is an issue of major concern for the UN (Addiscott et al. 1991), and a series of environmental policies have been implemented in the EU to decrease nitrogen (N) emissions from agriculture. The Nitrate Directive is one of the main policies to reduce nitrate leaching from agriculture primarily through good agricultural practise and the application of organic and mineral fertilizers according to regulations (Velthof et al. 2014).

As nitrogen is the key factor in a high and stable cultivated plant production, it is no surprise that modern agriculture affects nitrogen content in water, as it is a high nitrogen user (Addiscott et al. 1991; Bašić and Herceg 2010). Nitrate concentration in natural waters has been increasing during the last few decades, coinciding with the

✉ Jasna Nemčić-Jurec
nemcic-jurec@inet.hr; jnemcicj@gmail.com

Anamarija Jazbec
ajazbec@sumfak.hr

[1] Institute of Public Health of Koprivnica-Krizevci County, Trg Tomislava Bardeka 10/10, 48 000 Koprivnica, Croatia

[2] Department for Forest Inventory and Management, Faculty of Forestry, University of Zagreb, Svetošimunska 25, 10 000 Zagreb, Croatia

increased use of mineral fertilizers and a higher yield of cultivated plants in agriculture. This problem has become even more significant by introducing the stricter standards concerning the nitrate content of drinking water in the EU. The maximum acceptable value (MAV) is 50 mg/l NO_3^- (Regulation 2013), but the recommended limit value (RLV) is only 25 mg/l NO_3^- (Council Directive 1991).

Agriculture is the biggest source of pollution because of the so-called diffuse source pollution that does not originate from a single discrete source. Pollution is often a cumulative effect of small amounts of contaminants gathered from a large area (US EPA 1987). Numerous scientists have shown nitrate leaching from agricultural areas. Elhatip et al. (2003) showed that agricultural activities contribute to an increased nitrate concentration in groundwater either directly by nitrate leaching from agrochemicals or by agrochemicals that affect the processes in the soil and increase nitrate leaching from the soil. Nitrate leaching depends on a variety of environmental factors. Mild climate and favorable winter temperatures allow the extension of the growing season of plants, which reduces nitrate leaching (Hooker et al. 2008). Benson et al. (2006) also evaluated the importance of the excessive use of agrochemicals. They point out that the use of fertilizers in doses of 500 kg/ha increases the nitrate concentration in the water up to a distance of 2 km from the well.

Point source pollution refers to contaminants that enter a waterway from a single, identifiable source such as livestock farms, slurry lagoons or manure depots with an inappropriate building or location (US EPA 1987). Dense populations and discharges from point sources like septic systems or broken sewer systems contribute significantly to water pollution by nitrate in urban and suburban areas (Nemčić-Jurec et al. 2007, 2013; Drake and Bauder 2005; Melian et al. 1999; US EPA 1987).

Effects of point sources of pollution are usually localized and limited (Krapac et al. 2002; Nemčić-Jurec et al. 2013). Ciravolo et al. (1979) showed that point source (lagoon with liquid pig manure) affected the concentration of nitrate in the water at the distance of 36 meters from the well. However, they point out that the contamination occurred in the wells located on sandy soil, while at locations where there is clay soil, there was no significant pollution. In similar studies, Richards et al. (1996) showed that the concentration of nitrate in the water from the well in sandy soil, at the distance of 6 m from the point source pollution (animal feedlot or barnyard), was twice as high (3.6 mg/l) compared to the concentration of nitrate (1.8 mg/l) from the wells located 60 m from the pollution. Nemčić-Jurec et al. (2013) showed by comparison of nitrate concentration in two different geological areas that the nitrate concentration occasionally or regularly exceeds MAV in wells close to point sources of pollution (within 10 m of the well). In both areas of research, however, the frequency of unsafe samples was significantly higher in Prigorje area compared to Podravina area. Nitrate concentration in water depends on the vertical and horizontal transports of nitrate leaching (Houben et al. 2001). They point out that the highest nitrate concentrations are determined in the shallow aquifers that are richer in oxygen at a depth of 12 m. Katz et al. (2004) showed that the concentration of nitrate in the shallow aquifer is spatially and vertically highly variable. Although the average nitrate content between the shallow and deep zone was not significantly different (0.13 and 0.17 mg NO_3-N/l), the maximum content was much higher in the shallow zone (7.3 mg NO_3-N/l) compared to the deeper zone (1.6 mg NO_3-N/l). The variability of nitrate concentration was higher in shallower layers.

Podravina is an area of relatively intensive agriculture where, according to modern agricultural demands, higher doses of organic and mineral fertilizers are applied, causing leaching of the soil and nitrate input in aquatic ecosystems. Considering the large share of arable land in the River Drava basin, the highest possible nitrate pollution of water in this region can be expected. Due to the increasing trend of application of mineral and organic fertilizers in the world (Woo-Jung et al. 2007; Pacheco et al. 2001; Bouwman et al. 2005; Cetindag 2005; Widory et al. 2003) and, on the other hand, due to stricter criteria concerning reduction in nitrate pollutions in water (EU Nitrate Council Directive 1991), the aim of this research was to investigate the annual nitrate concentration variability in the area of an intensive agriculture and dense population of the middle Podravina region. The areas especially prone to pollution are easily permeable soils in drinking water protection zones in shallow unconfined aquifers (Burkart and Stoner 2007; Kraft and Stites 2003). Therefore, the aim was to determine to what extent point sources of pollution from agriculture and septic systems contribute to total pollution, as well as to determine the nitrate concentration variability in individual wells depending on the environmental factors.

Study areas

Location

The locations of the wells within the regions (R1–R6) are illustrated in Fig. 1. The regions are based and defined on political (municipality) and geographical boundaries. The study area (20 × 30 km) located in the water catchment of middle Podravina in northwest Croatia extends from Legrad in the east to Sokolovac in the west. From the north, the area is bordered by the state border with the Republic of Hungary and in the south with the northern slopes of Bilogora and Papuk. The old alluvial terraces area is

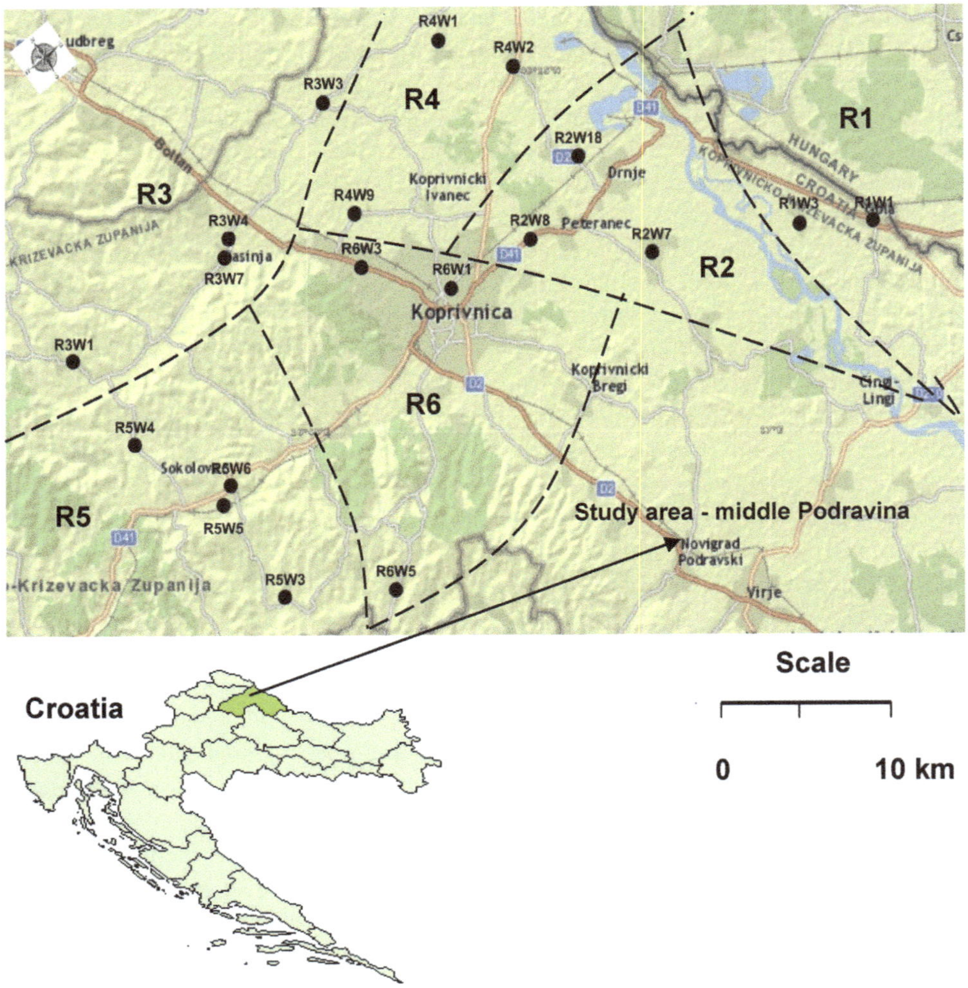

Fig. 1 Locations of the wells within the regions in middle Podravina (*R* region, *W* well)

composed of drift of the River Drava and its tributaries, as well as terraced sediment.

Climate and hydrogeology

From the hydrogeological point of view, the quaternary sand–gravel aquifer is the most important. The top part of the aquifer is built of sand and clay, with a significant proportion of quicksand and mostly of swampy loess in the southern and eastern part of water catchment (Mayer et al. 1996). A mild-continental climate dominates, and the average annual precipitation is about 820 mm. The average annual temperature is about 11 °C which indicates mild-warm climate (data obtained from the Metrological and Hydrological Service). These climate conditions favor agricultural production and increased population density.

Land use and point sources of pollution

The research area is inhabited by about 60,000 people with traditionally developed plant and animal production. The total research area covers 177,464 ha, out of which agricultural land covers 59 %. The remaining area consists of forests (32 %), arid land, settlements and water surfaces (9 %). Of the agricultural land, 99 % is arable (gardens, orchards, vineyards and meadows) and 1 % is pastures.

According to the data of the Faculty of Agriculture, average doses of organic and mineral fertilizers applied in middle Podravina (152 kg N/ha agricultural area) are higher in comparison with the average doses of fertilizers applied in Croatia (54 kg N/ha agricultural area). In the research area, point sources of pollution of agricultural origin such as farms and barns, and landfill of organic fertilizers depots are often very close to the wells (5–20 m). Most of the population in urban areas, up to 80 %, is connected to a sewerage system, while the population of suburban areas use septic systems which are also often very close to the wells.

On the sandstone outcrop area, the soil is freely draining so the nitrate input from point sources of pollution can be expected to accumulate in the aquatic ecosystems including the River Drava basin.

Materials and methods

Sampling

Nitrate concentration variability was tested in drinking water from 19 wells originating from the shallow aquifer. Characteristics of the wells are shown in Table 1. The locations of the wells were chosen in such a way to represent a range of end uses. The wells that supply water to public buildings, such as schools, kindergartens, shops and restaurants, in densely populated and agricultural areas of middle Podravina were selected. The surroundings of the wells were examined, the sources of pollution were identified, and the distance of the pollution source from the wells was measured by a digital meter (m). The wells were located at a greater or lesser distance from the point sources of pollution of agricultural origin (farms and organic fertilizers depots) and septic systems. Information about the location of septic systems was obtained from the owner of the well, and the distance was measured by a digital meter. During a 6-year study (2002–2007), a total of 253 samples of water were taken during all seasons (2–3 times per year) at the same time (through a week) from each well.

For quality control purposes, duplicate samples were taken and analyzed at a rate of 1 in 20. These were treated as completely separate samples. Precision judged (% RSD) from the duplicates is on average in the range of 2–8 % for nitrate concentration.

The sampling was done according to HRN ISO 5667-5:2000. The water sampling from wells was done after the releasing of water for 2–3 min on the outlet of a water supply system, in a clean polyethylene vessel of 1 l volume. The samples were stored in a portable fridge, in the dark at 1–5 °C for transportation. After delivery to the laboratory, the samples were stored at the same temperature until the analysis (within 24 h).

Water analysis

Nitrate concentration in drinking water was determined using the ion chromatography method (HRN ISO 10304-1:1998). Prior to the injection into the analyzer, ion chromatograph ICS 3000, Dionex (USA), samples were filtered through a membrane filter (of pore 0.45 μm) to remove any particulate matter. After establishing the calibration function, samples were injected into the ion chromatograph and the peaks were measured in accordance with the manufacturer's instructions.

The chromatograph system consists of the following components: auto sampler device, sample injection system (incorporating sample loop of 50 μl), gradient pump, liquid chromatography device, conductivity detector and eluent generator. The AS 15 anion separator column and appropriate precolumn were used for the separation. ASRS ULTRA II-4 mm performed chemical suppression. 38 mM NaOH was used as an eluent, and the flow was 0.5 ml/min.

Table 1 Characteristics of the wells by regions (R1–R6)

Region	Well	Well depth (m)	Pollution source	Proximity (m)	Soil type
R1	W1	6	Ag	>20	Gravel–sand
R1	W3	5	Ag, SS	>20	Gravel–sand
R2	W7	5	Ag	<10	Silt–gravel–sand
R2	W8	6	Ag	>20	Silt–gravel–sand
R2	W18	5	Ag	<10	Silt–gravel–sand
R3	W1	10	Ag	>20	Clay–silt–gravel
R3	W3	5	Ag	>20	Silt–gravel–sand
R3	W4	10	Ag	>20	Clay–silt–gravel
R3	W7	7	Ag, SS	10–20	Clay–silt–gravel
R4	W1	5	Ag	>20	Silt–gravel–sand
R4	W2	5	Ag	>20	Silt–gravel–sand
R4	W9	7	SS	>20	Gravel–sand
R5	W3	10	SS	10–20	Clay–gravel
R5	W4	10	Ag	>20	Clay–gravel
R5	W5	10	SS	10–20	Clay–gravel
R5	W6	15	Ag	<10	Clay–gravel
R6	W1	20	SS	>20	Gravel–sand
R6	W3	10	Ag, SS	<10	Gravel–sand
R6	W5	5	Ag, SS	<10	Clay–gravel

R region, *W* well, *Ag* agriculture, *SS* septic systems

The whole system was supervised by Dionex Chromeleon software.

Data analysis

Descriptive statistics was done for all the data ($N = 253$ samples; $N = 114$ data), and it was compared to the MAV Regulation (2013), i.e., recommended level value (RLV) of Nitrate Directive (Council Directive 1991). Average nitrate concentrations in each of 19 wells through the 6 years were calculated. Data from each year and location are averaged in order to determine the general water quality in a particular well due to nitrate. Box plots of annual nitrate concentration (mg/l) indicating MAV and RLV were log-transformed (ln) to provide a more symmetric (normalized) distribution.

The difference in the nitrate concentration during the years 2002–2007 among regions (R1–R6) was tested with an analysis of variance (ANOVA).

In order to compare the concentrations of nitrate by wells and depending on the distance from the point sources of pollution in the analyzed period, we used the repeated measure analysis of variance. Since we wanted to detect precisely which group differences were statistically significant, we used the Tukey multiple comparison post hoc test (Davis 2002).

In order to confirm the grouping of wells according to the average annual values and distance from the point source pollution, we used k-means clustering.

In order to analyze and present wells in respect of their annual average concentration of nitrate, we did principal component analysis (PCA).

The correlation between the concentrations of nitrate and depth of wells was done as well as the correlation between the concentrations of nitrate and precipitation.

Descriptive statistics and the repeated measure ANOVA were done using the SAS 9.2. PCA was done using the STATISTICA 8.0 (2011), and graphical presentations were done using both statistical packages.

Results

Nitrate concentration in groundwater

Concentration of nitrate ($N = 114$) in the research area of middle Podravina ranged from <0.1 to 367 mg/l in groundwater from shallow wells, and 29.8 % of the samples were above the MAV.

If we analyze the regions (R1–R6), shown in Fig. 1, we see that there is no statistically significant difference in the nitrate concentration during the years 2002–2007 among regions ($F = 1.98$, $p = 0.15$), and the interaction between regions and years is not statistically significant too ($F = 0.99, p = 0.49$).

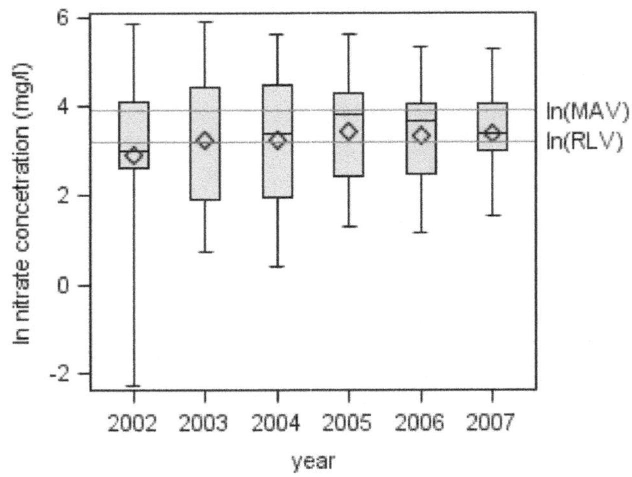

Fig. 2 Box plot of annual nitrate concentration (mg/l) indicating RLV (recommended level) and MAL (maximum allowed level) (*filled circle* mean, *hyphen* median, Q1–Q3 box, min–max whiskers)

The box plot of an annual nitrate concentration (mg/l) indicating MAV and RLV is shown in Fig. 2. Box plots, log-transformed to provide a more symmetric distribution, show that the average concentration of nitrate in all 19 wells for all the analyzed years was between RLV and MAV except in 2002 when the concentration was under the RLV. The distribution of the concentration of nitrate per year (mean, median, min and max) shows that the 2002 distribution was negatively asymmetric which means that most of the wells (over 50 %) had values below the RLV and almost 70 % below the MAV; 2003 and 2004 distribution was symmetric, i.e., about half of the wells had average concentration values up to the RLV. In 2005 and 2006, half of the wells had a concentration higher than the average and higher than the RLV, in 2005 almost reached the maximum (MAV), and after that value slightly decreased until 2007 (half of the wells had the average value that is slightly higher than the RLV). Nitrate pollution is present in wells of all regions, and during a 6-year study, there are year-to-year nitrate concentration differences. Figure 2 also shows that the highest variability of nitrate concentration was established in 2002. It is assumed that extremely low concentrations (detection limit—DL) of one well (R3W1) increase the total variability (negatively asymmetric) in 2002. Considering that the nitrate concentrations in the well are below the RLV and that the well is not exposed to contamination, it is difficult to explain the reason for the variability.

Point sources of nitrate pollution

The results of the repeated measure ANOVA show that there is a statistically significant difference in the concentration of nitrate due to the point source distance (Table 2).

Results of Tukey post hoc tests show that there is statistically significant difference between the wells within 10 m from the point sources of pollution compared to the wells within 20 m from the source of pollution ($F = 10.6$; $p < 0.001$), while the wells within 10–20 m from the point sources are not statistically different from wells within 10 m and within 20 m.

Average values of nitrate during the years are not statistically different, but the interaction between point source distance (proximity) and years is statistically significant ($F = 2.07$; $p = 0.04$), which indicates that the concentration of nitrate during the monitoring does not behave the same with respect to the distance (Fig. 3). Average values of nitrate concentration for wells whose distance is >20 m increase over the years, while those whose distance is <10 m decrease. In the group of the wells at the distance of point sources >20 m, nitrate concentration is generally below the MAV, and only in one well (R4W9), a concentration above the MAV in 2005 (74.2 mg/l) and 2006 (57.5 mg/l) was determined. After that, the concentration in that well decreases again.

Figure 3 also shows that the average concentration of nitrate at the distance of point sources <10 m is the highest in 2003. In a well (R2W7) from this group, concentration ranged from 43.6 to 186 mg/l. The maximum nitrate concentration (186 mg/l) was determined in 2003. In 2004, the concentration decreased (157 mg/l), and a decreasing trend continued until 2007. The variability of the above-mentioned well suggests an average increase in nitrate concentration in 2003. The figure shows that there are differences in the average concentration from year to year and that the variability of individual wells influences the average nitrate concentration through a period of research from 2002 to 2007. The reason behind the variability of the above-mentioned well is unknown.

It is evident that the proximity of point sources of pollution affects the average nitrate concentration. Figure 4 shows the grouping of the wells, considering the type of pollution (of agricultural origin—Ag, septic systems—SS) and a comparison with RLV and MAV. Results of k-means clustering confirmed a division into four clusters according to the pollution. Cluster 1 (R6W3AgSS) is the highest one with an average value ranging from 201 to 367 mg/l during the years and is far higher than the MAV. Cluster 2 (R2W7Ag, R2W8, R2W18Ag, R4W9SS, R5W5SS, R5W6Ag, 6W5AgSS) clustered wells whose average values are over the MAV, and it is evident that this includes

Table 2 Results of repeated measure analysis of variance for ln of nitrate concentration

Source of variation	DF	Mean square	*F* value	Pr > *F*
Between subject				
Point source distance	2	46.50	10.58	0.001
Error	16	4.39		
Within subject				
Year	5	0.06	0.23	0.956
Point source * year	10	0.52	2.07	0.036
Error	80	0.25		

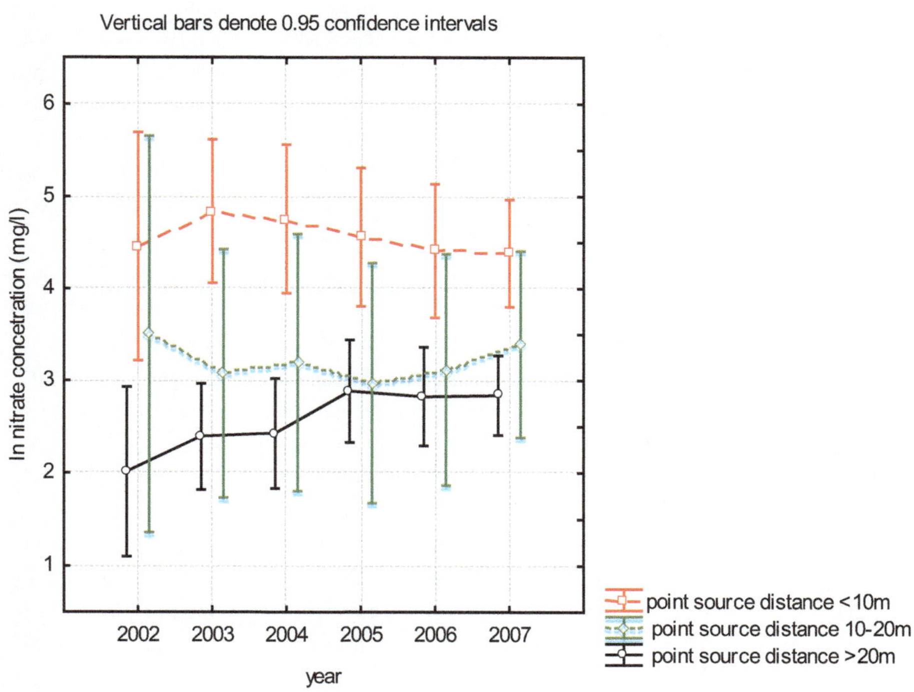

Fig. 3 Average of ln nitrate concentration (mg/l) and 95 % confidence intervals according to point source distance (m)

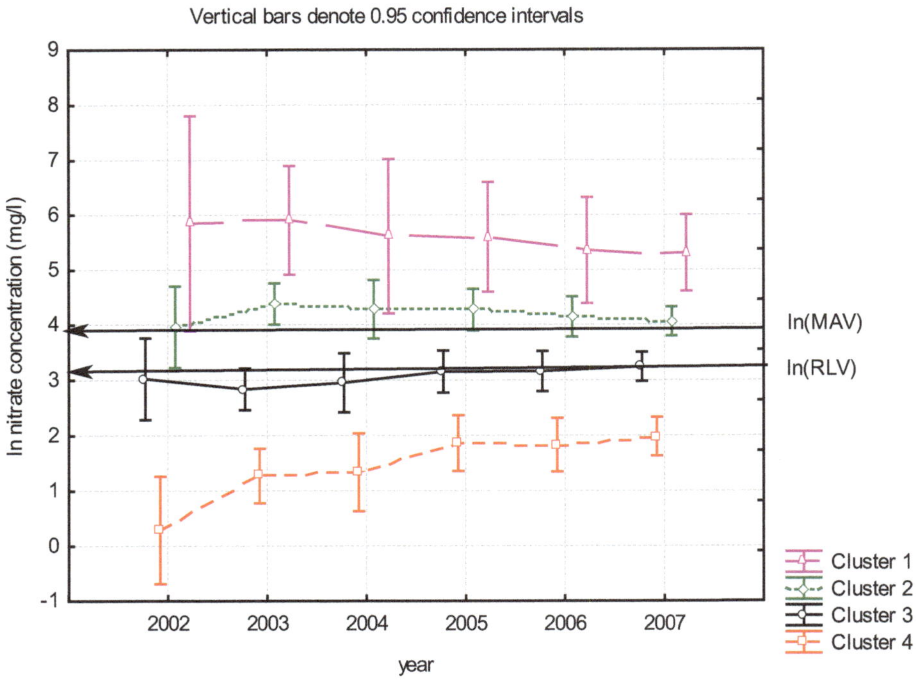

Fig. 4 Average of ln nitrate concentration (mg/l) of 4 clusters according to k-means clustering. RLV (recommend level) and MAL (maximum allowed level)

all wells that are exposed to Ag, SS or both. Cluster 3 (R1W1, R3W3, R3W4, R3W7, R5W3SS, R6W1, R4W1) has the nitrate concentration values a little less than the RLV, while cluster 4 (R1W3, R3W1, R4W2, R5W4) is the lowest.

According to the results, it is clear that the wells that are close to one type of point source of pollution (agriculture or septic systems) show an increased nitrate concentration, while the wells close to both types of pollution are most likely to have nitrate problems.

In the wells far from pollution (Cluster 3 and Cluster 4), established nitrate concentrations are lower (mostly below the RLV). However, according to Fig. 4, it can be seen that the difference in nitrate concentrations was also determined in these wells.

Given that the other environmental factors have not been studied in detail, we determined how many point sources of pollution contribute to the total variability of nitrate in the wells. Principal component analysis projections of the wells with respect to the first two factors according to the average annual nitrate concentration are shown in Fig. 5. Our results show that the first factor explains 91.6 % of the total variability, which means that the distance point source (proximity) is a very important predictor, and another factor accounts only for 5.67 %. We can see that the clusters are lined up one after the other and that the concentration of nitrate following point source distance varies from highest to the left to lowest to the right. The figure also shows that cluster includes two wells away from point sources of >20 m. In these wells, concentrations are close to the MAV (R2W8) or occasionally above the MAV (R4W9SS). Similar characteristics of these wells are such that nitrate concentrations are high even though the distance of point source is large or it is not determined at all.

Effects of multiple factors

The characteristics of the wells are very significant for contamination in shallow wells. For this reason, the correlation between the depth of the wells and nitrate concentrations was examined. There is no statistically significant correlation between the concentrations of nitrate and depth of wells (mean = 8.32 m, SD = 4.08 m) for any observed year ($r2002 = -0.02$, $r2003 = 0.14$, $r2004 = 0.11$, $r2005 = 0.11$, $r2006 = 0.07$, $r2007 = 0.07$).

In addition to depth, several other factors, for example precipitation, may be associated with more frequent contamination problems. The effect of some other factors (precipitation) can be related to the soil characteristics. According to the results, there was no statistically significant correlation ($r = -0.12$) between the average amount of rainfall (mm) per year and the average nitrate concentration.

Discussion

In the region of middle Podravina, a relatively high nitrate concentration (to 367 mg/l) was found in water wells from the shallow aquifer consistent with results of previous research. Kattan (2001) found a concentration of nitrate in

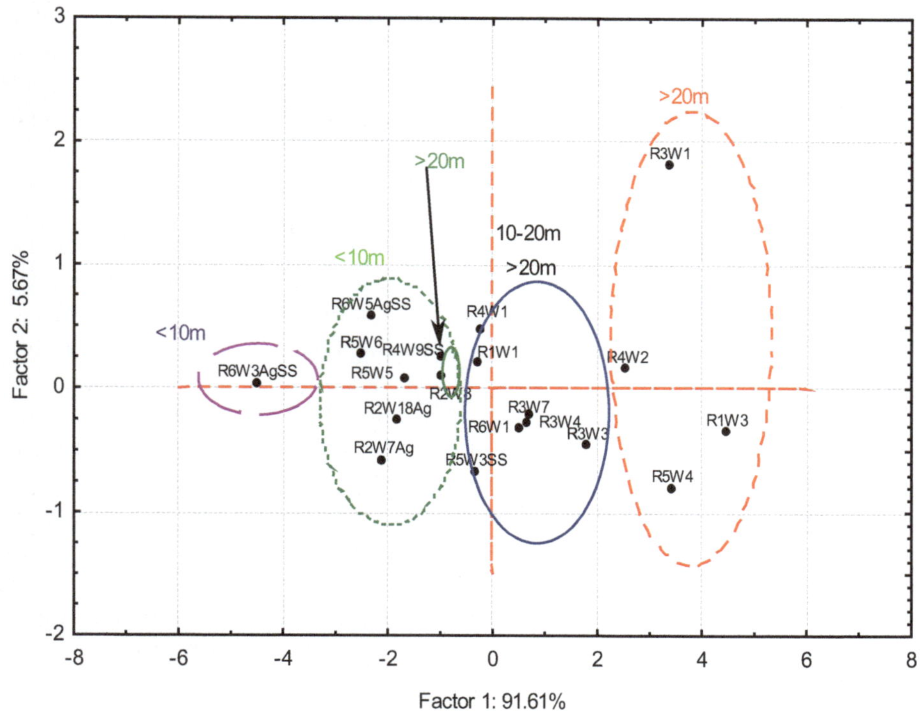

Fig. 5 Projections of the wells with respect to the first two factors (PCA) according to the ln of average annual nitrate concentrations indicating cluster membership and point source distance (Ag—agricultural source, SS—septic systems source, AgSS—both sources)

shallow wells ranging from 11 to 80 mg/l and some other researchers (Obeidat et al. 2007) ranging up to 330 mg/l in the wells, which was recognized as contamination. Other researchers reported an even lower nitrate concentration (8.5 mg/l) in drinking water (Hallberg 1989) that they already considered as pollution, while Kazemi (2004) and Jalali (2005) point out that the concentration of 14 mg/l is satisfactory because it is far below the MAV. Aelion and Conte (2004) found a low concentration (0.94 mg NO_3-N/l) in their research area. However, after an increase in the intensity of use of agricultural land, an increase in nitrate concentration up to 3 mg NO_3-N/l followed. Although scientists have different opinions about the concentrations considered as pollution, it is evident that 29.8 % of samples (above the MAV) in middle Podravina are contaminated by nitrate.

Contaminations of some wells do not necessarily reflect a generalized contamination of local or regional aquifers. Consistent local patterns of contamination probably indicate contamination of the aquifer locally (Richards et al. 1996; Bouwman et al. 2005). It is evident that the nitrate problem exists in all regions, but the nitrate concentration does not depend on the region but on the location of the wells within the region. This difference could be explained as a result of different specific habitat factors (Singh et al. 2015).

Although ammonia and nitrite can cause water quality problems, nitrate is most often associated with the contamination of drinking water (Follett and Hatfield 2001). In the earlier Podravina research, ammonia was not detected in the analyzed samples and it very rarely occurs in waters from shallow wells, whereas nitrate is very often a cause of water contamination. The water from the aquifer in the research area is rich in oxygen, and if ammonia occurs, it probably rapidly oxidizes to nitrate by redox processes (Houben et al. 2001; Follett and Hatfield 2001; Gates et al. 2008).

The most common sources of nitrate contamination include fertilizers, livestock manure, feedlots, septic tanks and land application of municipal sludge (Addiscott et al. 1991; Elhatip et al. 2003; Obeidat et al. 2007). Plant and livestock production in the middle Podravina region is well developed, but the infrastructure (water supply and sewage systems) in settlements that are supplied with water from shallow aquifers, i.e., wells, has not been built yet. Also, point sources of pollution like septic systems can often be found near the wells so these results were expected. Woo-Jung et al. (2007) compared the concentrations of nitrate during 3 years of monitoring in the agricultural area with various sources of pollution, such as sown agricultural land, farms of livestock or dense population, in order to determine types of source contaminations. They found a different variability and in some areas an increase in nitrate concentration above the MAV. In accordance with our results, they determined the lowest concentration of nitrate in the natural area where there are no sources of pollution close to the well, following the crop area. The largest concentrations were found in densely populated areas as well as in the areas with the farms close to the wells. The monitoring of the nitrate concentration in middle Podravina during a 6-year period has also shown that water from the wells far from pollution (determined by surveying of the well surroundings) was not contaminated by nitrate. The contamination by nitrate was determined in densely populated areas in the vicinity of point sources of agricultural origin (leaching from organic waste of private farms) or septic systems. Such results are shown by Krapac et al. (2002) as well, but alongside the type and vicinity of the source of contamination, the type of soil is also important for the leaching of nitrate (Ciravolo et al. 1979). By monitoring groundwater over 8 years, Ciravolo et al. showed contamination from the source of contamination in the area with sandy soil, while they did not determine significant water contamination in the area of clay soil. They emphasize the importance of the distance of point sources of pollution of the wells, but the effect of the impact depends on the permeability and the type of soil. Given that in Podravina there is sandy and gravelly soil (less clay soil), the obtained results are consistent with the hypotheses.

In this study, it was found that when the distance between the point sources and the wells was within 10 m, nitrate concentrations were above the MAV. In contrast, when the distance was greater (more than 10 m), effect on water quality was much rarer. Richards et al. (1996) also showed that the concentration of nitrate is significantly higher in well close to point source (within 6 m) of agricultural origin or septic systems from the well, compared to the concentration far from point source (within 60 m). It is obvious that the distance of the point source contributes to the final concentration of nitrate and nitrate variability. Gardner and Vogel (2005) also pointed out that distance and numbers of septic tanks and point sources of agricultural origin contribute to an increased nitrogen concentration in water even at distances up to 300 m and that the proximity of these types of pollution is a prerequisite for poor water quality due to nitrate.

In the study area, four clusters are shown according to the level of pollution. Water from wells close to point source pollution was worst quality. In wells that are far from point source, water quality is better, but nitrate concentrations are different. Factors that contribute to the difference in the nitrate concentration are not identified in this paper, but are assumed to be the result of the geological structure and porosity of the soil. The physical and chemical weathering of minerals, leaching and runoff are the main factors responsible for groundwater quality deterioration with some geogenic contribution from soil (Singh et al. 2015). The structure and depth of the wells and the proximity of diffuse pollution of agricultural origin or the use of agrochemicals on intensive agricultural areas affect the concentration of nitrate in groundwater in some locations of wells (Elhatip et al. 2003; Burkart and Stoner 2007; Kraft and Stites 2003). This result emphasizes the fact that vulnerability factors bear only a probabilistic relationship to the concentration of contaminants in a well and that even under conditions that combine several factors associated with a low concentration, a certain proportion of wells are contaminated.

Nemčić-Jurec et al. (2007), Melian et al. (1999), Nemčić-Jurec et al. (2013) identified multiple effects of different sources of pollution including agricultural point sources, since their samples had been mostly analyzed in connection with other point sources of pollution. However, some of the previous researches showed the pollution effect of different point sources like septic systems (Drake and Bauder 2005) and agricultural point sources (Widory et al. 2003). In this study, PCA showed that the distance is the main reason for the total variability, which means that it is a very important predictor in middle Podravina. Concentration of nitrate following point source distance ranges from highest (within 10 m, wells are exposed to Ag and SS) to lowest (wells are not exposed to contamination) which is in accordance with above-mentioned studies.

Concentrations of nitrate have shown different trends during the study. A higher nitrate concentration from year to year may be related to lower organic C content in older soils, and consequently less removal of nitrate by reduction or because of older wells, deteriorating structure and higher permeability (Richards et al. 1996). According to the results (data obtained from the Metrological and Hydrological Service) from 2004 to 2007, the annual precipitation is higher (753–991 mm) on average compared to the initial period of the study (740 mm–748 mm), which may contribute to the difference in the concentration of nitrate in the wells, in whose vicinity there are no point sources. However, there is a possibility of influencing underground water through diffuse sources of pollution, which can in general contribute to increasing concentrations in most of the wells. Nitrate concentration changes over time, and the interpretation of research results can be affected by the duration of the research period. Brian et al. (2004) emphasized that geocemical processes are too slow. They showed increase in nitrate concentration in groundwater during 3 years of monitoring. They pointed out that longer period of monitoring includes processes and the influence of potential hazards on groundwater quality. Variability or trends of water quality can be determined by 3 or more years of monitoring. The monitoring of trends during a period longer than 6 years was done by other researchers (Thorburn et al. 2003) reporting 4 categories of trends according to nitrate concentration: varying, steady, decreasing and increasing categories. They showed the effect on water quality during a longer period of time, as well as different effects of point sources near wells and influence of other agroecological factors.

Changes in the soil can affect the transport of nitrate. Alluvial areas which are usually formed by the deposition of soil or sediments have high porosity and permeability. They are considered good for groundwater recharge, and hence, they are also very vulnerable to groundwater pollution in comparison with confined aquifers (Singh et al. 2015). The concentration and variability of nitrate in water depend on the vertical and horizontal transports (Houben et al. 2001). According to Houben et al., vertical transport depends on the distribution of the reactive material in the aquifer, and it is estimated that the velocity of reactive vertical transport of pollutants is on average a few cm/year, whereas nonreactive is much faster, 1 m/year. They found the highest concentrations of nitrate in shallow layers of the aquifer that are richer in oxygen, and transport of nitrate is the fastest. According to Katz et al. (2004), the average concentrations of nitrate were similar in the shallow and deep zones, but the maximum content was much higher in the shallow zone compared to the deep zone. They point out that the concentration of nitrate is highly variable and that the variability of nitrate concentration is greater in shallow layers. As previously mentioned, the variability of nitrate concentrations in Podravina was also recorded in some wells close to pollution.

According to the studies of the Geological Survey, the hydraulic conductivity values range on average between 1×10^{-3} and 3×10^{-3} m/s. The analysis of anisotropy has found that the maximum anisotropy is determined in the shallowest part of the soil (2–7 m depth) made of gravel and sand. In the research area, the anisotropy coefficient usually varies from 10 to 35, while the vertical hydraulic conductivity is 10–35 times smaller than the horizontal. The results showed that the maximum horizontal conductivity is in a 5-m-deep layer (the most common depth of the monitored wells), which further explains the transfer of nitrate from the point source of pollution to the well. The nitrate of agricultural origin was leached for years from point sources into deeper layers, while septic systems are also often several meters deep and are often to be found at the level of the wells. For this reason, contamination of the wells through higher horizontal conductivity of nitrate was expected. Considering that the wells are shallow (mostly 5–10 m depth) and that in the research area, the soil is easily permeable, horizontal hydraulic conductivity values are high and a rapid transfer of nitrate from pollution sources to the well is expected, contributing to variability in individual wells. These shallow wells are more often irregularly, unsafely and spontaneously constructed. We assume that such kind of wells construction contributes to variability of water quality in the wells. Nemcic-Jurec et al. (2013) had examined nitrate concentration in part of Podravina not far away from the study area examined in this paper. Both areas showed similar geological structure and similar average nitrate concentrations (about 25 mg/l). However, proportions of samples above the MAV are different. According to Nemčić-Jurec et al. (2013), only 6 % was above the MAV, while at the nearby area examined in this paper, proportion of samples above MAV was significantly higher, about 30 %. We assume that similar geology and hydrogeology contribute to similar average nitrate concentrations in both areas, while higher proportion of samples above MAV shows periodically higher nitrate concentrations and higher variability. Differences are probably caused by micro-location and by safe environmental conditions, by proximity of source of pollution and by well construction itself. Bad and unsafe well construction and bad environmental conditions like proximity of source of pollution contribute to periodically higher nitrate concentrations, and they are an important prerequisite for poor water quality. Actually, bad and unsafe wells are indicators of proximity of pollution.

The perennial accumulation of nitrate in the soil near the source of pollution contributes to nitrate leaching into groundwater and drinking water in wells (Katz et al. 2004).

Richards et al. (1996) and McLay et al. (2001) have shown that the characteristics of the wells are very important for the contamination of groundwater from shallow wells. In the study area, the wells are usually shallow, and the difference in depths is not significant (90 % of the monitored wells are at 5–10 m depth), it is expected that there is no relationship between nitrate concentration and depth of the wells. Such a relationship should be investigated in different geological areas and in wells with significantly greater differences in depth (range 5–30 m or more). Previous research (Richards et al. 1996; McLay et al. 2001) determined the negative correlation between nitrate concentration and significantly different depth of wells (between wells <15 m deep and wells >30 m deep).

As previously noted, the effect of precipitation on wells depends on the characteristics of the soil and wells in sandy soils more frequently exceed the MAV. The alluvial aquifer is particularly sensitive to pollution (Burkart and Stoner 2007; Singh et al. 2015). Podravina predominantly has alluvial aquifers, but correlation between rainfall and nitrate concentration has not been established. Whatever the exact explanation for the unexpected relationships with well water quality shown by these two factors (precipitation and type of soil), it is clear that if they have any tendency to contribute to an increase in nitrate concentration, this tendency is small enough to be masked by confounding factors acting in the other directions. Similar results were obtained by other researchers (Burkart and Stoner 2007; Kraft and Stites 2003). The largest proportion of wells show varying trends over a longer period of time as a result of all environmental factors, and in these wells, the correlation between nitrate and precipitation was not found. They emphasize that groundwater is particularly sensitive to pollution in areas with a humid climate. More precipitation contributes to the variability of nitrate. However, according to some researchers, one of the more significant environmental factors which influence the concentration of nitrates in wells is precipitation. Pacheco et al. (2001) found the lowest nitrate concentration during the rain season due to dilution of water in wells caused by rain which led to low nitrate concentrations, and they also found a relationship between nitrate and precipitation.

Conclusions

Nitrate-contaminated groundwater has been observed in shallow wells in a typical agricultural region, middle Podravina. Nitrate concentration in some wells exceeded the maximum accepted value of 50 mg/l required by the EU Drinking Water Standards; 29.8 % of the samples were above the MAV.

There is no statistically significant difference in the nitrate concentration during the years 2002–2007 among regions (R1–R6) nor was the interaction between regions and years statistically significant, which indicates that during the monitoring, all regions behaved approximately equally. The nitrate concentration problem was observed in all regions with the key predictor of concentration being the well location within the region.

In the study area, there is a significant difference in the concentration of nitrate due to the point source distance (proximity). Water quality is better in the wells at the proximity of <20 m compared to the wells at the proximity of >10 m. Average values of nitrate during the years are not statistically different. In the wells close to one type of point source pollution (agricultural origin or septic systems), there is an increased nitrate concentration, while in the wells close to both types of contamination, the water is more frequently unsafe for health. Point source distance (proximity) explains 94 % of the total variability of nitrate, which means that the proximity is a very important predictor. Bad and unsafe well construction also contributes to periodically higher nitrate concentrations. It is an indicator of pollution proximity, and it is an important prerequisite for poor water quality.

Because high nitrate concentrations in water can cause different health problems, it is necessary to protect drinking water in the study area. High nitrate concentrations are associated with improperly located wells. It is necessary to locate new wells uphill and least 20 m from feedlots, septic systems, barnyards and chemical storage facilities. Monitoring of nitrate in the tested wells should be continued, and specific rehabilitation measures of wells to end users should be proposed. Wherever possible, it is necessary to build a public water supply system to ensure safe drinking water.

It is also important to manage nonpoint sources of water pollution (fields, lawns) to limit the loss of excess water and plant nutrients and match fertilizer applications to precise crop uptake needs in order to minimize groundwater contamination. Careful fertilizer management can reduce nitrate leaching to groundwater.

These measures will contribute to the implementation of the Nitrate Directive in Croatia and the EU, thus providing pollution control and agricultural product safety.

References

Addiscott TM, Whitmore AP, Powlson DS (1991) Farming, fertilizers and the nitrate problem. CAB International, Wallingford, Oxon, UK

Aelion CM, Conte BC (2004) Susceptibility of residential wells to VOC and nitrate contamination. Environ Sci Technol 38:1648–1653

Benson VS, VanLeeuwen JA, Sanchez J, Dohoo IR, Somers GH (2006) Spatial analysis of land use impact on ground water nitrate concentrations. J Environ Qual 35:421–432

Bouwman AF, Van Drecht G, Van der Hoek KW (2005) Global and regional surface nitrogen balances in intensive agricultural production systems for the period 1970–2030. Pedosphere 15:137–155

Brian GK, Angela RC, Thomas RP (2004) Use of chemical and isotopic tracers to assess nitrate contamination and ground-water age, Woodville Karst Plain, USA. J Hydrol 289:36–61

Burkart MR, Stoner JD (2007) Nitrate in aquifers beneath agricultural systems. Water Sci Technol 56:59–69

Cetindag B (2005) Investigation of discharge and groundwater contamination characteristics around the Karasu Spring, Mus Area, Turkey. Environ Geol 47:268–282

Ciravolo TG, Martens DC, Hallock DL, Collins ER, Kornegay ET, Thomas HR (1979) Pollutant movement to shallow ground water tables from anaerobic swine waste lagoons. J Environ Qual 8:126–130

Clean Water Act (CWA), Section 502 (1987). General definitions. US EPA

Davis CS (2002) Statistical methods for the analysis of repeated measurements. Springer, New York

Drake VM, Bauder JW (2005) Ground water nitrate-nitrogen trends in relation to urban development, Helena, Montana, 1971–2003. Ground Water Monit Remediat 25:118–130

Drinking water safety Regulation (2008) NN 47/2008

Elhatip H, Afsim M, Kuscu I, Dirik K, Kurmac Y, Kaviurmaci M (2003) Influences of human activities and agriculture on groundwater quality of Kayseri-Incesu-Dokuzpinar springs, central anatolian part of Turkey. Environ Geol 44:490–494

European Union(EU) (1991) Council Directive 91/676/EEC of 12 December 1991 concerning the protection of waters against pollution caused by nitrates from agricultural sources. Off J 375:0001–0008

European Union(EU) (2000) Council directive 2000/60/EEC of 23 October 2000 establishing a framework for the Community action in the field of water policy. Off J 327:0001–0073

Follett RF, Hatfield JL (2001) Nitrogen in the environment: sources, problems and management, 1st edn. Elsevier science, Amsterdam, The Netherlands

Gardner KK, Vogel RM (2005) Predicting ground water nitrate concentration from land use. Ground Water 43:343–352

Gates JB, Bohlke JK, Edmunds WM (2008) Ecohydrological factors affecting nitrate concentrations in a phreatic desert aquifer in northwestern China. Environ Sci Technol 15:3531–3537

Hallberg GR (1989) Nitrates in groundwater in the United States of America. In: Follet RF (ed) Nitrogen management and groundwater protection: developments in agriculture and managed-forest ecology. Elsevier, Dordrecht

Hooker KV, Coxon CE, Hackett R, Kirwan LE, O'Keeffe E, Richards KG (2008) Evaluation of cover crop and reduced cultivation for reducing nitrate leaching in Ireland. J Environ Qual 37:138–145

Houben GJ, Martiny A, Basler N, Langguth HR, Pluger WL (2001) Assessing the reactive transport of inorganic pollutants in groundwater of the Bourtanger Moor area (NW Germany). Environ Geol 41:480–488

HRN ISO 10304-1:1998. Water quality—Determination of dissolved anions by liquid chromatography of ions-Part 1: Determination of bromide, chloride, fluoride, nitrate, nitrite, phosphate and sulfate

HRN ISO 5667-5:2000. Water quality-Sampling-Part 5: Guidance on sampling of drinking water and water used for food and beverage processing

Jalali M (2005) Nitrates leaching from agricultural land in Hamadan, western Iran. Agric Ecosyst Environ 110:210–218

Kattan Z (2001) Use of hydrochemistry and environmental isotopes for evaluation of groundwater in the Paleogene limestone aquifer of the Ras Al-Ain area (Syrian Jezireh). Environ Geol 41:128–144

Katz BG, Chelette AR, Pratt TR (2004) Use of chemical and isotopic tracers to assess nitrate contamination and ground water age, Woodville Karst plain, USA. J Hydrol 289:36–61

Kazemi GA (2004) Temporal changes in the physical properties and chemical composition of the municipal water supply of Shahrood, northeastern Iran. Hydrogeol J 12:723–734

Kraft GJ, Stites W (2003) Nitrate impacts on groundwater from irrigated-vegetable systems in a humid north-central US sand plain. Agric Ecosyst Environ 100:63–74

Krapac IG, Dey WS, Roy WR, Smyth CA, Storment E, Sargent SL, Steele JD (2002) Impact of swine manure pits on groundwater quality. Environ Poll 120:475–492

McLay CDA, Dragten R, Sparling G, Selvarajah N (2001) Predicting groundwater nitrate concentrations in a region of mixed agricultural land use: a comparison of three approaches. Environ Poll 115:191–204

Melian R, Myrlian N, Gouriev A, Moraru C, Radstake F (1999) Groundwater quality and rural drinking-water supplies in the Republic of Moldova. Hydrogeol J 7:188–196

Nemčić-Jurec J, Mesić M, Bašić F, Kisić I, Zgorelec Ž (2007) Nitrate concentration in drinking water from wells at three different locations in northwest Croatia. Cereal Res Commun 35:845–848

Nemčić-Jurec J, Konjačić M, Jazbec J (2013) Monitoring of nitrates in drinking water from agricultural and residential areas of Podravina and Prigorje (Croatia). Environ Monit Assess 185:9509–9520

Obeidat MM, Massadeh AM, Al-Ajlouni AM, Athamneh FS (2007) Analysis and evaluation of nitrate levels in groundwater at Al-Hashimiya area, Jordan. Environ Monit Assess 135:1–3

Pacheco J, Marin L, Cabrera A, Steinich B, Escolero O (2001) Nitrate temporal and spatial patterns in 12 water-supply wells, Yucatan, Mexico. Environ Geol 40:708–715

Richards RP, Baker DB, Creamer NL, Kramer JW, Ewing DE, Merryfield BJ, Wallrabenstein LK (1996) Well water quality, well vulnerability and agricultural contamination in the Midwestern United States. J Environ Qual 25:389–402

SAS 9.2 (http://support.sas.com/documentation/92/index.html). Accessed 25 April 2008

Singh SK, Srivastava PK, Singh D, Han D, Gautam SK, Pandey AC (2015) Modeling groundwater quality over a humid subtropical region using numerical indices, earth observation datasets, and X-ray diffraction technique: a case study of Allahabad district, India. Environ Geochem Health 37:157–180

Statistica 8.0 (Electronic version): StatSoft, Inc (2011). Electronic statistics textbook. Tulsa, OK: StatSoft. WEB:http://www.statsoft.com/textbook/. Accessed 1 Oct 2011

Thorburn PJ, Biggs JS, Weier KL, Keating BA (2003) Nitrate in groundwaters of intensive agricultural areas in coastal Northeastern Australia. Agric Ecosyst Environ 94:49–58

Velthof GL, Lesschen JP, Webb J, Pietrzak S, Miatkowski Z, Pinto M, Kros J, Oenema O (2014) The impact of the nitrates directive on nitrogen emissions from agriculture in the EU-27 during 2000-2008. Sci Total Environ 468:1225–1233

Widory D, Kloppman W, Chery L, Bonnin J, Rochdi H, Guinamant JL (2003) Nitrate in groundwater: an isotopic multi-tracer approach. J Contam Hydrol 72:165–188

Woo-jung C, Gwang-Hyun H, Sang-mo L, Goon-Taek L, Kwang-sik Y, Soo-Myung C, Hee-Myong R (2007) Impact of land-use types on nitrate concentration and ó15N in unconfined groundwater in rural areas of Korea. Agric Ecosyst Environ 120:259–268

24

Characterization of a fluoride-resistant bacterium *Acinetobacter* sp. RH5 towards assessment of its water defluoridation capability

Shraboni Mukherjee[1] · Vaibhav Yadav[2] · Madhumanti Mondal[1] · Soumya Banerjee[1] · Gopinath Halder[1]

Abstract The present study investigates the defluoridation capability of fluoride-resistant bacteria from contaminated groundwater collected from Asanjola and Madhabpur, West Bengal, India. Seven strains of fluoride-resistant bacteria were isolated employing culture media containing 10–250 mg/L of fluoride to evaluate their ability in reducing fluoride concentration in water. Five isolates exhibited significant amount of reduction in fluoride. Isolate RH5 achieved a maximum fluoride removal of 25.7 % from the media at 30 °C and pH 7 after 8 days of incubation. Based on morphological, physiological characteristics and analysis of 16S rDNA gene sequence, isolate RH5 was identified as *Acinetobacter* sp. RH5. Growth of RH5 was analysed at a diverse pH range, and it could thrive at pH 5–10. The present investigation revealed that the selective pressure of fluoride results in growth of fluoride-resistant bacteria capable of secreting high-affinity anion-binding compounds. This bacterium played a dominant bioremediative role by concentrating the anions so that they become less available. Hence, the fluoride-resistant bacteria, *Acinetobacter* sp. RH5, could be used as a promising strain for application in water defluoridation from contaminated sites.

Keywords Bioremediation · Defluoridation · Xenobiotic · Characterization · *Acinetobacter*

✉ Gopinath Halder
gopinath_haldar@yahoo.co.in

[1] Department of Chemical Engineering, National Institute of Technology, Durgapur, West Bengal, India

[2] Department of Earth and Environmental Studies, National Institute of Technology , Durgapur, West Bengal, India

Introduction

Trace elements have toxic effects when their level of permissibility is exceeded. The permissible limits for the discharge of toxic metals into the water bodies have been prescribed by various regulatory bodies (Sud et al. 2008). The health hazards and environmental degradation caused due to addition of the metal ions at a much higher concentration than the permissible limits are given in Table 1. For example, the fluoride ion has got some side effects which are toxic to the protoplasmic content of the cell affecting its biochemical content when its levels exceed 1.5 mg/L in drinking water (Annadurai et al. 2014).

Fluorine accounts for almost 0.3 g/kg of the earth's crust and exists in the form of fluoride in minerals such as fluorspar, cryolite and fluorapatite (Ghosh et al. 2013). The permissible limit for fluoride in drinking water as prescribed by World Health Organization (WHO), Indian Standard (IS), US Environment Protection Agency (USEPA) and European Union (EU) is 1.5, 1.0, 4 and 1.5 mg/L, respectively, with a target of between 0.8 and 1.2 g/L to maximize benefits and minimize adverse effects (WHO 2006; Edmunds and Smedley 2013). The effects of varying concentrations of fluoride on human health are given in Table 2.

In recent years, high levels of fluoride in groundwater (McDonagh et al. 2000) have resulted in vitiation of water quality, which has led to widespread anxiety. In West Bengal, India, excess amount of fluoride in groundwater has been detected across seven districts, viz. Purulia, Birbhum, Bankura, Malda, South Dinajpur, North Dinajpur and South 24-Parganas. Literature review by Bhattacharya and Chakrabarti (2011) confirms the existence of fluoride as a complex ion in a naturally occurring mineral called apatite which is a fluorinated calcium phosphatic

Table 1 Permissible limits and health effects of various toxic heavy metals

Metal contaminants	IS: 2490 (1974) For surface water (mg/L)	IS: 3306 (1974) for public sewers (mg/L)	IS: 3307 (1974) for irrigation (mg/L)	WHO (μg/L)	USEPA (μg/L)	Adverse health effects
Arsenic	0.20	0.20	0.20	10.00	50.00	Carcinogenic, liver tumours, skin and gastrointestinal effects
Mercury	0.01	0.01	–	1.00	2.00	Dermatitis, anorexia, kidney damage and muscle pain
Cadmium	2.00	1.00	–	3.00	5.00	Carcinogenic, lung fibrosis, dyspnoea and weight loss
Lead	0.10	1.00	–	10.00	5.00	Loss of appetite, anaemia, sterility, kidney problem, high blood pressure
Chromium	0.10	2.00	–	50.00	100.00	Lung tumours, allergic dermatitis
Nickel	3.00	3.00	–	–	–	Chronic bronchitis, lung cancer, nasal sinus
Zinc	5.00	15.00	–	–	–	Metal fume fever, restlessness
Copper	3.00	3.00	–	–	1300.00	Irritation of nose, mouth, eyes, headache diarrhoea
Fluoride	1.0	1.0	–	0.0015	0.004	Dental fluorosis, skeletal fluorosis

Table 2 Effects of varying concentrations of fluoride on human health

Level in water	Effects
Nil	Limited growth and fertility
0.8–1.2 mg/L	Prevention of tooth decay, strengthening of skeleton
Above 1.5 mg/L	Fluorosis: pitting of tooth enamel and deposits in bones
Above about 10 mg/L	Crippling skeletal fluorosis

compound. It has been observed from experimental investigations that various physico-chemical factors are found to be responsible for fluoride contamination of water from the fluoride-bearing rock into the saturated zone of groundwater (Datta et al. 2014). Analysis of water samples revealed that the presence of Na–K–Ca–Mg–HCO_3 favours continuous erosion from the fluoride-rich host rock under alkaline pH condition of the circulating water as revealed by Saxena and Ahmed (2003). Due to highly fragile economic structure, the absence of any treatment or monitoring systems the rural populace is compelled to drink fluoride contaminated water in these places (Singh et al. 2001). Fluoride is an oligo-element indispensable to bone and teeth development when ingested below the specified limit of 1.5 mg/L (Li et al. 2001; Susheela 1999) but poses a threat to human health when the prescribed limit is breached. Fluoride intake in excess leads to dental, skeletal and non-skeletal fluorosis (Gopalakrishnan et al. 1991; Zhu et al. 2006). It has been reported to be endemic in 20 countries of the world (Messaitfa 2008) with 22 states of India bearing the brunt of fluoride contamination (Susheela 1999). Several methods have been reported to treat fluoride contaminated water such as osmosis, nanofiltration, electrodialysis, precipitation and adsorption. These methods have certain drawbacks like secondary pollution, high sludge generation and high capital investment. The advantages of biological processes are functional simplicity, cost-effectiveness, less sludge production, facile regeneration of the biomass. Microbes have the capability of developing resistance to their surroundings through several processes such as bioaccumulation, biotransformation and biosorption (Juwarkar and Yadav 2010) since the bacterial cell wall comprises metal binding groups like amines, carboxylates, sulfhydryl and phosphates which aid in the interaction of metal ions (Kleinubing et al. 2011; Wuertz and Mergeay 1997). Biological degradation of petroleum wastes has been reported by bacteria of the genus *Pseudomonas* (Chung et al. 2003), *Sphingomonas* (Liu et al. 2009) and *Acinetobacter* (Ahmad et al. 2012). *Acinetobacter* strains have been explored for degradation of persistent organic pollutants (POPs) like atrazine (Cail et al. 2003). In 2012, five fluoride-resistant strains were isolated and a maximum fluoride removal of 22.1 % was achieved by *Pseudomonas aerugenosa* (Chouhan et al. 2012). Table 3 gives a comparative study of fluoride removal by processes other than bioremediation.

A freshwater unicellular microbial species, *Acinetobacter* sp. RH5, was isolated from Asanjola, India. The aim was to characterize fluoride-resistant bacterial strains from oligotrophic and heterotrophic niches and to utilize these strains towards defluoridation of contaminated water.

Table 3 Comparative study of fluoride removal by various techniques

Sl. no.	Techniques	% Removal of fluoride/adsorption capacity	References
1.	Adsorption by activated alumina	69.5 %	Ghorai and Pant (2005)
2.	Adsorption by carbon materials $KMnO_4$ modified activated carbon	15.9 mg/g	Daifullah et al. (2007)
3.	Co-precipitation (Nalgonda technique)	2.1–0.7 mg/L	Dahi et al. (1996)
4.	Precipitation with calcium and phosphate compounds	15 mg/L	Saha (1993)
5.	Ion exchange (modified zeolite)	94 %	Samatya et al. (2007)
6.	Reverse osmosis	98.9 %	Diawara et al. (2011)
7.	Nanofiltration		
	Brackish water	63.3 %	Diawara et al. (2011)
	Fluorinated drinking water	71 %	
8.	Biosorption (powdered biomass *Tinospora cordifolia*)	25 mg/g	Pandey et al. (2012)
9.	Biosorption (powdered activated carbon of *Eichhornia crassipes*)	70 %	Halder et al. (2014)
10.	Bioremediation (*Pseudomonas aerugenosa*)	22.1 %	Chouhan et al. (2012)
11.	Bioremediation (*Acinetobacter* RH5)	25.7 %	Present study

Materials and methods

Sample collection and fluoride concentration measurement

Asanjola and Madhabpur villages in Rampurhat tehsil (24.17°N, 87.78°E), Birbhum district, West Bengal, India, were selected for collecting the samples as this district is known to be endemic to fluoride where the rural population is dependent only on groundwater resources for their daily consumption (Susheela 1999). Due to highly fragile economic structure, the absence of any treatment or monitoring systems causes the rural populace to drink fluoride contaminated water in these villages; 50 mL of water samples was collected in presterilized containers and analysed for their fluoride concentration.

A fluoride ion-selective electrode (Orion, Thermo Scientific, USA) was employed for measuring the fluoride concentration of water samples. Calibration of the instrument was performed with standard solutions containing fluoride concentrations of 0.1, 1, 10 and 100 ppm (mg/L). A total ionic strength adjustment buffer (TISAB) was added to each sample before measuring the fluoride concentration to equalize the pH and ionic strength of the samples and the standards. It also facilitates removal of any interference by Fe(III) and Al(III) by forming a complex with these ions, hence releasing fluoride ions into the solution (Schamschula et al. 1985; NIOSH 1994).

Isolation and adaptation of bacterial cultures on media containing sodium fluoride

All the glassware was acid washed with 10 % hydrochloric acid prior to use. The collected water samples were allowed to stand at room temperature for 3 h to permit sedimentation of debris. Luria–Bertani (LB) medium with 2 % agar was used for plating and was sterilized by autoclaving at 121 °C and 15 psi pressure for 15 min; 100 μL of the supernatant was inoculated on plates containing sterilized LB agar media using spread plate technique and incubated for 24 h at 30 °C. Petri plates exhibiting bacterial growth were selected, and distinct colonies were streaked on a new set of LB agar media and incubated for 24 h at 30 °C.

LB agar media containing 10 mg/L fluoride was prepared using sodium fluoride (NaF), and the bacterial cultures were streaked onto these plates. It was followed by incubation of these plates at 30 °C for 24 h. Resistant micro-organisms were inoculated in LB broth containing the same concentration of fluoride as in the source plate media and incubated for 24 h at 30 °C in a rotary shaker at 120 rpm. Low concentration of fluoride was initially selected to allow the culture micro-organisms to adapt on media containing fluoride and to prevent inhibition of growth (Zhang et al. 2013). The microbes were then subjected to an increased fluoride concentration when broth cultures were further subcultured on LB agar plates containing 50 mg/L fluoride concentration, followed by incubation and then inoculation in LB broth containing 50 mg/L of fluoride. The same procedure was repeated for media containing 150 mg/L and 250 mg/L fluoride. After three subcultures, isolated bacterial cultures were obtained.

Determination of strain performance and selection of initial pH of selective medium

In order to determine the fluoride degrading capability, the isolated strains were inoculated in 250-mL conical flasks in triplicates containing LB broth with 20 mg/L fluoride concentration and incubated at 30 °C on a rotary shaker at

120 rpm. Broth containing 20 mg/L fluoride was selected due to the reason that maximum fluoride concentration in contaminated groundwater was observed to be 19.2 mg/L. Samples were taken from these flasks and analysed for fluoride concentration and growth. Ten millilitres of sample was taken at a time for fluoride analysis and centrifuged at 4500 rpm for 15 min, and the supernatant was subjected to fluoride concentration analysis using a fluoride ion-selective electrode. The bacterial isolate designated as RH5 showing the maximum fluoride removal percentage was chosen, and effects of initial pH on growth in LB broth were studied for this isolate. Initial pH was adjusted to 5, 6, 7, 8, 9, and 10, respectively, by adding 0.1 mol/L sodium hydroxide or 0.1 mol/L hydrochloric acid solutions. Study of bacterial growth by means of spectrophotometric analysis at 600 nm gave the optimum pH for the isolate RH5.

Biochemical characterization and growth kinetics of isolate RH5

Methyl red test

The methyl red test was conducted to verify the ability of isolate RH5 to perform mixed acid fermentation (Palitzsch 1911; MacFaddin 1980); 100 μL of inoculum of isolate RH5 was inoculated in a tube containing MRVP (methyl red Voges–Proskauer) broth and incubated at 35 °C for 24 h. Five drops of methyl red were added after incubation, and the result was recorded. A red colour change indicated the presence of mixed acid fermentation.

Voges–Proskauer test

Voges–Proskauer test was performed to determine whether the isolate RH5 produced 2,3-butanediol as a fermentation product from glucose. Since 2,3-butanediol cannot be easily detected, the test targets acetoin, which is an intermediate in the pathway. An inoculum from a culture of isolate RH5 was inoculated in a tube containing MRVP (methyl red Voges–Proskauer) broth and incubated for 24 h at 35 °C. Five drops of Barritt's A reagent were added to the tube after incubation followed by addition of five drops of Barritt's B reagent. The tube was then allowed to stand in a slant position for half an hour. Development of red colour indicated a positive test (MacFaddin 1980).

Catalase test

This test was accomplished using the method of Evans and Kloos (1972). A sterile inoculating loop was used to collect a single 24-h-old colony of isolate RH5 culture from LB agar medium and placed on a slide. A single drop of 3 % H_2O_2 was placed on the smear and covered with a petri plate. The presence of catalase was signified by the formation of bubbles.

Oxidase test

Overnight grown individual colonies were removed using a sterile, plastic loop. The cells were rubbed onto a moistened strip impregnated with oxidase reagent (1 % *N,N,N,N*-tetramethyl *p*-phenylenediamine dihydrochloride). This chemical replaces oxygen as a recipient for the electrons from the oxidase cytochrome. The additional electrons turn the oxidase reagent from colourless to purple. If oxidase is not present, no colour change is observed.

Nitrate reduction test

Nitrate reduction test was executed to determine the ability of the isolate to reduce nitrate to nitrite using the enzyme nitrate reductase. Nitrate broth containing nutrients and potassium nitrate was inoculated with the test micro-organism. A second tube containing only uninoculated nitrate broth was used as control. Both inoculated and control tubes were incubated for 24 h at 30 °C. After incubation, a drop of reagent A (0.8 g sulfanilic acid and 100 mL 30 % acetic acid) and reagent B (500 mg *N,N*-dimethyl-1-naphthylamine and 100 mL 30 % acetic acid) each was added. Then a small amount of powdered zinc was added and results were recorded. Development of red colour after the addition of sulfanilic acid and *N,N*-dimethyl-1-naphthylamine, followed by its disappearance on addition of zinc, indicates a positive result. Appearance of no colour after the addition of the same reagents followed by appearance of red colour on addition of zinc indicates a negative result.

A spectrophotometer (UV-2300, Techcomp Ltd., Singapore) was employed to measure the growth of isolate RH5 in LB broth. Absorbance of the bacterial isolate was measured at 600 nm at intervals of 4 h. The absorbance readings were then plotted against time to determine the growth phases and to establish a link between fluoride removal and growth of the isolate RH5 (Campbell et al. 2006).

Identification and phylogenetic analysis of isolate RH5

Identification by means of 16S rDNA sequencing was carried out for identification of isolate RH5. Genomic DNA was isolated from the plate culture of isolate RH5 using AMpurE Bacterial Genomic DNA Mini kit (Amnion Biosciences Pvt. Ltd., India). Using consensus/universal oligos, the ~1.5-kb 16S rDNA fragment from isolated gDNA was amplified using *Taq* DNA polymerase, employing the primers 5′-CWG RCC TAN CAC ATG SAA GTC and 5′-GRC GGW GTG TAC NAG GC (R = A or G; S = C or

G; W = A or T; N = A or C or G or T; W = A or T). In brief, the PCRs (polymerase chain reaction) were performed under the following set of conditions: 35 cycles of denaturation at 94 °C for 30 s, annealing at 55 °C for 30 s and extension at 72 °C for 1 min 30 s. Sanger dideoxy sequencing was performed by using BigDye Terminator v.3.1 Cycle Sequencing kits (Applied Biosystems, USA). Purified cycle sequencing products were analysed on an ABI Prism 3130 xl Genetic Analyzer. The PCR product was bi-directionally sequenced using the forward, reverse and/or an internal oligos. The PCR product was loaded on 1.0 % agarose gel along with a 500-bp DNA ladder (Amnion Biosciences Pvt. Ltd., India). Sequence data were aligned and analysed for finding the closest homologous micro-organisms using nucleotide Basic Local Alignment Search Tool (NCBI-blastn).

Results and discussion

Fluoride concentration of samples

Water samples from Asanjola were highly contaminated with fluoride with 3 out of 4 groundwater samples containing fluoride concentration above 15 mg/L. The rock types present in the areas surrounding Asanjola are mainly composed of biotite which itself produces dissolved fluoride concentrations beyond 4 mg/L (Gupta et al. 2012). Madhabpur fared a little better with only one sample exceeding the prescribed fluoride limit. In the preliminary experiments, we found that isolate RH5 could grow in 100 mg/L of NaF and possess the ability of accumulating large amounts of fluoride. But, there are no systematic studies regarding the application of this strain for the defluoridation of water (Table 4).

Strain isolation and adaptation on media containing fluoride

A total of seven bacterial isolates were obtained from samples collected from Rampurhat. The isolates were labelled as VF1, VF2, VF3, VF4, VF5, R3 and RH5. These isolates were differentiated on the basis of their colony characteristics and subjected to Gram staining. The isolated strains were subjected to different fluoride concentrations and all exhibited growth till 100 mg/L of fluoride concentration after incubation at 30 °C for 24 h. The resistant isolates were then inoculated in LB broth with the same concentration of fluoride. Five isolates were able to withstand maximum fluoride concentrations up to 250 mg/L.

Table 4 Fluoride concentration of samples of Asanjola and Madhabpur

Area	Sites	Fluoride concentration (mg/L)
Asanjola	Hand pump1	19.2
	Primary school	15.4
	Hand pump2	19.0
	Pond	0.512
Madhabpur	Primary school	16.3
	Hand pump	0.418
	Pond	0.630

Identification of strain performance and determination of initial pH of selective medium

All of the isolates were analysed for fluoride reduction capability by measuring fluoride concentration of the broth after centrifugation at 4500 rpm for 15 min. Percentage reduction in fluoride content was determined for each alternate day, viz. days 0, 2, 4, 6, 8 and 10. Reduction in fluoride concentration was observed till day 8 after which there was no considerable defluoridation which can be attributed to reduction in cell viability. Isolates VF1, VF4, VF5, R3 and RH5 exhibited significant fluoride removal. Isolates VF2 and VF3 were not able to remove fluoride in significant amounts. Isolate RH5 obtained from water sample taken from Asanjola showed the maximum fluoride reducing capability of 25.7 % and was selected for identification and characterization. The growth of isolate RH5 can be due to high levels of fluoride which might produce a situation closer to that of using an antibacterial agent, in which a resistant organism is released from ecological restrictions of the community and can, therefore, grow in higher levels in the presence of fluoride. Previously, bacterial species have been reported to achieve a maximum of 22.1 % fluoride removal (Xu et al. 2011). Initial pH of medium was adjusted to 5, 6, 7, 8, 9 and 10, respectively, to study the growth of the isolate and uptake of fluoride by the isolated strain. The absorbance recorded at 600 nm was least for LB broth at pH 10 and maximum for pH 7 which is shown in Fig. 1. Therefore, pH 7 was deemed to be the optimum for growth of isolate RH5 (Table 5).

Biochemical characterization and growth kinetics of isolate RH5

Isolate RH5 obtained from water sample taken of Asanjola, Rampurhat, is a Gram negative, non-motile, coccobacillary bacterium. It is oxidative and aerobic in nature. The strain could grow at a diverse range of pH (pH 5–10) conditions, with optimum pH being 7. It could produce acid from galactose, glucose and maltose but not from mannitol and sucrose. The strain tested negative for methyl red, Voges–Proskauer, oxidase and nitrate reduction tests and positive for catalase test.

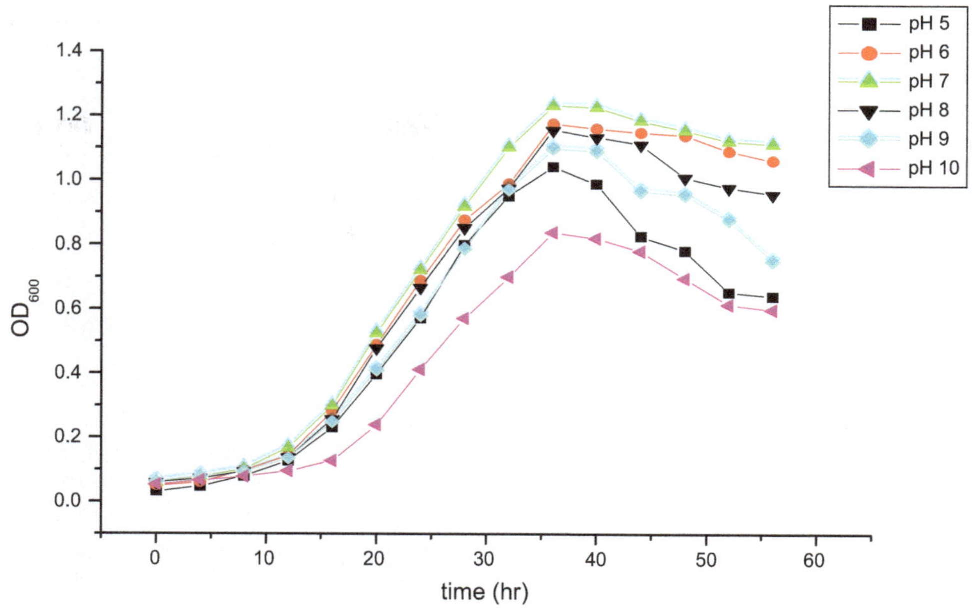

Fig. 1 Effect of pH on growth of *Acinetobacter* sp. RH5; the bacterium was able to thrive on a diverse pH range

Table 5 Percentage reduction in fluoride by isolates

Isolate	Fluoride reduction (%)				
	Day 0	Day 2	Day 4	Day 6	Day 8
VF1	0.0	4.5	7.2	9.1	15.7
VF2	0.0	1.6	2.7	3.7	5.1
VF3	0.0	2.1	3.9	5.1	7.4
VF4	0.0	8.5	14.7	16.8	20.5
VF5	0.0	7.6	12.6	14.2	19.1
R3	0.0	6	11.4	17.4	18.5
RH5	0.0	9.5	18.1	21	25.7

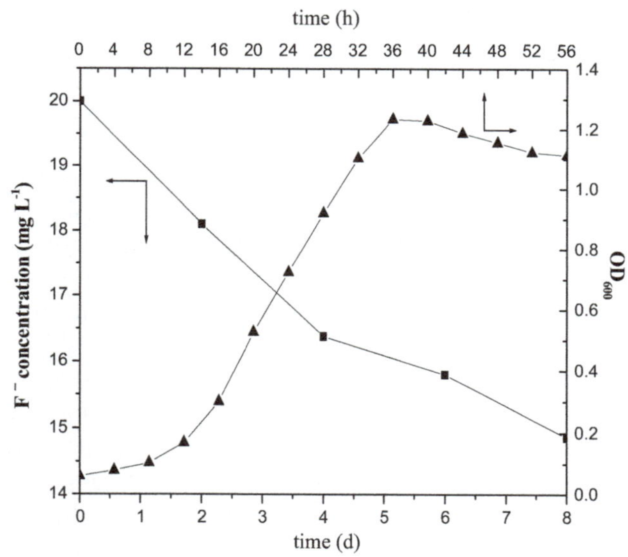

Fig. 2 Growth and fluoride removal efficiency of *Acinetobacter* sp. RH5; lag phase lasted for 6 h; fluoride concentration decreased rapidly till fourth day after which retardation in the fluoride removal rate was encountered

The strain RH5 exhibited a diminutive lag phase, lasting for only 6 h, implying its ability to adapt to the medium of high fluoride concentration. After 6 h, the strain reproduced at a rapid rate, which was indicated by a sharp increase in OD_{600} values. Figure 2 shows that the maximum growth rate was observed between 16 and 36 h.

PCR identification of 16S rDNA and phylogenetic analysis of isolate RH5

A 1.5-kb 16S rDNA fragment obtained by PCR was sequenced, and phylogenetic analysis based on 16S rDNA gene sequences suggested that isolate RH5 formed a phylogenetic lineage with members of the genus *Acinetobacter* and demonstrated the maximum 16S rDNA similarity of 99 % with *Acinetobacter calcoaceticus* as shown in Fig. 3. The isolate RH5 was, therefore, found to be a member of *Acinetobacter* genus and named *Acinetobacter* sp. RH5.

Bioremediation of toxic xenobiotics has many advantages over other techniques as it is cheap and non-destructive. Many studies have confirmed that microbes have high affinity for metals and non-metals through a variety of mechanisms. As reported by Doble and Kumar (2005), certain bacterial species secrete high-affinity anion-binding compounds called ionophores. These ionophores bind to certain specific forms of anions and make a complex that can be utilized by the bacteria. So, the reduction in the

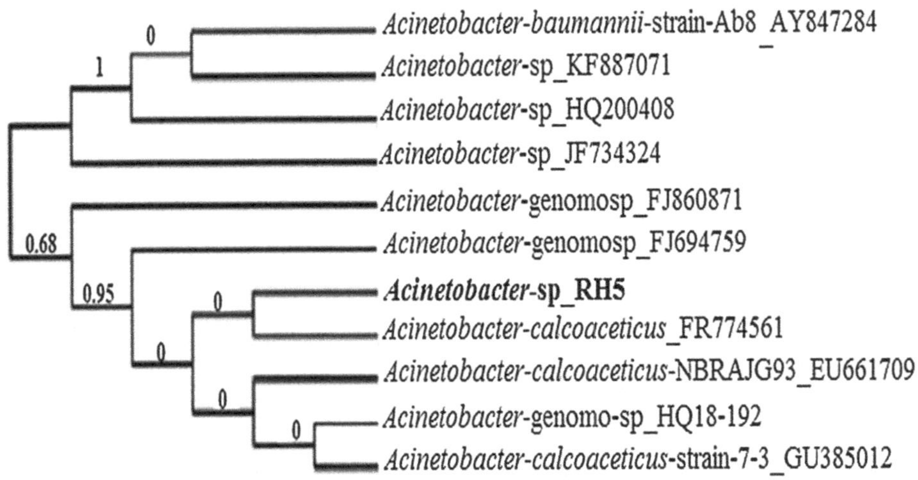

Fig. 3 Phylogenetic analysis of isolate RH5 and related species

fluoride concentration of the media in our present study can be attributed to this fact. Fluoride mostly exists with a proton (HF). Hydrogen fluoride can readily cross cell membrane and acts like a proton conductor. In this way, micro-organisms having the adapting mechanisms concentrate fluoride from the surroundings making it less available in the environment.

Conclusions

Through the current study, it can be convincingly implied that the problem of fluoride contamination can be alleviated by use of bioremediation technologies. *Acinetobacter* sp. RH5 had better defluoridation capability than other isolates obtained from natural and anthropogenic sources for defluoridation of water. The isolate showed maximum resistance to fluoride as well as the maximum fluoride removal capability of 25.7 % at 30 °C. The growth characteristics were tested for a wide pH range, wherein pH 7 exhibited maximum growth of isolate RH5. The strong adaptability of strain RH5 to high concentration of fluoride made it a promising candidate for treating fluoride contaminated water. Though this study dwelt only in the realm of identifying fluoride removal capabilities of bacteria, still it can be worthwhile to infer that development of a bioremediation system for defluoridation can be achieved either by immobilization of *Acinetobacter* sp. RH5 to prevent washing away by flowing water, or by use of sequence batch reactors to achieve longevity and maintenance of the bacteria. In order to determine the sustainability and pragmatism of the bioremediation process, further experiments are being conducted.

Acknowledgments The authors express their gratitude to Anoar Ali Khan and Sumit Hanspal Dhawane, PhD scholars of Department of Chemical Engineering, National Institute of Technology Durgapur, India, for their enthusiastic support and cooperation during execution of the work. The financial assistance from Department of Biotechnology, Government of West Bengal, India, through research project BT (Estt.) R & D-23/13 is gratefully acknowledged. The authors are also thankful to Department of Earth and Environmental Studies, National Institute of Durgapur, for providing the instrumental facilities towards successful completion of the project.

References

Ahmad S, Shamaan N, Arif N, Koon G, Shukor M, Syed M (2012) Enhanced phenol degradation by immobilized *Acinetobacter* sp. strain AQ5NOL 1. World J Microbiol Biotechnol 28:347–352

Annadurai ST, Rengasamy JK, Sundaram R, Munusamy AP (2014) Incidence and effects of fluoride in Indian natural ecosystem: a review. Adv Appl Sci Res 5(2):173–185

Bhattacharya HN, Chakrabarti S (2011) Incidence of fluoride in the groundwater of Purulia District, West Bengal: a geo-environmental appraisal. Sci Corresp 101(2):152

Cail B, Hanl Y, Liu B, Renl Y, Jiang S (2003) Isolation and characterization of an atrazine-degrading bacterium from industrial wastewater in China. Lett Appl Microbiol 36:272–276

Campbell W, Song HP, Barbier G (2006) Nitrate reductase for nitrate analysis in water. Environ Chem Lett 4:69–73

Chouhan S, Tuteja U, Flora SJS (2012) Isolation, identification and characterization of fluoride resistant bacteria: possible role in bioremediation. Appl Biochem Microbiol 48(1):43–50

Chung TP, Tseng HY, Juang RS (2003) Mass transfer effect and intermediate detection for phenol degradation in immobilized *Pseudomonas putida* systems. Process Biochem 38:1497–1507

Dahi E, Mtalo F, Njau B, Bregnhj H (1996) Defluoridation using the Nalgonda technique in Tanzania, reaching the unreached: challenges for the 21st century. In: 22nd WEDC conference, New Delhi

Daifullah AAM, Yakout SM, Elreefy SA (2007) Adsorption of fluoride in aqueous solutions using $KMnO_4$-modified activated carbon derived from steam pyrolysis of rice straw. J Hazard Mater 147:633–643

Datta AS, Chakrabortty AA, De Dalal SS, Lahiri SC (2014) Fluoride contamination of groundwater in West Bengal, India. 47(3):241–248

Diawara CK, Diop SN, Diallo MA, Farcy M, Deratani A (2011) Performance of nanofiltration (NF) and low pressure reverse osmosis (LPRO) membranes in the removal of fluorine and salinity from brackish drinking water. J Water Resour Protect 3:912–917

Doble M, Kumar A (2005) Biotreatment of industrial effluents. Elsevier, New York

Edmunds WM, Smedley PL (2013) Essentials of medical geology, revised edition. Springer, Berlin

Evans JB, Kloos EW (1972) Use of shake cultures in a semisolid thioglycolate medium for differentiating staphylococci from micrococci. Appl Microbiol 23(2):326–331

Ghorai S, Pant KK (2005) Equilibrium, kinetics and breakthrough studies for adsorption of fluoride on activated alumina. Sep Purif Technol 42(3):265–271

Ghosh A, Mukherjee K, Ghosh SK, Saha B (2013) Sources and toxicity of fluoride in environment. Res Chem Intermed 39(7):2881–2915

Gopalakrishnan S, Narasimhan S, Seenipandian C (1991) Incidence of fluorosis in 10 villages near Tuticorin—suggestions for prevention and control. Ind J Environ Protect 11(2):118–123

Gupta AB, Gupta SK (2001) Simultaneous carbon and nitrogen removal from high strength domestic wastewater in an aerobic RBC biofilm. Water Res 35:1714–1722

Gupta S, Mondal D, Bhardhan A (2012) Geochemical provenance and spatial distribution of fluoride in groundwater in parts of Raniganj coal field, West Bengal, India. Arch Appl Sci Res 4(1):292–306

Halder GN, Sinha K, Dhawane S (2014) Defluoridation of wastewater using powdered activated carbon developed from *Eichhornia crassipes* stem: optimization by Response surface methodology. Desalin Water Treat. doi:10.1080/19443994.2014.942375

Juwarkar AA, Yadav SK (2010) Bioaccumulation and biotransformation of heavy metals. In: Fulekar MH (ed) Bioremediation technology. Springer, Amsterdam, pp 266–284

Kleinubing SJ, da Silva EA, da Silva MGC, Guibal E (2011) Equilibrium of Cu (II) and Ni (II) biosorption by marine alga *Sargassum filipendula* in dynamic system: competitiveness and selectivity. Bioresour Technol 102(7):4610–4617

Li Y, Liang C, Slemenda CW, Ji R, Sun S, Cao J, Emsley C, Ma F, Wu Y, Ying P, Zhang Y, Gao S, Zhang W, Katz B, Niu S, Cao S, Johnston C (2001) Effect of long-term exposure to fluoride in drinking water on risks of bone fractures. J Bone Miner Res 16(5):932–939

Liu YJ, Zhang AN, Wang XC (2009) Biodegradation of phenol by using free and immobilized cells of *Acinetobacter* sp. XA05 and *Sphingomonas* sp. FG03. Biochem Eng J 44:187–192

MacFaddin JF (1980) Biochemical tests for identification of medical bacteria, 2nd edn. Lippincott Williams & Wilkins, Philadelphia

McDonagh MS, Whiting PF, Wilson PM, Sutton AJ, Chestnutt I, Cooper J, Misso K, Bradley M, Treasure E, Kleijnen J (2000) Systematic review of water fluoridation. Br Med J 321:855–859

Messaitfa A (2008) Fluoride contents in ground waters and the main consumed foods (dates and tea) in Southern Algeria region. Environ Geol 55:377–383

NIOSH (1994) Manual of analytical methods (NMAM), 4th edn, 8/15/94

Palitzsch S (1911) Application of methyl-red to the colorimetric estimation of hydrogen ion concentrations. Biochem Z 37:131–138

Pandey PK, Pandey M, Sharma R (2012) Defluoridation of water by a biomass: *Tinospora cordifolia*. J Environ Protect 3:610–616

Saha S (1993) Treatment of aqueous effluent for fluoride removal. Water Res 27(8):1347–1350

Samatya S, Yüksel U, Yüksel M, Kabay N (2007) Removal of fluoride from water by metal ions (Al^{3+}, La^{3+} and ZrO^{2+}) loaded natural zeolite. Sep Sci Technol 42:2033–2047

Saxena VK, Ahmed S (2003) Inferring the chemical parameters for the dissolution of fluoride in groundwater. Environ Geol 43:731–736

Schamschula RG, Sugar E, Un PSH, Toth K, Armes DEB, Adkins BL (1985) Physiological indicators of fluoride exposure and utilization: an epidemiological study. Community Dent Oral Epidemiol 13:104–107

Singh PP, Barjatiya MK, Dhing S, Bhatnagar R, Kothari S, Dhar V (2001) Evidence suggesting that high intake of fluoride provokes nephrolithiasis in tribal populations. Urol Res 29(4):238–244

Sud D, Mahajan G, Kaur MP (2008) Agricultural waste material as potential adsorbent for sequestering heavy metal ions from aqueous solutions—a review. Bioresour Technol 99:6017–6027

Susheela AK (1999) Fluorosis management programme in India. Curr Sci 77(10):1250–1256

WHO (2006) Chemical fact sheets: fluoride, guidelines for drinking water quality (electronic resource), 3rd edition. Incorporation first addendum, recommendations. WHO, Geneva, vol 1, pp 375–377

Wuertz S, Mergeay M (1997) The impact of heavy metals on soil microbial communities and their activities, Modern Soil Microbiology. Marcel Dekker Inc, New York, pp 607–643

Xu J, Song XA, Zhang Q, Pan H, Liang Y, Fan XW, Zhi Y (2011) Characterization of metal removal of immobilized *Bacillus* strain CR-7 biomass from aqueous solutions. J Hazard Mater 187:450–458

Zhang C, Zhang W, Xu J (2013) Isolation and identification of methanethiol-utilizing bacterium CZ05 and its application in bio-trickling filter of biogas. Bioresour Technol 150:338–343

Zhu C, Bai G, Liu X, Li Y (2006) Screening high fluoride and high arsenic drinking water and surveying endemic fluorosis and arsenism in Shaanxi province in western China. Water Res 40:3015–3022

25

New biosorbent in removing some metals from industrial wastewater in El Mex Bay, Egypt

Maha Ahmed Mohamed Abdallah[1] · Mohamed E. Mahmoud[2] · Maher M. Osman[2] · Somaia B. Ahmed[1]

Abstract Biosorption is an extensive technology applied for the removal of heavy metal ions and other pollutants from aqueous solutions. In the present study, the biosorption of cadmium, lead, chromium and mercury ions from polluted surface seawater in El-Max Bay was determined using hybrid active carbon sorbents. These sorbents were treated chemically by acid, base and redox reaction followed by surface loading of baker's yeast biomass for increasing their biosorption capacity and the highest metal uptake values. The surface function and morphology of the hybrid immobilized sorbents were studied by Fourier Transform Infrared analysis and scanning electron microscope imaging. Metal removal values proved that the vital role of baker's yeast as a significant high removable due to functional groups at baker's yeast cell wall surface that have the ability to forming various coordination complexes with metal ions. A noticeable increase in the removal of all studied metals was observed and reached to 100 %.

Keywords Activated carbon · Baker's yeast · Wastewater · El Mex Bay · Egypt

Introduction

The metal contaminations in the aquatic environments have received considerable concern due to their toxicity, abundance and persistence in the environment, and their subsequent accumulation in aquatic habitats. It has been recognized for many years that the concentrations of metals found in dissolved or particulate phase of coastal areas, may be derived from a variety of anthropogenic and natural sources (Abdallah et al. 2007). El-Mex Bay is a part of Alexandria coast on the Mediterranean Sea, it is adjacent to Alexandria City center that is populated with about six million inhabitants; in addition, it is considered as one of the main fishing sources in Egypt (Abdallah 2014). The contaminants are introduced through waterway and through several landbased sources into the Bay. El-Mex Bay has several industrial plants situated close to the coast and directly discharges its effluents into it. In addition this Bay is an estuarine zone of huge agricultural drain (Omoum Drain), which crosses areas of intensive agriculture and a county with rapidly growing population density, and industrial activities; its discharge rate is about 2547.7×10^6 m^3/year (Abdallah 2008; El-Rayis and Abdallah 2006). As a result, large amounts of pollutants including heavy metals were supplied to the bay. Besides aquatic inputs, El-Mex Bay is also exposed to atmospheric pollution from the same activities that cause aquatic pollution.

Removal of heavy metals can be mainly accomplished using various conventional methodologies which are generally known as inefficient and/or expensive especially at low metal concentrations (Abdallah 2008). Adsorption is one of the most efficient techniques for heavy metal removal, extraction and pre-concentration of low concentration levels from various aqueous and non-aqueous matrices (Mahmoud et al. 2010; Abdel-Fattah et al. 2014). A number of carbonaceous materials have been used and applied as sorbents due to their easy handling and high adsorption efficiency for water purification and waste water treatment of hazardous materials including organic and

✉ Maha Ahmed Mohamed Abdallah
mahaahmed72007@Yahoo.com; mahaahmed7@ymail.com

[1] Marine Pollution Lab, National Institute of Oceanography and Fisheries, Qait Bey, Alexandria, Egypt

[2] Chemistry Department, Faculty of Sciences, Alexandria University, Alexandria, Egypt

heavy metal species (Mahmoud et al. 2008; Tian et al. 2010; Faur- Brasquet et al. 2002; Mahmoud et al. 2012). The adsorption efficiency of active carbon sorbent toward heavy metals can be enhanced and improved by surface modification via immobilization and incorporation of certain chelating functional groups.

On the other hand, baker's yeast, known as *Saccharomyces cerevisiae*, is characterized by the presence of various chelating functional groups that were deployed in successful biosorption and removal of heavy metal ions such as Cr(III), Cr(VI), Hg(II), As(III) and As(v) from aqueous samples (Bag et al. 1998; Peez-Corona et al. 1998; Yavuz et al. 2006; Bag et al. 2002). These procedures were implemented using immobilized baker's yeast on several solid matrices such as silica gel, Dowex anion exchanger and other solid supports (Mahmoud et al. 2011; Baccara et al. 2009). In this study, a method was established and presented to enhance and improve the adsorptive efficiency of active carbon for pre-concentration of some toxic heavy metals such as Cd, Pb, Cr, and Hg from aqueous solutions by simple, direct and affordable chemical treatment and biological immobilization of baker's yeast, as a source of chelating functional groups. The metal-sorbent interaction processes and mechanisms were investigated in this study using FT-IR analysis which characterized the surface functional groups of the studied hybrid Ac sorbent–biosorbent.

Materials and methods

Study area

El-Mex bay bordering an industrial zone is located west of Alexandria City, one of the most densely populated cities with 6 million people. This bay extending for about 7 km between longitude 29° 45′ and 29° 54′E and latitude 31° 07′ and 31° 15′N (Fig. 1), from Agami headland (west) to the Western Harbor (east), occupies an area of 19.4 km^2 with a mean depth of 10 m and a water volume of 190.3 × 10^6 m^3 (Abdallah et al. 2007). As a consequence of growing heavy industries (chloro-alkali, cement, chemicals, textile, tanneries, industrial dyes, ink, petroleum refining, meat processing, fish production, and iron or steel industries) and the uncontrolled disposal of resulting wastes, the coastal water of El-Mex Bay receives huge amounts of untreated industrial wastes containing heavy metals like Hg, Cr, Cd and Pb dumped directly into the southern part of the bay via pipelines. In addition, El-Dekhaila Harbor (commercial harbour) has been recently constructed at the western side of El-Mex Bay.

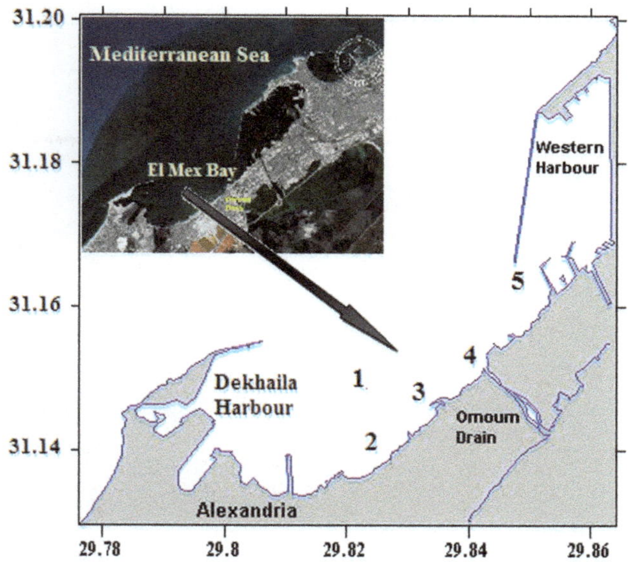

Fig. 1 Study area of El-Mex Bay

Sampling

Surface water samples were collected from five stations during June 2014, to provide a broad spatial coverage of the bay and allow assessment of the water quality throughout the Bay. The bay has been divided into two parts (inshore and middle regions). The inshore part represents the discharging points of different sources (stations 2, 3 and 4), heavily polluted by unauthorized discharges of industrial effluents. The middle part covers stations 1 (closed to the ethylene platform) and 5 (the entrance of the Western Harbour). Study area and positions of sampling stations are shown in Fig. 1. All the surface water samples, from all studied areas, were collected by hand from a rowing boat by submerging precleaned PE (5 L) bottles approximately 20 cm beneath the water surface. Procedural blanks were prepared in the same way as samples, and none of the target metals was detected.

Synthesis of hybrid active carbon-immobilized-baker's yeast sorbents

Commercial activated carbon (AC) was weighed and treated by HCl and stirred for 4 h to produce sorbent acid treated activated carbon (AT-AC). A normal weight of (AT-AC) sorbent was added to baker's yeast and the reaction mixture was mixed well and dried in an oven. Base treated active carbon (BT-AC) sorbent and base treated active carbon-immobilized-baker's yeast (BT-AC-BY) sorbent were also prepared in a similar way using NaOH. Oxidized active carbon sorbent (Ox-AC) sorbent and oxidized active carbon-immobilized-baker's yeast sorbent (Ox-AC-BY) were also prepared in a similar way using HNO_3 as an oxidizing agent.

Reduced active carbon (Rd-AC) sorbent and reduced active carbon-immobilized-baker's yeast sorbent (Rd-AC-BY)

Table 1 Hybrid chemically and biologically treated active carbon sorbents

Symbol sorbent	Active carbon
AC	Active carbon-modified-baker's yeast
AC-BY	Active carbon-immobilized-baker's yeast
AT-AC	Acid-treated-active carbon
AT-AC-BY	Acid treated active carbon- immobilized -baker's yeast
BT-AC	Base treated active carbon
BT-AC-BY	Base treated active carbon- immobilized -baker's yeast
Ox-AC	Oxidized active carbon by nitric acid
Ox-AC-BY	Oxidized active carbon by nitric acid and immobilized by baker's yeast
Rd-AC	Reduced active carbon by sodium sulfite
Rd-AC-BY	Reduced active carbon by sodium sulfite and immobilized by baker's yeast

were also prepared in a same manner using Na_2SO_3 solution as reducing agent (Grote and kettrup 1985). The list and specifications of all hybrids chemically and biologically modified active carbon biosorbents are given in Table 1.

Analysis

Dissolved heavy metals

In the laboratory, water samples were filtered through 0.45-μm membrane filter. The filtered samples (dissolved phase) were acidified using nitric acid and trace metals were pre-concentrated and measured using Atomic Absorption Spectrophotometry the procedures adopted followed (Abdullah and Royle 1974). Samples were analyzed using an air-acetylene flame atomic absorption spectrophotometer Shimadzu Model AA-6800, Duisburg, Germany, with D (subscript 2) background correction and an autosampler. Results were expressed as micrograms of element per liter of water samples. All absorbance readings were made in triplicate. Instrument settings were as recommended in the manufacturer's manual, with wavelengths (nm) of 228.8, 357.9 and 283.31 for Cd, Cr and Pb respectively. For Hg analysis, a cold vapor atomic absorption spectrophotometer flow injection Hg/hydride analyzer (MVu 1A–Mercury vaporizer unit) was used with wavelength of 253.7 nm and a quartz absorption cell was used for the metals determination according to the United Nations Environment Programme (UNEP 1993). The accuracy and analytical procedures were verified by analyzing a certified reference material (LGC-6010). All trace elements gave both accurate and precise results.

Potential of hybrid active carbon sorbents for removal of heavy metals from water samples

Surface functionality of carbon sorbents is characterized by its responsibility for all activity and reactivity as well as capability for all adsorption properties and processes. In this work, infrared spectroscopy was used to obtain information about the chemical structure and functional groups of the raw material and the hybrid activated carbon biosorbents. The FTIR spectrum were recorded using a Shimadzu Fourier Transform infrared spectrophotometer (FTIR system-BX 0.8009) was used in the range 200–4000 cm^{-1} to acquire the FT-IR spectra of all chemically and biologically treated active carbon sorbents. These sorbents were also imaged by the use of scanning electron microscope (JSM-5300, JEOL Ltd.). An ion sputtering coating device (JEOL-JFC-1100E) was used to coat the SEM specimens with gold to increase the conductivity. The Brunauer, Emmett and Teller (BET) technique is the most common method for determining the surface area of powders and porous materials. Nitrogen gas is generally employed as the probe molecule and is exposed to a solid under investigation at liquid nitrogen conditions (i.e. 77 K). The surface area of the solid is evaluated from the measured monolayer capacity and knowledge of the cross-sectional area of the molecule being used as a probe. For the case of nitrogen, the cross-sectional area is taken as 16.2 A2/molecule. Malvern Zetasizer Nano ZS90 was used to measure the particle size and the zeta potential of active carbon sorbents microspheres.

Pre-concentration of heavy metals was performed according to the following procedure. The wastewater samples were passed over a pre-concentration micro-column with a diameter of 30 mm and length: 70 mm packed with 50 mg (pack length: 5–7 mm) of the modified hybrid active carbon sorbents with a flow rate of 10 ml min^{-1} under air pressure. The adsorbed metal ions on the hybrid active carbon sorbents surface was eluted by the flow of 5.0 ml of conc. HNO_3 and determined by atomic absorption spectrophotometer, all values are triplicate (Grote and Kettrup 1985). The removal percentages were calculated as:

$$\text{Removal }\% = \frac{\text{Detected concentration of sample after passing over sorbent}}{\text{Initial concentration of sample before passing over the sorbent}} \times 100.$$

Fig. 2 FT-IR spectra of hybrid activated carbon sorbents

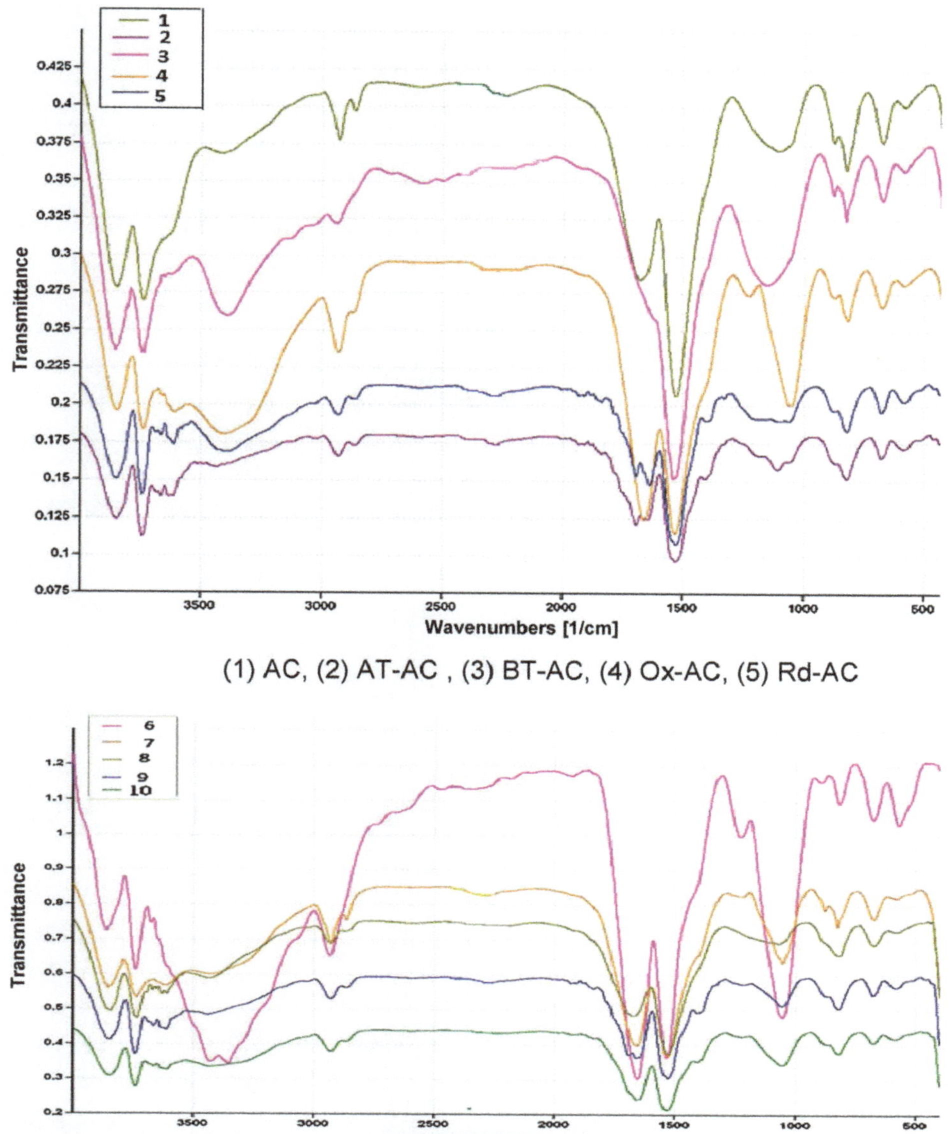

(1) AC, (2) AT-AC , (3) BT-AC, (4) Ox-AC, (5) Rd-AC

(6) AC-BY, (7) AT-AC-BY , (8) BT-AC-BY, (9) Ox-AC-BY, (10) Rd-AC-BY

Results and discussion

Surface characterization

The surface functional groups of the tested sorbent and biosorbents system were determined and characterized by FT-IR analysis. The infrared spectroscopy is a commonly used technique to figure out the various functional groups as well as to find information about the surface activity of chemically and biologically hybrid active carbon sorbents. Characteristics of infrared peaks are mainly identified and related to the nature of AC sorbent as shown in Fig. 2. However, upon chemical treatments using acid–base and redox reactions as well as surface immobilization of baker's yeast, several other IR peaks were produced. The peak centered at 1110–1120 cm^{-1} is mainly due to (C–O) of methoxy group and the strong peak appearing at 1520 and 1670 cm^{-1} are mainly ascribed to the formation of oxygen functional groups based on highly conjugated C–O stretching in carboxylic or carboxylate group as well as carbonyl group. The broad band in the region 3300–3600 cm^{-1} is typically attributed to O–H stretching or adsorbed water molecule. Thus, chemical and biological treatment processes of AC sorbent were found to improve and intensify the various functional groups of AC sorbent.

Scanning electron microscopy (SEM) is known as an efficient technique for investigation and evaluation of the surface morphology of modified sorbents, Fig. 3 represents the SEM-image of untreated AC sorbent and all treated AC forms, respectively. The surfaces of the chemically treated

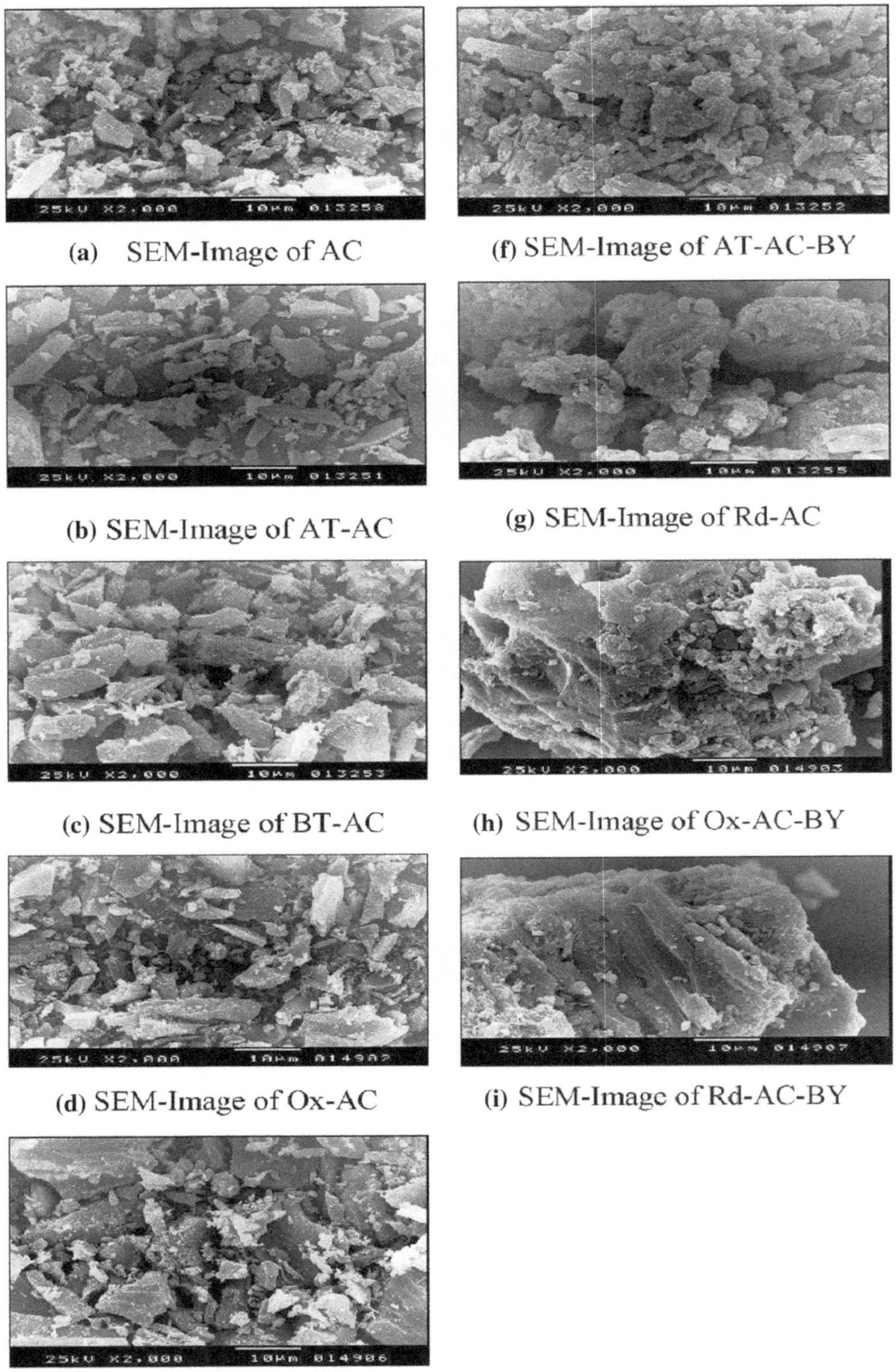

(a) SEM-Image of AC

(f) SEM-Image of AT-AC-BY

(b) SEM-Image of AT-AC

(g) SEM-Image of Rd-AC

(c) SEM-Image of BT-AC

(h) SEM-Image of Ox-AC-BY

(d) SEM-Image of Ox-AC

(i) SEM-Image of Rd-AC-BY

(e) SEM-Image of Rd-AC

Fig. 3 Scanning electron micrograph of hybrid activated carbon sorbents

AC sorbents have retained their original shape and distribution leading to little unnoticeable changes. On the other hand, biological treatment of chemically modified AC sorbents with baker's yeast for the formation of AC-BY, AT-AC-BY BT-AC-BY, Ox-AC-BY and Rd-AC-BY has resulted in pronounced surface changes of these four sorbents as represented by Fig. 3.

To estimate the importance of porous structure for two selective sorbents (AC-BY and Rd- AC-BY) during the adsorption process, the nitrogen adsorption–desorption

Fig. 4 Specific area characters of AC-BY by BET analysis

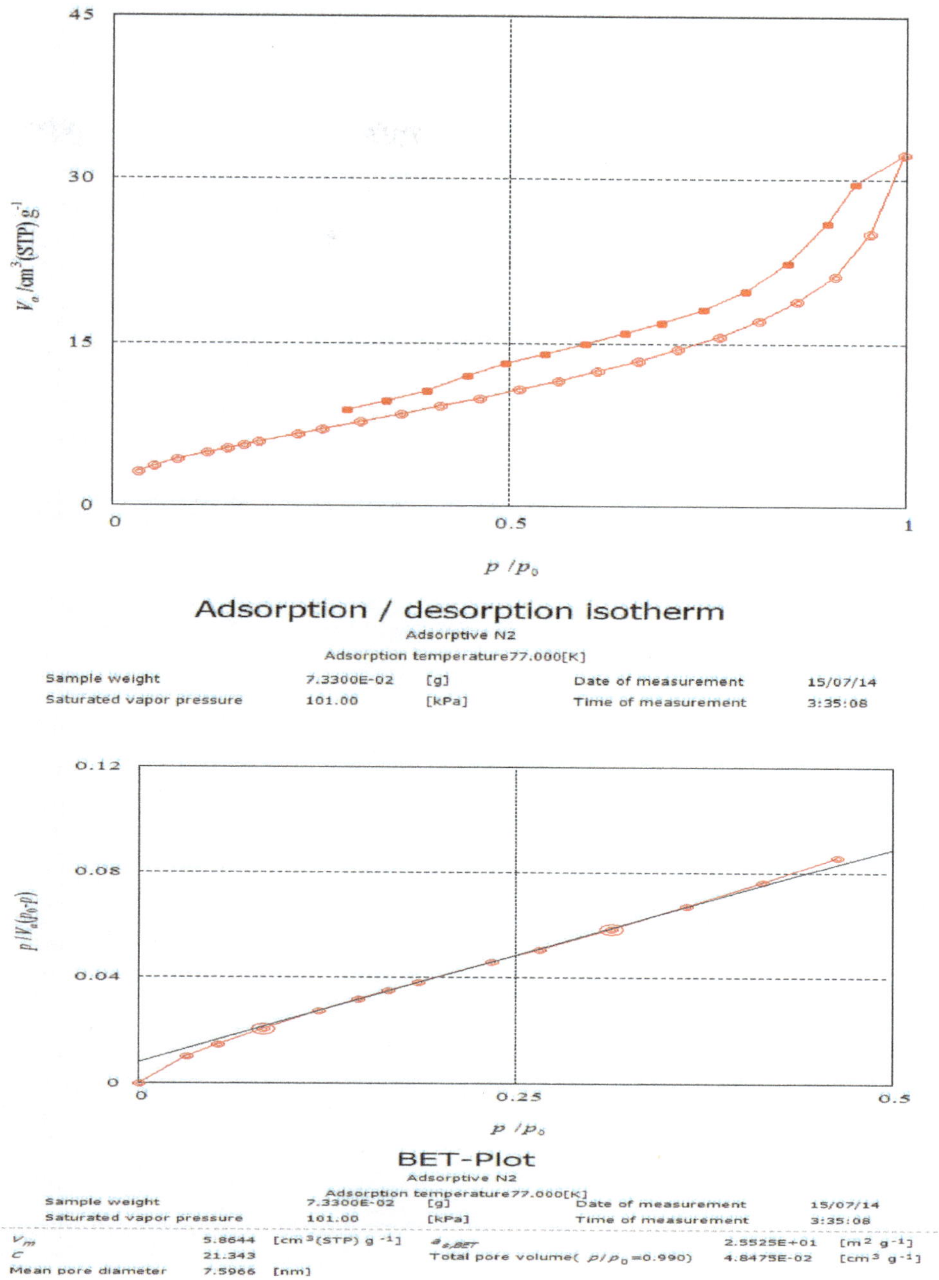

isotherm and the pore size distribution curve of the sorbents microspheres are also provided as contrast, as shown in Figs. 4 and 5. The BET surface area of the hollow sorbent Rd- AC-BY is 1.2927E + 02 m^2/g, which is larger than that of AC-BY sorbent (2.5525E + 01 m^2/g), and the total pore volume (0.1524 cm^3/g) of Rd- AC-BY is also larger than that of AC-BY sorbent (4.8475E − 02 cm^3/g). Furthermore, the mean pore diameter of AC-BY (7.5666 nm) is slightly large compared with Rd-AC-BY (4.7155 nm) sorbent which own a much richer porous structures, as shown in the inset of Fig. (5). The pre-concentration results indicates that adsorbent materials owning relatively high specific surface area, high porous volume and hollow structure will show a better adsorption ability, as these pore structures can provide more adsorption sites (Cui et al. 2015; Fu et al. 2015).

Removal of metals by different forms of the modified AC

Removal of Cd from wastewater samples of El-Max bay by hybrid AC sorbents

Removal of dissolved cadmium from industrial wastewaters by sorption or biosorption, as one of the most effective processes, was always planned and aimed. The traces

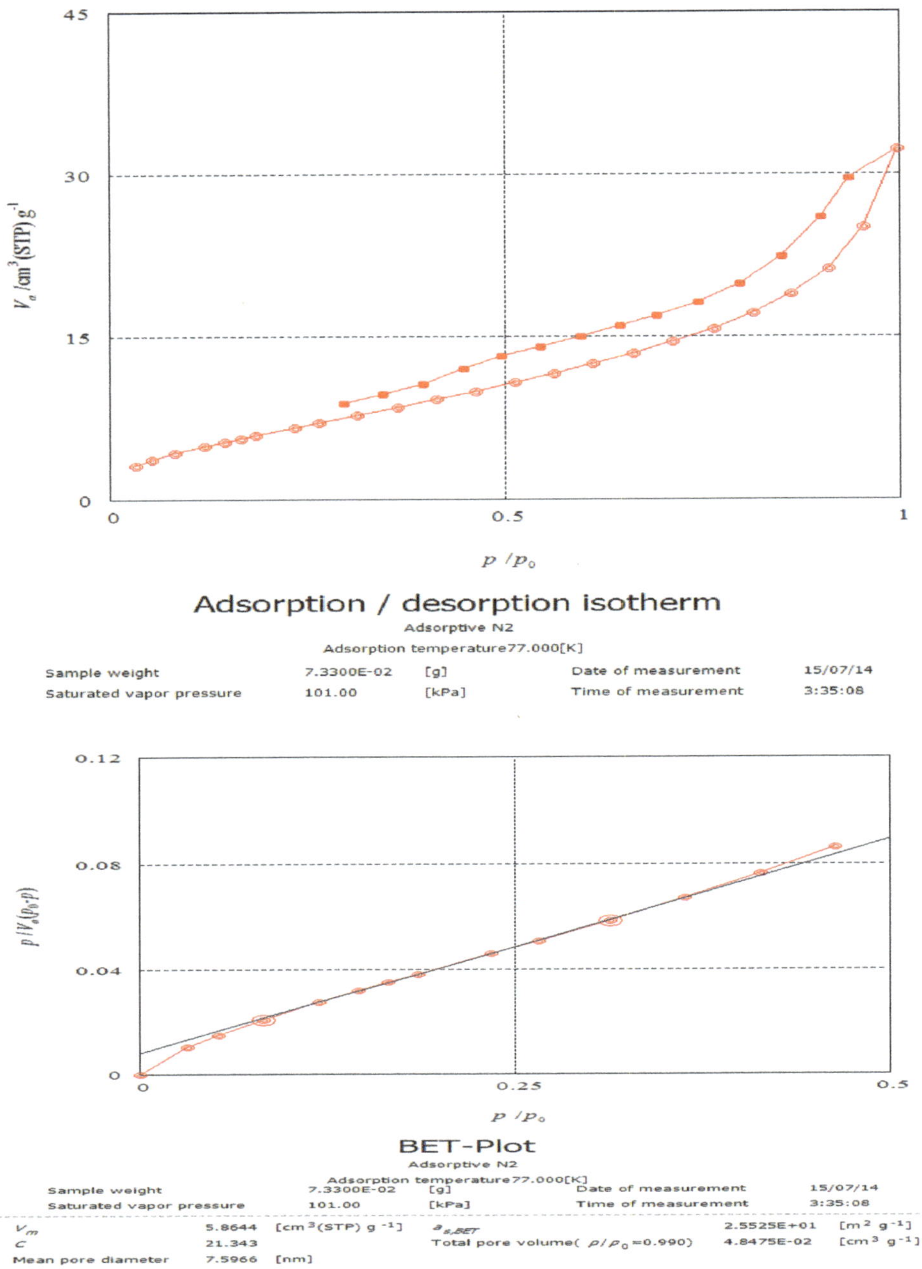

Fig. 5 Specific area characters of Rd- AC-BY by BET analysis

existence of Cd and other heavy metals in El-Mex bay leads to use pre-concentration technique as judging on the capacity of hybrid AC sorbents as a removal of cadmium ion. Therefore, the surface water samples were collected from five stations in El-Mex Bay and pre-concentrated for Cd via a multistage micro-column technique. The data show that the highest concentration (0.251 μg/L) was at station No. 5, at the iterance of the Western Harbour (Fig. 1), while the lowest contaminated areas were at stations 3 and 4 (0.051 and 0.077 μg/L), respectively. Stations 1 and 2 show moderate contamination by Cd (II) 0.122 and 0.141 μg/L, irrespectively. The results show that AC-treated-baker's yeast sorbents proved highly effective in the removal of the Cd than those of the chemically modified active carbon sorbents and both of these sorbents are generally higher than blank AC sorbent. The AC sorbent as a packing material in the micro-column was found to produce removal percentage values of 70–75 % after the first run (first passing of each sample through micro-column) and 82–85 % after the second one (second passing of each sample through micro-column). On the other hand, AC-BY, AT-AC, AT-AC-BY, BT-AC,BT-AC-BY, Ox-AC, Ox-AC-BY, Rd-AC and Rd-AC-BY were found to produce removal values in ranges of 90–94, 88–92, 95–99, 90–95, 97–100, 95–100, 100, 100 and 100 % (Table 2), respectively, all stations. Considering the results we can

Table 2 Removal of Cd (II) from wastewater samples of El-Max Bay

Sorbents	Removal %				
	Station 1	Station 2	Station 3	Station 4	Station 5
AC	85	84	82	82	85
AC-BY	93	94	93	90	94
AT-AC	92	91	88	90	92
AT-AC-BY	99	99	96	95	99
BT-AC	95	93	91	90	94
BT-AC-BY	99	100	98	97	100
Ox-AC	98	99	95	95	100
Ox-AC-BY	100	100	100	100	100
Rd-AC	100	100	100	100	100
Rd-AC-BY	100	100	100	100	100

find that,the three sorbents Ox-AC-BY, Rd-AC and Rd-AC-BY show the maximum removal of Cd (II) from all station by around 100 % after one run only.

The heavy metal uptake is attributed to different mechanisms of ion-exchange processes and/or metal complex formation as well as to the adsorption process. During the ion-exchange process, metal ions had to move through the pores of the hybrid AC mass, but also through channels of the lattice, and they had to replace exchangeable cations. Diffusion was faster through the pores and was retarded when the ions moved through the smaller diameter channels. In this case the metal ion uptake could mainly be attributed to ion-exchange reactions in the micro porous minerals of hybrid AC mass (Erdem et al. 2004).

The other mechanism of metal uptake could be forming various coordination complexes between heavy metals and the functional groups on the hybrid AC sorbent surface from chemical and biological modification. The data point out that the obvious increasing of Cd removal in all five stations done by increasing the chemical treated of AC, in order AT-AC > BT-AC > Ox-AC > Rd-AC. In another word the more donor atoms exist on the sorbent surface more binding sites available for directed interaction with metal ion. Further surface modification of AC, AT-AC, BT-AC, Ox-AC and Rd-AC sorbents via immobilization of baker's yeast for the formation of the corresponding hybrid carbon sorbents AC-BY, AT-AC-BY, BT-AC-BY, Ox-AC-BY and Rd-AC-BY, respectively were studied and compared for metals binding and sorption interaction processes. In general, the metal removals values are prove the vital role of baker's yeast as a significant high removable due to functional groups at baker's yeast cell wall surface that have the capable of forming various coordination complexes with metal ions. The functional groups responsible for binding with heavy metal ions are generally in the form of carboxyl (–COOH),phosphate(PO_4^{-3}), amine($-NH_2$), thiol (–SH), and hydroxyl (–OH) (Park and Kim 2005). Such noticed enhancement that cadmium sorption removal process are mainly due the contribution of both hybrid chemical and biological treatments of AC with Cd via coordination complex mechanism with minor role for ion exchange process.

Furthermore, all sorbents removal percentage increase as the concentration of Cd decreases in the collected samples due to exist of more available active sites on the sorbents surface, that trend excluded the samples from stations (3) and (4), which give removal values lower than the expected. That behavior could be explained by the existence of more sorbets in the samples to compete with Cd on the active sites of the sorbents or may be cause of the traces concentration of Cd in both stations decreases the chance of collation between cadmium ions and sorbents particles.

Removal of Pb from wastewater samples of El-Max bay by hybrid AC sorbents

The percentages of removal Pb from all five stations in the study area after two runs by AC sorbent as a packing material in a micro-column extraction were in range of 85–90 %. In other evidence, the different forms of the modified sorbents:AC-BY, AT-AC, AT-AC-BY, BTAC, BT-AC-BY, Ox-AC, Ox-AC-BY, Rd-AC and Rd-AC-BY were found to produce removal values of Pb in range of 93–97, 88–94, 97–99, 88–94, 93–100, 90–100, 99–100, 90–100 and 100 % for the same sorbents and biosorbents (Table 3). Two stations Nos. 3 and 5 were excluded from this trend and behavior when Ox-AC-BY, Rd-AC and Rd-AC-BY sorbents were applied for removal of Pb only in one run of micro-column enough to produce 100 % pre-concentration.

It is evident that the chemical treatment of AC improves the adsorption of all tested sorbents and biosorbents; an

Table 3 Removal of Pb(II) from wastewater samples of El-Max Bay

Sorbents	Removal %				
	Station 1	Station 2	Station 3	Station 4	Station 5
AC	85	89	90	88	89
AC-BY	93	95	97	93	97
AT-AC	88	89	94	89	92
AT-AC-BY	97	97	99	97	99
BT-AC	88	90	94	89	93
BT-AC-BY	93	97	100	95	99
Ox-AC	90	95	100	90	98
Ox-AC-BY	99	99	100	100	100
Rd-AC	90	95	100	92	100
Rd-AC-BY	100	100	100	100	100

increase in the percentage removal of Pb from the wastewater with increasing of chelating groups on the surface of hybrid AC was observed. Both the Ox-AC and Rd-AC sorbents with high number of chelating groups (Ox-AC(-NO_3) and Rd-AC(-SO_3)) shows higher removal percentage than those sorbent with less number of chelating groups (AT-AC (-Cl) and BT-AC (-OH)). Accordingly, the addition of baker's yeast on surface of hybrid AC gives a significant role in improving the adsorption strength of all tested sorbents (Vasudevan et al. 2002; Imamoglu and Tekir 2008). In the present study, a noticeable increase in the removal of Pb was observed due to the presence of more available chelating sites on the surface of hybrid AC from different functional groups of yeast cell wall (such as amine (–NH2), thiol (–SH), and hydroxyl (–OH) that are capable of forming various coordination complexes and/or ion exchange mechanisms with Pb ions. That evidence referring to the contribution of both chemical and biological treatments are responsible for providing huge number of biding sites available for chelating the metal ions on the surface of the hybrid sorbents and biosorbents.

Removal of Cr from wastewater samples of El-Max bay by hybrid AC sorbents

The ability of hybrid AC sorbents to remove total Cr from the study area in El-Max bay was further studied as the important step to explore the potential applications of hybrid AC sorbents and its results. Percentage recovery values of total Cr from station No. 4, as a highest contaminated area by Cr (2.260 μg/L), in the range of 85–94 % for second run by hybrid AC sorbents: AC, AC-BY, AT-AC, AT-AC-BY, BTAC, BT-AC-BY, Ox-AC, Ox-AC-BY, Rd-AC and Rd-AC-BY (Table 4). However, the percentage recovery values of the Cr from this station were in the range 88–100 % by the hybrid AC sorbents. The same trend and behavior were also observed for Cr species in stations (5), more contaminated area by Cr (1.076 μg/L), by hybrid AC sorbents. The percentage recovery values of Cr in this area were found to be in the range of 90–100 % as determined by hybrid AC sorbents(AC, AC-BY, AT-AC, AT-AC-BY, BTAC, BT-AC-BY, Ox-AC, Ox-AC-BY, Rd-AC and Rd-AC-BY) respectively (Table 4).

This trend were changed in the lowest contaminated areas by Cr species (Stations No. 1, 2 and 3), the recover percentage for Cr species were found in the range of 86–100 % for station No.1 and 87–100 % for stations No. 2 and 3 by all hybrid AC sorbents. This recovery percentages show increase reaching the maximum recovery around 100 % by all hybrid AC sorbents. This better trend excluded the Ox-AC, Ox-AC-BY, Rd-AC and Rd-AC-BY which needed only one run to reach 100 % recovery percentage. Maximum metal uptake values of Cr were obtained for the best performance hybrid AC sorbents (Ox-AC and Rd-AC) in all examined station especially at the lower contaminated stations (1, 2 and 3). the high sorption capacity for Cr may be explained by enhanced the direct chelating between metal ions and functional groups such as (–COOH, –NH and –OH) at the activated carbon surface and the activated carbon loaded with baker's yeast surface. The small size of Cr atom could be suitable for porous on the surface of hybrid AC sorbent that also enhance the capability of those sorbents and biosorbents for removal percent of Cr spices reach 100 % in some case after only first run. This explanation agrees with result gain from the above two metals (Cd and Pb), where we found that both small size atoms Cr and Cd show higher removal percent than the removal percent of the highest size Pb atom (Gupta et al. 2011).

Removal of Hg from wastewater samples of El-Max bay by hybrid AC sorbents

The recovery percentage of Hg from five locations in El-Max bay was performed using a multistage micro-column

Table 4 Removal of Cr from wastewater samples of El-Max Bay

Sorbents	Removal %				
	Station 1	Station 2	Station 3	Station 4	Station 5
AC	92	94	93	85	90
AC-BY	94	93	96	92	92
AT-AC	92	91	94	89	90
AT-AC-BY	97	96	96	95	95
BT-AC	94	95	97	92	90
BT-AC-BY	97	97	99	96	93
Ox-AC	96	100	100	94	94
Ox-AC-BY	100	100	100	100	100
Rd-AC	98	100	100	97	98
Rd-AC-BY	100	100	100	100	100

Table 5 Removal of Hg from wastewater samples of El-Max Bay

Sorbents	Removal %				
	Station 1	Station 2	Station 3	Station 4	Station 5
AC	80	85	65	99	73
AC-BY	90	91	72	100	78
AT-AC	84	90	75	100	77
AT-AC-BY	91	92	80	100	85
BT-AC	89	90	78	100	84
BT-AC-BY	91	94	84	100	89
Ox-AC	95	96	91	100	93
Ox-AC-BY	100	100	94	100	96
Rd-AC	98	100	96	100	99
Rd-AC-BY	100	100	97	100	99

technique. The data show high recovery percentage from stations (1, 2 and 4) where the percentages of recover were in the range of 70–89, 85–94 and 96–100 %, respectively, by all hybrid AC sorbents (AC-BY, AT-AC, AT-AC-BY, BTAC, BT-AC-BY, Ox-AC, Ox-AC-BY, Rd-AC and Rd-AC-BY). The second run for the same stations were found excellent recovery percentage reaching 80–100, 85–100 and 99–100 %, respectively (Table 5). On the contrary of this trend the higher pollutants area with high concentration of Hg (5.700 and 4.710 μg/L) represented by stations (3 and 5) show lower recovery percentage were 42–85 and 62–89 %, respectively of recovery by all hybrid AC sorbents (AC-BY, AT-AC, AT-AC-BY, BTAC, BT-AC-BY, Ox-AC, Ox-AC-BY, Rd-AC and Rd-AC-BY) which considered the lowest recovery percentage done by those sorbents all over tested metal ions or all over tested locations. However this values of recovery show good rising to reach 65–97.0 % and 73–99 % after 2nd run by all hybrid AC sorbents for these two stations. This data show that all hybrid AC sorbents couldn't reach 100 % of recovery even after second stage of pre-concentration.

The high affinity of Hg removal enhanced by hybrid AC sorbents that contained ligand groups (amine, amide, thiol, etc.) from chemical or biological treated can be explained well by the Pearson's theory (Knocke and Hemphill 1981), in which hard acids tend to co-ordinate with hard bases and soft acids with soft bases. Neutral mercury molecules are soft acids, therefore, the interactions of Hg species such as $HgCl_2$, $(HgCl_2)_2$, Hg $(OH)_2$ and HgOHCl with surface nitrogen and sulfur ligands (soft bases) are likely favored. The enhanced Hg(II) adsorption by hybrid AC may also result from the Hg reduction and $Hg(HS)_2$ and $Hg_2(HS)_2$ formations on sulfur-treated carbons, as suggested by Sinha and Walker (1972) and Lopez-Gonzalez et al. (1982). In an electronic microscopy study, Macias et al. (1995) and Adams (1991) confirmed the presence of Hg^+ and Cl − on the surface of activated carbons in contact with $HgCl_2$, supporting the reduction process. In addition, Hg ions could react with various ligands on the biological treated carbon via yeast loaded to form surface complexes. Therefore, a combination of mercury surface reduction, complexation, and/or ion exchange would contribute to the enhanced Hg sorption by hybrid AC sorbents and biosorbents. This study showed that nitric acid (oxidizing agent) together with sodium sulfite (reducing agent) treatment and ligand groups on the surface of the baker's yeast reaction would effectively in reach Ox-AC, Rd-AC, Ox-AC-BY and Rd-AC-BY to maximum removal percent for all five Hg contaminated stations.

Conclusion

The newly designed chemically and/or biologically hybrid activated carbon (AC) sorbents were synthesized as effective sorbents for heavy metals contaminated water in El Mex bay. These sorbents and biosorbents were successfully applied in removing cadmium, lead, chromium and mercury by dynamic micro-column technique. The chemical treatments of commercial active carbon via simple acid, base, oxidation and reduction reactions were found to improve the recovery efficiency of the chemically hybrid AC sorbents for the studied metals than AC blank in all case, this recovery efficiency give excellent values with addition of baker's yeast to all hybrid AC sorbents especially after 2nd run of micro-column. Reducing active carbon-immobilized-baker's yeast biosorbent (Rd-AC-BY) was characterized by its sorption superiority, under the effect of all studied contaminated areas a long of El-Max bay, compared to the other chemically and biologically immobilized active carbon sorbents and biosorbents, in almost cases Rd-AC-BY get 100 % recovery after only one run of micro-column. Finally, the presented work affords a number of environmental friendly hybrid AC sorbents and biosorbents characterized by high efficiency of removing different heavy metals from wastewater.

Acknowledgments The authors would like to thank the colleagues in the National Institute of Oceanography and Fisheries for their help and support during the collecting the samples from El Mex Bay.

References

Abdallah MAM (2008) Trace metal behavior in mediterranean-climate coastal bay: El-Mex Bay Egypt and its Coastal Environment. Glob J Environ Res 2:23–29

Abdallah MAM (2014) Chromium geochemistry in coastal environment of the Western Harbour, Egypt: water column, suspended matter and sediments. J Coast Conserv 18:1–10

Abdallah MAM, El Sayed NB, Saad M (2007) Distribution and enrichment evaluation of heavy metals in El-Mex Bay using normalization models. Fresenius Environ Bull 16:710–719

Abdel-Fattah TM, Mahmoud ME, Ahmed SB, Huff MD, Lee JW, Kumar S (2014). Biochar from woody biomass for removing metal contaminants and carbon sequestration, J Ind Eng Chem (in press)

Abdullah MI, Royle LG (1974) A study of the dissolved and particulate trace elements in the Bristol Channel. J Mar Biol Assess UK 54:581–597

Adams MD (1991) The mechanism of adsorption of Hg (CN)2 and HgCl2 onto activated carbon. Hydrometallurgy 26:201–210

Baccara R, Bouzida J, Fekib M, Montiela A (2009) Preparation of activated carbon from Tunisian olive-waste cakes and its application for adsorption of heavy metal ions. J Hazard Mater 162:1522–1529

Bag H, Lale M, Turker AR (1998) Determination of iron and nickel by flame atomic absorption spectrophotometry after preconcentration on Saccharomyces cerevisiae immobilized sepiolite. Talanta 47:689–696

Bag H, Turker AR, Lale M, Eli AT (2002) Separation and speciation of Cr(III) and Cr(VI) with Saccharomyces cerevisiae immobilized on sepiolite and determination of both species in water by FAAS. Talanta 51:895–902

Cui L, Wang Y, Gao L, Hu L, Yan L, Wei Q, Du B (2015) EDTA functionalized magnetic graphene oxide for removal of Pb (II), Hg (II) and Cu (II) in water treatment: adsorption mechanism and separation property. Chem Eng 281:1–10

El-Rayis OA, Abdallah MAM (2006) Contribution of Some Trace Elements from an Egyptian Huge; the Mediterranean Sea, east of Alexandria. Mediterr Mar Sci 7:79–86

Erdem E, Karapinar N, Donat R (2004) The removal of heavy metal cations by natural zeolites. J Colloid Interface Sci 280:309–314

Faur- Brasquet C, Reddad Z, Kadirvelu K, Le Cloirec P (2002) Modeling the adsorption of metal ions (Cu2 + , Ni2 + , Pb2 +) onto ACCs using surface complexation models. Appl Surf Sci 196:356–365

Fu J, Chen Z, Wu X, Wang M, Wang X, Zhang J, Zhang J, Xu Q (2015) Hollow poly (cyclotriphosphazene-co-phloroglucinol) microspheres: an effective and selective adsorbent for the removal of cationic dyes from aqueous solution. Chem Eng 281:42–52

Grote M, Kettrup A (1985) Ion exchange resins containing S-bonded dithizone and dehydrodithizone as functional groups-II: desorption properties and development of separation procedures for gold and platinum group metals. Anal Chim Acta 175:239

Gupta VK, Agarwal S, Saleh TA (2011) Chromium removal by combining the magnetic properties of iron oxide with adsorption properties of carbon nanotubes. Water Res 45:2207–2212

Imamoglu M, Tekir O (2008) Removal of copper (II) and lead (II) ions from aqueous solutions by adsorption on activated carbon from a new precursor hazelnut husks. Desalination 228:108–113

Knocke WR, Hemphill LH (1981) Mercury (II) sorption by waste rubber. Water Res 15:275–282

Lopez-Gonzalez JD, Moreno-Castilla C, Rodriguez-Reinoso F (1982) Effect of carbon–oxygen and carbon–sulfur surface complexes on the adsorption of mercuric chloride in aqueous solutions by activated carbons. J Chem Technol Biotechnol 32:575–579

Macias GA, Valenzuela CC, Espinosa MAY, Gomezserrano V (1995) Adsorption of Cd2 + by heat-treated and sulfurized activated carbon. Anal Quim 91:547–552

Mahmoud M, Yakout E, Ahmed SB, Osman MM (2008) Chromium speciation, selective extraction and preconcentration by aluminafunctionalised 2-pyridenecarboxyladehyde thiosemicarbazone. Int J Environ Anal Chem 88:1017–1031

Mahmoud ME, Hafez OF, Alrefaay A, Osman MM (2010) Performance evaluation of hybrid inorganic/organic adsorbents in removal and preconcentration of heavy metals from drinking and industrial waste water. Desalination 253:9–15

Mahmoud MM, Osman M, Ahmed SB, Abdel-Fattah TM (2011) Improved adsorptive removal of cadmium from water by hybrid chemically and biologically carbonaceous sorbents. ME Chem Eng J 175:84–94

Mahmoud MM, Osman M, Ahmed SB, Abdel-Fattah TM (2012). Enhanced removal of lead by chemically and biologically treated carbonaceous materials. Sci World J 2012

Park SJ, Kim YM (2005) Adsorption behaviors of heavy metal ions onto electrochemically oxidized activated carbon fibers. Mater Sci Eng A 39:121–123

Peez-Corona T, Madrid-Albamin Y, Camara C, Beceiro E (1998) Living organisms as an alternative to hyphenated techniques for metal speciation: evaluation of baker's yeast immobilized on silica gel for Hg speciation. Spectrochim Acta B 53:321–329

Sinha RK, Walker PL (1972) Removal of mercury by sulfurized carbons. Carbon 1:754–756

Tian Y, Yin P, Qu R, Wang C, Zheng H, Yu Z (2010) Removal of transition metal ions from aqueous solutions by adsorption using a novel hybrid material silica gel chemically modified by tri ethylene tetramino methylene phosphonic acid. Chem Eng J 162:573–579

United Nations Environment Programme (UNEP) (1993). United Nations Environment Programme, report of the governing council seventeenth session. Supplement No. 25 (A/48/25). New York pp 96

Vasudevan P, Padmavathy V, Dhingra SD (2002) Biosorption of monovalent and divalent ions on baker's yeast. Bioresour Technol 82:285–289

Yavuz H, Denizli A, Gungunes H, Safarikova M, Safarik I (2006) Biosorption of mercury on magnetically modified yeast cells. Sep Purif Technol 52:253–260

26

Pollutant source identification model for water pollution incidents in small straight rivers based on genetic algorithm

Shou-ping Zhang[1] · **Xiao-kang Xin**[2,3]

Abstract Identification of pollutant sources for river pollution incidents is an important and difficult task in the emergency rescue, and an intelligent optimization method can effectively compensate for the weakness of traditional methods. An intelligent model for pollutant source identification has been established using the basic genetic algorithm (BGA) as an optimization search tool and applying an analytic solution formula of one-dimensional unsteady water quality equation to construct the objective function. Experimental tests show that the identification model is effective and efficient: the model can accurately figure out the pollutant amounts or positions no matter single pollution source or multiple sources. Especially when the population size of BGA is set as 10, the computing results are sound agree with analytic results for a single source amount and position identification, the relative errors are no more than 5 %. For cases of multi-point sources and multi-variable, there are some errors in computing results for the reasons that there exist many possible combinations of the pollution sources. But, with the help of previous experience to narrow the search scope, the relative errors of the identification results are less than 5 %, which proves the established source identification model can be used to direct emergency responses.

✉ Xiao-kang Xin
xin.xiaokang@163.com

[1] Chongqing Water Resources and Electric Engineering College, Chongqing 402160, People's Republic of China

[2] State Key Laboratory of Hydraulics and Mountain River Engineering, Sichuan University, Chengdu 610065, People's Republic of China

[3] Changjiang Water Resources Protection Institute, Changjiang Water Resources Commission, Wuhan 430051, People's Republic of China

Keywords Water pollution incident · Basic genetic algorithm (BGA) · Pollutant amount · Pollution source position · Identification model

Introduction

In the past a few decades, thousands of water pollution incidents caused by accident or illegal emissions happened all over the world. Such as Tuojiang River pollution incident happened in 2004, Songhua River benzene pollution incident happened in 2005, Colorado River pollution incident happened in 2015, and so on. The pollution incident refers to a large amount of pollutant leaked into the water suddenly, resulting in a rapid increase in pollutant concentrations, which may serious impact on the safety of water supply. When water pollution incidents occur, the recognition of pollution sources (including source location and amount of pollutant) is a primary task in emergency response, which has attracted the attention of many researchers and governors. Xue and Wang identified the source location of a water pollution incident successfully by adopting the 'monitoring-judging' method (Xue and Wang 1997). With the help of an established pollution source database created by government agencies, Wang proposed the "Fingerprint Method" for identification of the source location of the pollution incident (Wang et al. 2009). Combining a one-dimensional water quality model in rivers with a geostatistical method, Fulvio inversely analyzed the pollutant amount on where the pollution source location was known (Fulvio et al. 2005). With the combination of a two-dimensional river water quality model and Adjoint Method, Nikolaos and Michael used an optimization procedure to predict the position of the pollution sources and the amount of pollutant (Nikolaos and Michael 1996). Combining MODFLOW and MT3DMS and BL-PDF, Neupauer and Wilson developed a groundwater pollution source identification model (Roseanna and Neupauer 2004). With the combination of the two-dimensional water quality mathematical model in rivers and BL–PDF, Cheng and Jia proposed a surface water pollution source inversion model (Cheng and Jia 2010). With the combination of water quality model and a Bayesian estimation method, Zhu obtained the location probability distribution of pollution source (Zhu et al. 2009). The inversion problem of water pollution sources is also studied by the researchers in the field of mathematics. Based on mathematical theory, the uniqueness of the pollution source (single source) position has been proved (Wang and Xu 2006), and the related identification method has been given (Wang and Qiu 2008).

To figure out the source for a pollution incident is of great difficult. Traditional identification models relied excessively on monitoring data, and are laborious and data intensive (Cheng and Jia 2010). With the high requirement for topographical, hydrological, water quality data, and needing a lot of incident environmental monitoring data as support, the methods that use numerical water quality models method may affect the efficiency. The recognition of pollution sources with the aid of an optimization algorithm combined with mathematical analysis methods is simple, practical and has been widely applied (Cao et al. 2010; Yin et al. 2011). Meanwhile, with the ability for global searching, robustness and high efficiency, the Genetic Algorithm (GA) is widely used on the inverse problems in the aquatic environment (Singh and Datta 2006). Combining GA and the mathematical analysis method, this paper proposed a pollution source identification model of sudden water pollution incidents in small straight rivers, which can concisely and effectively deal with the problem of the pollution source determination.

Water quality equations and its analytic solution

The motion and transportation process for pollutant in the river channel can be described as convection and diffusion processes, so the water quality equations are called as convection and diffusion equations. The three-dimensional equation is as follows (Wang and Shen 2005; Li et al. 2015):

$$\frac{\partial C}{\partial t}+u\frac{\partial C}{\partial x}+v\frac{\partial C}{\partial y}+w\frac{\partial C}{\partial z}=\frac{\partial}{\partial x}\left(E_x\frac{\partial C}{\partial x}\right)+\frac{\partial}{\partial y}\left(E_y\frac{\partial C}{\partial y}\right)+\frac{\partial}{\partial z}\left(E_z\frac{\partial C}{\partial z}\right)-\mathrm{KC}+S \tag{1}$$

where, C is the concentration of pollutant caused by the pollution incident (mg/L); t is the time (s); x, y, z are the lateral, transverse and vertical coordinates (m); u, v, w are the flow velocity in the above three coordinates, respectively (m/s); E_x is the diffusion coefficient, (m^2/s); K is the comprehensive decay coefficient, (1/s); S is the source item. In general, the concentration field can be computed by numerically models with Eq. (1). However, river terrain and hydrological data are usually insufficient where accident emergencies occur. Therefore, especially for small rivers, the three-dimensional convection–diffusion equation in rivers can be simplified as: (1) A river, whose discharge is less than 50 m^3/s, bend coefficient is less than 1.5 and water width is less than 100 m, can be assumed as small straight river (Ni and Wang 2000). For those rivers, the river courses can be assumed as homogeneous channel and the flow can be assumed as constant (Liu 2005). (2) Comparing to the quantity and concentration of pollutants leaked into the river, the background concentrations at two endpoints could be ignored, and the initial concentration could be assumed to be zero as well. (3) For a small river, the pollutant is considered to mix homogeneously across the transverse section, and have a constant attenuation. Therefore, the concentration distribution along the river can be simplified as the following one-dimensional convection–diffusion equations (Jiang and Wang 1997; Liu 2005; Abderrezzak et al. 2015):

$$\begin{cases}\frac{\partial C}{\partial t}+u\frac{\partial C}{\partial x}=E_x\frac{\partial^2 C}{\partial x^2}-\mathrm{KC}+\sum_{i=1}^{q}M_i\delta(x-x_i)\,0<x<l,\ t>0\\ C(x,0)=0,\quad 0<x<l\\ C(0,t)=0,\quad t>0\\ C(l,t)=0,\quad t>0\end{cases} \tag{2}$$

where l is the river length (m); M_i represents pollutant amount of the corresponding pollution source, (g/s); δ is the Dirac Function; x_i is the location (coordinate) of the pollution source, (m); q is the number of pollution sources in the incident.

When the location x_i of pollution sources and the amount of pollutant M_i released into the river are known, solving the distribution $C(x, t)$ of pollutant concentration in the river at different times is known as the "forward problem" for pollutant concentration simulation. With this "forward problem", we can simulate and predict the effect degree and scope of a water pollution incident. On the other hand, when the concentration monitoring values at different places at a particular time or the monitoring values at a place at different times are given, we need to find out the pollution source position or the pollutant amount, which is known as an "inverse problem". In the process of solving an "inverse problem", the first step is to build the optimization object function of the "inverse

Table 1 Parameters of river hydrodynamics and pollution sources

Parameter name	River length (l)	Velocity (u)	Diffusion coefficient (Ex)	Decay coefficient (K)	Single source	Double source
Unit	m	m/s	m^2/s	1/s	(m, g/s)	(m, g/s)
Value	1000	0.1	5.0	0.00001	(500, 3.0)	(300, 3.0), (500, 5.0)

problem" by applying the solution of the "forward problem". So we should use an appropriate method to solve Eq. (2). As we can see, the partial differential Eq. (2) belongs to the linear convection–diffusion equations. Some researchers used the method for combining Fourier transform with Laplace transform method (Kumar et al. 2010; Feng 2012), and others have used separation of variables to solve the equations (Min et al. 2004; Wu et al. 2008). Finally, it is concluded that the water quality equation's analytical solution of a river pollution incident is as follows:

$$C(x,t) = \sum_{n=1}^{\infty} \frac{\frac{2}{l}\sum_{i=1}^{q} M_i e^{-\frac{ux_i}{2E_x}} \sin\frac{n\pi x_i}{l}}{\frac{u^2}{4E_x} + \left(\frac{n\pi}{l}\right)^2 + K} \left[e^{\frac{u^2 t}{4E_x}} - e^{-t\left(\frac{n\pi}{l}\right)^2 + k}\right] \sin\frac{n\pi x}{l} \times e^{(ux/2E_x - u^2 t/4E_x)} \tag{3}$$

Supposing the experimental river length l is 1000 m, the flow velocity u is 0.1 m/s, the diffusion coefficient E_x is 5.0 m^2/s, and the comprehensive attenuation coefficient K is 0.00001/s, and the pollutant concentration can be calculated using the above Eq. (3). There are two supposed conditions: (1) a single pollutant source incident happens with the position $x = 500$ m and amount $M = 3.0$ g/s; (2) double pollutant sources incident occur with the position $x_1 = 300$ m, amount $M_1 = 3.0$ g/s, $x_1 = 500$ m, amount $M_1 = 5.0$ g/s, respectively. Hydrodynamic and pollution source characteristics parameters of the supposed conditions are listed in Table 1.

Calculating the infinite series from $n = 1$ to $n = 1000$ to obtain the pollutants concentration approximately at each moment, under the two kinds of working conditions, concentration distribution at each moment along the river channel is shown in Fig. 1, and concentration distribution of each section changing in time is shown in Fig. 2. As can be seen, the pollution causes the pollutant concentration jump in the adjacent river sections. And for a special position, the concentration increases at first and then tends to be stable, which did respond the truth

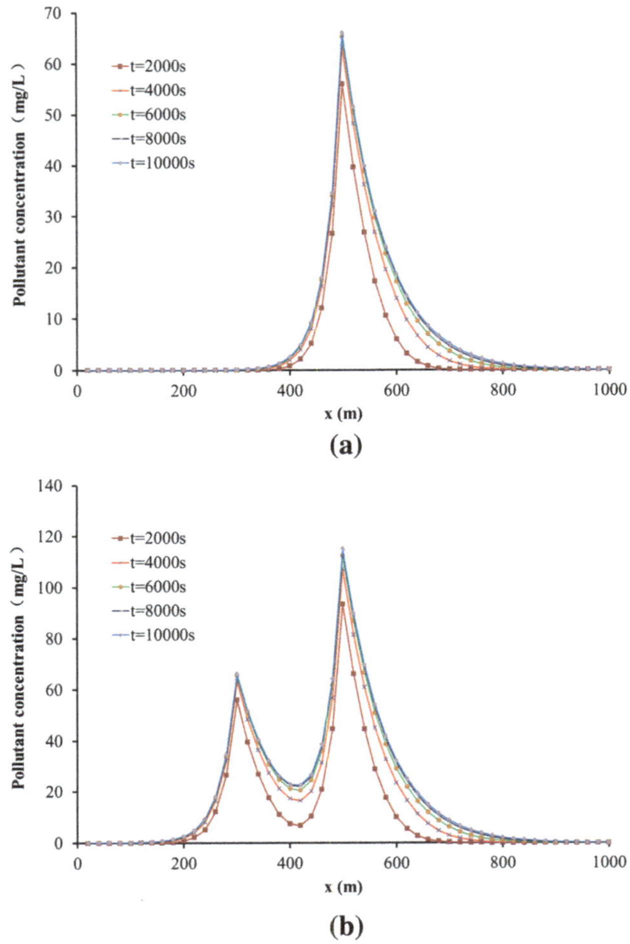

Fig. 1 Pollutant concentration distribution through time after incident occured. **a** Single pollutant source, **b** double pollutant sources

Genetic algorithm and the identification model of accident pollution source

Basic genetic algorithm

The basic genetic algorithm (BGA) is a kind of intelligent optimization method proposed by Holland (Holland 1975), which imitated the natural genetic phenomena of selection, crossover and mutation processes in nature. Jong carried out a great quantity of optimization tests with numerical functions using Holland's theory, which proofed BGA is an effective and efficient stochastic search method (Jong 1975). In the 1990s, BGA were widely used in the scope of engineering with the publication of 'Hand book of Genetic Algorithm', such as reservoir operation optimization (Jothiprakash and Shanthi 2006), numerical model parameter optimization (Tang et al. 2010), inverse problem research (Dai and Wang 2006), and so on.

BGA starts from an initial population that contains a number of individuals, and then new individuals are produced to be better adapted to the environment with random selection, crossover and mutation. The best individual is eventually achieved by a number of evolution steps

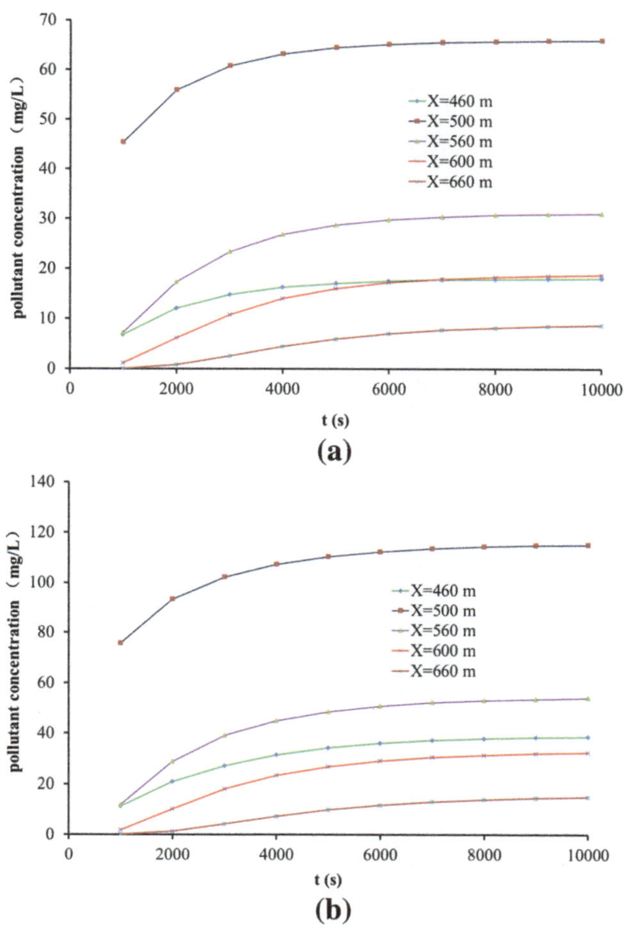

Fig. 2 Pollutant concentration profile at each cross-section following the incident. **a** Single pollutant source, **b** double pollutant sources

(generations). Every individual in a population is a feasible solution for the optimization problems, and the best individual is the optimal solution of the optimization problem. Compared with the other optimization algorithms, BGA has the following advantages: (1) the optimization objective function can be either a continuous function or a discrete function (Tang et al. 2010); (2) has the property of global search and automatic convergence to the optimal solution; (3) robust in dealing with complex nonlinear problems; (4) the principle is simple, easy to understand, versatile and highly maneuverable. Many improvements have been made to the BGA considering its wide application, including a niche technology of crossover operation (Srinivas and Patnaik 1994), uniform mutation operation method (Xin et al. 2009) and adaptive algorithm of crossover and mutation probability (Li et al. 2004; Yu et al. 2013).

Pollutant source identification model

It can be seen from the pollutant concentration Eq. (2) that solving $C(x, t)$ depends on the location x_i of sudden water pollution incident and the amount of pollutants M_i. Therefore, different x_i and M_i decide different pollutant concentration distributions. By setting up a monitoring cross section on an experimental river at time T and obtaining the concentration distribution $\overline{C}(\overline{x}_i, T)$ of the pollutant, where, i is the number of the monitoring section, $\overline{x}_i$ is the position coordinates of the ith monitoring section, then the pollution source inversion problem can be transformed into an optimization problem of a linear discrete function as follows:

$$\min \sum_{i=1}^{n} C(x_i, M_i, \overline{x}_i, T) - \overline{C}(\overline{x}_i, T) \tag{4}$$

If we obtain the x_i and M_i which are the closest to calculated values and measured values by optimal Eq. (4), and the location of incident pollution sources and pollutants amount are consequently confirmed. This optimization search process can be completed by the basic genetic algorithm. Hence, a pollution source identification model of a sudden water pollution incident based on the basic genetic algorithm can be established through the following steps:

1. Determining the parameters of the basic genetic algorithm, including the gene number N_g of individuals, and the value scope of the gene, the crossover probability p_c, the mutation probability p_m, the number of individuals in a population (population size) I_m, and the maximum evolution step T_m, among which the number of genes N_g is determined by the number of optimization variables. Take a single source identification problem as an example, if only the position or the amount need to be identified, the $N_g = 1$, else if both the position and amount needs to be optimized, the $N_g = 2$. The crossover probability, mutation probability and the population size are determined by experience (Zhang et al. 2009).
2. Determining the gene encoding. Since a water pollution incident may have q sources, and each source will have two optimization variables. So the number of optimization variables is $2q$. If we use binary encoding method, a dozen of gene may only represent an individual, which may waste a lot of storage space. So we choose the floating-point real number coding method which means the number of the gene is $2q$, accordingly. A floating-point real number coding method is adopted for the convenience of encoding and decoding. In the basic genetic algorithm, the genes of kth individual in the population is $X^k = (x_i^k, M_i^k)$, where x_i^k is the position and M_i^k represents pollutant amount for the ith pollution source.

3. Generating initial population $x_i^k = x_{i,\min} + \mathrm{rand}(x_{i,\max} - x_{i,\min})$,$M_i^k = M_{i,\min} + \mathrm{rand}(M_{i,\max} - M_{i,\min})$, where rand is a random function which creates a number between 0 and 1; $x_{i,\min}$ and $x_{i,\max}$ are the lower limit and upper limit for the position of the ith pollution source. $M_{i,\min}$ and $M_{i,\max}$ are the lower limit and upper limit for the pollutant amount of the ith pollution source.
4. Establishing the optimal objective function (fitness function of basic genetic algorithm). Because the basic genetic algorithm can only make the individual evolve towards the increasing fitness values, while Eq. (4) requires optimization variables changing in the decreasing direction of the objective function value, thus it is necessary to modify the objective function to the fitness function (Tang et al. 2010):

$$F = 1 \Big/ \left[C(x_i, M_i, \overline{x}_i, T) - \overline{C}(\overline{x}_i, T)\right]^2 \quad (5)$$

5. Optimization operation. The operators of selection, crossover and mutation are applied to the initial population, and it stops when the Genetic Algorithm converges or reaches the maximum evolution step. The best individual of the latest population represents the positions x_i and pollutant amount M_i.

With this, a pollutant source identification model of pollutant incidents based on BGA has been established, and the working flow chart of the model is shown in Fig. 3.

Tests of pollutant source identification model

Single variable identification for single source

In the reality, after some sudden water pollution incident happens, we may know the location of the pollutant incident by investigation with other methods, such as Chinese national pollution source information database (MEP 2010). But, we do not know the pollutant amount discharged into river. Under this condition, the pollutant amount is the optimization variable. Water pollution incident happened at Fujiang, China, in 2011 belongs to this situation (Wang et al. 2012). On the other hand, the pollutant amount can be figure out easily or the incident perpetrator know the amount of pollutant, but we cannot obtain relate information about pollution source position, so the position is the optimization variable. To forecast the pollution information and carry out rescue measures, using the established identification model of pollution source to identify the pollutant amount into river or pollutant source position is of great importance.

Take the following water pollution incident as an example, where the hydrodynamic parameters of the test river section are the same as in Table 1 (single source condition), the pollutant amount is 3.0 g/s and the pollution source location is at 500 m. For the sake of contrast analysis, the analytic solution to the "forward problem" at $t = 4000$ s is selected as the field monitoring value

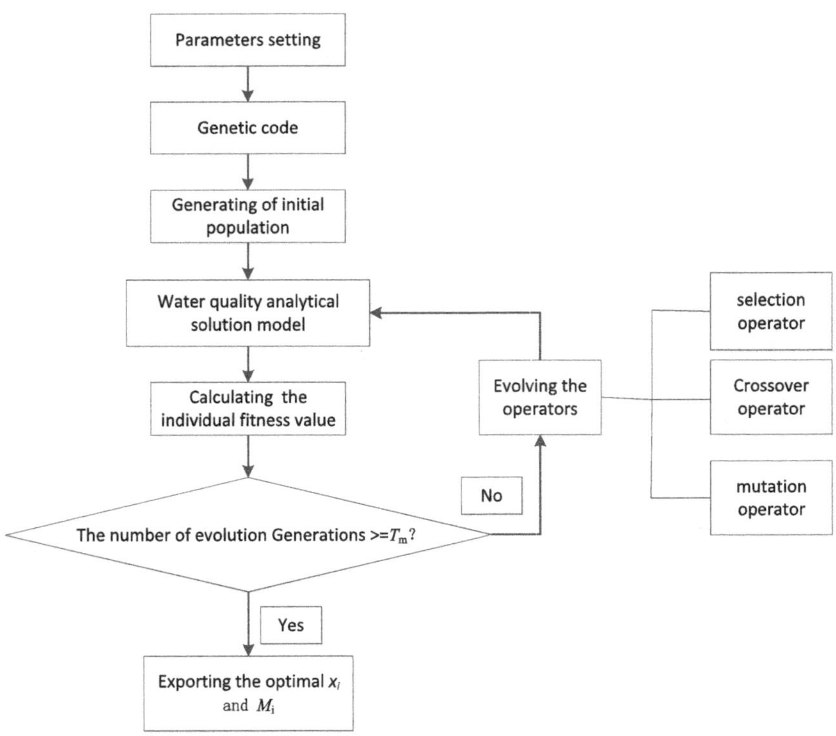

Fig. 3 Flow chart for pollutant source identification model based on GA

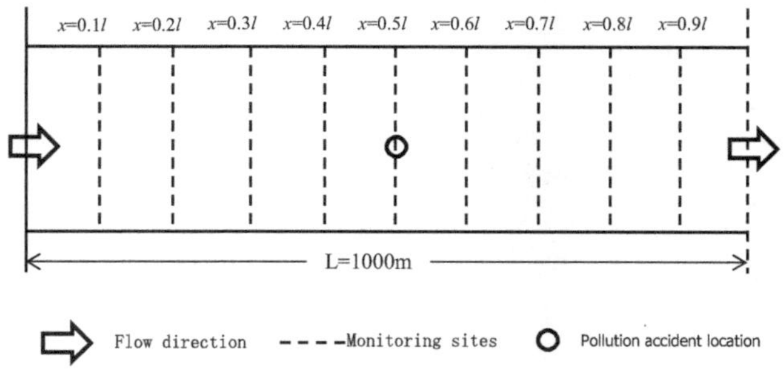

Fig. 4 Monitoring site locations for source identification of a water pollution incident

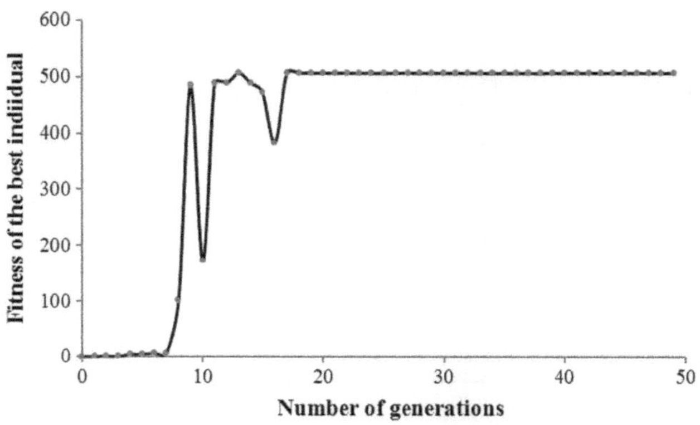

Fig. 5 Fitness curve of the identification model for multi-variable and single source

$\overline{C}(\overline{x}_i, T)$, and the monitoring sections (data collection points) are shown as Figs. 4 and 5, the "field monitoring data" are shown in Table 2.

Considering the fact that the population size I_m of the basic genetic algorithm has significant influence on the identification model, different computing conditions are set with different population sizes. The other parameters of computing conditions for BGA are shown in Table 3. The identification results of pollutant amount or the position of sources are shown in Table 4. As we can see from Table 4, this novel identification model based on the basic genetic algorithm has a relatively higher precision: the relative error between identification results of pollutant amount and analytic results varied from −1.7 to 5.0 %, and the relative error between identification results of source position and analytic results varied from 0 to −1.6 %. Since the crossover probability parameter p_m has been set as 0.8, so crossover operation is more fully undertaken with increasing population size, the "premature" phenomenon can be avoided and the model converges faster. When population size is set as 10, the model identification result and analytic result are exactly the same, where the relative error is 0.

Multi-variable identification for single source

For the other water pollution incidents, we not only do not know the locations of the pollution sources, but also do not know the pollutant amount discharged into the river, so we need to identify these two variables at the same time, which is a multi-variable identification for single source problems. Ammonia nitrogen pollution incident happened at Hanjiang, China, in 2014 belongs to this situation. (http://finance.sina.com.cn/consume/puguangtai/20140425/092218920579.shtml).

Table 2 Actual concentration value at each position of single source (t = 4000 s)

Monitoring points	1#	2#	3#	4#	5#	6#	7#	8#	9#	10#
Position (m)	0.1l	0.2l	0.3l	0.4l	0.5l	0.6l	0.7l	0.8l	0.9l	1.0l
Concentration (mg/L)	0.00	0.00	0.03	1.89	63.23	13.94	1.75	0.09	0.00	0.00

Table 3 Computing parameters for single variable identification model of single source

Condition/parameter	Optimization variable	Crossover probability (p_c)	Mutation probability (p_m)	Population size (I_m)	Maximum evolution step (T_m)
Condition 1	Amount	0.8	0.05	4	200
Condition 2	Amount	0.8	0.05	6	200
Condition 3	Amount	0.8	0.05	8	200
Condition 4	Amount	0.8	0.05	10	200
Condition 5	Position	0.8	0.05	4	200
Condition 6	Position	0.8	0.05	6	200
Condition 7	Position	0.8	0.05	8	200
Condition 8	Position	0.8	0.05	10	200

Table 4 Computing results for single variable identification model of single source

Parameter	Position (m)	Amount (g/s)	Relative error (%)	The best individual fitness value of the initial	convergence generations	The best individual fitness value of the last
Condition 1	–	3.03	1.0	0.12	40	178
Condition 2	–	2.95	−1.7	0.29	25	188
Condition 3	–	3.15	5.0	0.09	10	2
Condition 4	–	3.00	0	0.22	5	9613
Condition 5	492	–	−1.6	0.00	96	32
Condition 6	493	–	−1.4	0.00	118	78
Condition 7	493	–	−1.4	0.00	96	78
Condition 8	500	–	0	0.00	10	2150

Table 5 Computing parameters for multi-variable identification model for single source

parameter name	The inversion variable	Crossover probability (p_c)	Mutation probability (p_m)	Population size (I_m)	The biggest evolution algebra (T_m)
Condition 1	Position and amount	0.5	0.2	10	200

Table 6 Computing results for multi-variable identification model for single source

parameter name	Position (m)	amount (g/s)	The best individual fitness value of the initial	convergence generations	The best individual fitness value of the last
Condition 1	503	2.93	0.0	20	0.18

Table 7 Actual concentration value at each cross-section of double sources (t = 4000 s)

Monitoring point	1#	2#	3#	4#	5#	6#	7#	8#	9#	10#
Position (m)	0.1l	0.2l	0.3l	0.4l	0.5l	0.6l	0.7l	0.8l	0.9l	1.0l
Concentration (mg/L)	0.03	1.89	63.29	17.08	107.15	23.32	2.93	0.16	0.00	0.00

Continuing with the above water pollution incident as an example, where the hydrodynamic parameters of the test river section are the same as Table 1, the pollutant amount is 3.0 g/s and the pollution source location is at 500 m. For the sake of contrast analysis, the analytic solution to the "forward problem" at $t = 4000$ s is selected as the field monitoring value, and the monitoring sections (data collection points) are shown in Figs. 4 and 5, "field monitoring data" are shown in Table 2. An identification model is established to carry out a test experiment for the pollutant source position and the pollutant amount synchronously. Computing parameters of the source identification model are shown in Table 5. The computing results are shown in Table 6. Obviously, the pollution source location is 500 m calculated by the identification model (relative error is 0.0 %), pollutant total amount is 2.90 g/s (relative error is 2.34 %), which implies the precision is dropping in an acceptable range (relative error less than ±10 %). The model converged when calculated to the 20th generation, and the profile curve of fitness value changing along with the evolution steps is shown in Figs. 4 and 5.

Single variable identification for double sources

Although the probability of occurrence in practice is small, to test the applicability of the established identification model, taking a water pollutant incident with double sources as an example where pollutant amounts are 3.0 and 5.0 g/s, pollution sources' positions are 300 and 500 m, respectively. Analytic solution at $t = 4000$ s of "forward problem" is still used as the monitoring value $\overline{C}(\overline{x}_i, T)$ for the "inverse problem". The analytic results are shown in Table 7. On this basis, the identification model of a single variable (pollutant amounts or source positions) for multi-sources is established. The computing results are shown in Table 8. When the positions of double sources are known, computing results of pollutant amounts are 2.85 and 5.02 g/s, the relative errors are −5.0 and 0.4 %, respectively. When the pollutant amounts of double sources are

Table 8 Computing results for single variable identification model of double sources

Parameter	Positions (m)	Amounts (g/s)	The best individual fitness value of the initial	Stable convergence generations	The best individual fitness value of the last
Amounts identification	–	2.85, 5.02	0.03	333	72
Positions identification	310, 496	–	0.03	228	56

Table 9 Computing results for multi-variable identification model of double sources

Parameter	Positions (m)	Amounts (g/s)	The best individual fitness value of the initial	convergence generations	The best individual fitness value of the last
Positions and amounts	520, 295	9.15, 2.40	0.02	–	0.07
Positions and amounts[a]	293, 511	2.93, 5.22	0.02	145	154.5

[a] Narrow the search range with some previous experience

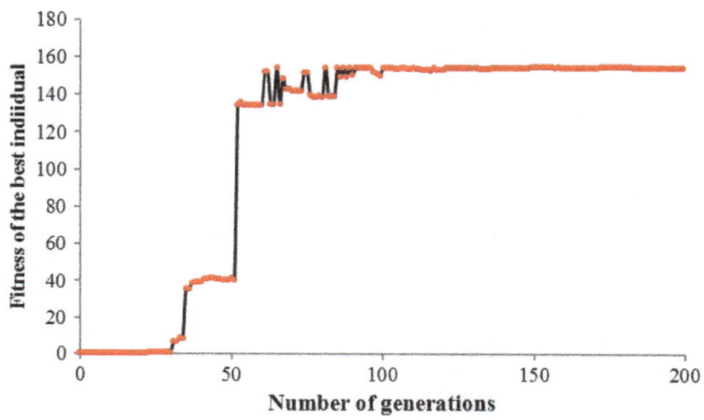

Fig. 6 Fitness curve of the identification model for multi-variable and double sources

known, the computing results of source positions are 310 and 496 m, the relative errors are, respectively, 3.3 and −0.8 %. So the precision is in an acceptable range (relative error less than ±10 %).

Multi-variable identification for double sources

If the amounts and the positions are not both known, continuing with the sudden water pollutant incident of double sources as an example where pollutant amounts are 3.0 and 5.0 g/s, and pollutant source locations are 300 and 500 m. The analytic results at $t = 4000$ s are still used as monitoring data $\overline{C}(\overline{x}_i, T)$. The identification results of pollutant amounts and positions are shown in Table 9. If without any experience to narrow the search scope of the positions or the amounts, the model is thoroughly unable to identify the pollution sources, with the computing positions $x_1 = 520$ and $x_2 = 295$, and the computing amounts $M_1 = 9.15$ and $M_2 = 2.40$, which draws a wrong results. However, if narrowed the positions x_1, x_2 to [100, 600] and the amounts M_1, M_2 to [1.0, 6.0], identification results of pollution source positions for double point sources are 293 and 511 m, the relative errors are −2.3 and 2.2 %, identification results of pollutant amounts are 2.93 and 5.22 g/s, the relative errors are −2.3 and −4.4 %, respectively. The model converges at an evolution step of 145, and the profile curve of fitness value changing along with the evolution steps is shown in Fig. 6.

Conclusions

1. To supply some help in the rescue of pollution incidents, coupling the BGA with the analytical solution of a one-dimensional convection–diffusion equation, a sources identification model for water pollution incidents in small straight rivers is established. Tests imply that for both single variable and multi-variables for single pollutant source, the inversion results are well satisfied when the population size is large enough, such as 10. The relative errors are less than 5 %.
2. For the pollutant source identification of multiple point sources in a water pollution incident, there exists certain error even mistake, because this kind of "inverse problem" is a so called "ill-posed problem" mathematically, which does not have a unique solution. So in this situation, we need some more previous experience to narrow the search scope of the BGA.
3. For the sources identification of a sudden water pollution incident, a water quality monitoring work is a necessary. The water quality monitoring scheme proposed in this paper is to set multiple monitoring points in the affected river section, and obtain the $\overline{C}(\overline{x}_i, T)$ in the same time. In fact, the field monitoring scheme can also be a single monitoring point and acquiring data $\overline{C}(X, t_i)$ by continuous time. According to the

numerical testing, ideal results can be obtained under both these conditions.

Acknowledgments This work was supported by the National Natural Science Foundation of China (Grant No. 51309031) and the "Chinese 12th Five-Year Science and Technology Support Program" (Grant No. 2012BAC06B00).

References

Abderrezzak KEK, Ata R, Zaoui F (2015) One-dimensional numerical modeling of solute transport in streams: the role of longitudinal dispersion coefficient. J Hydrol 527:978–989

Cao XQ, Song JQ, Zhang WM, Zhang LL (2010) MCMC method on an inverse problem of source term identification for convection–diffusion equation. Chin J Hydrodyn 25(2):127–136

Cheng WP, Jia YF (2010) Identification of contaminant point source in surface waters based on backward location probability density function method. Adv Water Resour 33:397–410

Dai HC, Wang LL (2006) Study and application of inverse problem on hydraulics. J Sichuan Univ (Engineering Science Edition) 38(1):15–19

Feng AZ (2012) An analytical solution to constant coefficients advection-diffusion equation in the semi-infinite domain. J Shaoyang Univ Sci Technol 9(1):19–22

Fulvio B, Roberto R, Luca R (2005) Source identification in river pollution problems: a geostatistical approach. Water Resour Res 41(7):23–30

Holland JH (1975) Adaptation in natural and artificial systems. An introductory analysis with application to biology, control, and artificial intelligence. The MIT Press, Cambridge

Jiang ZJ, Wang JH (1997) Study and modification on one dimensional water quality model in natural rivers. J Hunan Univ 24(6):90–94

Jong D (1975) Analysis of the behavior of a class of genetic adaptive systems. College of Literature, Science, and the Arts, The University of Michigan

Jothiprakash V, Shanthi G (2006) Single reservoir operating policies using genetic algorithm. Water Resour Manag 20(6):917–929

Kumar A, Kumar N, Kumar JD (2010) Analytical solutions to one-dimensional advection–diffusion equation with variable coefficients in semi-infinite media. J Hydrol 380(3):330–337

Li J, Feng ZP, Tsukamoto H (2004) Hydrodynamic optimization design of low solidity vaned diffuser for a centrifugal pump using genetic algorithm. J Hydrodyn Ser B 16(2):186–193

Li KQ, Li Z, Li Y, Zhang LJ, Wang XL (2015) A three-dimensional water quality model to evaluate the environmental capacity of nitrogen and phosphorus in Jiaozhou Bay, China. Mar Pollut Bull 91:306–316

Liu SY (2005) Simple simulation of river pollutant diffusion with one-dimensional water quality model. Water Transp Manag 27(4):33–35

Min T, Zhou XD, Zhang SM, Feng MQ (2004) Genetic algorithm to an inverse problem of source term identification for convection–diffusion equation. J Hydrodyn 19(4):520–524

Ministry of Environmental Protection of the People's Republic of China (MEP) (2010) The first national pollution source census bulletin. http://www.zhb.gov.cn/gkml/hbb/bgg/201002/t20100210_185698.htm. Accessed 18 Nov 2015

Ni JR, Wang SJ (2000) On straight river. Shuili Xuebao 12:14–20

Nikolaos DK, Michael P (1996) Site and optimization of contaminant sources in surface water system. J Environ Eng 122:917–923

Roseanna M, Neupauer JLW (2004) Numerical implementation of a backward probabilistic model of ground water contamination. Groundwater 42(2):175–188

Singh RM, Datta B (2006) Identification of groundwater pollution sources using GA-based linked simulation optimization model. J Hydrol Eng 11:101–109

Srinivas M, Patnaik LM (1994) Adaptive probabilities of crossover and mutation in genetic algorithm. IEEE Trans Syst Man Cybernet 24(4):656–667

Tang HW, Xin XK, Dai WH, Xiao Y (2010) Parameter identification for modeling river network using a Genetic Algorithm. J Hydrodyn Ser B 22(2):246–253

Wang ZW, Qiu SF (2008) Stability and numerical simulation of pollution point source identification in a watershed. Chin J Hydrodyn 23(4):364–371

Wang SD, Shen YM (2005) Three high-order splitting schemes for the 3D transport equation. Appl Math Mech 26(8):921–928

Wang ZW, Xu DH (2006) Uniqueness and numerical methods for point source pollution of watershed. J Ningxia Univ (Natural Science Edition) 27(2):124–129

Wang L, Wang Y, Wu CD, Xie QJ (2009) Pollution source fingerprinting system of accident water pollution based on GIS in chemical area. Environ Sci Technol 22(6):57–60

Wang Y, Wu SB, Wang CL, Liu YC (2012) Emergency monitoring and analysis for '721' water contamination incident in Fujiang River. Yangtze River 43(12):64–67

Wu ZK, Fan HM, Chen XR (2008) The numerical study of the inverse problem in reverse process of convection–diffusion equation with adjoint assimilation method. J Hydrodyn 23(2):121–125

Xin XK, Xiao Y, Zhu XD, Cao G, Lv SQ (2009) BP neural network based on DGA and its application in one-dimensional river network simulation. Adv Sci Technol Water Resour 29(32):9–13

Xue GP, Wang C (1997) The determination of source for drinking water pollution accident. Admin Tech Environ Monit 9(6):20–22

Yin FL, Li GS, Jia GZ (2011) An inverse problem of determining magnitudes of multi-point sources in the diffusion equation. J Shandong Univ Technol Sci Technol 25(2):1–5

Yu H, Fan Z, Du HD (2013) The optimal retrieval of ocean color constituent concentrations based on the vibrational method. J Hydrodyn Ser B 25(1):62–71

Zhang GL, Liu XX, Zhang T (2009) The impact of population size on the performance of GA. In: Proceeding of 2009 international conference on machine learning and cybernetics, pp 1866–1870

Zhu S, Liu GH, Wang LZ, Mao GH, Cheng WP, Huang YF (2009) A Bayesian approach for the identification of pollution source in water quality model coupled with hydrodynamics. J Sichuan Univ (Engineering Science Edition) 41(5):30–35

Permissions

All chapters in this book were first published in AWS, by Springer; hereby published with permission under the Creative Commons Attribution License or equivalent. Every chapter published in this book has been scrutinized by our experts. Their significance has been extensively debated. The topics covered herein carry significant findings which will fuel the growth of the discipline. They may even be implemented as practical applications or may be referred to as a beginning point for another development.

The contributors of this book come from diverse backgrounds, making this book a truly international effort. This book will bring forth new frontiers with its revolutionizing research information and detailed analysis of the nascent developments around the world.

We would like to thank all the contributing authors for lending their expertise to make the book truly unique. They have played a crucial role in the development of this book. Without their invaluable contributions this book wouldn't have been possible. They have made vital efforts to compile up to date information on the varied aspects of this subject to make this book a valuable addition to the collection of many professionals and students.

This book was conceptualized with the vision of imparting up-to-date information and advanced data in this field. To ensure the same, a matchless editorial board was set up. Every individual on the board went through rigorous rounds of assessment to prove their worth. After which they invested a large part of their time researching and compiling the most relevant data for our readers.

The editorial board has been involved in producing this book since its inception. They have spent rigorous hours researching and exploring the diverse topics which have resulted in the successful publishing of this book. They have passed on their knowledge of decades through this book. To expedite this challenging task, the publisher supported the team at every step. A small team of assistant editors was also appointed to further simplify the editing procedure and attain best results for the readers.

Apart from the editorial board, the designing team has also invested a significant amount of their time in understanding the subject and creating the most relevant covers. They scrutinized every image to scout for the most suitable representation of the subject and create an appropriate cover for the book.

The publishing team has been an ardent support to the editorial, designing and production team. Their endless efforts to recruit the best for this project, has resulted in the accomplishment of this book. They are a veteran in the field of academics and their pool of knowledge is as vast as their experience in printing. Their expertise and guidance has proved useful at every step. Their uncompromising quality standards have made this book an exceptional effort. Their encouragement from time to time has been an inspiration for everyone.

The publisher and the editorial board hope that this book will prove to be a valuable piece of knowledge for researchers, students, practitioners and scholars across the globe.

List of Contributors

Muheeb Awawdeh and Ghusun Zaiter
Department of Earth and Environmental Sciences, Yarmouk University, Irbid, Jordan

Mutewekil Obeidat
Faculty of Science and Arts, Jordan University of Science and Technology, Irbid 22110, Jordan

S. Krishna Kumar
Department of Geology, University of Madras, Guindy campus, Chennai 600025, Tamil Nadu, India

A. Logeshkumaran
Department of Civil Engineering, VKS College of Engineering and Technology, Desiyamagalam, Karur 639120, Tamil Nadu, India

N. S. Magesh, Prince S. Godson and N. Chandrasekar
Centre for GeoTechnology, Manonmaniam Sundaranar University, Tirunelveli 627 012, Tamil Nadu, India

R. Krishna, J. Iqbal, A. K. Gorai and G. Pathak
Environmental Science and Engineering Group, Birla Institute of Technology Mesra, Ranchi 835215, India

F. Tuluri
Department of Technology, Jackson State University, Jackson, MS 39217, USA

P. B. Tchounwou
Department of Biology, Jackson State University, Jackson, MS 39217, USA

Y. Srinivas, T. B. Aghil, C. Nithya Nair, N. Chandrasekar1
Centre for GeoTechnology, Manonmaniam Sundaranar University, Tirunelveli, Tamil Nadu 627012, India

D. Hudson Oliver
Senthamarai College of Arts and Science, Vadapalanji, Madurai, Tamil Nadu 625021, India

Vinod Atmaram Mendhe, Subhashree Mishra, Atul Kumar Varma and Awanindra Pratap Singh
Central Institute of Mining and Fuel Research, Dhanbad, Jharkhand, India

Priyanka Tiwari and Arun Goel
Department of Civil Engineering, National Institute of Technology (Institute of National Importance), Kurukshetra, Haryana 136119, India

G V. S. R. Pavan Kumar
Department of Chemistry, M V G R College Of Engineering, Vizianagaram, India

K. Rama Krishna
Department of Chemistry, GITAM Institute of Sciences, GITAM University, Visakhapatnam, India

Ouassil Ait El Mekki and Nour-Eddine Laftouhi
Department of Earth Sciences, Faculty of Sciences Semlalia, Cadi Ayyad University, BP 2390, Marrakesh, Morocco

Lahoucine Hanich
Department of Earth Sciences, Faculty of Sciences and Techniques Marrakech, Cadi Ayyad University, BP 549, Marrakesh, Morocco

N. Subba Rao, P. Surya Rao and A. Dinakar
Department of Geology, Andhra University, Visakhapatnam 530 003, India

P. V. Nageswara Rao
Department of Geology, Acharya Nagarjuna University, Nagarjunanagar, Guntur 522 510, India

Deepali Marghade
Priyadarshini Institute of Engineering and Technology, Nagpur 440 019, India

Palas Roy
Department of Chemistry, B.N. Mahavidyalaya (Itachuna), The University of Burdwan, Burdwan 713104, West Bengal, India
Environmental Chemistry Laboratory, Department of Environmental Science, The University of Burdwan, Burdwan, West Bengal 713104, India

Uttiya Dey, Soumya Chattoraj and Naba Kumar Mondal
Environmental Chemistry Laboratory, Department of Environmental Science, The University of Burdwan, Burdwan, West Bengal 713104, India

Debasis Mukhopadhyay
Department of Economics, B.N. Mahavidyalaya (Itachuna), The University of Burdwan, Burdwan, West Bengal 713104, India

Mukheled Al-Sameraiy
Environmental Research Center, University of Technology, Baghdad, Iraq

Alaa M. Al-Abadi
Department of Geology, College of Sciences, University of Basra, Basra, Iraq

Ayesha Kausar
Nanosciences and Catalysis Division, National Centre For Physics, Quaid-i-Azam University Campus, Islamabad 44000, Pakistan

Muhammad Siddiq
Department of Chemistry, Quaid-i-Azam University, Islamabad, Pakistan

Jasmit Kaur, Harpreet Walia, Samson Okongo Mabwoga and Saroj Arora
Department of Botanical and Environmental Sciences, Guru Nanak Dev University, Amritsar, Punjab 143005, India

Murugan Sangareswari and Mariappan Meenakshi Sundaram
Centre for Research and Post Graduate Studies in Chemistry, Ayya Nadar Janaki Ammal College (Autonomous), Sivakasi 626124, Tamil Nadu, India

Abhishek Banerjee and Prafull Singh
Amity Institute of Geo-Informatics and Remote Sensing, Amity University, Sector 125, Noida 201303, India

Kamleshwar Pratap
WAPCOS, Ministry of Water Resources, Government of India, New Delhi 110001, India

A. Shanmugasundharam, G. Kalpana, S. R. Mahapatra, E. R. Sudharson and M. Jayaprakash
Department of Applied Geology, Global and Environmental Change Research Group, University of Madras, Guindy Campus, Tamil Nadu, India

Bo Li
Key Laboratory of Karst Environment and Geohazard Prevention, Ministry of Education, Guizhou University, Guiyang 550000, Guizhou, China

Bo Li, Ganlu Wang, Hanghang Ding
College of Resource and Environmental Engineering, Guizhou University, Guiyang 550000, Guizhou, China

Yulong Chen
Department of Civil Engineering, University of Tokyo, Tokyo 113-8656, Japan

Mehdi Ebrahimi-Gatkash and Habibollah Younesi
Department of Environmental Science, Faculty of Natural Resources, Tarbiat Modares University, P.O. Box 46414-356, Noor, Iran

Afsaneh Shahbazi
Environmental Sciences Research Institute, Shahid Beheshti University, GC, Tehran, Iran

Ava Heidari
Natural Resources and Environment College, Ferdowsi University of Mashhad, Mashhad, Iran

Song Chen and Herong Gui
School of Resources and Civil Engineering, Suzhou University, Suzhou 234000, China
National Engineering Research Center of Coal Mine Water Hazard Controlling, Suzhou 234000, China

Song Chen
School of Earth and Space Science, University of Sciences and Technology of China, Hefei 230000, China

N. Prakash
Department of Chemistry, Thangavelu Engineering College, Chennai 600 097, India

M. Soundarrajan
Department of Chemistry, Valliammai Engineering College, Kattankulathur 603 203, India

S. Arungalai Vendan
School of Electrical Engineering, VIT University, Vellore, India

P. N. Sudha
Department of Chemistry, D.K.M. College for Women, Vellore 632 001, India

N. G. Renganathan
Department of Chemistry, Vel Tech University, Avadi, Chennai, India

Jasna Nemcic-Jurec
Institute of Public Health of Koprivnica-Krizevci County, Trg Tomislava Bardeka 10/10, 48 000 Koprivnica, Croatia

Anamarija Jazbec
Department for Forest Inventory and Management, Faculty of Forestry, University of Zagreb, Svetosimunska 25, 10 000 Zagreb, Croatia

Shraboni Mukherjee, Madhumanti Mondal, Soumya Banerjee and Gopinath Halder
Department of Chemical Engineering, National Institute of Technology, Durgapur, West Bengal, India

Vaibhav Yadav
Department of Earth and Environmental Studies, National Institute of Technology , Durgapur, West Bengal, India

Maha Ahmed Mohamed Abdallah and Somaia B. Ahmed
Marine Pollution Lab, National Institute of Oceanography and Fisheries, Qait Bey, Alexandria, Egypt

Mohamed E. Mahmoud and Maher M. Osman
Chemistry Department, Faculty of Sciences, Alexandria University, Alexandria, Egypt

Shou-ping Zhang
Chongqing Water Resources and Electric Engineering College, Chongqing 402160, People's Republic of China

Xiao-kang Xin
State Key Laboratory of Hydraulics and Mountain River Engineering, Sichuan University, Chengdu 610065, People's Republic of China
Changjiang Water Resources Protection Institute, Changjiang Water Resources Commission, Wuhan 430051, People's Republic of China